ISBN 978-0-266-98770-3
PIBN 11155898

English
Français
Deutsche
Italiano
Español
Português

www.forgottenbooks.com

Mythology Photography **Fiction**
Fishing Christianity **Art** Cooking
Essays Buddhism Freemasonry
Medicine **Biology** Music **Ancient
Egypt** Evolution Carpentry Physics
Dance Geology **Mathematics** Fitness
Shakespeare **Folklore** Yoga Marketing
Confidence Immortality Biographies
Poetry **Psychology** Witchcraft
Electronics Chemistry History **Law**
Accounting **Philosophy** Anthropology
Alchemy Drama Quantum Mechanics
Atheism Sexual Health **Ancient History**
Entrepreneurship Languages Sport
Paleontology Needlework Islam
Metaphysics Investment Archaeology
Parenting Statistics Criminology
Motivational

PURE AND APPLIED MATHEMATICS
A Series of Texts and Monographs

Edited by: R. COURANT · L. BERS · J. J. STOKER

Additional volumes in preparation

PURE AND APPLIED MATHEMATICS

A Series of Texts and Monographs

Edited by

R. COURANT · L. BERS · J. J. STOKER

VOLUME IV

Waves about a harbor

WATER WAVES

The Mathematical Theory
with Applications

J. J. STOKER

INSTITUTE OF MATHEMATICAL SCIENCES
NEW YORK UNIVERSITY, NEW YORK

INTERSCIENCE PUBLISHERS, INC., NEW YORK

To

NANCY

Introduction

1. Introduction

The purpose of this book is to present a connected account of the mathematical theory of wave motion in liquids with a free surface and subjected to gravitational and other forces, together with applications to a wide variety of concrete physical problems.

Surface wave problems have interested a considerable number of mathematicians beginning apparently with Lagrange, and continuing with Cauchy and Poisson in France.* Later the British school of mathematical physicists gave the problems a good deal of attention, and notable contributions were made by Airy, Stokes, Kelvin, Rayleigh, and Lamb, to mention only some of the better known. In the latter part of the nineteenth century the French once more took up the subject vigorously, and the work done by St. Venant and Boussinesq in this field has had a lasting effect: to this day the French have remained active and successful in the field, and particularly in that part of it which might be called mathematical hydraulics. Later, Poincaré made outstanding contributions particularly with regard to figures of equilibrium of rotating and gravitating liquids (a subject which will not be discussed in this book); in this same field notable contributions were made even earlier by Liapounoff. One of the most outstanding accomplishments in the field from the purely mathematical point of view — the proof of the existence of progressing waves of finite amplitude — was made by Nekrassov [N.1], [N.1a]† in 1921 and independently by a different means by Levi-Civita [L.7] in 1925.

The literature concerning surface waves in water is very extensive. In addition to a host of memoirs and papers in the scientific journals, there are a number of books which deal with the subject at length. First and foremost, of course, is the book of Lamb [L.3], almost a third of which is concerned with gravity wave problems. There are books by Bouasse [B.15], Thorade [T.4], and Sverdrup [S.39]

---

* This list would be considerably extended (to include Euler, the Bernoullis, and others) if hydrostatics were to be regarded as an essential part of our subject.
† Numbers in square brackets refer to the bibliography at the end of the book.

devoted exclusively to the subject. The book by Thorade consists almost entirely of relatively brief reviews of the literature up to 1931 — an indication of the extent and volume of the literature on the subject. The book by Sverdrup was written with the special needs of oceanographers in mind. One of the main purposes of the present book is to treat some of the more recent additions to our knowledge in the field of surface wave problems. In fact, a large part of the book deals with problems the solutions of which have been found during and since World War II; this material is not available in the books just now mentioned.

The subject of surface gravity waves has great variety whether regarded from the point of view of the types of physical problems which occur, or from the point of view of the mathematical ideas and methods needed to attack them. The physical problems range from discussion of wave motion over sloping beaches to flood waves in rivers, the motion of ships in a sea-way, free oscillations of enclosed bodies of water such as lakes and harbors, and the propagation of frontal discontinuities in the atmosphere, to mention just a few. The mathematical tools employed comprise just about the whole of the tools developed in the classical linear mathematical physics concerned with partial differential equations, as well as a good part of what has been learned about the nonlinear problems of mathematical physics. Thus potential theory and the theory of the linear wave equation, together with such tools as conformal mapping and complex variable methods in general, the Laplace and Fourier transform techniques, methods employing a Green's function, integral equations, etc. are used. The nonlinear problems are of both elliptic and hyperbolic type.

In spite of the diversity of the material, the book is not a collection of disconnected topics, written for specialists, and lacking unity and coherence. Instead, considerable pains have been taken to supply the fundamental background in hydrodynamics — and also in some of the mathematics needed — and to plan the book in order that it should be as much as possible a self-contained and readable whole. Though the contents of the book are outlined in detail below, it has some point to indicate briefly here its general plan. There are four main parts of the book:

Part I, comprising Chapters 1 and 2, presents the derivation of the basic hydrodynamic theory for non-viscous incompressible fluids, and also describes the two principal approximate theories which form the basis upon which most of the remainder of the book is built.

Part II, made up of Chapters 3 to 9 inclusive, is based on the approximate theory which results when the amplitude of the wave motions considered is small. The result is a linear theory which from the mathematical point of view is a highly interesting chapter in potential theory. On the physical side the problems treated include the propagation of waves from storms at sea, waves on sloping beaches, diffraction of waves around a breakwater, waves on a running stream, the motion of ships as floating rigid bodies in a seaway. Although this theory was known to Lagrange, it is often referred to as the Cauchy-Poisson theory, perhaps because these two mathematicians were the first to solve interesting problems by using it.

Part III, made up of Chapters 10 and 11, is concerned with problems involving waves in shallow water. The approximate theory which results from assuming the water to be shallow is not a linear theory, and wave motions with amplitudes which are not necessarily small can be studied by its aid. The theory is often attributed to Stokes and Airy, but was really known to Lagrange. If linearized by making the additional assumption that the wave amplitudes are small, the theory becomes the same as that employed as the mathematical basis for the theory of the tides in the oceans. In the lowest order of approximation the nonlinear shallow water theory results in a system of hyperbolic partial differential equations, which in important special cases can be treated in a most illuminating way with the aid of the method of characteristics. The mathematical methods are treated in detail in Chapter 10. The physical problems treated in Chapter 10 are quite varied; they include the propagation of unsteady waves due to local disturbances into still water, the breaking of waves, the solitary wave, floating breakwaters in shallow water. A lengthy section on the motions of frontal discontinuities in the atmosphere is included also in Chapter 10. In Chapter 11, entitled Mathematical Hydraulics, the shallow water theory is employed to study wave motions in rivers and other open channels which, unlike the problems of the preceding chapter, are largely conditioned by the necessity to consider resistances to the flow due to the rough sides and bottom of the channel. Steady flows, and steady progressing waves, including the problem of roll waves in steep channels, are first studied. This is followed by a treatment of numerical methods of solving problems concerning flood-waves in rivers, with the object of making flood predictions through the use of modern high speed digital computers. That such methods can be used to furnish accurate predictions has been verified for a flood in a 400-mile stretch of the Ohio River, and for a flood coming down the Ohio River and passing through its junction with the Mississippi River.

Part IV, consisting of Chapter 12, is concerned with problems solved in terms of the exact theory, in particular, with the use of the exact nonlinear free surface conditions. A proof of the existence of periodic waves of finite amplitude, following Levi-Civita in a general way, is included.

The amount of mathematical knowledge needed to read the book varies in different parts. For considerable portions of Part II the elements of the theory of functions of a complex variable are assumed known, together with some of the standard facts in potential theory. On the other hand Part III requires much less in the way of specific knowledge, and, as was mentioned above, the basic theory of the hyperbolic differential equations used there is developed in all detail in the hope that this part would thus be made accessible to engineers, for example, who have an interest in the mathematical treatment of problems concerning flows and wave motions in open channels.

In general, the author has made considerable efforts to try to achieve a reasonable balance between the mathematics and the mechanics of the problems treated. Usually a discussion of the physical factors and of the reasons for making simplified assumptions in each new type of concrete problem precedes the precise formulation of the mathematical problems. On the other hand, it is hoped that a clear distinction between physical assumptions and mathematical deductions — so often shadowy and vague in the literature concerned with the mechanics of continuous media — has always been maintained. Efforts also have been made to present important portions of the book in such a way that they can be read to a large extent independently of the rest of the book; this was done in some cases at the expense of a certain amount of repetition, but it seemed to the author more reasonable to save the time and efforts of the reader than to save paper. Thus the portion of Chapter 10 concerned with the dynamics of the motion of fronts in meteorology is largely self-contained. The same is true of Chapter 11 on mathematical hydraulics, and of Chapter 9 on the motion of ships.

Originally this book had been planned as a brief general introduction to the subject, but in the course of writing it many gaps and inadequacies in the literature were noticed and some of them have been filled in; thus a fair share of the material presented represents the result of researches carried out quite recently. A few topics which are even rather speculative have been dealt with at some length (the theory of the motion of fronts in dynamic meteorology, given in Chapter 10.12, for example); others (like the theory of waves on sloping beaches) have been treated at some length as much because the author had a special fondness for the material as for their intrinsic mathematical interest. Thus the author has written a book which is rather personal in character, and which contains a selection of material chosen, very often, simply because it interested him, and he has allowed his predilections and tastes free rein. In addition, the book has a personal flavor from still another point of view since a quite large proportion of the material presented is based on the work of individual members of the Institute of Mathematical Sciences of New York University, and on theses and reports written by students attending the Institute. No attempt at completeness in citing the literature, even the more recent literature, was made by the author; on the other hand, a glance at the Bibliography (which includes only works actually cited in the book) will indicate that the recent literature has not by any means been neglected.

In early youth by good luck the author came upon the writings of scientists of the British school of the latter half of the nineteenth century. The works of Tyndall, Huxley, and Darwin, in particular, made a lasting impression on him. This could happen, of course, only because the books were written in an understandable way and also in such a way as to create interest and enthusiasm: — but this was one of the principal objects of this school of British scientists. Naturally it is easier to write books on biological subjects for non-specialists than it is to write them on subjects concerned with the mathematical sciences — just because the time and effort needed to acquire a knowledge of modern mathematical tools is very great. That the task is not entirely hopeless, however, is indicated by John Tyndall's book on sound, which should be regarded as a great classic of scientific exposition. On the whole, the British school of popularizers of science wrote for people presumed to have little or no foreknow-ledge of the subjects treated. Now-a-days there exists a quite large potential audience for books on subjects requiring some knowledge of mathematics and physics, since a large number of specialists of all kinds must have a basic training in these disciplines. The author hopes that this book, which deals with so many phenomena of every day occurrence in nature, might perhaps be found interesting, and understandable in some parts at least, by readers who have some mathematical training but lack specific knowledge of hydrodynamics.* For example, the introductory discussion of waves on sloping beaches in Chapter 5, the purely geometrical discussion of the wave patterns created by moving ships in Chapter 8, great parts of Chapters 10 and 11 on waves in shallow water and flood waves in rivers, as well as the general discussion in Chapter 10 concerning the motion of fronts in the atmosphere, are in this category.

2. Outline of contents

It has already been stated that this book is planned as a coherent and unified whole in spite of the variety and diversity of its contents on both the mathematical and the physical sides. The possibility of achieving such a purpose lies in the fortunate fact that the material can be classified rather readily in terms of the types of mathematical problems which occur, and this classification also leads to a reasonably consistent ordering of the material with respect to the various types of physical problems. The book is divided into four main parts.

Part I begins with a brief, but it is hoped adequate, development of the hydrodynamics of perfect incompressible fluids in irrotational flow without viscosity, with emphasis on those aspects of the subject relevant to flows with a free surface. Unfortunately, the basic general theory is unmanageable for the most part as a basis for the solution of concrete problems because the nonlinear free surface conditions make for insurmountable difficulties from the mathematical point of view. It is therefore necessary to make restrictive assumptions which have the effect of yielding more tractable mathematical formulations. Fortunately there are at least two possibilities in this respect which are not so restrictive as to limit too drastically the physical interest, while at the same time they are such as to lead to mathematical problems about which a great deal of knowledge is available.

One of the two approximate theories results from the assumption that the wave amplitudes are small, the other from the assumption

---

* The book by Rachel Carson [C.16] should be referred to here. This book is entirely nonmathematical, but it is highly recommended for supplementary reading. Parts of it are particularly relevant to some of the material in Chapter 6 of the present book.

that it is the depth of the liquid which is small — in both cases, of course, the relevant quantities are supposed small in relation to some other significant length, such as a wave length, for example. Both of these approximate theories are derived as the lowest order terms of formal developments with respect to an appropriate small dimensionless parameter; by proceeding in this way, however, it can be seen how the approximations could be carried out to include higher order terms. The remainder of the book is largely devoted to the working out of consequences of these two theories, based on concrete physical problems: Part II is based on the small amplitude theory, and Part III deals with applications of the shallow water theory. In addition, there is a final chapter (Chapter 12) which makes up Part IV, in which a few problems are solved in terms of the basic general theory and the nonlinear boundary conditions are satisfied exactly; this includes a proof along lines due to Levi-Civita, of the existence, from the rigorous mathematical point of view, of progressing waves of finite amplitude.

Part II, which is concerned with the first of the possibilities, might be called the linearized exact theory, since it can be obtained from the basic exact theory simply by linearizing the free surface conditions on the assumption that the wave motions studied constitute a small deviation from a constant flow with a horizontal free surface. Since we deal only with irrotational flows, the result is a theory based on the determination of a velocity potential in the space variables (containing the time as a parameter, however) as a solution of the Laplace equation satisfying certain linear boundary and initial conditions. This linear theory thus belongs, generally speaking, to potential theory.

There is such a variety of material to be treated in Part II, which comprises Chapters 3 to 9, that a further division of it into subdivisions is useful, as follows: 1) subdivision A, dealing with wave motions that are simple harmonic oscillations in the time; 2) subdivision B, dealing with unsteady, or transient, motions that arise from initial disturbances starting from rest; and 3) subdivision C, dealing with waves created in various ways on a running stream, in contrast with subdivisions A and B in which all motions are assumed to be small oscillations near the rest position of equilibrium of the fluid.

Subdivision A is made up of Chapters 3, 4, and 5. In Chapter 3 the basically important standing and progressing waves in liquids of uniform depth and infinite lateral extent are treated; the important fact that these waves are subject to dispersion comes to light, and the notion of group velocity thus arises. The problem of the uniqueness of the solutions is considered — in fact, uniqueness questions are intentionally stressed throughout Part II because they are interesting mathematically and because they have been neglected for the most part until rather recently. It might seem strange that there could be any interesting unresolved uniqueness questions left in potential theory at this late date; the reason for it is that the boundary condition at a free surface is of the mixed type, i.e. it involves a linear combination of the potential function and its normal derivative, and this combination is such as to lead to the occurrence of non-trivial solutions of the homogeneous problems in cases which would in the more conventional problems of potential theory possess only identically constant solutions. In fact, it is this mixed boundary condition at a free surface which makes Part II a highly interesting chapter in potential theory — quite apart from the interest of the problems on the physical side. Chapter 4 goes on to treat certain simple harmonic forced oscillations, in contrast with the free oscillations treated in Chapter 3. Chapter 5 is a long chapter which deals with simple harmonic waves in cases in which the depth of the water is not constant. A large part of the chapter concerns the propagation of progressing waves over a uniformly sloping beach; various methods of treating the problem are explained — in part with the object of illustrating recently developed techniques useful for solving boundary problems (both for harmonic functions and functions satisfying the reduced wave equation) in which mixed boundary conditions occur. Another problem treated (in Chapter 5.5) is the diffraction of waves around a vertical wedge. This leads to a problem identical with the classical diffraction problem first solved by Sommerfeld [S.12] for the special case of a rigid half-plane barrier. Here again the uniqueness question comes to the fore, and, as in many of the problems of Part II, it involves consideration of so-called radiation conditions at infinity. A uniqueness theorem is derived and also a new, and quite simple and elementary, solution for Sommerfeld's diffraction problem is given. It is a curious fact that these gravity wave problems, the solutions of which are given in terms of functions satisfying the Laplace equation, nevertheless require for the uniqueness of the solutions that conditions at infinity of the radiation type, just as in the more familiar problems based on the linear wave equation, be imposed;

ordinarily in potential theory it is sufficient to require only boundedness conditions at infinity to ensure uniqueness.

In subdivision B of Part II, comprised of Chapter 6, a variety of problems involving transient motions is treated. Here initial conditions at the time $t = 0$ are imposed. The technique of the Fourier transform is explained and used to obtain solutions in the form of integral representations. The important classical cases (treated first by Cauchy and Poisson) of the circular waves due to disturbances at a point of the free surface in an infinite ocean are studied in detail. For this purpose it is very useful to discuss the integral representations by using an asymptotic approximation due to Kelvin (and, indeed, developed by him for the purpose of discussing the solutions of just such surface wave problems) and called the principle, or method, of stationary phase. These results then can be interpreted in a striking way in terms of the notion of group velocity. Recently there have been important applications of these results in oceanography: one of them concerns the type of waves called tsunamis, which are destructive waves in the ocean caused by earthquakes, another concerns the location of storms at sea by analyzing wave records on shore in the light of the theory at present under discussion. The question of uniqueness of the transient solutions — again a problem solved only recently — is treated in the final section of Chapter 6. An opportunity is also afforded for a discussion of radiation conditions (for simple harmonic waves) as limits as $t \to \infty$ in appropriate problems concerning transients, in which boundedness conditions at infinity suffice to ensure uniqueness.

The final subdivision of Part II, subdivision C, deals with small disturbances created in a stream flowing initially with uniform velocity and with a horizontal free surface. Chapter 7 treats waves in streams having a uniform depth. Again, in the case of steady motions, the question of appropriate conditions of the radiation type arises; the matter is made especially interesting here because the circumstances with respect to radiation conditions depend radically on the parameter U^2/gh, with U and h the velocity and depth at infinity, respectively. Thus if $U^2/gh > 1$, no radiation conditions need be imposed, if $U^2/gh < 1$ they are needed, while if $U^2/gh = 1$ something quite exceptional occurs. These matters are studied, and their physical interpretations are discussed in Chapter 7.3 and 7.4. In Chapter 8 Kelvin's theory of ship waves for the idealized case of a ship regarded as a point disturbance moving over the surface of the water is treated in considerable detail. The principle of stationary phase leads to a beautiful and elegant treatment of the nature of ship waves that is purely geometrical in character. The cases of curved as well as straight courses are considered, and photographs of ship waves taken from airplanes are reproduced to indicate the good accord with observations. Finally, in Chapter 9 a general theory (once more the result of quite recent investigations) for the motion of ships, regarded as floating rigid bodies, is presented. In this theory no restrictive assumptions — regarding, for example, the coupling (or lack of coupling, as in an old theory due to Krylov [K.20] between the motion of the sea and the motion of the ship, or between the various degrees of freedom of the ship — are made other than those needed to linearize the problem. This means essentially that the ship must be regarded as a thin disk so that it can slice its way through the water (or glide over the surface, perhaps) with a finite velocity and still create waves which do not have large amplitudes; in addition, it is necessary to suppose that the motion of the ship is a small oscillation relative to a motion of translation with uniform velocity. The theory is obtained by making a formal development of all conditions of the complete nonlinear boundary problem with respect to a parameter which is a thickness-length ratio of the ship. The resulting theory contains the classical Michell-Havelock theory for the wave resistance of a ship in terms of the shape of its hull as the simplest special case.

We turn next to Part III, which deals with applications of the approximate theory which results from the assumption that it is the depth of the liquid which is small, rather than the amplitude of the surface waves as in Part II. The theory, called here the shallow water theory, leads to a system of nonlinear partial differential equations which are analogous to the differential equations for the motion of compressible gases in certain cases. We proceed to outline the contents of Part III, which is composed of two long chapters.

In Chapter 10 the mathematical methods based on the theory of characteristics are developed in detail since they furnish the basis for the discussion of practically all problems in Part III; it is hoped that this preparatory discussion of the mathematical tools will make Part III of the book accessible to engineers and others who have not had advanced training in mathematical analysis and in the methods of mathematical physics. In preparing this part of the book the author's task was made relatively easy because of the existence of the book by Courant and Friedrichs [C.9], which deals with gas dynamics; the presentation of the basic theory given here is largely modeled on the presentation given in that book. The concrete problems dealt with in Chapter 10 are quite varied in character, including the propagation of disturbances into still water, conditions for the occurrence of a bore and a hydraulic jump (phenomena analogous to the occurrence of shock waves in gas dynamics), the motion resulting from the breaking of a dam, steady two dimensional motions at supercritical velocity, and the breaking of waves in shallow water. The famous problem of the solitary wave is discussed along the lines used recently by Friedrichs and Hyers [F.13] to prove rigorously the existence of the solitary wave from the mathematical point of view; this problem requires carrying the perturbation series which formulate the shallow water theory to terms of higher order. The problem of the motion of frontal discontinuities in the atmosphere, which lead to the development of cyclonic disturbances in middle latitudes, is given a formulation — on the basis of hypotheses which simplify the physical situation — which brings it within the scope of a more general "shallow water theory". Admittedly (as has already been noted earlier) this theory is somewhat speculative, but it is nevertheless believed to have potentialities for clarifying some of the mysteries concerning the dynamical causes for the development and deepening of frontal disturbances in the atmosphere, especially if modern high speed digital computing machines are used as an aid in solving concrete problems numerically.

Chapter 10 concludes with the discussion of a few applications of the linearized version of the shallow water theory. Such a linearization results from assuming that the amplitude of the waves is small. The most famous application of this theory is to the tides in the oceans (and also in the atmosphere, for that matter); strange though it seems at first sight, the oceans can be treated as shallow for this phenomenon since the wave lengths of the motions are very long because of the large periods of the disturbances caused by the moon and the sun. This theory, as applied to the tides, is dealt with only very summarily, since an extended treatment is given by Lamb [L.3]. Instead, some problems connected with the design of floating breakwaters in shallow water are discussed, together with brief treatments of the oscillations in certain lakes (the lake at Geneva in Switzerland, for example) called seiches, and oscillations in harbors.

Finally, Part III concludes with Chapter 11 on the subject of mathematical hydraulics, which is to be understood here as referring to flows and wave motions in rivers and other open channels with rough sides. The problems of this chapter are not essentially different, as far as mathematical formulations go, from the problems treated in the preceding Chapter 10. They differ, however, on the physical side because of the inclusion of a force which is just as important as gravity, namely a force of resistance caused by the rough sides and bottom of the channels. This force is dealt with empirically by adding a term to the equation expressing the law of conservation of momentum that is proportional to the square of the velocity and with a coefficient depending on the roughness and the so-called hydraulic radius of the channel. The differential equations remain of the same type as those dealt with in Chapter 10, and the same underlying theory based on the notion of the characteristics applies.

Steady motions in inclined channels are first dealt with. In particular, a method of solving the problem of the occurrence of roll waves in steep channels is given; this is done by constructing a progressing wave by piecing together continuous solutions through bores spaced at periodic intervals. This is followed by the solution of a problem of steady motion which is typical for the propagation of a flood down a long river; in fact, data were chosen in such a way as to approximate the case of a flood in the Ohio River. A treatment is next given for a flood problem so formulated as to correspond approximately to the case of a flood wave moving down the Ohio to its junction with the Mississippi, and with the result that disturbances are propagated both upstream and downstream in the Mississippi and a backwater effect is noticeable up the Ohio. In these problems it is necessary to solve the differential equations numerically (in contrast with most of the problems treated in Chapter 10, in which interesting explicit solutions could be given), and methods of doing so are explained in detail. In fact, a part of the elements of numerical analysis as applied to solving hyperbolic partial differential equations by the method of finite differences is developed. The results of a numerical prediction of a flood over a stretch of 400 miles in the Ohio River as it actually exists are given. The flood in question was the 1945 flood — one of the largest on record — and the predictions made (starting with the initial state of the river and using the known flows into it from tributaries and local drainage) by numerical integration on a high speed digital computer (the Univac) check quite closely with the actually observed flood. Numerical predictions were also made for the case of a flood (the 1947 flood in this case) coming down the Ohio and passing through its junction with the Mississippi; the accuracy of the prediction was good. This is a case in which the simplified methods of the civil engineers do not work well. These results, of course, have important implications for the practical applications.

Finally Part IV, made up of Chapter 12, closes the book with a few solutions based on the exact nonlinear theory. One class of problems is solved by assuming a solution in the form of power series in the time, which implies that initial motions and motions for a short time only can be determined in general. Nevertheless, some interesting cases can be dealt with, even rather easily, by using the so-called Lagrange representation, rather than the Euler representation which is used otherwise throughout the book. The problem of the breaking of a dam, and, more generally, problems of the collapse of columns of a liquid resting on a rigid horizontal plane can be treated in this way. The book ends with an exposition of the theory due to Levi-Civita concerning the problem of the existence of progressing waves of finite amplitude in water of infinite depth which satisfy exactly the nonlinear free surface conditions.

Acknowledgments

Without the support of the Mathematics Branch and the Mechanics Branch of the Office of Naval Research this book would not have been written. The author takes pleasure in acknowledging the help and encouragement given to him by the ONR in general, and by Dr. Joachim Weyl, Dr. Arthur Grad, and Dr. Philip Eisenberg in particular. Although she is no longer working in the ONR, it is nevertheless appropriate at this place to express special thanks to Dean Mina Rees, who was head of the Mathematics Branch when this book was begun.

Among those who collaborated with the author in the preparation of the manuscript, Dr. Andreas Troesch should be singled out for special thanks. His careful and critical reading of the manuscript resulted in many improvements and the uncovering and correction of errors and obscurities of all kinds. Another colleague, Professor E. Isaacson, gave almost as freely of his time and attention, and also aided materially in revising some of the more intricate portions of the book. To these fellow workers the author feels deeply indebted.

Miss Helen Samoraj typed the entire manuscript in a most efficient (and also good-humored) way, and uncovered many slips and inconsistencies in the process.

The drawings for the book were made by Mrs. Beulah Marx and Miss Larkin Joyner. The index was prepared by Dr. George Booth and Dr. Walter Littman with the assistance of Mrs. Halina Montvila.

A considerable part of the material in the present book is the result of researches carried out at the Institute of Mathematical Sciences of New York University as part of its work under contracts with the Office of Naval Research of the U.S. Department of Defense, and to a lesser extent under a contract with the Ohio River Division of the Corps of Engineers of the U.S. Army. The author wishes to express his thanks generally to the Institute; the cooperative and friendly spirit of its members, and the stimulating atmosphere it has provided have resulted in the carrying out of quite a large number of researches in the field of water waves. A good deal of these researches and new results have come about through the efforts of Professors K. O. Friedrichs, Fritz John, J. B. Keller, H. Lewy (of the University of California), and A. S. Peters, together with their students or with visitors at the Institute.

J. J. STOKER

New York, N.Y.
January, 1957.

Contents

PART I

PART II

Subdivision A

Waves Simple Harmonic in the Time

Subdivision B

Motions Starting from Rest. Transients

Subdivision C

Waves on a Running Stream. Ship Waves

PART III

PART IV

PART I

CHAPTER 1

Basic Hydrodynamics

1.1. The laws of conservation of momentum and mass

As has been stated in the introduction, we deal exclusively in this book with flows in water (and air) which are of such a nature as to make it unnecessary to take into account the effects of viscosity and compressibility. As a consequence of the neglect of internal friction, or in other words of neglect of shear stresses, it is well known that the stress system* in the liquid is a state of uniform compression at each point. The intensity of the compressive stress is called the pressure p.

The equation of motion of a fluid particle can then be obtained on the basis of Newton's law of conservation of momentum, as follows. A small rectangular element of the fluid is shown in Figure 1.1.1

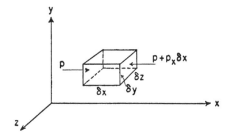

Fig. 1.1.1. Pressure on a fluid element

with the pressure acting on the faces normal to the x-axis. Newton's law for the x-direction is then

$$[-(p + p_x \, \delta x) + p]\delta y \, \delta z + X\varrho \, \delta x \, \delta y \, \delta z = \varrho a_{(x)} \, \delta x \, \delta y \, \delta z$$

* We assume that the usual concepts of the general mechanics of continuous media are known.

3

in which X is the external or body force component per unit mass
and $a_{(x)}$ is the acceleration component, both in the x-direction, and
ϱ is the density. The quantities p, X, and $a_{(x)}$ are in general functions
of x, y, z, and t. Here, as always, we shall use letter subscripts to
denote differentiation, and this accounts for the symbol $a_{(x)}$ to denote
the component of a vector in the x-direction. Upon passing to the
limit in allowing δx, δy, δz to approach zero we obtain the equation
of motion for the x-direction in the form $-p_x + \varrho X = \varrho a_{(x)}$, and
analogous expressions for the two other directions. Thus we have the
equations of motion

$$(1.1.1) \quad \begin{cases} -\dfrac{1}{\varrho} p_x + X = a_{(x)}, \\[2mm] -\dfrac{1}{\varrho} p_y + Y = a_{(y)}, \\[2mm] -\dfrac{1}{\varrho} p_z + Z = a_{(z)}, \end{cases}$$

or, in vector form:

$$(1.1.2) \quad -\frac{1}{\varrho} \operatorname{grad} p + \mathbf{F} = \mathbf{a},$$

with an obvious notation. The body force $\mathbf{F}$ plays a very important
role in our particular branch of hydrodynamics—in fact the main
results of the theory are entirely conditioned by the presence of the
gravitational force $\mathbf{F} = (0, -g, 0)$, in which g represents the acceleration of gravity. It should be observed that we consider *the positive
y-axis to be vertically upward, and the x, z-plane therefore to be horizontal*
(usually it will be taken as the undisturbed water surface). This convention regarding the disposition of the coordinate axes will be maintained, for the most part, throughout the book.

The differential equations (1.1.1) are in what is called the Lagrangian form, in which one has in mind a direct description of the motion
of each individual fluid particle as a function of the time. It is more
useful for most purposes to work with the equations of motion in the
so-called Eulerian form. In this form of the equations one concentrates attention on the determination of the velocity distribution in
the region occupied by the fluid without trying to follow the motion of
the individual fluid particles, but rather observing the velocity
distribution at fixed points in space as a function of the time. In

other words. the velocity field, with components u, v, w, is to be determined as a function of the space variables and the time. Afterwards, if that is desired, the motion of the individual particles can be obtained by integrating the system of ordinary differential equations $\dot{x} = u$, $\dot{y} = v$, $\dot{z} = w$, in which the dot over the quantities x, y, z means differentiation with respect to the time in following the motion of an individual· particle.

In order to restate the equations of motion (1.1.1) in terms of the Euler variables u, v, w, and in order to carry out other important operations as well, it is necessary to calculate time derivatives of various functions associated with a given fluid particle in following the motion of the particle. For example, we need to calculate the time derivative of the velocity of a particle in order to obtain the acceleration components occurring in (1.1.1), and quite a few other quantities will occur later on for which such *particle derivatives* will be needed. Suppose, then, that $F(x, y, z; t)$ is a function associated with a particle which follows the path given by the vector

$$\mathbf{x} = (x(t), y(t), z(t));$$

it follows that

$$\dot{\mathbf{x}} = (\dot{x}(t), \dot{y}(t), \dot{z}(t)) = (u, v, w)$$

is the velocity vector associated with the particle. For this particle the arguments x, y, z of the function F are of course the functions of t which characterize the motion of the particle; as a consequence we have

$$\frac{dF}{dt} = F_x \dot{x} + F_y \dot{y} + F_z \dot{z} + F_t$$

$$= u F_x + v F_y + w F_z + F_t,$$

and hence the operation of taking the particle derivative d/dt is defined as follows:

$$(1.1.3) \qquad \frac{d}{dt} (\ \ ) = u(\ \ )_x + v(\ \ )_y + w(\ \ )_z + (\ \ )_t.$$

The distinction between dF/dt and $\partial F/\partial t = F_t$ should be carefully noted.

Since the acceleration $\mathbf{a}$ of a particle is given by $\mathbf{a} = (du/dt, dv/dt, dw/dt)$, in which (u, v, w) are the components of the velocity $\mathbf{v}$ of the particle, it follows from (1.1.3) that the component $a_{(x)} = du/dt$ is given by

$$\frac{du}{dt} = uu_x + vu_y + wu_z + u_t,$$

with similar expressions for the other components. The equations of motion (1.1.1) are therefore given as follows in terms of the Euler variables:

(1.1.4)
$$\begin{cases} u_t + uu_x + vu_y + wu_z = -\frac{1}{\varrho} p_x, \\[2mm] v_t + uv_x + vv_y + wv_z = -\frac{1}{\varrho} p_y - g, \\[2mm] w_t + uw_x + vw_y + ww_z = -\frac{1}{\varrho} p_z \end{cases}$$

when we specify the external or body force to consist only of the force of gravity.

Equations (1.1.4) form a set of three nonlinear partial differential equations for the five quantities u, v, w, ϱ, and p. Since the fluid is assumed to be incompressible, the density ϱ can be taken as a known constant. At the same time, the assumption of incompressibility leads to a relatively simple differential equation expressing the law of conservation of mass, and this equation constitutes the needed fourth equation for the determination of the velocity components and the pressure. Perhaps the simplest way to derive the mass conservation law is to start from the relation

$$\iint_S \varrho v_n \, dS = 0,$$

which states that the mass flux outward through any fixed closed surface enclosing a region in which no liquid is created or destroyed is zero. (By v_n we mean the velocity component taken positive in the direction of the outward normal to the surface.) An application of Gauss's divergence theorem:

(1.1.5)
$$\iint_S \varrho v_n \, dS - \iiint_R \operatorname{div} (\varrho \mathbf{v}) \, d\tau$$

to the above integral leads to the relation

$$\iiint_R \operatorname{div} (\varrho \mathbf{v}) \, d\tau = 0$$

for any arbitrary region R. It follows therefore that div $(\varrho \mathbf{v}) = 0$ everywhere, and since $\varrho = $ constant, we have finally

$$(1.1.6) \qquad\qquad \text{div } \mathbf{v} = u_x + v_y + w_z = 0$$

as the expression of the law of conservation of mass. The equation (1.1.6) is also frequently called the *equation of continuity*.

Equations (1.1.4) and (1.1.6) are sufficient, once appropriate initial and boundary conditions (to be discussed shortly) are imposed, to determine the velocity components u, v, w, and the pressure p uniquely.

1.2. Helmholtz's theorem

Before discussing boundary conditions it is preferable to formulate a few additional conservation laws which are consequences of the assumptions made so far — in particular of the assumption that internal fluid friction can be neglected.

The first of these laws to be discussed is the law of conservation of *circulation*. The notion of circulation is defined as follows. Consider a closed curve C which moves with the fluid (that is, C consists always of the same particles of the fluid). The circulation $\Gamma = \Gamma(t)$ around C is defined by the line integral

$$(1.2.1) \qquad\qquad \Gamma(t) = \oint_C u\,dx + v\,dy + w\,dz$$

$$= \oint_C v_s\,ds$$

in which v_s is the velocity component of the fluid tangent to C, and ds is the element of arc length of C. The curve C is considered as given by the vector $\mathbf{x}(\sigma, t)$ with σ a parameter on C such that $0 \leq \sigma \leq 1$ and $\mathbf{x}(0, t) = \mathbf{x}(1, t)$. We are thus operating in terms of the Lagrange system of variables rather than in terms of the Euler system, and fixing a value of σ has the effect of picking out a specific particle on C.

We may write $\Gamma(t) = \int_0^1 \mathbf{v} \cdot \mathbf{x}_\sigma d\sigma$ in which $\mathbf{v} \cdot \mathbf{x}_\sigma$ is a scalar product and $\mathbf{x}_\sigma$, as usual, refers to differentiation with respect to σ. For the time derivative $\dot{\Gamma}$ we have therefore

$$\dot{\Gamma}(t) = \int_0^1 (\dot{\mathbf{v}} \cdot \mathbf{x}_\sigma + \mathbf{v} \cdot \dot{\mathbf{x}}_\sigma) d\sigma.$$

From the equation of motion (1.1.2) in the Lagrangian form with $\mathbf{a} = \dot{\mathbf{v}}$, $\mathbf{F} = (0, -g, 0) = -\operatorname{grad}(gy)$, and from $\dot{\mathbf{x}}_\sigma = \mathbf{v}_\sigma$, the last equation yields

$$(1.2.2) \quad \dot{\Gamma}(t) = \int_0^1 \left[-\frac{1}{\varrho} \mathbf{x}_\sigma \cdot \operatorname{grad} p - g\mathbf{x}_\sigma \cdot \operatorname{grad} y + \mathbf{v} \cdot \mathbf{v}_\sigma \right] d\sigma$$

$$= \int_0^1 \left[-\frac{1}{\varrho} p_o - gy_\sigma + \frac{1}{2} (\mathbf{v} \cdot \mathbf{v})_\sigma \right] d\sigma$$

$$= 0,$$

since the values of p, y, and $\mathbf{v}$ coincide at $\sigma = 0$ and $\sigma = 1$, and ϱ and g are constants. The last equation evidently states that *in a nonviscous fluid the circulation around any closed curve consisting of the same fluid particles is constant in time.* This is the theorem of Helmholtz. The assumption of zero viscosity entered into our derivation through the use of (1.1.2) as equation of motion.*

In this book we are interested in the special case in which the circulation for all closed curves is zero. This case is very important in the applications because it occurs whenever the fluid is assumed to have been at rest or to have been moving with a constant velocity at some particular time, so that $\mathbf{v} \equiv \text{const.}$ holds at that time, and hence Γ vanishes for all time. The cases in which the fluid motion begins from such states are obviously very important.

The assumption that Γ vanishes for all closed curves has a number of consequences which are basic for all that follows in this book. The first conclusion from $\Gamma \equiv 0$ follows almost immediately from Stokes's theorem:

$$(1.2.3) \qquad \Gamma = \oint_C v_s ds = \iint_S (\operatorname{curl} \mathbf{v})_n \, dA,$$

in which the surface integral is taken over any surface S spanning the curve C. If $\Gamma = 0$ for all curves C, as we assume, it follows easily by a well-known argument that the vector curl $\mathbf{v}$ vanishes everywhere:

* It should be added that the law of conservation of circulation holds under much wider conditions than were assumed here (cf. [C.9], p. 19).

(1.2.4) $\text{curl } \mathbf{v} = (w_y - v_z,\ u_z - w_x,\ v_x - u_y) = 0,$

and the flow is then said to be *irrotational*. In other words, a motion in a nonviscous fluid which is irrotational at one instant always remains irrotational. *Throughout the rest of this book we shall assume all flows to be irrotational.*

1.3. Potential flow and Bernoulli's law

The assumption of irrotational flow results in a number of simplifications in our theory which are of the greatest utility. In the first place, the fact that curl $\mathbf{v} = 0$ (cf. (1.2.4)) ensures, as is well known, the existence of a single-valued *velocity potential* $\Phi(x, y, z; t)$ in any *simply connected region*, from which the velocity field can be derived by taking the gradient:

(1.3.1) $\mathbf{v} = \text{grad } \Phi = (\Phi_x,\ \Phi_y,\ \Phi_z),$

or, in terms of the components of $\mathbf{v}$:

(1.3.2) $u = \Phi_x,\ v = \Phi_y,\ w = \Phi_z.$

The velocity potential is, indeed, given by the line integral

$$\Phi(x, y, z; t) = \int^{x, y, z} u\, dx + v\, dy + w\, dz.$$

The vanishing of curl $\mathbf{v}$ ensures that the expression to be integrated is an exact differential. Once it is known that the velocity components are determined by (1.3.2), it follows from the continuity equation (1.1.6), i.e. div $\mathbf{v} = 0$, that the velocity potential Φ is a solution of the Laplace equation

(1.3.3) $\nabla^2\Phi = \Phi_{xx} + \Phi_{yy} + \Phi_{zz} = 0,$

as one readily sees, and Φ is thus a harmonic function. This fact represents a great simplification, since the velocity field is derivable from a single function satisfying a linear differential equation which has been very much studied and about which a great deal is known.

Still another important consequence of the irrotational character of a flow can be obtained from the equations of motion (1.1.4). By making use of (1.2.4), it is readily verified that the equations of motion (1.1.4) can be written in the following vector form:

$$\text{grad } \Phi_t + \frac{1}{2}\text{grad }(u^2 + v^2 + w^2) = -\text{grad }\frac{p}{\varrho} - \text{grad }(gy),$$

use having been made of the fact that $\varrho = \text{constant}$. Integration of this relation leads to the important equation expressing what is called *Bernoulli's law*:

$$(1.3.4) \qquad \Phi_t + \frac{1}{2}(u^2 + v^2 + w^2) + \frac{p}{\varrho} + gy = C(t),$$

in which $C(t)$ may depend on t, but not on the space variables. There are two other forms of Bernoulli's law for the case of steady flows, one of which applies along stream lines even though the flow is not irrotational, but since we make no use of these laws in this book we refrain from formulating them.

The potential equation (1.3.3) together with Bernoulli's law (1.3.4) can be used to take the place of the equations of motion (1.1.4) and the continuity equation (1.1.6) as a means of determining the velocity components u, v, w, and the pressure p: in effect, u, v, and w are determined from the solution Φ of (1.3.3), after which the pressure p can be obtained from (1.3.4). It is true that the pressure appears to be determined only within a function which is the same at each instant throughout the fluid. On physical grounds it is, however, clear that a function of t alone added to the pressure p has no effect on the motion of the fluid since no pressure gradients result from such an addition to the pressure. In fact, if we set

$$\Phi = \Phi^* + \int^t C(\xi)\, d\xi, \quad \text{then} \quad \Phi^* \quad \text{is a harmonic function with}$$

grad $\Phi =$ grad Φ^* and the Bernoulli law with reference to it has a vanishing right hand side. Thus we may take $C(t) \equiv 0$ in (1.3.4) without any essential loss of generality.

While it is true that the Laplace equation is a linear differential equation, it does not follow that we shall be able to escape all of the difficulties arising from the nonlinear character of the basic differential equations of motion (1.1.4). As we shall see, the problems of interest here remain essentially nonlinear because the Bernoulli law (1.3.4), and another condition to be derived in the next section, give rise to nonlinear boundary conditions at free surfaces. In the next section we take up the important question of the boundary conditions appropriate to various physical situations.

1.4. Boundary conditions

We assume the fluid under consideration to have a boundary surface S, fixed or moving, which separates it from some other medium, and which has the property that any particle which is once on the surface remains on it.* Examples of such boundary surfaces of importance for us are those in which S is the surface of a fixed rigid body in contact with the fluid—the bottom of the sea, for example—or the free surface of the water in contact with the air.

If such a surface S were given, for example, by an equation $\zeta(x, y, z; t) = 0$, it follows from (1.1.3) that the condition

$$(1.4.1) \qquad \frac{d\zeta}{dt} = u\zeta_x + v\zeta_y + w\zeta_z + \zeta_t = 0$$

would hold on S. From (1.3.2) and the fact that the vector $(\zeta_x, \zeta_y, \zeta_z)$ is a normal vector to S it follows that the condition (1.4.1) can be written in the form

$$(1.4.2) \qquad \frac{\partial \Phi}{\partial n} = -\frac{\zeta_t}{\sqrt{\zeta_x^2 + \zeta_y^2 + \zeta_z^2}} = v_n,$$

in which $\partial/\partial n$ denotes differentiation in the direction of the normal to S and v_n means the common velocity of fluid and boundary surface in the direction normal to the surface.

In the important special case in which the boundary surface S is fixed, i.e. it is independent of the time t, we have the condition

$$(1.4.3) \qquad \frac{\partial \Phi}{\partial n} = 0 \text{ on } S.$$

This is the appropriate boundary condition at the bottom of the sea, or at the walls of a tank containing water.

Another extremely important special case is that in which S is a *free surface* of the liquid, i.e. a surface on which the pressure p is prescribed but the form of the surface is not prescribed a priori. We shall in general assume that such a free surface is given by the equation

$$(1.4.4) \qquad y = \eta(x, z; t).$$

On such a surface $\zeta = y - \eta(x, z; t) = 0$ for any particle, and hence (1.4.1) yields the condition

$$(1.4.5) \qquad \Phi_x\eta_x - \Phi_y + \Phi_z\eta_z + \eta_t = 0 \text{ on } S.$$

In addition, as remarked above, we assume that the pressure p is given on S; as a consequence the Bernoulli law (1.3.4) yields the condition:

---

* Actually, this property is a consequence of the basic assumption in continuum mechanics that the motion of the fluid can be described mathematically as a topological deformation which depends continuously on the time t.

$$(1.4.6) \qquad g\eta + \Phi_t + \frac{1}{2}(\Phi_x^2 + \Phi_y^2 + \Phi_z^2) + \frac{p}{\varrho} = 0 \text{ on } S.$$

(As remarked earlier, we may take the quantity $C(t) = 0$ in (1.3.4).) Thus the potential function Φ must satisfy the two nonlinear boundary conditions (1.4.5) and (1.4.6) on a free surface. This is in sharp contrast to the single linear boundary condition (1.4.3) for a fixed boundary surface, but it is not strange that two conditions should be prescribed in the case of the free surface since an additional unknown function $\eta(x, z; t)$, the vertical displacement of the free surface, is involved in the latter case.

Later on we shall also be concerned with problems involving rigid bodies floating in the water and S will be the portion of the rigid body in contact with the water. In such cases the function $\eta(x, z; t)$ will be determined by the motion of the rigid body, which in turn will be fixed (through the dynamical laws of rigid body mechanics) by the pressure p between the body and the water in accordance with (1.4.6). The detailed conditions for such cases will be worked out later on at an appropriate place.

1.5. Singularities of the velocity potential

In our discussion up to now it has been tacitly assumed that all quantities such as the pressure, velocity potential, velocity components, etc. are regular functions of their arguments. It is, however, often useful to permit singularities of one kind or another to occur as an idealization of, or an approximation to, certain physical situations. Perhaps the most useful such singularity is the point source or sink which is given by the harmonic function

$$(1.5.1) \qquad \Phi = \frac{-c}{4\pi r}, \qquad r^2 = x^2 + y^2 + z^2$$

in three dimensions, and by

$$(1.5.2) \qquad \Phi = \frac{c}{2\pi}\log r, \qquad r^2 = x^2 + y^2$$

in two dimensions. Both of these functions yield flows which are radially outward from the origin, and for which the flux per unit time across a closed surface (for (1.5.1)) or a closed curve (for (1.5.2)) surrounding the origin has the value c, as one readily verifies since $\partial\Phi/\partial n = d\Phi/dr$ for $r = $ constant. That these functions represent at best idealizations of the physical situations implied in the words source and sink is clear from the fact that they yield infinite velocities at $r = 0$. Nevertheless, it is very useful here—as in other branches of applied mathematics —to accept such infinities with the reservation that the results obtained are not to be taken too literally in the immediate vicinity of the singular point.

We shall have occasion to deal with other singularities than sources or sinks, such as dipoles and multipoles, but these will be introduced when needed.

1.6. Notions concerning energy and energy flux

In dealing with surface gravity waves in water it is important and useful to analyze in some detail the flow of energy in the fluid past a given surface S. Let R be the region occupied by water and bounded by a "geometric" surface S which may, or may not, move independently of the liquid. The energy E contained in R consists of the kinetic energy of the water particles in R and their potential energy due to gravity; hence E is given by

$$(1.6.1) \quad E = \varrho \iiint\limits_{R} \left[\frac{1}{2}(\Phi_x^2 + \Phi_y^2 + \Phi_z^2) + gy \right] dx\, dy\, dz,$$

or, alternatively, by

$$(1.6.2) \qquad E = - \iiint\limits_{R} (p + \varrho \Phi_t) dx\, dy\, dz$$

upon applying Bernoulli's law (1.3.4) with $C(t) = 0$.

We wish to calculate dE/dt, having in mind that the region R is not necessarily fixed, but may depend on the time t. Quite generally, if $E = \iiint\limits_{R} f(x, y, z; t) dx\, dy\, dz$, it is well known that

$$\frac{dE}{dt} = \iiint\limits_{R} f_t\, dx\, dy\, dz + \iint\limits_{S} f v_n\, dS$$

in which v_n denotes the normal velocity of the boundary S of R taken positive in the direction outward from R. In applying the formula for dE/dt we make use of the definition of the function f implied in (1.6.1) in the first term, but take f from (1.6.2) for the second term. The result is

$$\frac{dE}{dt} = \varrho \iiint\limits_{R} (\Phi_x \Phi_{xt} + \Phi_y \Phi_{yt} + \Phi_z \Phi_{zt})\, dx\, dy\, dz$$

$$- \iint\limits_{S} (p + \varrho \Phi_t) v_n\, dS.$$

The integrand in the first integral can be expressed in the form

$$\Phi_x (\Phi_t)_x + \Phi_y (\Phi_t)_y + \Phi_z (\Phi_t)_z = \operatorname{grad} \Phi \cdot \operatorname{grad} \Phi_t$$

and hence the integral can be written as the following surface integral:

$$\varrho \iint\limits_{S} \Phi_t \frac{\partial \Phi}{\partial n}\, dS,$$

in view of Green's formula and the fact that $\nabla^2 \Phi = 0$. Thus the expression for dE/dt, the rate of change of the energy in R, can be put into the following form:

(1.6.3) $$\frac{dE}{dt} = \iint\limits_{S} [\varrho \Phi_t (\Phi_n - v_n) - p v_n]\, dS.$$

We recall that v_n means the normal velocity component of S, and Φ_n refers to the velocity component of the fluid taken in the direction of the normal to S which points outward from R.

It happens frequently that the boundary surface S of R is made up of a number of different pieces which have different properties or for which various different conditions are prescribed. Suppose first that a portion S_P of S is a "physical" boundary containing always the same fluid particles. Then Φ_n and v_n are identical (cf. (1.4.2)) and

(1.6.4.) $$\left. \frac{dE}{dt} \right|_{S_P} = - \iint\limits_{S_P} p\, v_n\, dS.$$

If, in addition, the surface S_P is fixed in space, i.e. $v_n \equiv 0$, the contribution of S_P to dE/dt evidently vanishes, as it should, since no energy flows through a fixed boundary containing always the same fluid particles. Similarly, the contribution to the energy flux also vanishes in the important special case in which S_F is a free surface on which the pressure p vanishes; this result also accords with what one expects on physical grounds.

Suppose now that S_G is a "geometric" surface fixed in space, but not necessarily consisting of the same particles of water. In this case we have $v_n = 0$ and the flow of energy through S_G is given by

$$(1.6.5) \qquad \left.\frac{dE}{dt}\right|_{S_G} = \varrho \iint\limits_{S_G} \Phi_t \Phi_n dS.$$

An important special case for us is that in which Φ is the velocity potential for a plane progressing wave given, for example, by

$$(1.6.6) \qquad \Phi(x, y, z; t) = \varphi(x - ct, y, z),$$

which represents a wave moving with constant velocity c in the direction of the x-axis. The flux through a fixed plane surface S orthogonal to the x-axis is easily seen from (1.6.5) to be given by

$$(1.6.7) \qquad \frac{dE}{dt} = -\iint\limits_{S} \varrho c \Phi_x^2 dy \, dz.$$

The negative sign results since our stipulations amount to saying that the region R occupied by the fluid lies on the negative side of S (i.e. on the side away from the positive normal, the x-axis); and consequently the energy flux through S due to a progressing wave moving in the positive direction of the normal (so that c is positive) is such as to decrease the energy in R, as one would expect. It is to be noted that there is always a flow of energy through a surface S orthogonal to the direction of a progressing wave if $\Phi_n \not\equiv 0$—even though the motion of the individual particles of the fluid should happen, for example, to be such that the particles move in a direction opposite to that of the progressing wave.

1.7. Formulation of a surface wave problem

It is perhaps useful—although somewhat discouraging, it must be admitted—to sum up the above discussion concerning the fundamental mathematical basis for our later developments by formulating a rather general, but typical, problem in the hydrodynamics of surface waves. The physical situation is indicated in Figure 1.7.1; what is intended is a situation like that on any ocean beach. The water is assumed to be initially at rest and to fill the space R defined by

$$- h(x, z) \leqq y \leqq 0, \; -\infty < z < \infty,$$

and extending to $+\infty$ in the x-direction. At the time $t = 0$, a given disturbance is created on the surface of the water over a region D (by the wind, perhaps), and one wishes to determine mathematically the subsequent motion of the water; in particular, the form of the

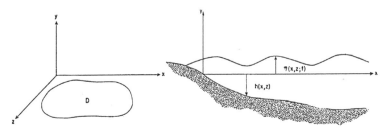

Fig. 1.7.1. A very general surface wave problem

free surface $y = \eta(x, z; t)$ is to be determined. On the basis of these assumptions the following conditions should be satisfied: First of all, the differential equation to be satisfied by Φ is, of course, the Laplace equation

$$(1.7.1) \quad \nabla^2\Phi = \Phi_{xx} + \Phi_{yy} + \Phi_{zz} = 0 \text{ for } \begin{cases} x^s(z; t) \leqq x < \infty \\ -h(x, z) \leqq y \leqq \eta(x, z; t) \\ -\infty < z < \infty \end{cases}$$

It is to be noted that $x^s(z; t)$, the abscissae of the water line on shore, and $\eta(x, z; t)$, the free surface elevation, are not known in advance but are rather to be determined as an integral part of the solution. As boundary condition to be satisfied at the bottom of the sea we have

$$(1.7.2) \qquad\qquad \frac{\partial\Phi}{\partial n} = 0 \text{ for } y = -h(x, z),$$

while the free surface conditions are the kinematic condition (cf. (1.4.5))

$$(1.7.3) \qquad \Phi_x\eta_x - \Phi_y + \Phi_z\eta_z + \eta_t = 0 \text{ for } y = \eta(x, z; t),$$

and the dynamic condition

$$(1.7.4) \quad g\eta + \Phi_t + \tfrac{1}{2}(\Phi_x^2 + \Phi_y^2 + \Phi_z^2) = F(x, z; t) \text{ on } y = \eta(x, z; t),$$

with $F(x, z; t) \equiv 0$ everywhere except over the region D where the disturbance is created. At ∞, i.e. for $x \to \infty$ and $|z| \to \infty$, we might prescribe that Φ and η remain bounded, or perhaps even that they and certain of their derivatives tend to zero. Next we have the initial conditions

$$(1.7.5) \qquad\qquad \eta(x, z; t) = 0 \text{ for } t = 0,$$

$$(1.7.6) \qquad\qquad \Phi_x = \Phi_y = \Phi_z = 0 \text{ for } t = 0,$$

appropriate to the condition of rest in an equilibrium position. Finally we must prescribe conditions fixing the disturbance; this could be done, for example, by giving the pressure p over the disturbed region D of the surface, in other words by prescribing the function F in (1.7.4) appropriately there.

One has only to write down the above formulation of our problem to realize how difficult it is to solve it. In the first place the problem is nonlinear, but what makes for perhaps even greater difficulties is the fact that the free surface ·is not known a priori and hence the domain in which the velocity potential is to be determined is not known in advance—aside from the fact that its boundary varies with the time.

These are, however, not the only difficulties in the above problem. If we assume that the function Φ is regular throughout the interior of R and uniformly bounded (together with some of its derivatives, perhaps) in R, the formulation of the problem given above would seem to be reasonable from the point of view of mechanics. However, the solution would probably not exist for all $t > 0$ for the following reason: everyone who has visited an ocean beach is well aware that the waves do not come in smoothly all the way to the shore (except possibly in very calm weather), but, rather, they steepen in front, curl over, and eventually break. In other words, any mathematical formulation of the problem which would fit the commonly observed facts even for a limited time would necessitate postulating the existence of singularities of unknown location in both space and time.

Because of the difficulty of the general nonlinear theory very little progress has been made in solving concrete problems which employ it. An exception is the problem of proving the existence of two-dimensional periodic progressing waves in water of uniform depth. This was done first by Nekrassov [N.1], [N.1a] and by Levi-Civita [L.7] for water of infinite depth, and later by Struik [S.29] for water of constant finite depth. In Chapter 12 an account of Levi-Civita's theory is given. In both cases the authors prove rigorously the existence of waves having amplitudes near to zero by showing that perturbation series in the amplitude converge. Another exception to the above statement is the problem of the solitary wave, the existence of which, from the mathematical point of view, has been proved recently by Lavrentieff [L.4] and by Friedrichs and Hyers [F.13]; an account of the work of the latter two authors is given in Chapter 10.9.

It seems likely that solutions of problems in the full nonlinear version of the theory will, for a long time to come, continue to be of the nature of existence theorems for motions of a rather special nature.

In order to make progress with the theory of surface waves it is in general necessary to simplify the theory by making special hypotheses of one kind or another which suggest themselves on the basis of the general physical circumstances contemplated in a given class of problems. As we have already explained in the introduction, up to now attention has been concentrated almost exclusively upon the two approximate theories which result when either a) the amplitude of the surface waves is considered small (with respect to wave length, for example), or b) the depth of the water is considered small (again with respect, say, to wave length). The first hypothesis leads to a linear theory and to boundary value problems more or less of classical type; while the second leads to a nonlinear theory for initial value problems, which in lowest order is of the type employed in wave propagation in compressible gases. If both hypotheses are made, the result is a linear theory involving essentially the classical linear wave equation; the present theory of the tides belongs in this class of problems.

In the next chapter we derive the approximate theories arising from the two hypotheses by starting from the general theory and then developing formally with respect to an appropriate parameter— essentially the surface wave amplitude in one case and the depth of the water in the other—and in subsequent chapters we continue by treating a variety of special problems in each of the two classes.

CHAPTER 2

The Two Basic Approximate Theories

2.1. Theory of waves of small amplitude

It has already been stated that the theory of waves of small amplitude can be derived as an approximation to the general theory presented in Chapter I on the basis of the assumption that the velocity of the water particles, the free surface elevation $y = \eta(x, z; t)$, and their derivatives, are all small quantities. We assume, in fact, that the velocity potential Φ and the surface elevation η possess the following power series expansions with respect to a parameter ε:

$$(2.1.1) \qquad \Phi = \varepsilon \Phi^{(1)} + \varepsilon^2 \Phi^{(2)} + \varepsilon^3 \Phi^{(3)} + \ldots, \text{ and}$$

$$(2.1.2) \quad \eta(x, z; t) = \eta^{(0)}(x, z; t) + \varepsilon \eta^{(1)}(x, z; t) + \varepsilon^2 \eta^{(2)}(x, z; t) + \ldots$$

It follows first of all that each of the functions $\Phi^{(k)}(x, y, z; t)$ is a solution of the Laplace equation, i.e.

$$(2.1.3) \qquad \nabla^2 \Phi^{(k)} = 0.$$

We turn next to the discussion of the boundary conditions. At a fixed physical boundary (cf. section 1.4) of the fluid we have clearly the conditions

$$(2.1.4) \qquad \frac{\partial \Phi^{(k)}}{\partial n} = 0,$$

in which $\partial/\partial n$ represents differentiation along the normal to the boundary surface.

At a free surface S: $y = \eta(x, z; t)$ on which the pressure is zero we have two boundary conditions. One of them arises from the Bernoulli law and has the form

$$g\eta + \Phi_t + \tfrac{1}{2}(\Phi_x^2 + \Phi_y^2 + \Phi_z^2) = 0 \text{ on } S.$$

Upon insertion of (2.1.1) and (2.1.2) in this condition and developing Φ_t, Φ_x^2, etc. systematically in powers of ε (due regard being paid to the fact that the functions $\Phi_t^{(k)}$, $\Phi_x^{(k)}$, etc. are to be evaluated for

$y = \eta(x, z; t)$ and that η in its turn is given in terms of ε by $\eta = \eta^{(0)} + \varepsilon\eta^{(1)} + \ldots$) one finds readily the conditions

(2.1.5) $$\eta^{(0)} = 0,$$

(2.1.6) $$g\eta^{(1)} + \Phi_t^{(1)} = 0,$$

(2.1.7) $$g\eta^{(2)} + \Phi_t^{(2)} + \tfrac{1}{2}[(\Phi_x^{(1)})^2 + (\Phi_y^{(1)})^2 + (\Phi_z^{(1)})^2] + \eta^{(1)}\Phi_{ty}^{(1)} = 0$$
$$\cdot \quad \cdot ,$$

to be satisfied for $y = \eta^{(0)}$, and since $\eta^{(0)} = 0$ from (2.1.5) it follows that the conditions (2.1.6), (2.1.7), etc. *are all to be satisfied on the originally undisturbed surface of the water* $y = 0$. The other boundary condition on S arises from the fact that the water particles stay on S (cf. section 1.4); it is expressed in the form

$$\Phi_x\eta_x + \Phi_z\eta_z + \eta_t = \Phi_y \text{ on } S.$$

Insertion of the power series for Φ and η in this expression leads to the conditions

(2.1.8) $$\eta_t^{(0)} = 0,$$

(2.1.9) $$\Phi_x^{(1)}\eta_x^{(0)} + \Phi_z^{(1)}\eta_z^{(0)} + \eta_t^{(1)} = \Phi_y^{(1)},$$

(2.1.10) $$\Phi_x^{(2)}\eta_x^{(0)} + \Phi_z^{(2)}\eta_z^{(0)} + \eta_t^{(2)} = \Phi_y^{(2)} - \Phi_x^{(1)}\eta_x^{(1)} - \Phi_z^{(1)}\eta_z^{(1)}$$
$$- \eta^{(1)}(\Phi_{xy}^{(1)}\eta_x^{(0)} + \Phi_{zy}^{(1)}\eta_z^{(0)} - \Phi_{yy}^{(1)}),$$
$$\cdot \quad \cdot \quad \cdot \quad \cdot \quad \cdot \quad \cdot \quad \cdot \quad \cdot \quad \cdot \quad \cdot \quad \cdot \quad \cdot \quad \cdot \quad \cdot \quad \cdot \quad \cdot \quad \cdot \quad \cdot \quad \cdot \quad \cdot ,$$

which are also to be satisfied for $y = 0$.

In view of the fact that $\eta^{(0)} = 0$, the free surface conditions can be put in the form

(2.1.11) $$g\eta^{(1)} + \Phi_t^{(1)} = 0,$$

(2.1.12) $$g\eta^{(2)} + \Phi_t^{(2)} = -\tfrac{1}{2}[(\Phi_x^{(1)})^2 + (\Phi_y^{(1)})^2 + (\Phi_z^{(1)})^2] - \eta^{(1)}\Phi_{ty}^{(1)},$$
$$\cdot \quad \cdot \quad \cdot \quad \cdot \quad \cdot \quad \cdot \quad \cdot \quad \cdot \quad \cdot \quad \cdot \quad \cdot \quad \cdot \quad \cdot \quad \cdot \quad \cdot \quad \cdot \quad \cdot \quad \cdot$$

(2.1.13) $$g\eta^{(n)} + \Phi_t^{(n)} = F^{(n-1)},$$

in which the symbol $F^{(n-1)}$ refers to a certain combination of the functions $\eta^{(k)}$ and $\Phi^{(k)}$ with $k \leq n - 1$, and all conditions are to be satisfied for $y = 0$. Similarly, the other set of free surface conditions becomes

$(2.1.14) \qquad \eta_t^{(1)} = \Phi_y^{(1)},$

$(2.1.15) \qquad \eta_t^{(2)} = \Phi_y^{(2)} - \Phi_x^{(1)} \eta_x^{(1)} - \Phi_z^{(1)} \eta_z^{(1)} + \Phi_{yy}^{(1)} \eta^{(1)},$

. . . .

$(2.1.16) \qquad \eta_t^{(n)} = \Phi_y^{(n)} + G^{(n-1)},$

in which $G^{(n-1)}$ depends only upon functions $\eta^{(k)}$ and $\Phi^{(k)}$ with $k \leq n - 1$, and once more all conditions are to be satisfied for $y = 0$. This theory therefore is a development in the neighborhood of the rest position of equilibrium of the water.

The relations (2.1.11) to (2.1.16) thus, in principle, furnish a means of calculating successively the coefficients of the series (2.1.1) and (2.1.2), assuming that such series exist: The conditions (2.1.11) and (2.1.14) at the free surface together with appropriate conditions at other boundaries, and initial conditions for $t = 0$, would in conjunction with $\nabla^2 \Phi^{(1)} = 0$ lead to unique solutions $\eta^{(1)}$ and $\Phi^{(1)}$. Once $\eta^{(1)}$ and $\Phi^{(1)}$ are determined, they can be inserted in the conditions (2.1.12) and (2.1.15) to yield two conditions for $\eta^{(2)}$ and $\Phi^{(2)}$ which with the subsidiary boundary and other conditions on $\Phi^{(2)}$ serve to determine them, etc. One could interpret the work of Levi-Civita [L.7] and Struik [S.29] referred to in section 1.7 as a method of proving the existence of progressing waves which are periodic in x by showing that the functions Φ and η can indeed be represented as convergent power series in ε for ε sufficiently small.

In what follows in Part II of this book we shall content ourselves in the main with the degree of approximation implied in breaking off the perturbation series after the terms $\varepsilon\Phi^{(1)}$ and $\varepsilon\eta^{(1)}$ in the series (2.1.1) and (2.1.2), i.e. we set $\Phi \equiv \varepsilon\Phi^{(1)}$ and $\eta \equiv \varepsilon\eta^{(1)}$. With this stipulation the free surface conditions (2.1.11) and 2.1.14) yield

$(2.1.17) \qquad\qquad g\eta + \Phi_t = 0 \left.\begin{array}{c} \\ \\ \end{array}\right\}$ for $y = 0$.

$(2.1.18) \qquad\qquad \eta_t - \Phi_y = 0$

By elimination of η between these two relations the single condition on Φ:

$(2.1.19) \qquad\qquad \Phi_{tt} + g\Phi_y = 0 \qquad$ for $y = 0$

is obtained; this condition is the one which will be used mainly in Part II in order to determine Φ from $\nabla^2 \Phi = 0$, after which the free surface elevation η can be determined from (2.1.17). The usual method of obtaining the last three conditions is to reject all but the linear terms in η and Φ and their derivatives in the kinematic (cf. (1.4.5)) and dynamic (cf. (1.4.6)) free surface boundary conditions. By proceeding in this way we can obtain a first approximation to the pressure p (which was not considered in the above general perturbation scheme) in the form:

$$(2.1.20) \qquad \frac{p}{\varrho} = -gy - \Phi_t. *$$

We can now see the great simplifications which result through the linearization of the free surface conditions: not only does the problem become linear, but also the domain in which its solution is to be determined becomes fixed a priori and consequently the surface wave problems in this formulation belong, from the mathematical point of view, to the classical boundary problems of potential theory.

2.2. Shallow water theory to lowest order. Tidal theory

A different kind of approximation from the foregoing linear theory of waves of small amplitude results when it is assumed that the *depth* of the water is sufficiently small compared with some other significant length, such as, for example, the radius of curvature of the water surface. In this theory it is not necessary to assume that the displacement and slope of the water surface are small, and the resulting theory is as a consequence not a linear theory. There are many circumstances in nature under which such a theory leads to a good approximation to the actual occurrences, as has already been mentioned in the introduction. Among such occurrences are the tides in the oceans, the "solitary wave" in sufficiently shallow water, and the breaking of waves on shallow beaches. In addition, many phenomena met with in hydraulics concerning flows in open channels such as roll waves, flood waves in rivers, surges in channels due to sudden influx of water, and other kindred phenomena, belong in the nonlinear shallow water theory. Chapters 10 and 11 are devoted to the working out of consequences of the shallow water theory.

The shallow water theory is, in its lowest approximation, the basic theory used in hydraulics by engineers in dealing with flows in open

---

* In case the surface pressure $p_0(x, z; t)$ is not zero one finds readily that (2.1.17) is replaced by

$$(2.1.20)_1 \qquad g\eta + \Phi_t = -p_0/\varrho,$$

while (2.1.18) remains unaltered.

channels, and also the theory commonly referred to in the standard treatises on hydrodynamics as the theory of long waves. We begin by giving first a derivation of the theory for two-dimensional motion along essentially the lines followed by Lamb [L.8], p. 254. As usual, the undisturbed free surface of the water is taken as the x-axis and the y-axis is taken vertically upwards. The bottom is given by $y = -h(x)$, so that h represents the variable depth of the undisturbed water. The surface displacement is given by $y = \eta(x, t)$. The velocity components are denoted by $u(x, y, t)$ and $v(x, y, t)$.

The equation of continuity is

$$(2.2.1) \qquad u_x + v_y = 0.$$

The conditions to be satisfied at the free surface are the kinematical condition:

$$(2.2.2) \qquad (\eta_t + u\eta_x - v)\,|_{y=\eta} = 0;$$

and the dynamical condition on the pressure:

$$(2.2.3) \qquad p\,|_{y=\eta} = 0.$$

At the bottom the condition is

$$(2.2.4) \qquad (uh_x + v)\,|_{y=-h} = 0.$$

Integration of (2.2.1) with respect to y yields

$$(2.2.5) \qquad \int_{-h}^{\eta} (u_x)\,dy + v\,|_{-h}^{\eta} = 0.$$

Use of the condition (2.2.2) at $y = \eta$ and (2.2.4) at $y = -h$ yields the relation

$$(2.2.6) \qquad \int_{-h}^{\eta} u_x\,dy + \eta_t + u\,|_{\eta} \cdot \eta_x + u\,|_{-h} \cdot h_x = 0.$$

We introduce the relation

$$(2.2.7) \qquad \frac{\partial}{\partial x} \int_{-h(x)}^{\eta(x)} u\,dy = u\,|_{y=\eta} \cdot \eta_x + u\,_{y=-h} \cdot h_x + \int_{-h}^{\eta} u_x\,dy.$$

and combine it with (2.2.6) to obtain

$$(2.2.8) \qquad \frac{\partial}{\partial x} \int_{-h}^{\eta} u\,dy = -\eta_t.$$

Up to this point no approximations have been introduced.

The shallow water theory is an approximate theory which results from the assumption that the y-component of the acceleration of the water particles has a negligible effect on the pressure p, or, what amounts to the same thing, that the pressure p is given as in hydrostatics * by

(2.2.9) $p = g\varrho(\eta - y).$

The quantity ϱ is the density of the water. A number of consequences of (2.2.9) are useful for our purposes. To begin with, we observe that

(2.2.10) $p_x = g\varrho\eta_x,$

so that p_x is independent of y. It follows that the x-component of the acceleration of the water particles is also independent of y; and hence u, the x-component of the velocity, is also independent of y for all t if it was at any time, say at $t = 0$. We shall assume this to be true in all cases—it is true for example in the important special case in which the water was at rest at $t = 0$—so that $u = u(x, t)$ depends only on x and t from now on. As equation of motion in the x-direction we may write, therefore, in view of (2.2.10):

(2.2.11) $u_t + uu_x = - g\eta_x.$

This is simply the usual equation of motion in the Eulerian form, use having been made of $u_y = 0$. In addition, (2.2.8) may now be written

(2.2.12) $[u(\eta + h)]_x = - \eta_t,$

since $\int_{-h}^{\eta} u\,dy = u\int_{-h}^{\eta} dy$ on account of the fact that u is independent of y. The two first order differential equations (2.2.11) and (2.2.12) for the functions $u(x, t)$ and $\eta(x, t)$ are the differential equations of the nonlinear shallow water theory. Once the initial state of the fluid is prescribed, i.e. once the values of u and η at the time $t = 0$ are given, the equations (2.2.11) and (2.2.12) yield the subsequent motion.

If in addition to the basic assumption of the shallow water theory expressed by (2.2.9) we assume that u and η, the particle velocity and free surface elevation, and their derivatives are small quantities whose squares and products can be neglected in comparison with linear terms, it follows at once that equations (2.2.11) and (2.2.12) simplify to

(2.2.13) $u_t = - g\eta_x,$

(2.2.14) $(uh)_x = - \eta_t,$

* We have $p_y = - g\varrho$ and (2.2.9) results through the use of $p = 0$ for $y = \eta$.

from which η can be eliminated to yield for u the equation

$$(2.2.15) \qquad (uh)_{xx} - \frac{1}{g} u_{tt} = 0.$$

If, in addition, the depth h is constant it follows readily that u satisfies the *linear wave equation*

$$(2.2.16) \qquad u_{xx} - \frac{1}{gh} u_{tt} = 0.$$

In this case η satisfies the same equation. One observes therefore the important result that the propagation speed of a disturbance is given by $\sqrt{gh}$. In principle, this linearized version of the shallow water theory is the one which has always been used as the basis for the theory of the tides. Of course, the tidal theory for the oceans requires for its complete formulation the introduction of the external forces acting on the water due to the gravitational attraction of the moon and the sun, and also the Coriolis forces due to the rotation of the earth, but nevertheless the basic fact about the tidal theory from the standpoint of mathematics is that it belongs to the linear shallow water theory. The actual oceans do not from most points of view impress one as being shallow; in the present connection, however, the depth is actually very small compared with the curvature of the tidal wave surface so that the shallow water approximation is an excellent one. That the tidal phenomena should be linear to a good approximation would also seem rather obvious on account of the small amplitudes of the tides compared with the dimensions of the oceans. A few additional remarks about tidal theory and some other applications of the linearized version of the shallow water theory to concrete problems (seiches in lakes, and floating breakwaters, for example) are given in Chapter 10.13.

2.3. Gas dynamics analogy

It is possible to introduce a different set of dependent variables in such a way that the equations of the shallow water theory become *analogous to the fundamental differential equations of gas dynamics for the case of a compressible flow involving only one space variable x.* (This seems to have been noticed first by Riabouchinsky [R.8].) To this end we introduce the mass per unit area given by

$$(2.3.1) \qquad \bar{\varrho} = \varrho(\eta + h).$$

Since h depends only on x we have

$$(2.3.2) \qquad\qquad \bar{\varrho}_t = \varrho \eta_t .$$

We next define the force $\bar{p}$ per unit width:

$$(2.3.3) \qquad\qquad \bar{p} = \int_{-h}^{\eta} p \, dy,$$

which, in view of (2.2.9) and (2.3.1), leads to

$$(2.3.4) \qquad\qquad \bar{p} = \frac{g\varrho}{2} (\eta + h)^2 = \frac{g}{2\varrho} \bar{\varrho}^2.$$

The relation between $\bar{p}$ and $\bar{\varrho}$ is thus of the form $\bar{p} = A\bar{\varrho}^{\gamma}$ with $\gamma = 2$, that is, the "pressure" $\bar{p}$ and the "density" $\bar{\varrho}$ are connected by an "adiabatic" relation with the fixed exponent 2.

Equation (2.2.11) may now be written

$$\varrho(\eta + h)(u_t + uu_x) = - g\varrho(\eta + h)\eta_x$$

and this, in turn, may be expressed through use of (2.3.1) and (2.3.4) as follows:

$$(2.3.5) \qquad\qquad \bar{\varrho}(u_t + uu_x) = - \bar{p}_x + g\bar{\varrho}h_x ,$$

as one can readily verify.

The equation (2.2.12) may be written as

$$(2.3.6) \qquad\qquad (\bar{\varrho}u)_x = - \bar{\varrho}_t ,$$

in view of (2.3.2) as well as (2.3.1). The differential equations (2.3.5) and (2.3.6), together with the "adiabatic" law $\bar{p} = g\bar{\varrho}^2/2\varrho$ given by (2.3.4), are identical in form with the equations of compressible gas dynamics for a one-dimensional flow except for the term $g\bar{\varrho}h_x$ on the right hand side of (2.3.5), and this term vanishes if the original undisturbed depth h of the water is constant. The "sound speed" c corresponding to our equations (2.3.5) and (2.3.6) is, in analogy with gas dynamics, given by $c = \sqrt{d\bar{p}/d\bar{\varrho}}$, and this from (2.3.4) and (2.3.1) has the value

$$(2.3.7) \qquad\qquad c = \sqrt{\frac{g\bar{\varrho}}{\varrho}} = \sqrt{g(\eta + h)}.$$

It will be seen later that $c(x, t)$ represents the local speed at which a small disturbance advances relative to the water.

2.4. Systematic derivation of the shallow water theory

It is of course a matter of importance to know under what circumstances the shallow water theory can be expected to furnish sufficiently accurate results. The only assumption made above in addition to the customary assumptions of hydrodynamics was that the pressure is given as in hydrostatics by (2.2.9), but no assumption was made regarding the magnitude of the surface elevation or the velocity components. Consequently the shallow water theory may be accurate for waves whose amplitude is not necessarily small, provided that the hydrostatic pressure relation is not invalidated. The above derivation of the shallow water theory is, however, open to the objection that the role played by the undisturbed depth of the water in determining the accuracy of the approximation is not put in evidence. In fact, since we shall see later on that all motions die out rather rapidly in the depth, it would at first sight seem reasonable to expect that the hydrostatic law for the pressure would be, on the whole, more accurate the deeper the water. That this is not the case in general is well known, since the solutions for steady progressing waves of small amplitude (i.e. for solutions obtained by the linearized theory) in water of uniform but finite depth are approximated accurately by the solutions of the shallow water theory (when it also is linearized) only when the depth of the water is small compared with the wave length (cf. Lamb [L.3], p. 368). It is possible to give a quite different derivation of the shallow water theory in which the equations (2.2.11) and (2.2.12) result from the exact hydrodynamical equations as the approximation of lowest order in a perturbation procedure involving a formal development of all quantities in powers of the ratio of the original depth of the water to some other characteristic length associated with the horizontal direction.* The relation (2.2.9) is then found to be correct within quadratic terms in this ratio. In this section we give such a systematic derivation of the shallow water theory, following K. O. Friedrichs (see the appendix to [S.19]), which, unlike the derivation given in section

* In this book the parameter of the shallow water theory is defined in two different ways: in dealing with the breaking of waves in Chapter 10.10 it is the ratio of the depth to a significant radius of curvature of the free surface; in dealing with the solitary wave, however, it is essentially the ratio of the depth to the quantity U^2/g, with U the propagation speed of the wave, and in this case the development is carried out for U^2/gh near to one. In still other problems it might well be defined differently in terms of parameters that are characteristic for such problems.

2.2 above, is capable of yielding higher order approximations.

The disposition of the coordinate axes is taken in the usual manner, with the x, z-plane the undisturbed water surface and the y-axis positive upward. The free surface elevation is given by $y = \eta(x, z, t)$ and the bottom surface by $y = -h(x, z)$. We recapitulate for the sake of convenience the differential equations and boundary conditions in terms of the Euler variables, that is, the equations of continuity and motion, the vanishing of the rotation, and the boundary conditions:

$$(2.4.1) \qquad u_x + v_y + w_z = 0,$$

$$(2.4.2) \quad \begin{cases} u_t + uu_x + vu_y + wu_z = -\dfrac{1}{\varrho}p_x, \\[2mm] v_t + uv_x + vv_y + wv_z = -\dfrac{1}{\varrho}p_y - g, \\[2mm] w_t + uw_x + vw_y + ww_z = -\dfrac{1}{\varrho}p_z, \end{cases}$$

$$(2.4.3) \qquad w_y = v_z, \ u_z = w_x, \ v_x = u_y,$$

$$(2.4.4) \qquad \eta_t + u\eta_x + w\eta_z = v \text{ at } y = \eta,$$

$$(2.4.5) \qquad p = 0 \text{ at } y = \eta,$$

$$(2.4.6) \qquad uh_x + v + wh_z = 0 \text{ at } y = -h.$$

We now introduce dimensionless variables through the use of two lengths d and k, with d intended to represent a typical depth and k a typical length in the horizontal direction—it is characteristic of the procedure followed here that the horizontal and vertical directions are not treated in the same way. The new independent variables are as follows:

$$(2.4.7) \qquad \bar{x} = x/k, \ \bar{y} = y/d, \ \bar{z} = z/k, \ \tau = t\sqrt{gd}/k,$$

while the new dimensionless dependent variables are

$$(2.4.8) \quad \begin{cases} \bar{u} = (\sqrt{gd})^{-1}u, \ \ \bar{v} = (k\sqrt{gd}/d)^{-1}v, \ \ \bar{w} = (\sqrt{gd})^{-1}w \\[2mm] \bar{p} = \dfrac{1}{\varrho gd}p, \\[2mm] \bar{\eta} = \eta/d, \ \ \bar{h} = h/d. \end{cases}$$

In addition, we introduce the important parameter

$$(2.4.9) \qquad \sigma = d^2/k^2$$

in terms of which all quantities will be developed; *when this parameter is small the water is considered to be shallow.* This means, of course, that d is small compared with k, and hence that the x and z coordinates (cf. (2.4.7)) are stretched differently from the y coordinate and in a fashion which depends upon the development parameter. Since it is the horizontal coordinate which is strongly stretched relative to the depth coordinate, it seems reasonable to refer to the resulting theory as a shallow water theory. The stretching process combined with a development with respect to σ is the characteristic feature of what we call the shallow water theory throughout this book. The dimensionless development parameter σ has a physical significance, of course, but its interpretation will vary depending on the circumstances in individual cases, as has already been noted above. For example, consider a problem in which the motion is to be predicted starting from rest with initial elevation $y = \eta_0(x, y, z)$ prescribed; from (2.4.8) we have

$$y = \eta_0 = d\bar{\eta}_0(\bar{x}, \bar{y}, \bar{z})$$

$$= d\bar{\eta}_0\left(\frac{x}{k}, \frac{y}{d}, \frac{z}{k}\right)$$

from which we obtain

$$dy_{xx} = \frac{d^2}{k^2}\bar{\eta}_{0\bar{x}\bar{x}} = \sigma\bar{\eta}_{0\bar{x}\bar{x}}.$$

It is natural to assume that the dimensionless second derivative $\bar{\eta}_{0\bar{x}\bar{x}}$ will be at least bounded and consequently one sees that the assumption that σ is small might be interpreted in this case as meaning that the product of the curvature of the free surface of the water and a typical depth is a small quantity.

The object now is to consider a sequence of problems depending on the small parameter σ and then develop in powers of σ. Introduction of the new variables in the equations (2.4.1) to (2.4.6) yields

(2.4.1)′ $$\sigma u_x + v_y + \sigma w_z = 0,$$

(2.4.2)′ $$\begin{cases} \sigma[u_t + uu_x + wu_z + p_x] + vu_y = 0, \\ \sigma[v_t + uv_x + wv_z + p_y + 1] + vv_y = 0, \\ \sigma[w_t + uw_x + ww_z + p_z] + vw_y = 0, \end{cases}$$

(2.4.3)′ $$w_y = v_z, \quad u_z = w_x, \quad v_x = u_y,$$

(2.4.4)′ $$\sigma[\eta_t + u\eta_x + w\eta_z] = v \text{ at } y = \eta,$$

$(2.4.5)'$ $p = 0$ at $y = \eta$,

$(2.4.6)'$ $\sigma[uh_x + wh_z] + v = 0$ at $y = -h$,

when bars over all quantities are dropped and τ is replaced by t.

The next step is to assume power series developments for u, v, w, η, and p:

$(2.4.10)$
$$\begin{cases} u = u^{(0)} + u^{(1)}\sigma + u^{(2)}\sigma^2 + \cdots, \\ v = v^{(0)} + v^{(1)}\sigma + v^{(2)}\sigma^2 + \cdots, \\ w = w^{(0)} + w^{(1)}\sigma + w^{(2)}\sigma^2 + \cdots, \\ \eta = \eta^{(0)} + \eta^{(1)}\sigma + \eta^{(2)}\sigma^2 + \cdots, \\ p = p^{(0)} + p^{(1)}\sigma + p^{(2)}\sigma^2 + \cdots, \end{cases}$$

and insert them in the equations $(2.4.1)'$ to $(2.4.6)'$ to obtain, by equating coefficients of like powers of σ, equations for the successive coefficients in the series, which are of course functions of x, y, z, and t. The terms of zero order yield the equations

$(2.4.1)'_0$ $v_y^{(0)} = 0$,

$(2.4.2)'_0$
$$\begin{cases} v^{(0)}u_y^{(0)} = 0, \\ v^{(0)}v_y^{(0)} = 0, \\ v^{(0)}w_y^{(0)} = 0, \end{cases}$$

$(2.4.3)'_0$ $w_y^{(0)} = v_z^{(0)}, \ u_z^{(0)} = w_x^{(0)}, \ v_x^{(0)} = u_y^{(0)}$,

$(2.4.4)'_0$ $v^{(0)} = 0$ at $y = \eta^{(0)}$,

$(2.4.5)'_0$ $p^{(0)} = 0$ at $y = \eta^{(0)}$,

$(2.4.6)'_0$ $v^{(0)} = 0$ at $y = -h$.

These equations yield the following:

$(2.4.11)$ $v^{(0)} \equiv 0$,

$(2.4.12)$ $w^{(0)} = w^{(0)}(x, z, t)$,

$(2.4.13)$ $u^{(0)} = u^{(0)}(x, z, t)$,

$(2.4.14)$ $p^{(0)}(x, \eta^{(0)}, z, t) = 0$,

which contain the important results that the vertical velocity component is zero and the horizontal velocity components are independent of the vertical coordinate y in lowest order.

The first order terms arising from $(2.4.1)'$ to $(2.4.6)'$ in their turn yield the equations

$(2.4.1)'_1$ $$u_x^{(0)} + w_z^{(0)} = -v_y^{(1)},$$

$(2.4.2)'_1$ $$\begin{cases} u_t^{(0)} + u^{(0)}u_x^{(0)} + w^{(0)}u_z^{(0)} + p_x^{(0)} = 0, \\ p_y^{(0)} + 1 = 0, \\ w_t^{(0)} + u^{(0)}w_x^{(0)} + w^{(0)}w_z^{(0)} + p_z^{(0)} = 0, \end{cases}$$

$(2.4.4)'_1$ $$\eta_t^{(0)} + u^{(0)}\eta_x^{(0)} + w^{(0)}\eta_z^{(0)} = v^{(1)} \text{ at } y = \eta^{(0)},$$

$(2.4.6)'_1$ $$u^{(0)}h_x + w^{(0)}h_z + v^{(1)} = 0 \text{ at } y = -h,$$

upon making use of (2.4.11), (2.4.12), and (2.4.13). Equation $(2.4.1)'_1$ can be integrated at once since $u^{(0)}$ and $w^{(0)}$ are independent of y to yield

$(2.4.15)$ $$v^{(1)} = -(u_x^{(0)} + w_z^{(0)})y + F(x, z, t),$$

with F an arbitrary function which can be determined by using $(2.4.6)'_1$; the result for $v^{(1)}$ is then

$(2.4.16)$ $$v^{(1)} = -(u_x^{(0)} + w_z^{(0)})y - [(u^{(0)}h)_x + (w^{(0)}h)_z]_{y=-h}.$$

To second order the vertical component of the velocity is thus linear in the depth coordinate. In similar fashion the second of the equations $(2.4.2)'_1$ can be integrated and the additive arbitrary function of x, z, t determined from (2.4.14); the result is

$(2.4.17)$ $$p^{(0)}(x, y, z, t) = \eta^{(0)}(x, z, t) - y$$

which is obviously the hydrostatic pressure relation (in dimensionless form).

In the derivation of the shallow water theory given in the preceding section this relation was taken as the starting point; here, it is derived as the lowest order approximation in a formal perturbation scheme. However, it is of course not true that we have *proved* that (2.4.17) is in some sense an appropriate assumption: instead, it should be admitted frankly that our dimensionless variables were introduced in just such a way that (2.4.17) would result. If it could be shown that our perturbation procedure really does yield a correct asymptotic development (that the development converges seems unlikely since the equations (2.4.1)' to (2.4.6)' degenerate in order so greatly for $\sigma = 0$) then the hydrostatic pressure assumption could be considered as having been justified mathematically. A proof that this is the case would be of great interest, since it would give a mathematical justification for the shallow water theory; to do so in a general way would seem to be a very difficult task, but Friedrichs and Hyers [F.13] have shown that the development does yield the existence of the solution in the important special case of the solitary wave (cf. Chapter 10.9). (Keller [K.6] had shown earlier that the formal procedure yields the solitary wave.) The problem is of considerable mathematical interest also because of the following intriguing circumstance: the approximation of lowest order to the solution of a problem in potential theory is sought in the form of a solution of a nonlinear wave equation, and this means that the solution of a problem of elliptic type is approximated (at least in the lowest order) by the solution of a problem of hyperbolic type.

The values of $v^{(1)}$ and $p^{(0)}$ given by (2.4.16) and (2.4.17) are now inserted in the first and third equations of $(2.4.2)'_1$ and in $(2.4.4)'_1$ to yield finally

$$(2.4.18) \qquad u_t^{(0)} + u^{(0)} u_x^{(0)} + w^{(0)} u_z^{(0)} + \eta_x^{(0)} = 0,$$

$$(2.4.19) \qquad w_t^{(0)} + u^{(0)} w_x^{(0)} + w^{(0)} w_z^{(0)} + \eta_z^{(0)} = 0,$$

$$(2.4.20) \qquad \eta_t^{(0)} + [u^{(0)}(\eta^{(0)} + h)]_x + [w^{(0)}(\eta^{(0)} + h)]_z = 0,$$

as definitive equations for $u^{(0)}$, $w^{(0)}$, and $\eta^{(0)}$ —all of which, we repeat, depend only upon x, z, and t. If the superscript is dropped, $w^{(0)}$ is taken to be zero, and it is assumed that all quantities are independent of z, one finds readily that these equations become identical with equations (2.2.11) and (2.2.12) except for the factor g in (2.2.11) which is missing here because of our introduction of a dimensionless pressure.

It is clear that the above process can be continued to obtain the higher order approximations. An example of such a calculation will be given later in Chapter 10.9, where we shall see that the first non-trivial term in the development which yields the solitary wave is of second order.

PART II

Summary

In Part II we treat a variety of problems in terms of the theory which arises through linearization of the free surface condition (cf. the preceding chapter); thus the problems refer to waves of small amplitude. To this theory the names of Cauchy and Poisson are usually attached. The material falls into three different types, or classes, of problems, as follows: A) *Waves that are simple harmonic in the time.* These problems are treated in Chapters 3, 4, and 5 and they include a study of the classical standing and progressing wave solutions in water of uniform depth, and waves over sloping beaches and past obstacles of one kind or another. The mathematical tools employed here comprise, aside from classical methods in potential theory, a thorough-going use of integrals in the complex domain. B) *Waves created by disturbances initiated at an instant when the water is at rest.* These problems, which are treated in Chapter 6, comprise a variety of unsteady motions, including the propagation of waves from a point impulse and from an oscillatory source. Uniqueness theorems for the unsteady motions are derived. The principle mathematical tool used in solving these problems is the Fourier transform. The method of stationary phase is justified and used. C) *Waves arising from obstacles immersed in a running stream.* This category of problems differs from the first two in that the motion to be investigated is a small oscillation in the neighborhood of a uniform flow, while the former cases concern small oscillations near the state of rest. This difference is in one respect rather significant since the problems of the first two types require no restriction on the shape of immersed bodies, or obstacles, while the third type of problem requires that the immersed bodies should be in the form of thin disks, since otherwise the flow velocity would be changed by a finite amount, and a linearization of the free surface condition would not then be justified. In other words, the problems of this third type require a linearization based on assuming a small thickness for any immersed bodies, as well as a linearization with respect to the amplitude of the surface waves. These problems are treated in Chapters 7, 8, and 9. The classical case of the waves created by a small obstacle in a running stream of uniform depth is first treated. This includes the classical shipwave problem, discussed in Chapter 8, in which the "ship" is treated as though it could be replaced by a point singularity. A treatment is given in Chapter 9 of the problem of the waves created by a ship moving through a sea of arbitrary waves, assuming the ship to be a floating rigid body with six degrees of freedom and with its motion determined by the propeller thrust and the pressure of the water on its hull.

Finally, in an Appendix to Part II a brief summary of some of the more recent literature concerned with the above types of problems is given, since the cases selected for detailed treatment here do not by any means exhaust the interesting problems which have been solved.

WAVES SIMPLE HARMONIC IN THE TIME

CHAPTER 3

Simple Harmonic Oscillations in Water of Constant Depth

3.1. Standing waves

In Chapter 2 we have derived the basic theory of irrotational waves of small amplitude with the following results (in the lowest order, that is). Assuming the x, z-plane to coincide with the free surface in its undisturbed position, with the y-axis positive upward, the velocity potential $\Phi(x, y, z; t)$ satisfies the following conditions:

$$(3.1.1) \qquad \nabla^2\Phi = \Phi_{xx} + \Phi_{yy} + \Phi_{zz} = 0$$

in the region bounded above by the plane $y = 0$ and elsewhere by any other given boundary surfaces. The free surface condition under the assumption of zero pressure there is

$$(3.1.2) \qquad \Phi_{tt} + g\Phi_y = 0 \text{ for } y = 0.$$

The condition at fixed boundary surfaces is that $\partial\Phi/\partial n = 0$; for water of uniform depth $h = \text{const.}$ we have therefore the condition

$$(3.1.3) \qquad \Phi_y = 0 \text{ for } y = -h.$$

Once the velocity potential Φ has been determined the elevation $\eta(x, z; t)$ of the free surface is given by

$$(3.1.4) \qquad \eta = -\frac{1}{g}\Phi_t(x, 0, z; t).$$

Conditions at ∞ as well as appropriate initial conditions at $t = 0$ must also be prescribed.

In this section we are interested in those special types of *standing waves* which are simple harmonic in the time; we therefore write

(3.1.5) $\Phi(x, y, z; t) = e^{i\sigma t}\varphi(x, y, z)$ *

with φ a real function, and with the understanding that either the
real or the imaginary part of the right hand side is to be taken.
The problems to be treated here thus belong to the theory of small
oscillations of dynamical systems in the neighborhood of an equilibrium position.

The conditions on Φ given above translate into the following
conditions on φ:

(3.1.6) $\nabla^2\varphi = 0$, $- h < y < 0$, $- \infty < x, z < \infty$,

(3.1.7) $\varphi_y - \dfrac{\sigma^2}{g}\varphi = 0$, $y = 0$,

(3.1.8) $\varphi_y = 0$, $y = - h$.

As conditions at ∞ we assume that φ and φ_y are uniformly bounded.**
Arbitrary initial conditions cannot now be prescribed, of course,
since we have assumed the behavior of our system to be simple
harmonic in the time. The free surface elevation is given by

(3.1.9) $\eta = - \dfrac{i\sigma}{g} e^{i\sigma t} \cdot \varphi(x, 0, z).$

We look first for standing wave motions which are *two-dimensional*,
so that φ depends only upon x and y: $\varphi = \varphi(x, y)$, and also consider
first the case of *water of infinite depth*, i.e. $h = \infty$. One verifies
readily that the functions

(3.1.10) $\begin{cases} \varphi = e^{my} \cos mx \\ \varphi = e^{my} \sin mx \end{cases}$

are harmonic functions which satisfy the free surface condition
(3.1.7) provided that the constant m satisfies the relation

(3.1.11) $m = \sigma^2/g.$

In addition, the conditions at ∞ are satisfied. In particular, it is of
interest to observe that the oscillations die out exponentially in the
depth. The free surface elevation is then given by

---

* The most general standing wave would be given by $\Phi = f(t)\varphi(x, y, z)$. This
means, of course, that the shape of the wave in space is fixed within a multiplying
factor depending only on the time. Thus nodes, maxima and minima, etc. occur
at the same points independent of the time.
** This means that the vertical components of the displacement and velocity
are bounded at ∞. One could prescribe more general conditions at ∞ without
impairing the uniqueness of the solutions of our boundary value problems, but
it does not seem worth while to do so in this case.

$$(3.1.12) \qquad \eta = - \frac{i\sigma}{g} e^{i\sigma t} \cdot \begin{cases} \cos mx \\ \sin mx \end{cases}.$$

It should be pointed out specifically that our boundary problem, though it is linear and homogeneous, has in addition to the solution $\varphi \equiv 0$ a two-parameter set of "non trivial" solutions obtained by taking linear combinations of the two solutions given in (3.1.10).

The surface waves given by (3.1.10) are thus simple harmonic in x as well as in t. The relation (3.1.11) furthermore states the very important fact that the wave length λ given by

$$(3.1.13) \qquad \lambda = 2\pi/m = 2\pi g/\sigma^2$$

is not independent of the frequency of the oscillation, but varies inversely as its square.

The above discussion yields standing wave solutions of physically reasonable type, but one nevertheless wonders whether there might not be others—for example, standing waves which are not simple sine or cosine functions of x, but rather waves with amplitudes which, for example, die out as x tends to infinity. Such waves do not occur in two dimensions,* however, in the sense that *all solutions for water of infinite depth*, except $\varphi \equiv 0$, *of the homogeneous boundary problem formulated in (3.1.6) and (3.1.7)* together with the condition that φ and φ_y are uniformly bounded at ∞ *are given by (3.1.10)* with m satisfying (3.1.11). This is a point worth pausing to prove, especially since the method of proof foreshadows a mode of attack on our problems which will be used in a more essential way later on. The first step in the uniqueness proof is to introduce the function $\psi(x, y)$ defined by

$$(3.1.14) \qquad \psi = \varphi_y - m\varphi, \qquad m > 0.$$

Since φ is a harmonic function, obviously ψ is also a harmonic function. In addition, ψ vanishes for $y = 0$ on account of its definition and (3.1.7). Hence ψ can be continued by reflection over the x-axis into a potential function which is regular and defined as a single-valued function in the entire x, y-plane. Since φ and φ_y were assumed to be uniformly bounded in the entire lower half plane it follows that ψ is bounded in the entire x, y-plane since reflection in the x-axis does not destroy boundedness properties. Thus ψ is a potential function which is regular and bounded in the entire x, y-plane. By Liou-

---

* This statement is not valid in three dimensions as we shall see later on in this section.

ville's theorem it is therefore a constant, and since $\psi = 0$ for $y = 0$, the constant must be zero. Hence ψ vanishes identically. From (3.1.14) it therefore follows that any solutions $\varphi(x, y)$ of our boundary value problem are also solutions of the differential equation

(3.1.15) $\qquad \varphi_y - m\varphi = 0, \qquad -\infty < y < 0.$

The most general solution of this differential equation is given by

(3.1.16) $\qquad\qquad \varphi = c(x)e^{my}$

with $c(x)$ an arbitrary function of x alone. However, $\varphi(x, y)$ is a harmonic function and hence $c(x)$ is a solution of

(3.1.17) $\qquad\qquad \dfrac{d^2c}{dx^2} + m^2c = 0$

which, in turn, has as its general solution the linear combinations of $\sin mx$ and $\cos mx$. It follows, therefore, that the standing wave solutions of our problem are indeed all of the form $Ae^{my} \cos (mx+\alpha)$,* with α and A arbitrary constants fixing the "phase" and the amplitude of the wave, while m is a fixed constant which determines the wave length λ in terms of the given frequency σ through (3.1.13).

In water of uniform finite depth h it is also quite easy to obtain two-dimensional standing wave solutions of our boundary value problem. One has, corresponding to the solutions (3.1.10) for water of infinite depth, the harmonic functions

(3.1.18) $\qquad \begin{cases} \varphi = \cosh m(y + h) \cos mx, \\ \varphi = \cosh m(y + h) \sin mx, \end{cases}$

as solutions which satisfy the boundary condition at the bottom, while the free surface condition is satisfied provided that the constant m satisfies the relation

(3.1.19) $\qquad\qquad \sigma^2 = gm \tanh mh$

instead of the relation (3.1.11), as one readily sees. Since $\tanh mh \to 1$ as $h \to \infty$ it is clear that the relation (3.1.19) yields (3.1.11) as limit relation for water of infinite depth. The uniqueness of the solutions (3.1.18) for the two-dimensional case under the condition of boundedness at ∞ was first proved by A. Weinstein [W.7] by a method

---

* It can now be seen that the negative sign in the free surface condition (3.1.7) is decisive for our results: if this sign were reversed one would find that the solution φ analogous to (3.1.16) would be bounded at ∞ only for $c(x) \equiv 0$, because (3.1.16) would now be replaced by $c(x)e^{-my}$, with $m > 0$.

different from the method used above for water of infinite depth which can not be employed in this case (see [B. 12]).

It is of interest to calculate the motion of the individual water particles. To this end let δx and δy represent the displacements from the mean position (x, y) of a given particle. Our basic assumptions mean that δx, δy and their derivatives are small quantities; it follows therefore that we may write

$$\frac{d\delta x}{dt} = u(x, y) = \Phi_x = - m A \cos \sigma t \cosh m(y + h) \sin mx$$

$$\frac{d\delta y}{dt} = v(x, y) = \Phi_y = mA \cos \sigma t \sinh m(y + h) \cos mx$$

within the accuracy of our basic approximation. The constant A is an arbitrary factor fixing the amplitude of the wave. Hence we have upon integration

$$(3.1.20) \quad \begin{cases} \delta x = - \dfrac{mA}{\sigma} \sin \sigma t \cosh m(y + h) \sin mx, \\[2mm] \delta y = \dfrac{mA}{\sigma} \sin \sigma t \sinh m(y + h) \cos mx. \end{cases}$$

The motion of each particle takes place in a straight line the direction of which varies from vertical under the wave crests ($\cos mx = 1$) to horizontal under the nodes ($\cos mx = 0$). The motion also naturally becomes purely horizontal on approaching the bottom $y = - h$. These consequences of the theory are verified in practice, as indicated in Fig. 3.1.1, taken from a paper by Ruellan and Wallet (cf. [R.12]). The photograph at the bottom makes the particle trajectories visible in a standing wave; this is the final specimen in a series of photographs of particle trajectories for a range of cases beginning with a pure progressing wave (cf. sec. 3.2), and continuing with superpositions of progressing waves traveling in opposite directions and having the same wave length but not the same amplitudes, finally ending with a standing wave when the wave amplitudes of the two trains are equal.

We proceed next to study the special class of *three-dimensional* standing waves that are simple harmonic in the time, and which depend only on the distance r from the y-axis. In other words, we seek standing waves having cylindrical symmetry. Again we seek solutions of (3.1.6) which satisfy (3.1.7). Only the case of water of infinite depth will be treated here, and hence (3.1.8) is replaced by

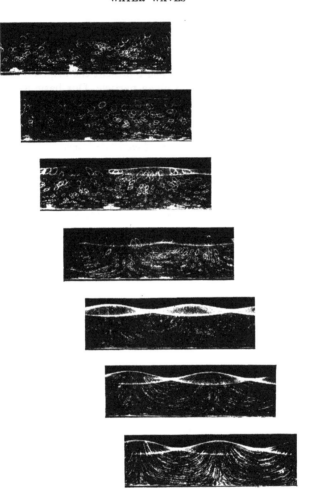

Fig. 3.1.1. Particle trajectories in progressing and standing waves

the condition that the solutions be bounded at ∞ in the negative y-direction as well as in the x- and z-directions. It is once more of interest to derive *all* possible standing wave solutions which are everywhere regular and bounded at ∞ because of the fact that the solutions in the present case behave quite differently from those obtained above for motions that are independent of the z-coordinate.

In particular, we shall see that all bounded standing waves with cylindrical symmetry die out at ∞ like the inverse square root of the distance, while in two dimensions we have seen that the assumption that the wave amplitude dies out at ∞ leads to waves of zero amplitude everywhere.

It is natural to make use of cylindrical coordinates in deriving our uniqueness theorem. Thus we write (3.1.6) in the form

$$(3.1.21) \quad \frac{1}{r}\frac{\partial}{\partial r}\left(r\frac{\partial \varphi}{\partial r}\right) + \frac{\partial^2 \varphi}{\partial y^2} = 0, \ 0 \geq y > -\infty, \ 0 \leq r < \infty$$

with r the distance from the y-axis. The assumption that φ depends only upon r and y and not on the angle θ has already been used. For our purposes it is useful to introduce a new independent variable ϱ replacing r by means of the relation

$$(3.1.22) \quad \varrho = \log r,$$

in terms of which (3.1.21) becomes

$$(3.1.23) \quad e^{-2\varrho}\frac{\partial^2 \varphi}{\partial \varrho^2} + \frac{\partial^2 \varphi}{\partial y^2} = 0, \ y < 0, \ -\infty < \varrho < \infty.$$

This equation holds, we observe, in the half-plane $y < 0$ of the y, ϱ-plane. The boundary condition to be satisfied at $y = 0$ is (cf. (3.1.7)):

$$(3.1.24) \quad \varphi_y - m\varphi = 0, \ m = \sigma^2/g.$$

We wish to find all regular solutions of (3.1.23) satisfying (3.1.24) for which φ and φ_y are bounded at ∞. To this end we proceed along much the same lines as above (cf. (3.1.14) and the reasoning immediately following it) for the case of two dimensions, and introduce the function $\psi(\varrho, y)$ by the identity

$$(3.1.25) \quad \psi = \varphi_y - m\varphi, \ y < 0, \quad -\infty < \varrho < \infty.$$

Since ψ involves only a derivative of φ with respect to y and not with respect to ϱ it follows at once that ψ is a solution of (3.1.23). Since ψ vanishes at $y = 0$ from (3.1.24) it follows easily that it can be continued analytically into the upper half-plane $y > 0$ by setting $\psi(\varrho, y) = -\psi(\varrho, -y)$ and that the resulting function will be a solution of (3.1.23) in the entire ϱ, y-plane. The function ψ thus obtained will be bounded in the entire plane, since it was bounded in the lower half plane by virtue of the boundedness assumptions with respect to φ. A theorem of S. Bernstein now yields the result that ψ is everywhere constant* if it is a uniformly bounded solution of (3.1.23) in the entire ϱ, y-plane. Since ψ vanishes on the y-axis it follows that ψ vanishes identically. Consequently we conclude from (3.1.25) that φ satisfies the relation

$$(3.1.26) \qquad \varphi_y - m\varphi = 0, \qquad y < 0.$$

The most general function $\varphi(\varrho, y)$ satisfying this equation is

$$(3.1.27) \qquad \varphi = e^{m y} f(\varrho) = e^{m y} f(\log r) = e^{m y} g(r)$$

with $g(r)$ an arbitrary function. But $\varphi(r, y)$ is also a solution of (3.1.21) and hence $g(r)$ is a solution of the ordinary differential equation

$$(3.1.28) \qquad \frac{1}{r}\frac{d}{dr}\left(r\frac{dg}{dr}\right) + m^2 g = 0,$$

or, in other words, $g(r)$ is a Bessel function of order zero:

$$(3.1.29) \qquad g(r) = AJ_0(mr) + BY_0(mr).$$

Since we restricted ourselves to bounded solutions only it follows that all solutions $\varphi(r, y)$ of our problem are given by

$$(3.1.30) \qquad g(r, y) = Ae^{m y} J_0(mr), \qquad m = \sigma^2/g,$$

with A an arbitrary constant. Upon reintroduction of the time factor we have, therefore, as the only bounded velocity potentials the functions

$$(3.1.31) \qquad \Phi(r, y; t) = Ae^{i\sigma t} e^{m y} J_0(mr).$$

As is well known, these functions behave for large values of r as follows:

$$(3.1.32) \qquad \Phi(r, y; t) \simeq Ae^{i\sigma t} e^{m y} \cdot \sqrt{\frac{2}{\pi m r}}\cos\left(mr - \frac{\pi}{4}\right)$$

and thus they die out like $1/\sqrt{r}$, as stated above.

In two dimensions we were able to find bounded standing waves of arbitrary phase (in the space variable) at ∞. In the present case of circular waves we have found bounded waves with only one phase at ∞. However, if we were to permit a logarithmic singularity at the

* What is needed is evidently a generalization of Liouville's theorem to the elliptic equation (3.1.23) which has a variable coefficient. The theorem of Bernstein referred to is much more general than is required for this special purpose, but it is also not entirely easy to prove (cf. E. Hopf [H.17] for a proof of it).

axis $r = 0$ and thus admit the singular Bessel function $Y_0(mr)$ as a solution of (3.1.28), we would have as possible velocity potentials the functions

(3.1.33) $$\Phi(r, y; t) = Be^{i\sigma t}e^{my}Y_0(mr)$$

which behave for large r as follows:

(3.1.34) $$\Phi(r, y; t) \simeq Be^{i\sigma t}e^{my} \sqrt{\frac{2}{\pi m r}} \sin\left(mr - \frac{\pi}{4}\right).$$

Admitting solutions with a logarithmic singularity on the y-axis thus leads to standing waves which behave at ∞ in the same way as those which are everywhere bounded, except that they differ by $90°$ in phase at ∞. Thus waves having an arbitrary phase at ∞ can be constructed, but not without allowing a singularity. It has, however not been shown that (3.1.31) and (3.1.33) yield all solutions with this property.

3.2. Simple harmonic progressing waves

Since our boundary problem is linear and homogeneous we can reintroduce the time factors $\cos \sigma t$ and $\sin \sigma t$ and take appropriate linear combinations of the standing waves (3.1.5) to obtain simple harmonic *progressing wave* solutions in water of uniform depth of the form

(3.2.1) $$\Phi = A \cosh m(y + h) \cos (mx \pm \sigma t + \alpha)$$

with m and σ satisfying

(3.2.2) $$\sigma^2 = gm \tanh mh,$$

as before.

The *wave*, or *phase*, *speed* c is of course given by

(3.2.3) $$c = \sigma/m,$$

or, in terms of the wave length $\lambda = 2\pi/m$ by

(3.2.3)$_1$ $$c = \sqrt{\frac{g\lambda}{2\pi} \tanh \frac{2\pi h}{\lambda}}.$$

It is useful to write the relation (3.2.2) in terms of the wave length $\lambda = 2\pi/m$ and then expand the function $\tanh mh$ in a power series to obtain

(3.2.4) $$\sigma^2 = \frac{2\pi g}{\lambda}\left[\frac{2\pi h}{\lambda} - \frac{1}{3}\left(\frac{2\pi h}{\lambda}\right)^3 + \cdots\right].$$

We see therefore that

(3.2.5) $$\sigma^2 \to \left(\frac{2\pi}{\lambda}\right)^2 gh = m^2 gh \quad \text{as} \quad \frac{h}{\lambda} \to 0,$$

and hence that

(3.2.6) $$c \simeq \sqrt{gh} \quad \text{if } h/\lambda \text{ is small.}$$

This last relation embodies the important fact that *the wave speed becomes independent of the wave length when the depth is small compared with the wave length, but varies as the square root of the depth.* This fact is in accord with what resulted in Chapter 2 upon linearizing the shallow water theory (cf. (2.2.16)) and the sentence immediately following), which led to the linear wave equation and to $c = \sqrt{gh}$ as the propagation speed for disturbances. We can gain at least a rough idea of the limits of accuracy of the linear shallow water theory by comparing the values of c given by $c^2 = gh$ with those given by the exact formula

(3.2.7) $$c^2 = \frac{g\lambda}{2\pi} \tanh \frac{2\pi h}{\lambda}$$

for various values of the ratio h/λ. One finds that c as given by $\sqrt{gh}$ is in error by about 6 % if the wave length is ten times the depth and by less than 2 % if the wave length is twenty times the depth. The error of course increases or decreases with increase or decrease in h/λ.

In water of infinite depth, on the other hand, we have already observed (cf. (3.2.2)) that

(3.2.8) $$c^2 = g\lambda/2\pi.$$

Actually, the error in c as computed by the formula $c^2 = g\lambda/2\pi$ is already less than 1/2 % if $h/\lambda > \frac{1}{2}$. One might therefore feel justified in concluding that variations in the bottom elevation will have but slight effect on a progressing wave provided that they do not result in depths which are less than half of the wave length, and observations seem to bear this out. In other words, the wave would not "feel" the bottom until the depth becomes less than about half a wave length.

It is of interest to determine the paths of the individual water particles as the result of the passage of a progressing wave. As in the preceding section we take δx and δy to represent the displacements of a particle from its average position, and determine those displacements from the equations

$$\frac{d\delta x}{dt} = \Phi_x = -Am \cosh m(y+h) \sin (mx + \sigma t + \alpha),$$

$$\frac{d\delta y}{dt} = \Phi_y = Am \sinh m(y+h) \cos (mx + \sigma t + \alpha),$$

since Φ is given by (3.2.1) in the present case. Integration of these equations yields

(3.2.9)
$$\begin{cases} \delta x = \dfrac{Am}{\sigma} \cosh m(y+h) \cos (mx + \sigma t + \alpha), \\[2mm] \delta y = \dfrac{Am}{\sigma} \sinh m(y+h) \sin (mx + \sigma t + \alpha), \end{cases}$$

so that the path of a particle at depth y is an ellipse

$$\frac{\delta x^2}{a^2} + \frac{\delta y^2}{b^2} = 1$$

with semi-axes a and b given by

$$a = \frac{Am}{\sigma} \cosh m(y+h)$$

$$b = \frac{Am}{\sigma} \sinh m(y+h).$$

On the bottom, $y = -h$, the ellipse degenerates into a horizontal straight line, as one would expect. Both axes of the ellipse shorten with increase in the depth. For experimental verification of these results, the discussion with reference to Fig. 3.1.1 should be consulted. In water of infinite depth the particle paths would be circles, as one can readily verify. The fact that the displacement of the particles dies out exponentially in the depth explains why a submarine need only submerge a slight distance below the surface—a half wave length, say—in order to remain practically unaffected even by severe storms.

3.3. Energy transmission for simple harmonic waves of small amplitude

In Chapter 1 the general formulas for the energy E stored in a fluid and its flux or rate of transfer F across given surfaces were derived for the most general types of motion. In this section we apply these formulas to the special motions considered in the present chapter, that is, under the assumption that the free surface conditions are linearized. The formula for the energy E stored in a region R is (cf. (1.6.1)):

$$(3.3.1) \qquad E = \varrho \iiint_R [\tfrac{1}{2}(\Phi_x^2 + \Phi_y^2 + \Phi_z^2) + gy]\,dxdydz;$$

while the flux of energy F in a time T across a surface S_G fixed in space is given by (cf. (1.6.5)):

$$(3.3.2) \qquad F = \varrho \int_t^{t+T} \left(\iint_{S_G} \Phi_t \frac{\partial \Phi}{\partial n}\,dS \right) dt.$$

We suppose first that the motion considered is the superposition of two standing waves which are simple harmonic in the time, as follows:

$$(3.3.3) \quad \Phi(x, y, z; t) = \varphi_1(x, y, z) \cos \sigma t + \varphi_2(x, y, z) \sin \sigma t.$$

Insertion of this in (3.3.2) with $T = 2\pi/\sigma$, i.e. for a time interval equal to the period of the oscillation, leads at once to the following expression for the energy flux F through S_G:

$$(3.3.4) \qquad F = \varrho \pi \iint_{S_G} \left(\varphi_2 \frac{\partial \varphi_1}{\partial n} - \varphi_1 \frac{\partial \varphi_2}{\partial n} \right) dS.$$

One observes that the energy flux over a period is zero if either φ_1 or φ_2 vanishes, i.e. if the motion is a standing wave: a fact which is not surprising since one expects an actual transport of energy only if the motion has the character of a progressing wave. Still another fact can be verified from (3.3.4) in our present cases, in which φ_1 and φ_2 are, as we know, harmonic functions: if S_G is a fixed *closed* surface in the fluid enclosing a region R Green's formula states that

$$\iint_{S_G} \left(\varphi_2 \frac{\partial \varphi_1}{\partial n} - \varphi_1 \frac{\partial \varphi_2}{\partial n} \right) dS = \iiint_R \left(\varphi_2 \nabla^2 \varphi_1 - \varphi_1 \nabla^2 \varphi_2 \right) dxdydz$$

provided that φ_1 and φ_2 have no singularities—sources or sinks for example—in R. In this case the energy flux F clearly vanishes since φ_1 and φ_2 are harmonic. Also one sees by a similar reasoning that the flux F over a period remains constant if S_G is deformed without passing over singularities. In particular, the energy flux through a vertical plane passing from the bottom to the free surface of the water in a two-dimensional motion would be the same (per unit width of the plane) for all positions of the plane provided that no singularities are passed over. This fact makes it possible, if one wishes, to consider the energy in the fluid as though the energy itself were an incompressible fluid, and to speak of its rate of flow.

In the literature dealing with waves in all sorts of media, but particularly in dispersive media, it is indeed commonly the custom to introduce the notion of the velocity of the flow of energy accompanying a progressing wave, and to bring this velocity in relation to the kinematic notion of the group velocity (to be discussed in the next section). The author has found it difficult to reconcile himself to these discussions, and feels that it would be better to discard the difficult concept of the velocity of transmission of energy, since this notion is not of primary importance, and nothing can be accomplished by its use which cannot be done just as well by using the well-founded and clear-cut concept of the flux of energy through a given surface. On the other hand, the notion is used in the literature (and probably will continue to be used) and consequently a discussion of it is included here, following pretty much the derivation given by Rayleigh in an appendix to the first volume of his Sound [R.4]. In the next section, where the notion of group velocity is introduced, some further comments about the concept of the velocity of transmission of energy will be made.

We consider the energy flux per unit breadth across a vertical plane in the case of a simple harmonic progressing wave in water of uniform depth (or, in view of the above remarks, across any surface of unit breadth extending from the bottom to the free surface). The velocity potential Φ is given by (cf. (8.2.1))

$$(8.8.5) \qquad \Phi = A \cosh m(y + h) \cos (mx + \sigma t + \alpha)$$

and (8.8.2) yields

$$(8.8.6) \quad F = A^2 \varrho \sigma m \int_t^{t+2\pi/\sigma} \left(\int_{-h}^{\eta} \cosh^2 m(y + h) dy \right) \sin^2 (mx + \sigma t) \, dt$$

for the flux across a strip of unit breadth in the time $T = 2\pi/\sigma$, the period of the oscillation. Hence the average flux per unit time is given by

$$(8.8.7) \qquad F_{av} = \frac{F}{T} = \frac{A^2 \varrho \sigma m h}{4} \left(1 + \frac{\sinh 2mh}{2mh} \right)$$

since the average of $\sin^2 \theta$ over a period is $1/2$. We have also taken

$\eta = 0$ in the upper limit of the integral in (3.3.6) and thus neglected a term of higher order in the amplitude. It is useful to rewrite the formula (3.3.7) in the following form through use of the relations $\sigma^2 = gm \tanh mh$ and $c = \sigma/m$:

$$(3.3.8) \qquad F_{av} = \frac{A^2 \varrho \sigma^2}{2g} \cosh^2 mh \cdot U,$$

with U a quantity having the dimensions of a velocity and given by the relation

$$(3.3.9) \qquad U = \frac{1}{2} c \left(1 + \frac{2mh}{\sinh 2mh} \right).$$

Next we calculate the average energy stored in the water as a result of the wave motion with respect to the *length* in the direction of propagation of the wave. This is obtained from (3.3.1) by calculating first the energy E_λ over a wave length $\lambda = 2\pi/m$ at any arbitrary fixed time. In the present case we have

$$
\begin{aligned}
E_\lambda - E_0 = m^2 \varrho \int_{-h}^{\eta} \int_0^\lambda [\tfrac{1}{2} A^2 \sinh^2 m(y+h) \cos^2 (mx + \sigma t + \alpha) \\
(3.3.10) \qquad + \tfrac{1}{2} A^2 \cosh^2 m(y+h) \sin^2 (mx + \sigma t + \alpha)] \, dx dy \\
+ \varrho \int_0^\eta \int_0^\lambda gy \, dx dy,
\end{aligned}
$$

in which the constant E_0 refers to the potential energy of the water of depth h when at rest. On evaluating the integrals, and ignoring certain terms of higher order, we obtain for the energy between two planes a wave length apart arising from the passage of the wave the expression

$$(3.3.11) \qquad E_\lambda - E_0 = \frac{A^2 \varrho \sigma^2}{2g} \lambda \cosh^2 mh,$$

as one finds without difficulty. Thus the average energy E_{av} in the fluid per unit length in the x-direction which results from the motion is given by

$$(3.3.12) \qquad E_{av} = \frac{A^2 \varrho \sigma^2}{2g} \cosh^2 mh.$$

Upon comparison with equation (3.3.8) we observe that E_{av} is exactly the coefficient of U in the formula (3.3.8) for the average energy flux per unit time across a vertical plane. It therefore follows, assuming that no energy is created or destroyed within the fluid itself, that *the energy is transmitted in the direction of propagation of the wave on the average with the velocity U.* As we see from (3.3.9) the velocity U is not the same as the phase or propagation velocity c; in fact, U is always less than c: for water of infinite depth it has the value $c/2$ and it increases with decrease in depth, approaching the phase velocity c as the depth approaches zero.

3.4. Group velocity. Dispersion

In any body of water the motion of the water in general consists of a superposition of waves of various amplitudes and wave lengths. For example, the motion of the water due to a disturbance over a restricted area of the surface can be analyzed in terms of the super-position of infinitely many simple harmonic wave trains of varying amplitude and wave length; such an analysis will in fact be carried out in Chapter 6. However, we know from our previous discussion (cf. (3.2.7)) that the propagation speed of a train of waves is an increasing function of the wave length—in other words, the wave phenomena with which we are concerned are subject to *dispersion*— and thus one might expect that the waves would be sorted out as time goes on into various groups of waves such that each group would consist of waves having about the same wave length. We wish to study the properties of such groups of waves having approximately the same wave length.

Suppose, for example, that the motion can be described by the superposition of two progressing waves given by

(3.4.1)
$$\begin{cases} \Phi_1 = A \sin (mx - \sigma t) \\ \Phi_2 = A \sin ([m + \delta m]x - [\sigma + \delta \sigma]t) \end{cases}$$

with δm and $\delta \sigma$ considered to be small quantities. The superposition of the two wave trains yields

(3.4.2)
$$\Phi = 2A \cos \frac{1}{2} (x\delta m - t\delta \sigma) \sin \left(\left[m + \frac{\delta m}{2} \right] x - \left[\sigma + \frac{\delta \sigma}{2} \right] t \right)$$
$$= B \sin (m'x - \sigma't)$$

with $m' = m + \delta m/2$, $\sigma' = \sigma + \delta \sigma/2$. Since δm and $\delta \sigma$ are small it follows that the function B varies slowly in both x and t so that Φ is an amplitude-modulated sine curve at each instant of time, as indicated schematically in Figure 3.4.1. In addition, the "groups" of waves thus defined—in other words the configuration represented by the dashed curves of Figure **3.4.1**.—advance with the velocity $\delta\sigma/\delta m$ in the x-direction. In our problem σ will in general be a function

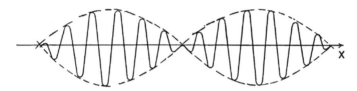

Fig. 3.4.1. Wave groups

of m so that the velocity U of the group is given approximately by $d\sigma/dm$, or, in terms of the wave length $\lambda = 2\pi/m$ and wave velocity $c = \sigma/m$, by

$$(3.4.3) \qquad U = \frac{d(mc)}{dm} = c - \lambda\frac{dc}{d\lambda}.$$

The matter can also be approached in the following way (cf. Sommerfeld [S.13]), which comes closer to the more usual circumstances. Instead of considering the superposition of only two progressing waves, consider rather the superposition, by means of an integral, of infinitely many waves with amplitudes and wave lengths which vary over a small range:

$$(3.4.4) \qquad \Phi = \int_{m_0-\varepsilon}^{m_0+\varepsilon} A(m)\exp\left\{i(mx - \sigma t)\right\}dm.$$

The quantity $mx - \sigma t$ can be written in the form

$$(3.4.5) \qquad mx - \sigma t = m_0 x - \sigma_0 t + (m - m_0)x - (\sigma - \sigma_0)t.$$

From (3.4.4) one then finds

$$(3.4.6) \qquad \Phi = C\exp\left\{i(m_0 x - \sigma_0 t)\right\},$$

in which the amplitude factor C is given by

$$(3.4.7) \quad C = \int_{m_0-\varepsilon}^{m_0+\varepsilon} A(m)\exp\left\{i[(m - m_0)x - (\sigma - \sigma_0)t]\right\}dm.$$

We are interested here in seeking out those places and times (if any) where the function C represents a wave progressing with little change in form, since (3.4.6) will then furnish what we call a group of waves. Since x and t occur only in the exponential term in (3.4.7), it follows that the values of interest are those for which this term must be nearly constant, i.e. those for which $(m - m_0)x - (\sigma - \sigma_0)t \simeq$ const.

It follows that the propagation speed of such a group is given by $(\sigma - \sigma_0)/(m - m_0)$, and if $(m - m_0)$ is small enough we obtain again the formula (3.4.3).

Evidently, it is important for this discussion of the notion of group velocity that the motion considered should consist of a superposition of waves differing only slightly in frequency and amplitude. In practice, the motions obtained in most cases —through use of the Fourier integral technique, for example,— are the result of super-position of waves whose frequencies vary from zero to infinity and whose amplitudes also vary widely. However, as we shall see in Chapter 6, it happens very frequently that the motion at certain places and times is approximated with good accuracy by integrals of the type given in (3.4.4) with ε arbitrarily small. (This is, indeed, the sense of the principle of stationary phase, to be treated in Chapter 6.) In such cases, then, groups of waves do exist and the discussion above is pertinent.

In our problems the relation between wave speed and wave length is given by (3.2.2) and consequently the velocity U of a group is readily found, from (3.4.3), to be

$$(3.4.8) \qquad U = \frac{1}{2} c \left(1 + \frac{2mh}{\sinh 2mh} \right).$$

We observe that the group velocity has the same value as was given in the preceding section for the average rate of propagation of energy in a uniform train of waves having the same wave length as those of the group. In other words, the rate at which energy is propagated is given by the group velocity and not the phase velocity. This is often considered as the salient fact with respect to the notion of group velocity. As indicated already in the preceding section, the author does not share this view, but feels rather that the kinematic concept of group velocity is of primary significance, while the notion of velocity of propagation of energy might better be discarded. It is true that the two velocities, in spite of the fact that one is derived from dynamics while the other is of purely kinematic origin, turn out to be the same—not only in this case, but in many others as well*—but it is also true that they are not always the same—for example, the two velocities are not the same if there is dissipation of energy in the medium. In addition, we have seen in the preceding section that the notion of velocity of energy can be derived when no

---

* A general analysis of the reason for this has been given by Broer [B.18].

wave group exists at all—we in fact derived this velocity for the case of a wave having but one harmonic component.

In Chapter 6 we shall have occasion to see how illuminating the kinematic concept of a group and its velocity can be in interpreting and understanding the complicated unsteady wave motions which arise when local disturbances propagate into still water.

CHAPTER 4

Waves Maintained by Simple Harmonic Surface Pressure in Water of Uniform Depth. Forced Oscillations

4.1. Introduction

In our previous discussions we have considered always that the pressure at the free surface was constant (usually zero) in both space and time. In other words, only the *free* oscillations were treated and the problems were, correspondingly, linear and homogeneous boundary value problems. Here we wish to consider two problems in which the surface pressure p_0 is simple harmonic in the time and the resulting motions are thus forced oscillations; the problems then also have a nonhomogeneous boundary condition. In the first such problem we assume that the motion is two-dimensional and that the surface pressure is a periodic function of the space coordinate x over the entire x-axis; in the second problem the surface pressure is assumed to be zero except over a segment of finite length of the x-axis. In these problems the depth of the water is assumed to be everywhere infinite, but the corresponding problems in water of constant finite depth can be, and have been, solved by much the same methods.

The formulation of the first two problems is as follows. A velocity potential $\Phi(x, y; t)$ is to be determined which is simple harmonic in the time t and satisfies

$$(4.1.1) \qquad \nabla^2 \Phi = 0 \qquad \text{for } y < 0.$$

The surface pressure $p(x; t)$ is given by

$$(4.1.2) \qquad p(x; t) = \bar{p}(x) \sin \sigma t,$$

and the boundary conditions at the free surface are the dynamical condition (cf. $(2.1.20)_1$)

$$(4.1.3) \qquad \eta = -\frac{1}{g} \Phi_t - p/\varrho g,$$

55

and the kinematic condition

(4.1.4) $$\eta_t = \Phi_y.$$

The last condition means that no kinematic constraint is imposed on the surface—it can deform freely subject to the given pressure distribution. In addition, we require that Φ_t and Φ_y should be uniformly bounded at ∞. This means effectively that the vertical displacement and vertical velocity components are bounded. In section 4.3, the amplitude $\bar{p}(x)$ of the surface pressure p will have discontinuities at two points and we shall impose appropriate conditions on Φ at these points when we consider this case.

We seek the most general simple harmonic solutions of our problem; they have the form

(4.1.5) $$\Phi = \varphi(x, y) \cos \sigma t + \psi(x, y) \sin \sigma t.$$

The functions φ and ψ are of course harmonic in the lower half plane. The conditions (4.1.2), (4.1.3), and (4.1.4) are easily seen to yield for the function φ the boundary condition

(4.1.6) $$\varphi_y - m\varphi = -\frac{\sigma}{\varrho g}\bar{p}(x) \qquad \text{for } y = 0$$

with the constant m defined by

(4.1.7) $$m = \sigma^2/g;$$

while for ψ they yield the condition

(4.1.8) $$\psi_y - m\psi = 0 \qquad \text{for } y = 0.$$

The phase $\sin \sigma t$ assumed for p in (4.1.2) has the effect that ψ satisfies the homogeneous free surface condition, as one sees.

We know from the first section of the preceding chapter that the only bounded and regular harmonic functions ψ which satisfy the condition (4.1.8) are given by

(4.1.9) $$\psi(x, y) = Ae^{my} \cdot \begin{cases} \cos mx \\ \sin mx \end{cases}.$$

In the next two sections we shall determine the function $\varphi(x, y)$, i.e. that part of Φ which has the phase $\cos \sigma t$, in accordance with two different choices for the amplitude $\bar{p}(x)$ of the surface pressure p.

4.2. The surface pressure is periodic for all values of x

We consider now the case in which the surface pressure is periodic in x such that $\overline{p}(x)$ in (4.1.2) and (4.1.6) is given by

$$(4.2.1) \qquad \overline{p}(x) = P \sin \lambda x, \qquad -\infty < x < \infty.$$

One verifies at once that the following function $\varphi(x, y)$:

$$(4.2.2) \qquad \varphi(x, y) = \frac{\sigma P}{\varrho g} \cdot \frac{e^{\lambda y}}{m - \lambda} \sin \lambda x$$

is a harmonic function which satisfies the free surface boundary condition (4.1.6) imposed in the present case. Since the difference χ of two solutions φ_1, φ_2 both satisfying all of our conditions would satisfy the homogeneous boundary condition $\chi_y - m\chi = 0$, it follows that *all* solutions φ of our boundary value problem can be obtained by adding to the special solution given by (4.2.2) any solution of the homogeneous problem, and these latter solutions are the functions ψ given by (4.1.9) since χ satisfies the same conditions as ψ. Therefore the most general simple harmonic solutions of the type (4.1.5) are given in the present case by

$$(4.2.3) \quad \Phi(x, y; t) = \left[\frac{\sigma P}{\varrho g} \frac{e^{\lambda y}}{m - \lambda} \sin \lambda x + A e^{my} \begin{Bmatrix} \cos mx \\ \sin mx \end{Bmatrix} \right] \cos \sigma t$$

$$+ B e^{my} \begin{Bmatrix} \cos mx \\ \sin mx \end{Bmatrix} \sin \sigma t,$$

with A and B constants which are at our disposal. In other words, the resulting motions are, as usual in linear vibrating systems, a linear combination of the forced oscillation and the free oscillations. These solutions—without the uniqueness proof—seem to have been given first by Lamb [L.2].

We observe that the case $\lambda = m$ must be excluded, and that if λ is near to m large amplitudes of the surface waves are to be expected. This means physically, as one sees immediately, that waves of large amplitude are created if the periodic surface pressure distribution has nearly the wave length which belongs to a surface wave of the same frequency for pressure zero at the surface—that is, the wave length of the corresponding free oscillation.

If instead of (4.1.2) we take the surface pressure as a *progressing wave* of the form

$$(4.2.4) \qquad p(x; t) = H \sin (\sigma t - \lambda x)$$

it is readily found that progressing surface waves result which are given by

$$(4.2.5) \qquad \Phi(x, y; t) = \frac{H\sigma}{\varrho g} \frac{e^{\lambda y}}{m - \lambda} \cos (\sigma t - \lambda x).$$

To this one may, of course, add any of the wave solutions which occur under zero surface pressure. Again one observes an odd kind of "resonance" phenomenon: large amplitudes are conditioned by the wave length in *space* of the applied pressure once the frequency has been fixed.

4.3. The variable surface pressure is confined to a segment of the surface

In this section we consider the case in which the surface pressure p

$$(4.3.1) \qquad p(x; t) = \begin{cases} P \sin \sigma t, & | x | \leq a \\ 0, & | x | > a \end{cases}, \quad y = 0$$

with P à constant. Some of the motions which can arise under such circumstances are discussed by Lamb [L.2] in the paper quoted above. However, here as elsewhere, Lamb assumes fictitious damping forces* in order to be rid of the free oscillations and thus achieve a unique solution, and he also makes use of the Fourier integral technique which we prefer to replace by a different procedure. In fact, the present problem is a key problem for this Part II and a peg upon which a variety of observations important for other discussions in later chapters will be hung. As we shall see, the present problem is also decidedly interesting for its own sake, although Lamb strangely enough made no attempt in his paper to point out the really striking results.

In addition to prescribing the pressure p through (4.3.1) it is necessary to add to the conditions imposed in section 4.1 appropriate conditions at the points $(\pm a, 0)$ where p has discontinuities. In view of (4.1.3) it is clear that a finite discontinuity in Φ_t or η or both must be admitted and it seems also likely that the derivatives Φ_x and Φ_y of Φ would be unbounded near these points. We shall make

---

* Lamb assumes resistances which are proportional to the velocity. In this way the irrotational character of the flow is preserved, but it is difficult to see how such resistances can be justified mechanically. It would seem preferable to secure the uniqueness of the solution in unbounded domains by imposing physically reasonable conditions on the behavior of the waves at infinity.

the following requirements

(4.3.2) Φ_t bounded; $\Phi_y = O(\varrho^{-1+\varepsilon})$, $\varepsilon > 0$

in a neighborhood of the points $(\pm a, 0)$ with ϱ the distance from these points. This means, in particular, that the surface elevation is bounded near these points and that the singularity admitted is not as strong as that of a source or sink. We recall that Φ_t and Φ_y were required to be uniformly bounded at ∞.

The stipulations made so far do not ensure the uniqueness of the solution Φ of our problem any more than similar conditions ensured uniqueness of the solution of the problem treated in the preceding section. However, we have in mind now a physical situation in which we expect the solution to be unique: We imagine the motion resulting from the applied surface pressure p given by (4.3.1) to be the limit approached after a long time subsequent to the application of p to the water *when initially at rest.* Under these circumstances one feels instinctively that the motion of the water far away from the source of the disturbance should have the character of a progressing wave moving away from the source of the disturbance, since at no time is there any reason why waves should initiate at infinity. (We shall show (cf. (6.7)) that the motion of the water arising from such initial conditions actually does approach, as the time increases without limit, the motion to be obtained here.) Consequently we add to our conditions on Φ the condition—often called the Sommerfeld condition in problems concerning electromagnetic wave propagation—that *the waves should behave at ∞ like progressing waves moving away from the source of the disturbance.* As we shall see, this qualitative condition leads to a unique solution of our problem.

In solving our problem there are some advantages to be gained by not stipulating at the outset that the Sommerfeld condition should be satisfied, but to obtain first *all* possible solutions of the form (4.1.5), and only afterwards impose the condition. We have therefore to find the harmonic functions φ which satisfy the condition (cf. (4.1.6) and (4.3.1))

(4.3.3) $\varphi_y - m\varphi - \begin{cases} c, & |x| \leqq a \\ 0, & |x| > a \end{cases}, \quad y = 0$

with

(4.3.4) $m = \sigma^2/g, \qquad c = -\dfrac{P\sigma}{\varrho g}$

on the free surface, and the boundedness conditions which follow from those imposed on Φ:

$$(4.3.5) \quad \begin{cases} \varphi \text{ and } \varphi_y \text{ bounded at } \infty, \\ \varphi \text{ bounded and } \varphi_y = O(\varrho^{-1+\varepsilon}), \quad \varepsilon > 0, \quad \text{at } x = \pm a. \end{cases}$$

The functions ψ in (4.1.5), i.e. those which yield the waves of phase $\sin \sigma t$ in Φ, satisfy the same conditions as in section 4.1 and are therefore given by (4.1.9). We have therefore only to determine the functions φ. To this end it is convenient to introduce new dimensionless quantities

$$(4.3.6) \qquad x_1 = mx, \qquad y_1 = my, \qquad a_1 = ma$$

together with $c_1 = c/m$ so that the free surface condition (4.3.3) takes the form

$$(4.3.7) \qquad \varphi_{y_1} - \varphi = \begin{cases} c_1, & |x_1| \le a_1 \\ 0, & |x_1| > a_1 \end{cases}, \qquad y_1 = 0.$$

In what follows we use the condition in this form but drop the subscripts for the sake of convenience.

In most of the two-dimensional problems treated in the remainder of Part II we make use of the fact that any harmonic function $\varphi(x, y)$ can be taken as the real part of an analytic function $f(z)$ of the complex variable $z = x + iy$ and write

$$(4.3.8) \qquad f(z) = \varphi(x, y) + i\gamma(x, y) = f(x + iy).$$

In our present problem $f(z)$ is defined and analytic in the lower half plane. To express the surface condition (4.3.7) in terms of $f(z)$ we write

$$\varphi_y - \varphi = \left(\frac{\partial}{\partial y} - 1\right)\varphi = \mathscr{R}e\left(\frac{\partial}{\partial y} - 1\right)(\varphi + i\gamma) = \mathscr{R}e\left(\frac{\partial}{\partial y} - 1\right)f(z)$$

$$= \mathscr{R}e\left(i\frac{df}{dz} - f\right) = \mathscr{R}e\,(if' - f),$$

in which the symbol $\mathscr{R}e$ means that the real part of what follows is to be taken. Consequently the free surface condition has the form:

$$(4.3.9) \quad \varphi_y - \varphi = \mathscr{R}e\,(if' - f) = \begin{cases} c, & |x| \le a \\ 0, & |x| > a \end{cases}, \qquad y = 0.$$

We now introduce a new analytic function $F(z)$ by the equation*

(4.3.10) $$F(z) = if'(z) - f(z)$$

and seek to determine $F(z)$ uniquely through the conditions imposed on $\varphi = \mathscr{Re}\, f(z)$. We observe to begin with that $F(z)$ satisfies the condition

(4.3.11) $$\mathscr{Re}\, F(z) = \begin{cases} c, & |x| \leq a \\ 0, & |x| > a \end{cases}, \quad y = 0,$$

in view of (4.3.9). We show now that $F(z)$ is uniquely determined within an additive pure imaginary constant, as follows: Suppose that $G(z) = F_1(z) - F_2(z)$ is the difference of two functions $F(z)$ satisfying the conditions resulting from (4.3.10) through those on $f(z)$. Then $\mathscr{Re}\, G(z)$ would vanish on the entire real axis, except possibly at $x = \pm\, a$, as one sees from (4.3.11). Hence $\mathscr{Re}\, G(z)$ is a potential function which can be continued analytically by reflection over the real axis into the entire upper half plane; it will then be defined and single-valued in the whole plane except for the points $(\pm\, a, 0)$. At ∞, $\mathscr{Re}\, G(z)$ is bounded in the lower half plane, while $\mathscr{Re}\, G(z) = O(\varrho^{-1+\varepsilon})$, $\varepsilon > 0$, at $x = \pm\, a$ in view of the regularity conditions and the definition of $G(z)$. These boundedness properties are evidently preserved in the analytic continuation into the upper half plane. Consequently $\mathscr{Re}\, G(z)$ has a removable singularity at the points $x = \pm\, a$ on the real axis since the singularity is weaker than a pole of first order and the function is single-valued in the neighborhood of these points. Thus $\mathscr{Re}\, G(z)$ is a potential function which is regular and bounded in the entire plane, and is zero on the real axis; by Liouville's theorem it is therefore zero everywhere. Consequently the analytic function $G(z)$ is a pure imaginary constant, and the result we want is obtained. On the other hand it is rather easy to find a function $F(z)$ which has the prescribed properties—for example by first finding its real part from (4.3.11) through use of the Poisson integral formula. We simply give it:

(4.3.12) $$F(z) = \frac{ic}{\pi} \log \frac{z-a}{z+a} ;$$

one verifies readily that it has all of the required properties. We take that branch of the logarithm which is real for $(z - a)/(z + a)$

* This device has been used by Kotschin [K.14], and it was exploited by Lewy [L.8] and the author [S.18] in studying waves on sloping beaches.

real and positive.

Once $F(z)$ has been uniquely determined, the complex velocity potential $f(z)$ is restricted to the solutions of the first order ordinary differential equation (4.3.10), which means that the solutions depend only on the arbitrary constant which multiplies the non-vanishing solution e^{-iz} of the homogeneous equation $if'(z) - f = 0$. But $\mathscr{Re} (A + iB)e^{-iz} = e^y(A \cos x + B \sin x)$ and these are the standing wave solutions for the case of surface pressure $p \equiv 0$. The most general solution of (4.3.10), with $F(z)$ given by (4.3.12), can be written, as one can readily verify, in the form

$$(4.3.13) \qquad f(z) = \frac{c}{\pi} e^{-iz} \int_{z_0}^{z} e^{it} \log \frac{t - a}{t + a}\, dt,$$

with the initial point z_0 and the path of integration any arbitrary path in the slit plane. Changing z_0 obviously would have the effect of changing the additive solution of the homogeneous equation. It is convenient to replace (4.3.13) by the following expression, obtained through an integration by parts:

$$(4.3.14)\ f(z) = \frac{ci}{\pi} \left[- \log \frac{z - a}{z + a} + e^{-iz} \int_{+i\infty}^{z} e^{it} \left(\frac{1}{t - a} - \frac{1}{t + a} \right) dt \right],$$

and at the same time to fix the path of integration as indicated in

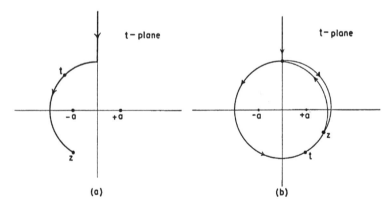

Fig. 4.3.1a,b. Path of integration in t-plane

Figure 4.3.1. This path comes from ∞ along the positive imaginary axis and encircles the origin, leaving it and the point $(-a, 0)$ to the left. Use has been made of the fact that $\log(z - a)/(z + a) \to 0$ when $z \to \infty$; we observe also that the integrals converge on account of the exponential factor.

That $\varphi(x, y) = \mathscr{R}e\, f(z)$ as given through (4.3.14) satisfies the boundary conditions imposed at the free surface and the regularity condition at the points $(\pm\, a,\, 0)$ is easy to verify. We proceed to discuss the behavior of $f(z)$ at ∞ (always for z in the lower half plane). For this purpose it suffices to discuss the integrals

$$I(z) = e^{-iz} \int_{i\infty}^{z} \frac{e^{it}}{t \pm a}\, dt \text{ since the function } \log\left(\frac{z - a}{z + a}\right)$$

behaves like $1/z$ at ∞ (as one readily sees). To this end we integrate once by parts to obtain

$$I(z) = -\frac{i}{z \pm a} - i \int_{i\infty}^{z} \frac{e^{i(t-z)}}{(t \pm a)^2}\, dt.$$

We suppose that the curved part of the path of integration in Figure 4.3.1a is an arc of a circle. It follows at once that the complex number $t - z$ has a positive imaginary part on the path of integration as long as the real part of z is negative, and hence we have

$$|\,I(z)\,| \leq \left|\frac{1}{z \pm a}\right| + \left|\int_{i\infty}^{z} \frac{|\,dt\,|}{|\,t \pm a\,|^2}\right| \leq \left|\frac{1}{z \pm a}\right| + \left|\int_{\sigma}^{\infty} \frac{|\,dt\,|}{|\,t \pm a\,|^2}\right|.$$

Consequently $I(z)$ behaves like $1/z$ at infinity when the real part of z is negative, and $f(z)$ likewise. The situation is different, however, if the real part of z is positive. To study this case, we add and subtract circular arcs, as indicated in Figure 4.3.1b, in order to have an integral over the entire circle enclosing the singularities at $\pm\, a$ as well as over a path symmetrical to the path in Figure 4.3.1a. By the same argument as above, the contribution of the integral over the latter path behaves like $1/z$ at ∞, and hence the non-vanishing contribution arises as a sum of residues at the points $\pm\, a$. These contributions to $I(z)$ are at once seen to have the values $2\pi i e^{-iz} e^{\mp ia}$. Thus we may describe the behavior of $f(z)$ as given by (4.3.14) at ∞ as follows:

$$(4.3.15) \quad f(z) = \begin{cases} O\left(\dfrac{1}{z}\right) & \text{for } \mathscr{R}e\, z < 0 \\[2mm] -4ci\,(\sin a)\,e^{-iz} + O\left(\dfrac{1}{z}\right) & \text{for } \mathscr{R}e\, z > 0. \end{cases}$$

From (4.3.10) and (4.3.12) one sees that $f'(z)$ has the same behavior at ∞ as $f(z)$, except for a factor $-i$. Hence $f(z)$, and with it $\varphi(x, y) = \mathscr{R}e\, f(z)$, has the postulated behavior at ∞. It is convenient to write down explicitly the behavior of $\varphi(x, y)$ at ∞:

$$(4.3.16) \quad \varphi(x, y) = \mathscr{R}e\, f(z) = \begin{cases} O\left(\dfrac{1}{r}\right) & \text{for } x < 0, \\ -4c \sin a\, e^y \sin x + O\left(\dfrac{1}{r}\right) & \text{for } x > 0. \end{cases}$$

It follows that *all* simple harmonic solutions of our problem are given by linear combinations of

$$(4.3.17) \quad \Phi(x, y; t) = (\mathscr{R}e\, f(z) + Ae^y \sin x + Be^y \cos x) \cos \sigma t$$

and

$$(4.3.18) \quad \Phi(x, y; t) = (Ce^y \sin x + De^y \cos x) \sin \sigma t$$

in which A, B, C, and D are arbitrary constants, and $f(z)$ is given by (4.3.14). In other words, the standing waves $\varphi(x, y) \cos \sigma t$ just found above, together with the standing waves which exist for vanishing free surface pressure, constitute all possible standing waves.

We now impose the condition that the wave $\Phi(x, y; t)$ we seek behaves like an outgoing progressing wave at ∞, i.e. that it behaves like

$$S_-: \quad e^y(H \sin (x + \sigma t) + K \cos (x + \sigma t)) \text{ at } x = -\infty$$

and like

$$S_+: \quad e^y(L \sin (x - \sigma t) + M \cos (x - \sigma t)) \text{ at } x = +\infty.$$

In view of the behavior of $\varphi(x, y) = \mathscr{R}e\, f(z)$ at $x = -\infty$ (cf. (4.3.16)), i.e. the fact that it dies out there, it is clear that we may combine the standing wave solutions (4.3.17) and (4.3.18) in such a way as to obtain a progressing wave solution

$$(4.3.19) \quad \Phi(x, y; t) = e^y(H \sin (x + \sigma t) + K \cos (x + \sigma t)) \\ + \varphi(x, y) \cos \sigma t$$

valid everywhere and which satisfies the condition S_-, with the two constants H and K still arbitrary. At $x = +\infty$ this wave has the behavior

$$(4.3.20) \quad \Phi(x, y; t) = e^y[(H \sin (x + \sigma t) + K \cos (x + \sigma t)) \\ - 4c \sin a \sin x \cos \sigma t] + O\left(\frac{1}{r}\right)$$

in view of (4.3.16). In order that S_+ should hold for this solution for all t one sees readily that the constants H and K must satisfy the linear equations

(4.3.21)
$$\begin{cases} L = H - 4c \sin a \\ L = - H \\ M = K, \ M = - K, \end{cases}$$

from which we conclude that

(4.3.22)
$$\begin{cases} L = - 2c \sin a, \quad H = 2c \sin a \\ M = K = 0. \end{cases}$$

Thus the solution is now uniquely determined through imposition of the Sommerfeld condition, and can be expressed as follows:

(4.3.23) $\quad \Phi(x, y; t) = \dfrac{2P\sigma}{\varrho gm} \sin ma \, e^{my} \sin (mx - \sigma t) + O\left(\dfrac{1}{r}\right), \quad x > 0$

upon reintroduction of the original variables and parameters (cf. (4.3.6)), with $O(1/r)$ representing a function which dies out at infinity like $1/r$. The function Φ of course yields a wave with symmetrical properties with respect to the y-axis. We observe that the wave length $\lambda = 2\pi/m$ of these waves at ∞ is the same as that of free oscillations of the same frequency, as one would expect.

The most striking thing about the solution is the fact that for certain frequencies and certain lengths of the segment over which the periodic pressure differs from zero, the amplitude of the progressing wave is zero at ∞; this occurs obviously for $\sin ma = 0$, i.e. for $ma = k\pi$, $k = 1, 2, 3, \ldots$. Since $m = 2\pi/\lambda$, with λ the wave length of a free oscillation of frequency σ, it follows that the amplitude of the progressing wave at ∞ vanishes when

(4.3.24) $\qquad\qquad 2a = k\lambda, \qquad k = 1, 2, \ldots,$

i.e. when the length of the segment on which the pressure is applied is an integral multiple of the wave length of the free oscillation having the same frequency as the periodic pressure. This does not of course mean that the entire disturbance vanishes, but only that the motion in this case is a standing wave given by

(4.3.25) $\qquad\qquad \Phi(x, y; t) = \varphi(x, y) \cos \sigma t,$

since the quantities H and K in (4.3.19) are now both zero. Since φ now behaves like $1/r$ at both infinities, the amplitude of the standing wave tends to zero at infinity. A wave generating device based on the physical situation considered here would thus be ineffective at certain frequencies. It is clear that no energy is carried off to infinity in this case, and hence that the surface pressure p on the segment $-a \leq x \leq +a$ can do no net work on the water on the average. Since $\eta_t = \varPhi_y$ it follows that the rate at which work is done by the pressure p (per unit width at right angles to the x, y-plane) is $\int_{-a}^{a} p \varphi_y \cos \sigma t \, dx$, and since p has the phase $\sin \sigma t$ it is indeed clear that the average rate of doing work is zero in this case.

There is a limit case of the present problem which has considerable interest for us. It is the limit case in which the length of the segment over which p is applied shrinks to zero while the amplitude P of p increases without limit in such a way that the product $2aP$ approaches a finite limit. In this way we obtain the solution for an oscillating pressure point. One sees easily that the function $f(z)$ given by (4.3.18), which yields the forced oscillation in our problem, takes the following form in the limit:

$$(4.3.26) \qquad f(z) = \frac{C}{\pi} e^{-iz} \int^z \frac{e^{it}}{t} \, dt,$$

with C the real constant $2aP\sigma/\varrho g$. At ∞ this function behaves as follows

$$(4.3.27) \qquad f(z) = \begin{cases} O\left(\dfrac{1}{z}\right) & \text{for } \mathcal{R}e \, z < 0, \\[2ex] 2Ci \, e^{-iz} + O\left(\dfrac{1}{z}\right) & \text{for } \mathcal{R}e \, z > 0. \end{cases}$$

In this limit case of an oscillating pressure point we see that there are no exceptional frequencies: application of the external force always leads to transmission of energy through progressing waves at ∞. The singularity of $f(z)$ at the origin is clearly a *logarithmic* singularity since $f(z)$ behaves near the origin like

$$(4.3.28) \qquad f(z) = \frac{C}{\pi} e^{-iz} \int^z \frac{dt}{t} + \ldots .$$

We see that a logarithmic singularity is appropriate at a source or sink of energy when the motion is periodic in the time.

4.4. Periodic progressing waves against a vertical clift

With the aid of the complex velocity potential defined by (4.3.13) we can discuss a problem which is different from the one treated in the preceding section. The problem in question is that of the determination of two-dimensional progressing waves moving toward a vertical cliff, as indicated in Figure 4.4.1. The cliff is the vertical

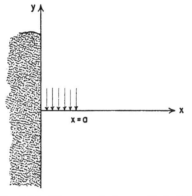

Fig. 4.4.1. Waves against a vertical cliff

plane containing the y-axis. As in the preceding section, we assume also that a periodic pressure (cf. (4.3.1)) is applied over the segment $0 \leq x \leq a$ at the free surface. To solve the problem we need only combine the standing waves given by (4.3.17) and (4.3.18) in such a way as to obtain progressing waves which move *inward* from the two infinities, and this can be done in the same way as in section 4.3. The result will be again a wave symmetrical with respect to the y-axis, and hence one for which $\Phi_x = 0$ along the y-axis; thus such a wave satisfies the boundary condition appropriate to the vertical cliff. We would find for the velocity potential Φ the expression, valid for $x > 0$:

$$(4.4.1) \quad \Phi(x, y; t) = \frac{2P\sigma}{\varrho g m} \sin ma\, e^{my}[\sin{(mx + \sigma t)}] + O\left(\frac{1}{r}\right)$$

with $O(1/r)$ a function behaving like $1/r$ at ∞ but with a singularity at $(a, 0)$. In order to obtain a system of waves which are not reflected back to ∞ by the vertical cliff it was necessary to employ a mechanism —the oscillating pressure over the segment $0 \leq x \leq a$ on the free surface—which absorbs the energy brought toward shore by the in coming wave. However, the particular mechanism chosen here, i.e. one involving an oscillatory pressure having the same frequency as the wave, will not always serve the purpose since the amplitude A of the surface elevation of the progressing wave at ∞ is given, from (4.4.1) and (4.1.4), by

$$(4.4.2) \qquad\qquad A = \frac{2P}{\varrho g} \sin ma.$$

Thus the ratio of the pressure amplitude P applied on the water surface near shore to the amplitude of the wave at ∞ would obviously become ∞ when $\sin ma = 0$. In other words, such a mechanism would achieve its purpose for waves whose wave length λ at ∞ satisfies the relation $a = k\,\lambda/2$, with k any integer, only if infinite pressure fluctuations at the shore occur. Presumably this should be interpreted as meaning that for these wave lengths the mechanism at shore is not capable of absorbing all of the incoming energy, or in other words, some reflection back to ∞ would occur. This remark has a certain practical aspect: a device to obtain power from waves coming toward a shore based on the mechanism considered here would function differently at different wave lengths.

It is of interest in the present connection to consider the same limit case as was discussed at the end of the preceding section, in which the segment of length a shrinks to zero while Pa remains finite. In this case no exceptional wave lengths or frequencies occur. However, the limit complex potential now has a logarithmic singularity at the shore line, as we noticed in the preceding section, and the amplitude of the surface would therefore also be infinite at the shore line. What would really happen, of course, is that the waves would break along the shore line if no reflection of wave energy back to ∞ occurred, and the infinite amplitude obtained with our theory represents the best approximation to such an essentially nonlinear phenomenon that the linear theory can furnish.

This limit case represents the simplest special case of the problem of progressing waves over uniformly sloping beaches which will be treated more generally in the next chapter. However, the present case has furnished one important insight: a singularity of the complex velocity potential is to be expected at the shore line if the condition at ∞ forbids reflection of the waves back to ∞, and the singularity should be at least logarithmic in character.

CHAPTER 5

Waves on Sloping Beaches and Past Obstacles

5.1. Introduction and summary

Perhaps the most striking—and perhaps also the most fascinating—single occurrence among all water wave phenomena encountered in nature is the breaking of ocean waves on a gently sloping beach. The purpose of the present chapter is to analyze mathematically the behavior of progressing waves over a uniformly sloping beach insofar as that is possible within the accuracy of the linearized theory for waves of small amplitude; that is, within the accuracy of the theory with which we are concerned in the present Part II. Later, in Chapter 10.10, we shall discuss the breaking of waves from the point of view of the nonlinear shallow water theory.

To begin with, it is well to recall the main features of what is often observed on almost any ocean beach in not too stormy weather. Some distance out from the shore line a train of nearly uniform progressing waves exists having wave lengths of the order of say fifty to several hundred feet. These waves can be considered as simple sine or cosine waves of small amplitude. As the waves move toward shore, the line of the wave crests and troughs becomes more and more nearly parallel to the shore line (no matter whether this was the case in deep water or not), and the distance between successive wave crests shortens slightly. At the same time the height of the waves increases somewhat and their shape deviates more and more from that given by a sine or cosine—in fact the water in the vicinity of the crests tends to steepen and in the troughs to flatten out until finally the front of the wave becomes nearly vertical and eventually the water curls over at the crest and the wave breaks. These observations are all clearly borne out in Figures 5.1.1, 5.1.2, which are photographs, given to the author by Walter Munk of the Scripps Institution of Oceanography, of waves on actual beaches. At the same time, it should be stated here that the breaking of waves also occurs in a manner different from this—a fact which will be discussed

69

Fig. 5.1.1. Waves breaking on a beach

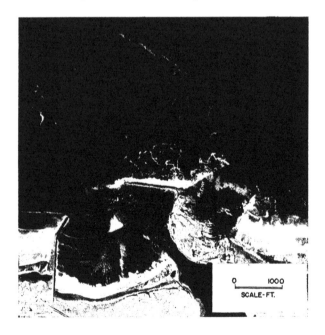

in Chapter 10.10 on the basis of other photographs of actual waves and a nonlinear treatment of the problem.

It is clear that the linear theory we apply here can not in principle yield large departures from the sine or cosine form of the waves in deep water, and still less can it yield the actual breaking phenomena: obviously these are nonlinear in character. On the other hand the linear theory is to be applied and should yield a good approximation for deep water and for the intermediate zone between deep water and the actual surf region. However, the fact that breakers do in general occur in nature cannot by any means be neglected even in formulating the problems in terms of the linear theory, for the following reasons. Suppose we consider a train of progressing waves coming from deep water in toward shore. As we know from Chapter 3, such a train of waves is accompanied by a flow of energy in the direction toward the shore. If we assume that there is little or no reflection of the waves from the shore—which observations show to be largely the case for a gently sloping beach* —it follows that there must exist some mechanism which absorbs the incoming energy; and that mechanism is of course the breaking of the waves which converts the incoming wave energy partially into heat through turbulence and partially into the energy of a different type of flow, i.e. the undertow. In terms of the linear theory about the only expedient which we have at our disposal to take account of such an effect in a rough general way is to permit that the wave amplitude may become very large at the shore line, or, in mathematical terms, that the velocity potential should be permitted to have an appropriate singularity at the shore line. As we have already hinted at the end of the preceding chapter, the appropriate singularity for a two-dimensional motion seems to be logarithmic, and hence the wave amplitude would be logarithmically infinite at the shore line. Indeed, it turns out that no progressing wave solutions without reflection from the shore line exist at all within the framework of the linear theory unless a singularity at least as strong as a logarithmic singularity is admitted at the shore line.

The actual procedure works out as follows: Once the frequency of the wave motion has been fixed, two different types of *standing*

* This fact is also used in laboratory experiments with water waves: the experimental tanks are often equipped with a sloping "beach" at one or more of the ends in order to absorb the energy of the incoming waves through breaking, and thus prevent reflection from the ends of the tank. This makes it possible to perform successive experiments without long waits for the motions to subside.

waves are obtained, one of which has finite amplitude, the other infinite amplitude, at the shore line. These two different types of standing waves behave at ∞ like the simple standing wave solutions for water of infinite depth obtained in Chapter 3; i.e. one of them behaves like $e^{mv} \sin(mx + \alpha)$ while the other behaves like $e^{mv} \cos(mx + \alpha)$; hence the two may be combined with appropriate time factors to yield arbitrary simple harmonic progressing waves at ∞. If the amplitude at ∞ is prescribed, and also the condition (cf. the last two sections of the preceding chapter) requiring that the wave at ∞ be a progressing wave moving toward shore, then the solution is uniquely determined; in particular, the strength of the logarithmic singularity at the shore line is uniquely fixed once the amplitude of the incoming wave is prescribed at ∞.

The fact that progressing waves over uniformly sloping beaches can be uniquely characterized in the simple way just stated is not a thing which has been known for a long time, but represents rather an insight gained in relatively recent years (cf. the author's paper [S.18] of 1947 and the other references given there). The method employed in the author's paper makes essential use of an idea due to H. Lewy to obtain the actual solutions for the case of two-dimensional waves over beaches sloping at the angles $\pi/2n$, with n an integer; H. Lewy [L.8] extended his method also to solve the problem for slope angles $(p/2n)\pi$, with p an odd integer and n any integer such that $p < 2n$. For the special slope angles $\pi/2n$ the progressing wave solutions were obtained first by Miche [M.8] (unknown to the author at the time because of lack of communications during World War II), and somewhat later by Bondi [B.14], but without uniqueness statements. Actually, the special standing wave solutions for these same slope angles which are finite at the shore line had already been obtained by Hanson [H.3].

All of these solutions for the slope angles $\omega = \pi/2n$, become more complicated and cumbersome as n becomes larger, that is, as the beach slope becomes smaller. In fact, the solutions consist of finite sums of complex exponentials and exponential integrals, and the number of the terms in these sums increases with n. Actual ocean beaches usually slope rather gently, so that many of the interesting cases are just those in which the slope angle is small—of the order of a few degrees, say. It is therefore important to give at least an approximate representation of the solution of the problem valid for small angles ω independent of the integer n. Such a representation has been given by Friedrichs [F.14]. To derive it the exact solution is first obtained for integer n in the form of a single complex integral, which can in turn be treated by the saddle point method to yield asymptotic solutions valid for large n, that is, for beaches with small slopes. The resulting asymptotic representation turns out to be very accurate. A comparison with the exact numerical solution for $\omega = 6^0$ shows the asymptotic solution to be practically identical with the exact solution all the way from infinity to within a distance of less than a wave length from the shore line. Eckart [E.2, 3] has devised an approximate theory which gives good results in both deep and shallow water.

For slope angles which are rational multiples of a right angle of the special form $\omega = p\pi/2n$ with p any odd integer smaller than $2n$, the problem of progressing waves has been treated by Lewy, as was mentioned above. Thus the theory is available for cases in which ω is greater than $\pi/2$, so that the "beach" becomes an overhanging cliff. The solution for a special case of this kind, i.e. for $\omega = 135^0$ or $p = 3$, $n = 2$, has been carried out numerically by E. Isaacson [I.2]. It turns out that there is at least one interesting contrast with the solutions for waves over beaches in which $\omega < \pi/2$. In the latter case it has been found that as a progressing wave moves in toward shore the amplitude first decreases to a value below the value at ∞, before it increases and becomes very large at the shore line. (This fact has also often been verified experimentally in wave tanks). The same thing holds for standing waves: at a certain distance from shore there exists always a crest which is lower than the crests at ∞. In the case of the overhanging cliff with $\omega = 135^0$, however, the reverse is found to be true: the first maximum going outward from the shore line is about 1 % higher than the height of the crests at ∞. Still another fact regarding the behavior of the solutions near the shore line is interesting. In all cases there exists just one standing wave solution which has a finite amplitude at the shore line; Lewy [L.8] has shown that the ratio of the amplitude there to the amplitude at ∞ is given in terms of the angle ω by the formula $(\pi/2\omega)^{1/2}$. Thus for angles ω less than $\pi/2$ the amplitude of the standing wave with finite amplitude is greater on shore than it is at infinity (becoming very large as ω becomes small) while for angles ω greater than $\pi/2$ the amplitude on shore is less than it is at ∞. Since the observations indicate that the standing wave of finite amplitude is likely to be the wave which actually occurs in nature for angles ω greater than about 40⁰, the above results can be used to give a rational explanation for what might be called the "wine glass" effect: wine is much more apt to spill over the edge of a glass with an edge which is flared outward than from a glass with an edge turned over slightly toward the inside of the glass.

A limit case of the problem of the overhanging cliff has a special interest, namely the case in which ω approaches the value π and the problem becomes what might be called the "dock problem": the water surface is free up to a certain point but from there on it is covered by a rigid horizontal plane. The solutions given by Lewy are so complicated as p and n become large that it seems hopeless to consider the limit of his solutions as $\omega \to \pi$. Friedrichs and Lewy [F.12] have, however, attacked and solved the dock problem directly for two-dimensional waves. For *three-dimensional waves* in water of constant finite depth the problem has been solved by Heins [H.13] (also see [H.12]).

It would be somewhat unsatisfying to have solutions for the sloping beach problem only for slope angles which are rational multiples of π: it is clear that this limitation is imposed by the methods used to solve the problem and not by any inherent characteristics of the problem itself. The two-dimensional problem has, in fact, been solved for all slope angles by Isaacson [I.1]. Isaacson obtained an integral representation of Lewy's solutions for the angles $p\pi/2n$ analogous to the representation obtained by Friedrichs for the angles $\pi/2n$, and then observed that his representation depended only upon the ratio of p to n and not on these quantities separately. Thus the solutions for all angles are given by this representation. Peters [P.5] has solved the same problem by an entirely different method, which makes no use of solutions for the special slope angles $p\pi/2n$.

The problem of two-dimensional progressing waves over sloping beaches thus has been completely solved as far as the theory of waves of small amplitude is concerned. Only one solution for three-dimensional motion has been mentioned so far, i.e. the solution by Heins for three-dimensional motion in the case of the dock problem. For certain slope angles $\omega = \pi/2n$ the method used by the author [S.18] can be extended in such a way as to solve the problem of three-dimensional waves on sloping beaches; in the paper cited the solution is carried out for the case $\omega = \pi/2$, i.e. for the case of waves approaching at an angle and breaking on a vertical cliff. Roseau [R.9] has used the same method for the case $\omega = \pi/4$. Subsequently the problem of three-dimensional waves on sloping beaches has been solved by Peters [P.6] and Roseau [R.9], who make use of a certain functional equation derived from a representation of the solution by a Laplace integral. In section 5.4. we shall give an account of this method of attack. Roseau [R.9] has solved the problem of waves in an ocean having different constant depths at the two different infinities in the x-direction which are connected by a bottom of variable depth.

Before outlining the actual contents of the present chapter, it may be well to summarize the conclusions which have been obtained from studying numerical solutions of the problems being considered here, which have been carried out (cf. [S.18]) for two-dimensional waves for slope angles $\omega = 135^0$, 90^0, 45^0, and 6^0, and for three-dimensional waves for the case $\omega = 90^0$. The results for the case of an overhanging cliff with $\omega = 135^0$ have already been discussed earlier. In the other three cases the most striking and important result is the following: The wave lengths and amplitudes change very little from their values at ∞ until points about a wave length from shore have been reached. Closer inshore the amplitude becomes large, as it must in accord with our theory. It is a curious fact (already mentioned earlier) that the amplitude of a progressing wave becomes less (for $\omega = 6^0$ about 10 % less) at a point near shore than its value at ∞, although it becomes infinite as the shore is approached. This effect has often been observed experimentally. This statement holds for the three-dimensional waves against a vertical cliff (with an amplitude decrease of about 2 %), as well as for the two-dimensional cases.

The exact numerical solution for the case of a beach sloping at 6^0 is useful for the purpose of a comparison with the results obtained from the linear shallow water theory (treated in Chapter (10.13) of Part III) and from the asymptotic approximation to the exact theory obtained by Friedrichs [F.14]. The linear shallow water theory, as its name indicates, can in principle not furnish a good approximation to the waves on sloping beaches in the deep water portion since it yields waves whose amplitude tends to zero at ∞. For a beach sloping at 6^0, for example, it is found that the shallow water theory furnishes a good approximation to the exact solution for a distance of two or three wave lengths outward from the shore line if the wave length is, say, about eight times the maximum depth of the water in this range; but the amplitudes furnished by the shallow water theory would be 50 to 60 percent too small at about 15 wave lengths away from the shore line. One of the asymptotic approximations to the exact theory given by Friedrichs yields a good approximation over practically the whole range from the shore line to infinity (it is inaccurate only very close to shore); this approximation, which even yields the decrease in amplitude under the value at ∞ mentioned above, is almost identical with one obtained by Rankine (cf. Miche [M.8, p. 287]) which is based upon an argument using energy flux considerations in connection with the assumption that the speed of the energy flux can be computed at each point in water of slowly varying depth by using the formula (cf. (8.3.9)) which is appropriate in water having everywhere the depth at the point in question. Friedrichs thus gives a mathematical justification for such a procedure on beaches of small slope.

It has already been made clear that the discussion in this chapter cannot yield information about the breaking of waves, which is an essentially nonlinear phenomenon. However, it is possible to analyze the breaking phenomena in certain cases and within certain limitations by making use of the nonlinear shallow water theory, as we shall see in Part III. For this purpose, one needs to know in advance the motion at some point in shallow water, and this presumably could be done by using the methods of the present chapter, combined possibly with the methods provided by the linear shallow water theory.

The material in the subsequent sections of this chapter is ordered as follows. In section 5.2. the problem of two-dimensional progressing waves over beaches sloping at the angles $\pi/2n$, n an integer, is discussed following the method of Lewy [L.8] and the author [S.18]. In section 5.3 the problem of three-dimensional waves against a vertical cliff is treated, also using the author's method. The reasons for including these treatments in spite of the fact that they yield results that are included in the more general treatments of Peters [P.6] and Roseau [R.9] is that they are interesting in themselves as an example of method, and also they can be applied to other problems, such as the problem of plane barriers inclined at the angles $\pi/2n$ (cf. F. John [J.4]), which have not been treated by other methods. In section 5.4, the general problem of three-dimensional waves on beaches sloping at any angle is treated following essentially the ideas of Peters.

In section 5.5 the problem of diffraction of waves around a rigid vertical wedge is treated; in case the wedge reduces to a plane the problem becomes the classical diffraction problem of Sommerfeld [S.12] for the case of diffraction of plane waves in two dimensions around a half-plane barrier. A new uniqueness theorem and a new and elementary solution for the problem are given. Methods of analyzing the solution are also discussed; photographs of the waves in such cases and comparisons of theory and experiment are made.

Finally, in section 5.6 a brief survey of a variety of solved and unsolved problems which might have been included in this chapter, with references to the literature, is given. Included are brief references to researches in oceanography, seismology, and to a selection of papers dealing with simple harmonic waves by using mathematical methods different from those employed otherwise in this chapter. In particular, a number of papers employing integral equations as a basic mathematical tool are mentioned and occasion is taken to explain the Wiener-Hopf technique of solving certain singular integral equations.

5.2. Two-dimensional waves over beaches sloping at angles $\omega = \pi/2n$

We consider first the problem of two-dimensional progressing waves over a beach sloping at the angle $\omega = \pi/2n$ with n an integer

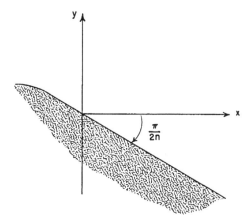

Fig. 5.2.1. Sloping beach problem

(cf. Figure 5.2.1), in spite of the fact that the problem can be solved, as was mentioned in the preceding section, by a method which is not restricted to special angles (cf. Peters [P.6], and Roseau [R.9]). The problem is solved here by a method which makes essential use of the fact that the slope angle has the special values indicated because the method has some interest in itself, and it yields representations which have been evaluated numerically in certain cases. In addition, the relevant uniqueness theorems are obtained in a very natural way.

We assume that the velocity potential $\Phi(x, y; t)$ is taken in the form $\Phi = e^{i\sigma t}\varphi(x, y)$. Hence $\varphi(x, y)$ is a harmonic function in the sector of angle $\omega = \pi/2n$. The free surface boundary condition then takes the form

$$(5.2.1) \qquad \varphi_y - \frac{\sigma^2}{g}\varphi = 0, \qquad \text{for } y = 0, \qquad x > 0,$$

as we have often seen (cf. (3.1.7)), while the condition at the bottom is

$$(5.2.2) \qquad \frac{\partial \varphi}{\partial n} = 0.$$

It is useful to introduce the same dimensionless independent variables as were used in the preceding chapter:

$$(5.2.3) \qquad x_1 = mx, \qquad y_1 = my, \qquad m = \sigma^2/g.$$

The function $\varphi(x, y)$ obviously remains harmonic in these variables, and conditions (5.2.1) and (5.2.2) become

$$(5.2.1)' \qquad \varphi_y - \varphi = 0, \qquad y = 0, \qquad x > 0,$$
$$(5.2.2)' \qquad \varphi_n = 0,$$

after dropping subscripts.

The simple harmonic standing waves in water of infinite depth everywhere are given by

$$(5.2.4) \qquad \Phi(x, y; t) = e^{i\sigma t} \cdot e^y \begin{cases} \cos(x + \alpha) \\ \sin(x + \alpha) \end{cases};$$

we write these down because we expect that they will represent the behavior of the standing waves in our case at large distances from the origin, that is, far away from the shore line.

The solution of the problem is obtained in terms of the complex potential $f(z)$ defined by

$$(5.2.5) \qquad f(z) = f(x + iy) = \varphi(x, y) + i\chi(x, y).$$

The function $f(z)$ should, like φ, be regular and analytic in the entire sector (including the boundaries,* except for the origin). The boundary conditions (5.2.1)' and (5.2.2)' are given in terms of $f(z)$ by

(5.2.6) $\varphi_y - \varphi = \mathscr{R}e \left(\dfrac{\partial}{\partial y} - 1 \right) (\varphi + i\chi) = \mathscr{R}e \left(\dfrac{\partial}{\partial y} - 1 \right) f(z)$

$= \mathscr{R}e \, (if' - f) = 0$ for z real and positive,

(5.2.7) $\varphi_n = \mathscr{R}e \, \dfrac{\partial}{\partial n} \, (f(z)) = \mathscr{R}e \, (- i \exp{(- i\pi/2n)} \, f')$

$= 0$ for $z = r \exp{\{- i\pi/2n\}}, \quad r > 0.$

The second condition results from

$$\mathscr{R}e \, \frac{\partial}{\partial n} \, (f(z)) = \mathscr{R}e \left\{ - \frac{1}{r} \frac{d}{d\theta} \, (f(z)) \right\} = \mathscr{R}e \, \{-ie^{i\theta} f'(z)\}.$$

We introduce the two following linear differential operators:

(5.2.8) $L_1(D) = - i \exp{\{- i\pi/2n\}} \, D,$

(5.2.9) $L_{2n}(D) = iD - 1$

with D meaning d/dz. The basic idea of the method invented by H. Lewy is to find additional linear operators, $L_2, L_3, \ldots, L_{2n-1}$ such that the operation $L_1 \cdot L_2 \cdot \ldots \cdot L_{2n}$ applied on $f(z)$ yields a function $F(z)$ *whose real part vanishes on both boundaries of our sector.* Once this has been done, the function $F(z)$ can be continued analytically over the boundaries of the sector by successive reflections to yield a single-valued function defined in the entire complex plane except possibly the origin. It can then be shown (see [S.18]), essentially by using Liouville's theorem, that the function $F(z)$ is *uniquely* determined within a constant multiplying factor by boundedness conditions on the complex potential $f(z)$ at ∞ together with the order of the singularity admitted at the origin. After $F(z)$ has been thus determined, the complex potential $f(z)$ is obtained as a solution of the ordinary differential equation $L_1 L_2 \ldots L_{2n} f(z) = F(z)$. Of course, it is necessary in the end to determine the arbitrary constants in the general solution of this differential equation in such a way as to satisfy all conditions of the problem, and this can in fact be done explicitly. It turns out that the resulting solution behaves at ∞ like

* Far less stringent conditions at the boundaries could be prescribed, since analytic continuations over the boundaries can easily be obtained explicitly in the present case.

the known solutions for waves in water having infinite depth everywhere and that it is uniquely determined by prescribing the amplitude
of the wave at ∞ together with the assumption that it should be,
say, an *incoming wave*.

We proceed to carry out this program, without however giving
all of the details (which can be found in the author's paper [S.18]).
To begin with, the ordinary differential equation for $f(z)$ and the
operators L_i are given by

$$(5.2.10) \quad L(D)f = L_1 \cdot L_2 \cdot L_3 \cdot \ldots \cdot L_{2n}f$$
$$= (\alpha_1 D)(\alpha_2 D - 1)(\alpha_3 D)(\alpha_4 D - 1) \ldots (\alpha_{2n-1}D)(\alpha_{2n}D - 1)f$$
$$= F(z)$$

with the complex constants α_k defined by

$$(5.2.11) \qquad \alpha_k = e^{-i\pi \left(\frac{k}{2n} + \frac{1}{2}\right)}, \qquad k = 1, 2, \ldots, 2n.$$

One observes that $L_1(D)$ and $L_{2n}(D)$ coincide with the definitions
given in (5.2.8) and (5.2.9). It is, in fact, not difficult to verify that

$$(5.2.12) \qquad\qquad \mathscr{R}e \; F(z) = 0$$

on both boundaries of the sector, by making use of the properties
of the numbers α_k and of the fact that $\mathscr{R}e \; L_1(D)$ and $\mathscr{R}e \; L_{2n}(D)f$
vanish on the bottom and the free surface, respectively, by virtue
of the boundary conditions (5.2.7) and (5.2.6).

So far we have not prescribed conditions on $f(z)$ at ∞ and at the
origin, and we now proceed to do so. At the origin we assume, in
accordance with the remarks made in section 5.1 and the discussion
in the last section of the preceding chapter, that $f(z)$ has at most a
logarithmic singularity; we interpret this to mean that $|d^k f(z)/dz^k| <
M_k/|z|^k$ in a neighborhood of the origin for $k = 1, 2, \ldots, 2n$, with
M_k certain constants. At ∞ we require that $\varphi = \mathscr{R}e \; f(z)$ together
with $|d^k f(z)/dz^k|$ for $k = 1, 2, \ldots, 2n$ be uniformly bounded when
$z \to \infty$ in the sector. (These conditions could be weakened considerably, but they are convenient and are satisfied by the solutions
we obtain.) In other words, although we expect the solutions of our
problem to behave at ∞ in accordance with (5.2.4) it is not necessary
to prescribe the behavior at ∞ so precisely since the boundedness
conditions yield solutions having this property automatically. Once
these conditions on $f(z)$ have been prescribed we see that the function
$F(z)$ defined by (5.2.10) has the following properties: 1) $|F(z)|$ is

uniformly bounded in the sector, and 2) $|F(z)| = O(1/z^{2n})$ in the neighborhood of the origin.

We have already observed that $\mathscr{Re}\, F(z) = 0$ on both boundaries of the sector and that $F(z)$ can therefore be continued as a single-valued function into the whole plane, except the origin, by the reflection process. Here we make decisive use of the assumption that ω, the angle of the sector, is $\pi/2n$ with n an integer. Since the boundedness properties of $F(z)$ at ∞ and the origin are preserved in the reflection process, it is clear from well-known results concerning analytic functions that $F(z)$ is an analytic function over the whole plane having a pole of order at most $2n$ at the origin. Since in addition the real part of $F(z)$ vanishes on all rays $z = r \exp\{ik\pi/2n\}$, $k = 1, 2, \ldots, 4n$, it follows that $F(z)$ is given uniquely by

$$(5.2.13) \qquad F(z) = \frac{A_{2n}i}{z^{2n}}$$

with A_{2n} an arbitrary *real* constant which may in particular have the value zero. Thus the complex potential $f(z)$ we seek satisfies the differential equation

$$(5.2.14) \quad (\alpha_1 D)(\alpha_2 D - 1) \ldots (\alpha_{2n-1} D)(\alpha_{2n} D - 1)f = \frac{A_{2n}i}{z^{2n}}.$$

Our problem is reduced to finding a solution $f(z)$ of this differential equation which satisfies all of the conditions imposed on $f(z)$. From the discussion of section 5.1 we expect to find *two* solutions $f_1(z)$ and $f_2(z)$ of our problem which behave differently at the origin and at ∞; at the origin, in particular, we expect to find one solution, say $f_1(z)$, to be bounded and the other, $f_2(z)$, to have a logarithmic singularity.

The regular solution $f_1(z)$ is the solution of (5.2.14) which one obtains by taking for the real constant A_{2n} the value zero, while $f_2(z)$ results for $A_{2n} \neq 0$. In other words the solution of the non-homogeneous equation contains the desired singularity at the origin. One finds for $f_1(z)$ the solution

$$(5.2.15) \qquad f_1(z) = \frac{\pi}{(n-1)!\sqrt{n}} \sum_{k=1}^{n} c_k e^{z\beta_k},$$

in which the constants c_k and β_k are the following complex numbers:

$$(5.2.16) \quad \begin{cases} \beta_k = \exp\left\{i\pi\left(\dfrac{k}{n} + \dfrac{1}{2}\right)\right\} \\[2mm] c_k = \exp\left\{i\pi\left(\dfrac{n+1}{4} - \dfrac{k}{2}\right)\right\} \cot\dfrac{\pi}{2n}\cot\dfrac{2\pi}{2n}\ldots\cot\dfrac{(k-1)\pi}{2n}, \\[2mm] \hspace{6cm} k = 2, 3, \ldots, n \\[2mm] c_1 = \bar{c}_n. \end{cases}$$

The constants c_k are obtained by adjusting the arbitrary constants in the solution of (5.2.14) so that the boundary conditions on $f(z)$ at the free surface and the bottom are satisfied; that such a result can be achieved by choosing a finite number of constants is at first sight rather startling, but it must be possible if it is true that a function $f(z)$ having the postulated properties exists since such a function must satisfy the differential equation (5.2.14). The calculation of the constants c_k is straightforward, but not entirely trivial. The function $f_1(z)$ is uniquely given by (5.2.15) within a real multiplying factor. As $|z| \to \infty$ in the sector, all terms clearly die out exponentially except the term for $k = n$, which is $c_n \exp\{-iz\}$, since all β_k's except β_n have negative real parts. Even the term for $k = n$ dies out exponentially except along lines parallel to the real axis. (The value of c_n, by the way, is $\exp\{-i\pi(n-1)/4\}$ since the cotangents in (5.2.16) cancel each other for $k = n$.) This term thus yields the asymptotic behavior of $f_1(z)$:

$$(5.2.17) \qquad f_1(z) \sim \frac{\pi}{(n-1)!\sqrt{n}} \cdot c_n e^{-iz}.$$

The solution $f_2(z)$ of the nonhomogeneous equation (5.2.14) which satisfies the boundary conditions is as follows:

$$(5.2.18) \qquad f_2(z) = \sum_{k=1}^{n} a_k\left[e^{z\beta_k}\int_{i\infty}^{iz\beta_k}\frac{e^{it}}{t}\,dt - \pi i e^{z\beta_k}\right],$$

for the case in which the real constant A_{2n} is set equal to one. The constants β_k are defined in (5.2.16); and the constants a_k are defined by

$$(5.2.19) \qquad a_k = c_k/\{(n-1)!\sqrt{n}\},$$

that is, they are a fixed multiple (for given n) of the constants c_k defined in (5.2.16). The constants a_k, like the c_k, are uniquely determined within a real multiplying factor. The path of integration for all integrals in (5.2.18) is indicated in Figure 5.2.2. That the points $iz\beta_k$ lie in the lower half of the complex plane (as indicated in the figure) can be seen from our definition of the constants β_k and the fact that z is restricted to the sector $-\pi/2n \leqq \arg z \leqq 0$.

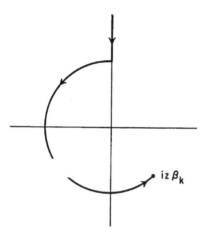

Fig. 5.2.2. Path of integration in t-plane

The behavior of $f_2(z)$ at ∞ of course depends on the behavior of the functions in (5.2.18). It is not hard to show—for example, by the procedure used in arriving at the result given by (4.3.15) in the preceding chapter—that these functions behave asymptotically as follows:

$$(5.2.20) \quad e^{z\beta_k}\int_{i\infty}^{iz\beta_k} \frac{e^{it}}{t}\,dt \sim \begin{cases} o\left(\frac{1}{z}\right), & \mathscr{R}e\,(iz\beta_k) < 0, \quad \mathscr{I}m\,(iz\beta_k) \leqq 0, \\ 2\pi i - o\left(\frac{1}{z}\right), & \mathscr{R}e\,(iz\beta_k) > 0, \quad \mathscr{I}m\,(iz\beta_k) \leqq 0. \end{cases}$$

Once this fact is established it is clear from (5.2.19) and (5.2.18) that $f_2(z)$ behaves asymptotically as follows:

$$(5.2.21) \qquad f_2(z) \sim \frac{\pi}{(n-1)!\sqrt{n}} \cdot c_n i e^{-iz},$$

since the term for $k = n$ dominates all others (cf. (5.2.20)) and $\mathscr{R}e(iz\beta_k) > 0$ in this case. Comparison of (5.2.21) with (5.2.17) shows that the real parts of $f_1(z)$ and $f_2(z)$ would be 90^0 out of phase at ∞.

That the derivatives of $f_2(z)$ behave asymptotically in the same fashion as $f_2(z)$ itself is easily seen, since the only terms in the derivatives of (5.2.18) of a type different from those in (5.2.18) itself are of the form b_k/z^k, k an integer $\geqq 1$. Finally, it is clear that $f_2(z)$ has a logarithmic singularity at the origin. Hence $f_1(z)$ and $f_2(z)$ satisfy all requirements. Just as in the 90^0 case (cf. the last section of the preceding chapter) it is now clear that $f(z) = b_1 f_1(z) + b_2 f_2(z)$, with b_1 and b_2 any real constants, yields *all* standing wave solutions of our problem.

The relations (5.2.17) and (5.2.21) yield for the asymptotic behavior of the real potential functions φ_1 and φ_2 the relations:

$$(5.2.22) \quad \varphi_1(x, y) = \mathscr{R}e\, f_1 \sim \frac{\pi}{(n-1)!\sqrt{n}}\, e^y \cos\left(x + \frac{n-1}{4}\pi\right)$$

$$(5.2.23) \quad \varphi_2(x, y) = \mathscr{R}e\, f_2 \sim \frac{\pi}{(n-1)!\sqrt{n}}\, e^y \sin\left(x + \frac{n-1}{4}\pi\right)$$

when it is observed that $c_n = \exp\{- i\pi(n-1)/4\}$. It is now possible to construct either standing wave or progressing wave solutions which behave at ∞ like the known solutions for steady progressing waves in water which is everywhere infinite in depth. In particular we observe that it makes sense to speak of the wave length at ∞ in our cases and that the relation between wave length and frequency satisfies asymptotically the relation which holds everywhere in water of infinite depth. For this, it is only necessary to reintroduce the original space variables by replacing x and y by mx and my, with $m = \sigma^2/g$ (cf. (5.2.3)), and to take note of (5.2.22) and (5.2.23).

Finally, we write down a solution $\Phi(x, y; t)$ which behaves at ∞ like $e^y \cos(x + t + \alpha)$, i.e. like a steady progressing wave moving toward shore:

$$(5.2.24) \quad \Phi(x, y; t) = A[\varphi_1(x, y) \cos(t + \alpha) - \varphi_2(x, y) \sin(t + \alpha)].$$

As our discussion shows, this solution is uniquely determined as soon as the amplitude is prescribed at ∞ (i.e. as soon as A is fixed) since $\varphi_1(x, y)$ and $\varphi_2(x, y)$ yield the *only* standing wave solutions of our problem and they are determined also within a real factor. As we have already stated in the preceding section, the progressing wave solutions (5.2.24) have been determined numerically (cf. [S.18]) for slope angles $\omega = 90^0$, 45^0, and 6^0, with results whose general features were already discussed in that section.

5.3. Three-dimensional waves against a vertical cliff

It is possible to treat some three-dimensional problems of waves over sloping beaches by a method similar to the method used in the preceding section for two-dimensional waves, in spite of the fact that it is now no longer possible to make use of the theory of analytic functions of a complex variable. In this section we illustrate the method by treating the problem of progressing waves in an infinite ocean bounded on one side by a vertical cliff when the wave crests at ∞ may make any angle with the shore line (cf. [S.18]).

We seek solutions $\Phi(x, y, z; t)$ of $\nabla^2_{(x,y,z)}\Phi = 0$ in the region $x \geq 0$, $y \leq 0$, $-\infty < z < \infty$ with the y-axis taken normal to the undisturbed free surface of the water and the z-axis* taken along the "shore", i.e. at the water line on the vertical cliff $x = 0$. Progressing waves moving toward shore are to be found such that the wave crests (or other curves of constant phase) at large distances from shore tend to a straight line which makes an arbitrary angle with the shore line. For this purpose we seek solutions of the form

(5.3.1) $\Phi(x, y, z; t) = \exp\{i(\sigma t + kz + \beta_1)\}\varphi(x, y)$

that is, solutions in which periodic factors in both z and t are split off.

As in the preceding section, we introduce new variables and parameters through the relations $x_1 = mx$, $y_1 = my$, $z_1 = mz$, $k_1 = k/m$, $m = \sigma^2/g$ and obtain for φ the differential equation

(5.3.2) $$\nabla^2_{(x,y)}\varphi - k^2\varphi = 0$$

and the free surface condition

(5.3.3) $$\varphi_y - \varphi = 0 \qquad \text{for } y = 0,$$

after dropping the subscript 1 on all quantities. The condition at the cliff is, of course,

(5.3.4) $$\frac{\partial \varphi}{\partial x} = 0 \qquad \text{for } x = 0.$$

At the origin $x = 0$, $y = 0$ (i.e. at the shore line on the cliff) we require, as in former cases, that φ should be of the form

(5.3.5) $$\varphi = \bar{\varphi} \log r + \bar{\bar{\varphi}}, \qquad r \ll 1,$$

for sufficiently small values of $r = (x^2 + y^2)^{1/2}$. with $\bar{\varphi}$ and $\bar{\bar{\varphi}}$ certain bounded functions with bounded first and second derivatives in a neighborhood of the origin. The functions $\bar{\varphi}$ and $\bar{\bar{\varphi}}$ should be considered at present as certain given functions; later on, they will be chosen specifically.

For large values of r we wish to have $\Phi(x, y, z; t)$ behave like

---

* It has already been pointed out that functions of a complex variable are not used in this section, so that the reintroduction of the letter z to represent a space coordinate should cause no confusion with the use of the letter z as a complex variable in earlier sections.

$e^y \exp \{i(\sigma t + kz + \alpha x + \beta)\}$ with $k^2 + \alpha^2 = 1$ but k and α otherwise arbitrary constants, so that progressing waves tending to an arbitrary plane wave at ∞ can be obtained. This requires that $\varphi(x, y)$ should behave at ∞ like $e^y \exp \{i(\alpha x + \beta_2)\}$ because of (5.3.1). However, it is no more necessary here than it was in our former cases to require that φ should behave in this specific way at ∞; it suffices in fact to require that

$$(5.3.6) \qquad |\varphi| + |\varphi_x| + |\varphi_{xy}| < M \qquad \text{for } r > R_0,$$

i.e. that φ and the two derivatives of φ occurring in (5.3.6) should be uniformly bounded at ∞. As we shall see, this requirement leads to solutions of the desired type.

We proceed to solve the boundary value problem formulated in equations (5.3.2) to (5.3.6). The procedure we follow is analogous to that used in the two-dimensional cases in every respect. To begin with, we observe that

$$(5.3.7) \qquad \frac{\partial}{\partial x}\left(\frac{\partial}{\partial y} - 1\right)\varphi = 0 \qquad \text{for both } x = 0 \text{ and } y = 0,$$

because of the special form of the linear operator on the left hand side together with the fact that (5.3.3) and (5.3.4) are to be satisfied. A function $\psi(x, y)$ is introduced by the relation

$$(5.3.8) \qquad \psi = \frac{\partial}{\partial x}\left(\frac{\partial}{\partial y} - 1\right)\varphi.$$

The essential point of our method is that the function ψ is determined uniquely within an arbitrary factor if our function φ, having the properties postulated, exists. Furthermore, ψ can then be given explicitly without difficulty. The properties of ψ are as follows.

1. ψ satisfies the same differential equation as φ, i.e. equation (5.3.2), as one sees from the definition (5.3.8) of ψ.

2. ψ is regular in the quadrant $x > 0$, $y < 0$ and vanishes, in view of (5.3.7), on $x = 0$, $y < 0$ and $y = 0$, $x > 0$. Hence ψ can be continued over the boundaries by the reflection process to yield a continuous and single-valued function having continuous second derivatives ψ_{xx} and ψ_{yy} (as one can readily see since $\nabla^2 \psi - k^2 \psi = 0$, and $\psi = 0$ on the boundaries) in the entire x, y-plane with the exception of the origin. (Here we use the fact that our domain is a sector of angle $\pi/2$.)

3. At the origin, ψ has a possible singularity which is of the form $\overline{\varphi}(x, y)/r^2$, with $\overline{\varphi}$ regular, as one can see from (5.3.5) and (5.3.8). This statement clearly holds for the function ψ when it has been extended by reflection to a full neighborhood of the origin.

4. The condition (5.3.6) on φ clearly yields for ψ the condition that ψ is uniformly bounded at ∞ after ψ has been extended to the whole plane.

Thus ψ is a solution of $\nabla^2\psi - k^2\psi = 0$ in the entire plane which is uniformly bounded at ∞. At the origin $\psi = \overline{\varphi}/r^2 + \overline{\overline{\varphi}}$ with $\overline{\varphi}$ and $\overline{\overline{\varphi}}$ certain regular functions ($\overline{\varphi} = 0$ not excluded). In addition, $\psi = 0$ on the entire x and y axes. We shall show, following Weinstein [W.5],* that the function

$$(5.3.9) \quad \psi(x, y) = AiH_2^{(1)}(ikr) \sin 2\theta, \; r = \sqrt{x^2 + y^2}, \; 0 \le k \le 1$$

is the unique solution for ψ in polar coordinates (r, θ) with A an arbitrary real constant, and $H_2^{(1)}$ the Hankel function of order two which tends to zero as $r \to \infty$. The function ψ has real values for r real. (The notation given in Jahnke-Emde, Tables of Functions, is used.)

The solution ψ is obtained by Weinstein in the following way. In polar coordinates (r, θ) the differential equation for ψ is

$$\frac{\partial^2\psi}{\partial r^2} + \frac{1}{r}\frac{\partial\psi}{\partial r} + \frac{1}{r^2}\frac{\partial^2\psi}{\partial\theta^2} - k^2\psi = 0.$$

For any fixed value of r the function ψ can be developed in the following sine series:

$$\psi = \sum_{n=1}^{\infty} c_n(r) \sin 2n\theta$$

since ψ vanishes for $\theta = 0, \pi/2, \pi, 3\pi/2$; and the coefficients $c_n(r)$ are given by

$$c_n(r) = C_n \int_0^{\pi/2} \psi(r, \theta) \sin 2n\theta \, d\theta, \qquad n = 1, 2, \ldots,$$

with C_n a normalizing factor. From this formula one finds by differentiations with respect to r and use of the differential equation for ψ that $c_n(r)$ satisfies the equation

$$c_n''(r) + \frac{1}{r}c_n'(r) - \left(k^2 + \frac{4n^2}{r^2}\right)c_n(r) = -\frac{C_n}{r^2}\int_0^{\pi/2}\left(\frac{\partial^2\psi}{\partial\theta^2} + 4n^2\psi\right)\sin 2n\theta \, d\theta.$$

* In the author's paper the solution ψ was obtained, but with a less general uniqueness statement.

The right hand side of this equation vanishes, as can be seen by integrating the first term twice by parts and making use of the boundary conditions $\psi = 0$ for $\theta = 0$ and $\theta = \pi/2$. Thus the functions $c_n(r)$ are Bessel functions, as follows:

$$c_n(r) = A_{2n} i^{2n+1} H^{(1)}_{2^i_i}(ikr) + B_{2n} I_{2n}(kr),$$

with A_{2n} and B_{2n} arbitrary real constants. The functions I_{2n} are unbounded at ∞; the Hankel functions $H^{(1)}_{2n}$ behave like r^{-2n} for $r \to 0$ and tend to zero exponentially at ∞. It follows therefore that the Fourier series for ψ in our case reduces to the single term given by (5.3.9) because of the boundedness assumptions on ψ.

For our purposes it is of advantage to write the solution ψ in the following form:

$$(5.3.10) \qquad \psi = Ai \frac{\partial^2}{\partial x \partial y} H^{(1)}_0 (ikr), \qquad r = \sqrt{x^2 + y^2},$$

in which A is any real constant and $H^{(1)}_0$ is the Hankel function of order zero which is bounded as $r \to \infty$. It is readily verified that this solution differs from that given by (5.3.9) only by a constant multiplier: for example, by using the well-known identities involving the derivatives of Bessel functions of different orders.

Once ψ is determined we may write (5.3.8) in the form

$$(5.3.11) \qquad \frac{\partial}{\partial x} \left(\frac{\partial}{\partial y} - 1 \right) \varphi = Ai \frac{\partial^2}{\partial x \partial y} H^{(1)}_0 (ikr), \qquad A \text{ arbitrary.}$$

This means that our function φ, if it exists, must satisfy (5.3.11) as well as (5.3.2). By integration of (5.3.11) it turns out that we are able to determine φ explicitly without great difficulty on account of the simple form of the left hand side of (5.3.11). This we proceed to do.

Integration of both sides of (5.3.11) with respect to x leads to

$$(5.3.12) \qquad \left(\frac{\partial}{\partial y} - 1 \right) \varphi = Ai \frac{\partial}{\partial y} H^{(1)}_0 (ikr) + g(y),$$

in which $g(y)$ is an arbitrary function. But $g(y)$ must satisfy (5.3.2), since all other terms in (5.3.12) satisfy it. Hence $d^2g/dy^2 - k^2 g = 0$. In addition $g(0) = 0$, since the other terms in (5.3.12) vanish for $y=0$ because of (5.3.3) and the fact that $\partial H^{(1)}_0 /\partial y = (ik)^{-1}(y/r)dH^{(1)}_0 /dr$. Finally, $g(y)$ is bounded as $y \to -\infty$ because of condition (5.3.6) and the fact that $\partial H^{(1)}_0 /\partial y$ tends to zero as $r \to \infty$. The function

$g(y)$ is therefore readily seen to be identically zero. By integration of (5.3.12) we obtain (after setting $g(y) = 0$):

$$(5.3.13) \quad \varphi = Aie^y \int_{+\infty}^{y} e^{-t} \frac{\partial}{\partial t} [H_0^{(1)} (ik\sqrt{x^2 + t^2})] dt + B(x)e^y.$$

The function $B(x)$ and the real constant A are arbitrary. The integral converges, since $\partial(H_0^{(1)})/\partial t$ dies out exponentially as $t \to \infty$.

We shall see that two solutions $\varphi_1(x, y)$ and $\varphi_2(x, y)$ satisfying all conditions of our problem can be obtained from (5.3.13) by taking $A = 0$ in one case and $A \neq 0$ in the other case, and that these solutions will be 90° "out of phase" at ∞. (This is exactly analogous to the behavior of the solutions in our previous two-dimensional cases.) Consider first the case $A = 0$. The function φ given by (5.3.13) satisfies (5.3.2) only if

$$(5.3.14) \quad \frac{d^2 B(x)}{dx^2} + (1 - k^2)B(x) = 0.$$

It is important to recall that $k^2 < 1$. The boundary condition $\varphi_x = 0$ for $x = 0$ requires that $B_x(0) = 0$. The condition $\varphi_y - \varphi = 0$ for $y = 0$ is automatically satisfied because of (5.3.12) and $g(y) \equiv 0$. Hence $B(x) = A_1 \cos \sqrt{1 - k^2} x$, with A_1 arbitrary, and the solution $\varphi_1(x, y)$ is

$$(5.3.15) \quad \varphi_1(x, y) = A_1 e^y \cos \sqrt{1 - k^2} x.$$

This leads to solutions Φ_1 in the form of standing waves,* as follows:

$$(5.3.15)' \quad \Phi_1(x, y, z; t) = A_1 e^{i\sigma t} e^y \cos \sqrt{1 - k^2} x \cdot \begin{cases} \cos kz \\ \sin kz \end{cases}$$

for $k^2 < 1$. If $k = 1$, the solution Φ_1 given by (5.3.15)' continues to be valid.

As we have already stated, we obtain solutions $\varphi_2(x, y)$ from (5.3.13) for $A \neq 0$ which behave for large x like $\sin \sqrt{1 - k^2} x$ rather than like $\cos \sqrt{1 - k^2} x$, and with these two types of solutions progressing waves approaching an arbitrary plane wave at ∞ can be constructed by superposition.

We begin by showing that (5.3.2) is satisfied for all $x > 0$, $y < 0$ by φ as given in (5.3.13) with $A \neq 0$, provided only that $B(x)$

---

* The standing wave solutions of this type (but not of the type with a singularity) for beaches sloping at angles $\pi/2n$ were obtained by Hanson [H.3] by a quite different method.

satisfies (5.3.14). Since $x > 0$, it is permissible to differentiate under
the integral sign in (5.3.13), even though t takes on the value zero
(since the upper limit y is negative). By differentiating we obtain

$$(5.3.16) \quad \nabla^2 \varphi - k^2 \varphi = Ai \left\{ e^y \int_\infty^y e^{-t} \frac{\partial}{\partial t} \left[\frac{\partial^2}{\partial x^2} + (1 - k^2) \right] H_0^{(1)} \, dt \right.$$

$$\left. + \frac{\partial H^{(1)}}{\partial y} + \frac{\partial^2 H_0^{(1)}}{\partial y^2} \right\} + \{ B''(x) + (1 - k^2) B(x) \} \, e^y.$$

Since $H_0^{(1)}$ is a solution of (5.3.2) the operator $(\partial^2/\partial x^2 - k^2)$ occurring under the integral sign can be replaced by $- \partial^2/\partial y^2$ and hence
the integral can be written in the form

$$\int_\infty^y e^{-t} \left[-\frac{\partial^3}{\partial t^3} + \frac{\partial}{\partial t} \right] H_0^{(1)} (ikr) dt.$$

We introduce the following notation

$$I_m(x, y) = e^y \int_\infty^y e^{-t} \frac{\partial^m}{\partial t^m} H_0^{(1)} (ikr) dt,$$

and obtain through two integrations by parts the result

$$I_m(x, y) = \left[\frac{\partial^{m-1}}{\partial y^{m-1}} + \frac{\partial^{m-2}}{\partial y^{m-2}} \right] H_0^{(1)} + e^y \int_{+\infty}^y e^{-t} \frac{\partial^{m-2} H_0^{(1)}}{\partial t^{m-2}} \, dt,$$

in which we have made use of the fact that the boundary terms are
zero at the lower limit $+ \infty$, since all derivatives of $H_0^{(1)}$ (ikr) tend
to zero as $r \to + \infty$. The integral of interest to us is given obviously
by $I_1 - I_3$ and this in turn is given by

$$- I_3 + I_1 = - \frac{\partial^2 H_0^{(1)}}{\partial y^2} - \frac{\partial H_0^{(1)}}{\partial y} - e^y \int_\infty^y e^{-t} \frac{\partial H_0^{(1)}}{\partial t} \, dt$$

$$+ e^y \int_\infty^y e^{-t} \frac{\partial H_0^{(1)}}{\partial t} \, dt = - \frac{\partial^2 H_0^{(1)}}{\partial y^2} - \frac{\partial H_0^{(1)}}{\partial y}$$

by use of the above relations for I_m. Hence the quantity in the first
bracket in (5.3.16) is identically zero—in other words the term
containing the integral on the right hand side of (5.3.13) is a solution
of (5.3.2). Hence φ is a solution of (5.3.2) in the case $A \neq 0$ if
$B(x)$ satisfies (5.3.14). Since (5.3.12) holds and $g(y) \equiv 0$ it follows
that the free surface condition (5.3.3) is satisfied by φ in view of
the fact that $\partial H_0^{(1)} (ikr)/\partial y = 0$ for $y = 0$.

We have still to show that a solution $B(x)$ of (5.3.14) can be chosen so that $\varphi_x = 0$ for $x = 0$, and that φ has the desired behavior for large values of r. Actually, these two things go hand in hand. An integration by parts in (5.3.13) yields the following for φ:

$$(5.3.17) \quad \varphi = Aie^y \int_\infty^y e^{-t}H_0^{(1)}(ik\sqrt{x^2+t^2})dt + AiH_0^{(1)}(ik\sqrt{x^2+y^2}) + B(x)e^y,$$

provided that $x > 0$. It should be recalled that the upper limit y of the integral is negative; thus the integrand has a singularity for $x = 0$ since $t = 0$ is included in the interval of integration and $H_0^{(1)}(ikr)$ is singular for $r = 0$. We shall show that $\lim_{x\to 0}\partial\varphi/\partial x = 0$ provided that $B_x(0) = -2A \neq 0$. We have, for $x > 0$ and $y < 0$:

$$\frac{\partial\varphi}{\partial x} = Aie^y \int_\infty^y e^{-t}\frac{\partial}{\partial x}[H_0^{(1)}(ik\sqrt{x^2+t^2})]dt$$

$$+ Ai\frac{\partial}{\partial x}[H_0^{(1)}(ik\sqrt{x^2+y^2})] + B_x(x)e^y.$$

The second term on the right hand side is readily seen to approach zero as $x \to 0$ since this term can be written as the product of x and a factor which is bounded for $y < 0$. For the same reason it is clear that the only contribution furnished by the integral in the limit as $x \to 0$ arises from a neighborhood of $t = 0$ since the factor x may be taken outside of the integral sign. We therefore consider the limit

$$\lim_{x\to 0}\int_\varepsilon^{-\varepsilon} e^{-t}\frac{\partial}{\partial x}[iH_0^{(1)}(ik\sqrt{x^2+t^2})]dt, \qquad \varepsilon > 0.$$

The function $iH_0^{(1)}(ikr)$ has the following development valid near $r = 0$:

$$iH_0^{(1)}(ikr) = -\frac{2}{\pi}[J_0(ikr)\log r + p(r)]$$

in which $p(r)$ represents a convergent power series containing only even powers of r, and J_0 is the regular Bessel function with the following development

$$J_0(ikr) = 1 + \frac{(kr)^2}{2^2} + \ldots.$$

It follows that

$$\frac{\partial}{\partial x}[iH_0^{(1)}(ikr)] = -\frac{2}{\pi}\left[\frac{x}{r^2}J_0(ikr) + J_0'(ikr)\frac{x}{r}\log r + xg(r)\right]$$

$$= -\frac{2}{\pi}\left[\frac{x}{r^2}J_0(ikr) + \frac{1}{2}k^2x\log r + x\bar{g}(r)\right]$$

in which $g(r) = (1/r)dp/dr$ is bounded as $x \to 0$ since $y < 0$. The contribution of our integral in the limit is therefore easily seen to be given by

$$\lim_{x \to 0} -\frac{2}{\pi} \int_\varepsilon^{-\varepsilon} e^{-t} \frac{x}{x^2 + t^2}\, dt = \lim_{x \to 0} -\frac{2}{\pi} \int_\varepsilon^{-\varepsilon} \frac{x}{x^2 + t^2}\, dt.$$

By introducing $u = t/x$ as new integration variable and passing to the limit we may write

$$\lim_{\substack{x \to 0}} -\frac{2}{\pi} \int_\varepsilon^{-\varepsilon} \frac{x}{x^2 + t^2}\, dt = -\frac{2}{\pi} \int_\infty^{-\infty} \frac{du}{1 + u^2} = 2.$$

It therefore follows that $\lim_{x \to 0} \partial\varphi/\partial x = 0$ provided that

(5.3.18) $B_x(0) = -2A.$

The function $B(x)$ which satisfies this condition and the differential equation (5.3.14) is

(5.3.19) $B(x) = -\dfrac{2A}{\sqrt{1 - k^2}} \sin \sqrt{1 - k^2}\,x.$

Since $H_0^{(1)}(ikr)$ dies out exponentially as $r \to \infty$ it follows that the solution φ given by (5.3.17) with $B(x)$ defined by (5.3.19) behaves at ∞ like $e^y \sin[(1 - k^2)^{1/2}x]$.

A solution φ_2 of our problem which is out of phase with φ_1 (cf. (5.3.15)) is therefore given by

(5.3.20) $\varphi_2(x, y) = A_2 \left[ie^y \int_\infty^y e^{-t} H_0^{(1)}(ik\sqrt{x^2 + t^2})dt \right.$

$$\left. + iH_0^{(1)}(ik\sqrt{x^2 + y^2}) - \frac{2e^y}{\sqrt{1 - k^2}} \sin \sqrt{1 - k^2}\, x \right],$$

with A_2 an arbitrary real constant. Standing wave solutions Φ_2 are then given by

(5.3.20)' $\Phi_2 = A_2 e^{i\sigma t} \varphi_2(x, y) \cdot \begin{Bmatrix} \cos kz \\ \sin kz \end{Bmatrix}.$

By taking appropriate values of k progressing waves tending at ∞ to any arbitrary plane wave solution for water of infinite depth can be obtained by forming proper linear combinations of solutions of the type (5.3.15)' and (5.3.20)'. For a progressing wave traveling toward shore, for example, we may write

$$(5.3.21)\ \ \Phi(x, y, z; t) = A\left[\varphi_1(x, y)\cos kz + \frac{\sqrt{1-k^2}}{2}\varphi_2(x, y)\sin kz\right]\cos \sigma t$$

$$- A\left[\varphi_1(x, y)\sin kz - \frac{\sqrt{1-k^2}}{2}\varphi_2(x, y)\cos kz\right]\sin \sigma t$$

in which A_1 and A_2 in (5.3.15) and (5.3.20) are both taken equal to A. The solution (5.3.21) behaves at ∞ like $Ae^y \cos(\sqrt{1-k^2}x + kz + \sigma t)$ as one can readily verify by making use of the asymptotic behavior of $\varphi_1(x, y)$ and $\varphi_2(x, y)$,* and it is the only such solution since φ_1 and φ_2 are uniquely determined.

The special case $k = 1$ has a certain interest. It corresponds to waves which at ∞ have their crests at right angles to the shore. One readily sees from (5.3.15) and (5.3.20) that as $k \to 1$ the progressing wave solution (5.3.21) tends to

$$(5.3.22)\qquad\qquad \Phi(y, z; t) = Ae^y \cos(z + \sigma t)$$

that is, the progressing wave solution for this case is independent of x, is free of a singularity at the origin, and the curves of constant phase are straight lines at right angles to the shore line—all properties that are to be expected.

The progressing wave solution (5.3.21) was studied numerically

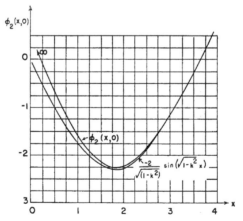

Fig. 5.3.1. Standing wave solution for a vertical cliff (with crests at an angle of 30° to shore)

* We remark once more that the original space and time variables can be reintroduced simply by replacing x, y, z by mx, my, mz and k by k/m.

for $k = 1/2$, i.e. for the case in which the wave crests tend at ∞ to a straight line inclined at $30°$ to the shore line. The function $\varphi_2(x, 0)$ is plotted in Figure 5.3.1. With the aid of these values the contours for Φ were calculated and are given in Figure 5.3.2. These are also essentially contour lines for the free surface elevation η, in accordance with the formula $\eta = -\dfrac{1}{g} \Phi_t|_{y=0}$. The water surface is shown between a pair of successive "nodes" of Φ, that is, curves for which $\Phi = 0$. These curves go into the z-axis (the shore line) under zero angle, as do all other contour lines. This is seen at once from their equation (cf. (5.3.21) with $\sigma t = \pi/2$)

$$(5.3.23) \quad \varphi_1(x, 0) \cos kz + \frac{\sqrt{1 - k^2}}{2} \varphi_2(x, 0) \sin kz = \text{const.}$$

Since $\varphi_2 \to \infty$ as $x \to 0$ while φ_1 remains bounded, it is clear that $\sin kz$ must approach zero as $x \to 0$ on any such curve. That the contours are all tangent to the z-axis at the points $z = 2\pi n$, n an integer, is also readily seen. It is interesting to observe that the

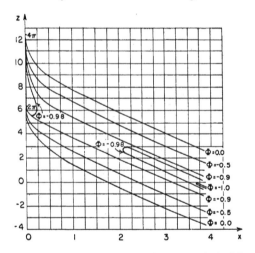

Fig. 5.3.2. Level lines for a wave approaching a vertical cliff at an angle

height of the wave crest is lower at some points near to the cliff than it is at ∞. It may be that the wave crest is a ridge with a number of saddle points.

It should be pointed out that we are no more able to decide in the present case than we were in the two-dimensional cases whether the waves are reflected back to infinity from the shore, and if so to what extent. Our numerical solution was obtained on the assumption that no reflection takes place, which is probably not well justified for the case of a vertical cliff, but would be for a beach of small slope.

5.4. Waves on sloping beaches. General case

We discuss here the most general case of periodic waves on sloping beaches which behave at ∞ like an arbitrary progressing wave—in particular, a wave with crests at an arbitrary angle to the shore line—and for a beach sloping at any angle. As has been mentioned earlier, this problem was first solved by Peters and Roseau (cf. the remarks in section 5.1).

We seek a harmonic function $\Phi(x, y, z; t)$ of the form $\exp \{i(\sigma t + kz)\}$ $\cdot \varphi(x, y)$ in the region indicated in cross section in Figure 5.4.1. At

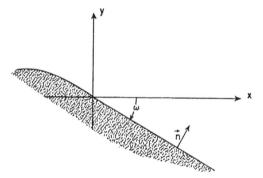

Fig. 5.4.1. Sloping beach of arbitrary angle

∞ the function Φ should behave like $\exp \{i(\sigma t + kz + \alpha x)\} \cdot \exp \{\sigma^2 y/g\}$ with k and α arbitrary. The function $\varphi(x, y)$ is not a harmonic function, but satisfies, as one readily sees, the differential equation

$$(5.4.1) \qquad \varphi_{xx} + \varphi_{yy} - k^2\varphi = 0,$$

the free surface condition

$$(5.4.2) \qquad \varphi_y - m\varphi = 0, \qquad y = 0, \qquad m = \frac{\sigma^2}{g},$$

and the condition at the bottom*

(5.4.3) $\varphi_n = 0$, $y = - x \tan \omega$.

By introducing (as we have done before) the new dimensionless
quantities $x_1 = mx$, $y_1 = my$, $\alpha_1 = \alpha/m$, $k_1 = k/m$ the conditions of
the problem for $\varphi(x, y)$ can be put in the form

$(5.4.1)_1$ $\varphi_{xx} + \varphi_{yy} - k^2 \varphi = 0$, $0 \leq k \leq 1$,

$(5.4.2)_1$ $\varphi_y - \varphi = 0$, $y = 0$,

$(5.4.3)_1$ $\varphi_n = 0$, $y = - x \tan \omega$

after dropping subscripts. Since we require $\varphi(x, y)$ to behave like
$e^{i\alpha x} e^{my} = \exp \{i\alpha_1 x_1 + y_1\}$ at ∞, it follows from (5.4.1) that
$- \alpha^2 + m^2 - k^2 = 0$ and hence that $\alpha_1^2 + k_1^2 = 1$. Thus k in $(5.4.1)_1$
(really it is k_1) is, as indicated, restricted to the range $0 \leq k \leq 1$,
and this fact is of importance in what follows.** Finally, we know
from past experience that a singularity must be permitted at the
origin. (In the problems treated earlier in this chapter we have
prescribed only boundedness conditions at ∞ in a way which led to
a statement concerning the uniqueness of the solution. In the present
case we do not obtain a similar uniqueness theorem—in fact, as has
been pointed out by Ursell [U.7, 8], Stokes showed that there exist
motions different from the state of rest and which die out at ∞.
For these motions, however, the quantity k is larger than unity).

We seek functions $\varphi(x, y)$ satisfying the above conditions as the
real or the imaginary part of a complex function $f(z, \bar{z})$ which is
analytic in each of the variables $z = x + iy$ and its conjugate
$\bar{z} = x - iy$. In the two-dimensional cases, it was sufficient to consider
analytic functions $f(z)$ of one complex variable, but in the present
case it is necessary to take more general functions since $\varphi(x, y)$ is
not a harmonic function. Note that we now use the variable z in a
different sense than above, where it is one of the space variables;
no confusion should result since the space variable z hardly occurs
again in the discussion to follow. It is useful to calculate some of the
derivatives of such functions with respect to x and y; we have,
clearly:

* Peters [P.6] solves the problem when the condition (5.4.3) is replaced by
the more general mixed boundary condition $\varphi_n + a\varphi = 0$, $a = $ const.
** Involved in this remark is the assumption that the derivatives of the solution
behave asymptotically the same as the derivatives of its asymptotic development;
but this is indeed the case, as we could verify on the basis of our final representation of the solution.

$$f_x = f_z z_x + f_{\bar{z}} \bar{z}_x = f_z + f_{\bar{z}},$$
$$f_y = i(f_z - f_{\bar{z}}),$$
$$f_{xx} + f_{yy} = 4 f_{z\bar{z}}.$$

Consequently our differential equation $(5.4.1)_1$ can be replaced by the differential equation

$$f_{z\bar{z}} - \frac{k^2}{4} f = 0$$

since the real or the imaginary part of any solution of it is clearly a solution of $(5.4.1)_1$.

Among the solutions of the last equation are the following simple special solutions (obtained, for example, by separating the variables in writing $f = f_1(z) \cdot f_2(\bar{z})$):

$$f(z, \bar{z}) = e^{\zeta z + \frac{k^2}{4} \frac{\bar{z}}{\zeta}}, \qquad \zeta = \text{const.},$$

which, when $\zeta = -i$ for example, is of the form

$$\exp\left\{\left(1 + \frac{k^2}{4}\right) y\right\} \cdot \exp\left\{-i\left(1 - \frac{k^2}{4}\right) x\right\}, \text{ and this is a solution of}$$

$(5.4.1)_1$ which has the proper behavior at ∞, at least. (Actually, when combined with the factor e^{ikz}, with z once more the space variable, the result is a harmonic function yielding a plane wave in water of infinite depth and satisfying the free surface condition).

One can obtain a great many more solutions by multiplying the above special solution by an analytic function $g(\zeta)$ and integrating along a path P in the complex ζ-plane:

$$(5.4.4) \qquad f(z, \bar{z}) = \frac{1}{2\pi i} \int_P e^{z\zeta + \frac{k^2}{4} \frac{\bar{z}}{\zeta}} \cdot g(\zeta) d\zeta.$$

By appropriate choices of the analytic function $g(\zeta)$ and the path P, we might hope to satisfy the boundary conditions and the condition at ∞. This does, indeed, turn out to be the case.

Still another way to motivate taking (5.4.4) as the starting point of our investigation is the following. It would seem reasonable to look for solutions of (5.4.1) in the form of the exponential functions $\varphi = \exp\{mx + ly\}$. However, since we wish to work with analytic functions of complex variables it would also seem reasonable to express x and y in terms of $z = x + iy$ and $\bar{z} = x - iy$, and this would lead to

$$\varphi = \exp\left\{m\left(\frac{z + \bar{z}}{2}\right) - li\left(\frac{z - \bar{z}}{2}\right)\right\}. \text{ In order that this function}$$

(which is clearly analytic in z and $\bar{z}$ separately) be a solution of
$(5.4.1)_1$ we must require that $m^2 + l^2 - k^2 = 0$, and this leads at
once to a solution of the form $\exp\left\{\zeta z + \dfrac{k^2}{4}\dfrac{\bar{z}}{\zeta}\right\}$, with ζ an arbitrary parameter, as one can readily verify. The method used by
Peters [P.6] to arrive at a representation of the form (5.4.4) is better
motivated though perhaps more complicated, since he operates with
(5.4.1) in polar coordinates, applies the Laplace transform with
respect to the radius vector, transforms the resulting equation to
the Laplace equation, and eventually arrives at (5.4.4).

One of the paths of integration used later on is indicated in Figure
5.4.2. The essential properties of this parth are: it is symmetrical
with respect to the real axis, goes to infinity in the negative direction

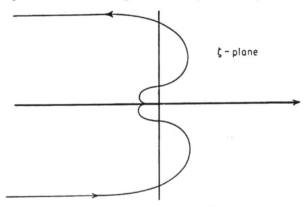

Fig. 5.4.2. The path P in the ζ-plane

of the real axis, enters the origin tangentially to the real axis and
from the left, and contains in the region lying to the left of it a
number of poles of $g(\zeta)$. (The path is assumed to enter the origin
in the manner indicated so that the term $\bar{z}/\zeta$ in the exponential
factor will not make the integral diverge). Our discussion will take
the following course: We shall assume $g(\zeta)$ to be defined in the ζ-plane
slit along the negative real axis (and also on occasion on a Riemann
surface obtained by continuing analytically over the slit). The choice
of the symmetrical path P leads to a functional equation for $g(\zeta)$
through use of the boundary conditions $(5.4.2)_1$ and $(5.4.3)_1$, and vice
versa a solution $g(\zeta)$ of the functional equation leads to a function

$\varphi(x, y) = \mathscr{R}e\, f(z, \bar{z})$ satisfying the boundary conditions. (By the symbols $\mathscr{I}m$ and $\mathscr{R}e$ we mean, of course, that the imaginary, or real, part of what follows is to be taken.) We seek a solution of the functional equation which is defined and regular in the slit ζ-plane, with at most poles in the left half-plane (including certain first order poles on the negative imaginary axis), and dying out at ∞ like $1/\zeta$. Once such a function has been found, the prescribed conditions at ∞ will be seen to follow by deforming the path P over the poles into a path on the two edges of the slit along the negative real axis: the residues at the poles on the negative imaginary axis clearly would yield contributions of the type

$$g(-ri) \cdot \exp\left\{- irz + \frac{k^2}{4}\frac{\bar{z}}{(-ir)}\right\}, r > 0,\ \text{which are easily seen to be}$$

of the desired type at ∞, while the remaining poles and the integral over the deformed path will be found to yield contributions that tend to zero when $\mathscr{R}e\, z \to +\infty$.

We begin this program by expressing the boundary conditions $(5.4.2)_1$ and $(5.4.3)_1$ in terms of the function $f(z, \bar{z})$. The first of these conditions will be satisfied if the following condition holds:

$(5.4.2)'_1$ $\qquad \mathscr{I}m(f_z - f_{\bar{z}} + if) = 0,$ $\qquad z$ real, positive,

as one readily sees. The condition $(5.4.3)_1$ will be satisfied if $\mathbf{n} \cdot \text{grad}\, \varphi = 0$, with $\mathbf{n}$ the unit normal at the bottom surface, i.e. if $\mathscr{R}e\, \{\mathbf{n} \cdot \text{grad}\, f\} = 0$, and the latter is given by

$$\mathscr{R}e\, \{(f_z + f_{\bar{z}})\sin \omega + i(f_z - f_{\bar{z}})\cos \omega\} = 0,$$

or finally, in the form

$(5.4.3)'_1$ $\qquad \mathscr{I}m\, \{f_z e^{-i\omega} - f_{\bar{z}} e^{i\omega}\} = 0,$ $\qquad z = re^{-i\omega},$ $\qquad r > 0.$

Upon making use of (5.4.4) in $(5.4.2)'_1$ the result is

$$(5.4.5) \qquad \mathscr{I}m\, \frac{1}{2\pi i}\int_P e^{z\zeta + \frac{k^2\bar{z}}{4\zeta}} \cdot \left[\zeta - \frac{k^2}{4\zeta} + i\right] g(\zeta)d\zeta = 0,$$
$$z\ \text{real, positive,}$$

while $(5.4.3)'_1$ yields

$$(5.4.6) \qquad \mathscr{I}m\, \frac{1}{2\pi i}\int_P e^{z\zeta + \frac{k^2\bar{z}}{4\zeta}} \cdot \left[\zeta e^{-i\omega} - \frac{k^2}{4\zeta}e^{i\omega}\right] g(\zeta)d\zeta = 0,$$
$$z = re^{-i\omega}, \qquad r > 0.$$

To satisfy the boundary condition (5.4.5) it is sufficient to require that $g(\zeta)$ satisfies the condition

$$(5.4.7) \qquad \mathscr{I}m \left[\zeta - \frac{k^2}{4\zeta} + i \right] g(\zeta) = 0, \qquad \zeta \text{ real, positive.}$$

The proof is as follows: If (5.4.7) holds, then the integrand $G(z, \bar{z}, \zeta)$ in (5.4.5) is real for real z and real positive ζ. Hence G takes on values G, $\bar{G}$ at conjugate points ζ, $\bar{\zeta}$ which are themselves conjugate, by the Schwarz reflection principle. Since the path P is symmetrical, as shown in Figure 5.4.2, it follows that $d\zeta$ takes on values at ζ, $\bar{\zeta}$ that are negative conjugates. Thus the integral $(1/2\pi i) \int_P G \, d\zeta$ is real when z is real and (5.4.7) holds. In considering next (5.4.6) we first introduce a new variable $s = \zeta e^{-i\omega}$ to obtain for $z = re^{-i\omega}$ the condition. replacing (5.4.6):

$$(5.4.8) \qquad \mathscr{I}m \frac{1}{2\pi i} \int_{P'} e^{rs + \frac{k^2 r}{4s}} \cdot \left[s - \frac{k^2}{4s} \right] g(se^{i\omega}) e^{i\omega} \, ds = 0. \qquad r \text{ real.}$$

Here P' is the path obtained by rotating P (and the slit in the ζ-plane as well, of course) clockwise about the origin through the angle ω. If g behaves properly at ∞, and if the rotation of P' can be accomplished without passing over any poles of the integrand, we may deform P' back to P and obtain

$$(5.4.8)' \qquad \mathscr{I}m \frac{1}{2\pi i} \int_{P} e^{rs + \frac{k^2 r}{s}} \cdot \left[s - \frac{k^2}{4s} \right] g(se^{i\omega}) e^{i\omega} \, ds = 0, \qquad r \text{ real.}$$

By the same argument as before we now see that the condition (5.4.6) will be satisfied provided that $g(\zeta)$ satisfies the condition

$$(5.4.9) \qquad \mathscr{I}m \, g(\zeta e^{i\omega}) e^{i\omega} = 0, \qquad \zeta \text{ real, positive.}$$

Thus if the function $g(\zeta)$ satisfies the conditions (5.4.7) and (5.4.9), the function $f(z, \bar{z})$ constructed by its aid will satisfy the boundary conditions. As we have already remarked, $g(\zeta)$ must satisfy still other conditions —at ∞, for example. In addition, we know from earlier discussions in this and the preceding chapter that it is necessary to find two solutions $\varphi(x, y)$ and $\varphi_1(x, y)$ of our problem which are "out of phase at ∞", in order that a linear combination of them with appropriate time factors will lead to a solution having the form of an arbitrary progressing wave at ∞. In this connection we observe that if the path P_1 of integration (as shown in Figure 5.4.3) is taken instead of the path P (it differs from P only in reversal of direction of the portion in the upper half-plane), and if we define $\varphi_1(x, y)$ as the *imaginary part* of $f_1(z, \bar{z})$ instead of its real part:

$$(5.4.10) \quad \varphi_1(x, y) = \mathscr{I}m \, \frac{1}{2\pi i} \int_{P_1} G(z, \bar{z}, \zeta) d\zeta = \mathscr{I}m \, f_1(z, \bar{z}),$$

with G the same integrand as before, then $\varphi_1(x, y)$, by the same argument as above, will satisfy the boundary conditions provided that the function $g(\zeta)$ also in this case satisfies the conditions (5.4.7)

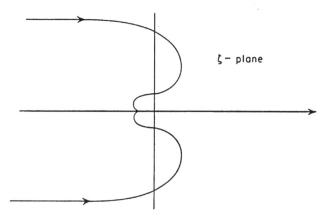

ζ - plane

Fig. 5.4.3. The path P_1 in the ζ-plane

and (5.4.9). It seems reasonable to expect that the integral over P_1 will behave the same as the integral over P when $\mathscr{R}e \, z$ is large and positive (since the poles in the lower half-plane alone determine this behavior and the paths P and P_1 differ only in the upper half-plane) except that a factor i will appear, and hence that φ and φ_1 will differ in phase at $+ \infty$ (in the variable x, that is) by 90°. This does indeed turn out to be the case.

Thus to satisfy the boundary conditions for both types of standing wave solutions we have only to find a function $g(\zeta)$ satisfying the conditions (5.4.7) and (5.4.9) which behaves properly at ∞—the last condition being needed in order that the path of integration can be rotated in the manner specified in deriving (5.4.8)'. To this end we derive a functional equation for $g(\zeta)$ by making use of these conditions. From (5.4.7) we have, clearly:

$$(5.4.11) \quad \left(\zeta - \frac{k^2}{4\zeta} + i\right) g(\zeta) = \left(\zeta - \frac{k^2}{4\zeta} - i\right) \overline{g(\zeta)}, \quad \zeta \text{ real, positive,}$$

while from (5.4.9) we have

(5.4.12) $\qquad \overline{g(\zeta)}e^{-i\omega} = g(\zeta e^{2i\omega})e^{i\omega}, \qquad \zeta$ real, positive,

both by virtue of the reflection principle. Eliminating $\overline{g(\zeta)}$ from the two equations we obtain

(5.4.13) $\qquad \left(\zeta - \dfrac{k^2}{4\zeta} + i\right)g(\zeta) = e^{2i\omega}\left(\zeta - \dfrac{k^2}{4\zeta} - i\right)g(\zeta e^{2i\omega}).$

This functional equation was derived for ζ real and positive, but since $g(\zeta)$ is analytic it is clear that it holds throughout the domain of regularity of $g(\zeta)$; it is the basic functional equation for $g(\zeta)$, a solution of which will yield the solution of our problem. Of course, this equation is only a necessary condition that must be fulfilled if the boundary conditions are satisfied; later on we shall show that the solution of it we choose also satisfies the condition (5.4.11), and hence the condition (5.4.12) will also be satisfied since (5.4.13) holds.

We proceed now to find a solution $g(\zeta)$ of (5.4.13) which has all of the desired properties needed to identify (5.4.4) and (5.4.10) as functions furnishing the solution of our problem, as has been done by Peters in the paper cited above.

We therefore proceed to treat the functional equation (5.4.13), which is easily put in the form:

(5.4.14) $\qquad \dfrac{g(e^{2i\omega}\zeta)}{g(\zeta)} = e^{-2i\omega}\dfrac{\zeta^2 + i\zeta - \dfrac{k^2}{4}}{\zeta^2 - i\zeta - \dfrac{k^2}{4}}$

$$= e^{-2i\omega}\dfrac{(\zeta + ir_1)(\zeta + ir_2)}{(\zeta - ir_1)(\zeta - ir_2)}$$

$$\text{with } r_{1,2} = \dfrac{1 \pm \sqrt{1 - k^2}}{2}.$$

The numbers $r_{1,2}$ are real since we know that k lies between 0 and 1. It is convenient to set

(5.4.15) $\qquad g(\zeta) = \dfrac{\zeta h(\zeta)}{(\zeta + ir_1)(\zeta + ir_2)}$

in which $h(\zeta)$, like $g(\zeta)$, is defined in the ζ-plane slit along the negative real axis. The function $h(\zeta)$ will have poles in the left half-plane, but only the poles at $\zeta = -ir_1$ and $\zeta = -ir_2$ of $g(\zeta)$ will be found to contribute a non-vanishing residue of $f(z)$ for $\mathscr{Re}\, z \to +\infty$, and this in turn would guarantee that $f(z)$ behaves at ∞ on the free surface like Ae^{-iz}. For $h(\zeta)$ we have from (5.4.14) and (5.4.15) the equation

(5.4.16) $\quad \dfrac{h(e^{2i\omega}\zeta)}{h(\zeta)} = \dfrac{(\zeta + ir_1 e^{-2i\omega})(\zeta + ir_2 e^{-2i\omega})}{(\zeta - ir_1)(\zeta - ir_2)} = m(\zeta).$

This equation is solved by introducing the function $l(\zeta)$ by

(5.4.17) $\quad\quad\quad\quad\quad \log h(\zeta) = l(\zeta),$

and one finds at once that $l(\zeta)$ satisfies the difference equation

(5.4.18) $\quad\quad\quad l(e^{2i\omega}\zeta) - l(\zeta) = \log m(\zeta) = w(\zeta).$

In solving this equation we shall begin by producing a solution $l(\zeta)$ free of singularities in the sector $-\omega \leqq \arg \zeta \leqq \omega$, after which the function $h(\zeta)$—which is (cf. (5.4.17)) then also regular in the same sector—can be continued analytically into the whole ζ-plane slit along the negative real axis (or, if desired, into a Riemann surface having the origin as its only branch point) by using (5.4.16). As an aid in solving equation (5.4.18) we set

(5.4.19) $\quad \begin{cases} \omega = \alpha\pi, & 0 < \alpha \leqq 1, \\ \zeta = \tau^\alpha, & l(\tau^\alpha) = L(\tau), \quad w(\tau^\alpha) = W(\tau), \end{cases}$

and operate now in a τ-plane. One observes that the sector $-\omega < \arg \zeta < \omega$ in the ζ-plane corresponds to the τ-plane slit along its negative real axis. For $L(\tau)$ one then finds at once from (5.4.18) the equation

(5.4.20) $\quad\quad\quad\quad L(\tau e^{2\pi i}) - L(\tau) = W(\tau).$

Our object in putting the functional equation into this form (following Peters) is that a solution is now readily found by making use of the Cauchy integral formula. Let us assume for this purpose that $L(\tau)$ is an analytic function in the closed τ-plane slit along its negative real axis* (which would imply that $l(\zeta)$ is regular in the sector $-\omega \leqq \arg \zeta \leqq \omega$, as we see from (5.4.19)); in such a case $L(\tau)$ can be represented by the Cauchy integral formula:

(5.4.21) $\quad\quad\quad\quad L(\tau) = \dfrac{1}{2\pi i} \oint_C \dfrac{L(\xi)}{\xi - \tau} \, d\xi,$

with C the path in the ξ-plane indicated by Figure 5.4.4. If we suppose in addition that $L(\xi)$ dies out at least as rapidly as, say, $1/\xi$ at ∞, it is clear that we can let the radius R of the circular part of C tend to infinity, draw the path of integration into the two edges of the slit and, in the limit, find for $L(\tau)$ the representation

* We shall actually produce such a regular solution shortly.

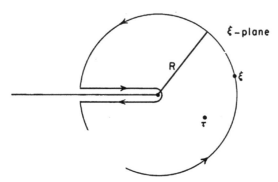

Fig. 5.4.4. Path C in the ξ-plane

$$(5.4.22) \qquad L(\tau) = \frac{1}{2\pi i}\int_{\rightarrow} + \frac{1}{2\pi i}\int_{\leftarrow},$$

in readily understandable notation. On making use of (5.4.20), and drawing the two integrals together, it is readily seen that $L(\tau)$ is given by

$$(5.4.23) \qquad L(\tau) = \frac{1}{2\pi i}\int_{-\infty}^{0} \frac{W(\xi)}{\xi - \tau}\, d\xi.$$

The path of integration is the negative real ξ-axis, and $W(\xi)$ is to be evaluated for $\arg \xi = -\pi$. Since $W(\xi)$ has no singularities (cf. (5.4.18)), it follows that $L(\tau)$ as given by (5.4.23) is indeed regular in the slit τ-plane. $L(\tau)$ also has no singularity on the slit except at the origin, where it has a logarithmic singularity. Since the numerator in the integrand behaves like $1/\xi^{\alpha}$, $\alpha > 0$, at ∞ (cf. (5.4.19), (5.4.18), (5.4.16)), it is clear that the function $L(\tau)$ dies out like $1/\tau$ at ∞ in the τ-plane. This function therefore has all of the properties postulated in deriving (5.4.23) from (5.4.21), and hence is a solution of the difference equation (5.4.20) in the slit plane including the lower edge of the slit.

A solution of (5.4.18) can now be written down through use of (5.4.19); the result is:

$$(5.4.24) \qquad l(\zeta) = \frac{1}{2\pi i}\int_{-\infty}^{0} \frac{\log m(\xi^{\alpha})}{\xi - \zeta^{1/\alpha}}\, d\xi,$$

with $m(\xi^{\alpha})$ to be evaluated for $\arg \xi = -\pi$. This solution is valid so far only for ζ in the sector $-\omega \leqq \arg \zeta \leqq \omega$, where it is regular, as we know from the discussion above. However, it is necessary to define the function $h(\zeta) = e^{l(\zeta)}$ (cf. (5.4.17)) in the entire slit ζ-plane, and this can be done by analytic continuation with the aid of the functional equation (5.4.16). In the process of analytic continuation, starting with the original sector in which $l(\zeta)$, and hence $h(\zeta)$, is free of singularities, one sees that the only singularities which could occur in continuing into the upper half-plane, say, would arise from the function on the right hand side of the equation (5.4.16). The only singularities of this function occur obviously at $\zeta = ir_{1,2}$. Consequently no singularity of $h(\zeta)$ appears in the analytic continuation into the upper half-plane, through widening of the sector in which $h(\zeta)$ is defined, until the points $\zeta = ir_1$ and $\zeta = ir_2$ have been covered, and one sees readily from (5.4.16) that the first such singularities of $h(\zeta)$—poles of first order—appear at the points $r_{1,2} \exp \{i(2\omega + \pi/2)\}$, the next at $r_{1,2} \exp \{i(4\omega + \pi/2)\}$, etc., though some, or all, of these poles may not appear on the first sheet of the slit ζ-plane, depending on the value of the angle ω. The continuation into the lower half-plane is accomplished by writing (5.4.16) in the equivalent form

$$(5.4.16)' \qquad \frac{h(\zeta e^{-2i\omega})}{h(\zeta)} = \frac{1}{m(\zeta e^{-2i\omega})} .$$

Again we see that poles will occur in the lower half-plane in the course of the analytic continuation, this time at $r_{1,2} \exp \{-i(2\omega + \pi/2)\}$ $r_{1,2} \exp \{-i(4\omega + \pi/2)\}$, etc. The situation is indicated in Figure 5.4.5; $h(\zeta)$ lacks the singularities of $g(\zeta)$ at the points $-ir_1$ and $-ir_2$ (cf. (5.4.15)). Thus the function $h(\zeta)$ is defined in the slit ζ-plane. (It can also be continued analytically over the slit which permits a rotation of the path of integration.) We see that $h(\zeta)$ may have poles in the open left half-plane, on two circles of radii r_1 and r_2, but the poles closest to the imaginary axis are at the angular distance 2ω from it. There is also a simple pole of $h(\zeta)$ at the origin, but $g(\zeta)$ (cf. (5.4.15)) is regular there.

The behavior of $h(\zeta)$ at ∞ in the slit plane is now easily discussed: In the original sector we know from (5.4.24) that $l(\zeta)$ dies out at ∞ like $1/\zeta^{1/\alpha}$. Hence $h(\zeta) = e^{l(\zeta)}$ is bounded in the sector, and since the right hand side of (5.4.16) is clearly bounded at ∞ it follows that $h(\zeta)$ is bounded at ∞ in the ζ-plane.

The function $g(\zeta) = \zeta h(\zeta)/(\zeta + ir_1)(\zeta + ir_2)$ (cf. (5.4.15)) can now be seen to have all of the properties needed to identify the functions

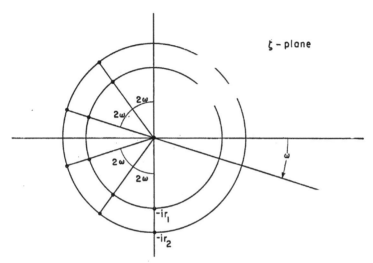

Fig. 5.4.5. The singularities of $h(\zeta)$ and $g(\zeta)$

$f(z)$ in (5.4.4) and $f_1(z)$ in (5.4.10) as functions whose real part and imaginary part, respectively, yield the desired standing wave solutions of our problem. To this end we write down the integrals

$$(5.4.25) \quad f, f_1 = \frac{1}{2\pi i} \int_{P, P_1} e^{\left(z\zeta + \frac{k^2 z}{4\zeta}\right)} \frac{\zeta h(\zeta)}{(\zeta + ir_1)(\zeta + ir_2)} \, d\zeta$$

over the paths indicated in Figure 5.4.6, where the direction is indicated only on the part of the path in the lower half-plane, since the paths P, P_1 differ only in the direction in which the remainder of the path is traversed.

Since $h(\zeta)$ is bounded at ∞, and $\mathscr{R}e \, z > 0$, the integrals clearly converge. One sees also that the paths of integration can be rotated through the angle ω about the origin without passing over singularities of the integrand, and also without changing the value of the integrals. (This was needed in deriving (5.4.8)'.) We prove next that $g(\zeta)$ satisfies the boundary condition (5.4.11). To begin with, we shall show that $l(\zeta)$ as defined by (5.4.24) is real when ζ is real and positive. Once this is admitted to be true, then $h(\zeta)$ as given by (5.4.17) would have the same property, and the function $g(\zeta)$ defined by (5.4.15) would easily be seen to satisfy the condition (5.4.11). We have, then, only to show that $l(\zeta)$ is real for real ζ, and this can be seen as follows:

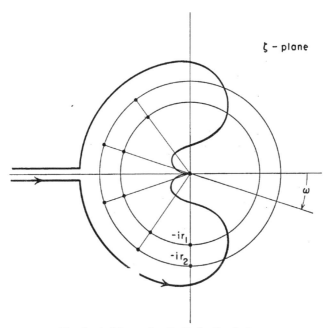

Fig. 5.4.6. The paths P, P_1 in the ζ-plane

In (5.4.24) $\log m(\xi^\alpha)$ is to be evaluated for $\arg \xi = -\pi$. But in this case one sees easily from the equation (5.4.16) defining $m(\xi^\alpha)$ (with $\alpha = \omega/\pi$, cf. (5.4.19)) that $m(\xi^\alpha)$ has its values on the unit circle when $\arg \xi = -\pi$, and hence its logarithm is pure imaginary on the path of integration; it follows at once from (5.4.24) that $l(\zeta)$ is real for ζ real and positive. Since $g(\zeta)$ was constructed in such a way as to satisfy (5.4.13) we know that (5.4.12) is satisfied automatically. Thus our standing wave solutions satisfy the boundary conditions.

Finally, we observe that the behavior of f and f_1 for $\mathscr{R}e\, z \to \infty$ is what was prescribed. To this end we deform the path of integration into a path running along the two banks of the slitted negative real axis. The residues at $\zeta = -ir_{1,2}$ contribute terms already discussed above which furnish the desired behavior for $\mathscr{R}e\, z \to +\infty$. We have, then, only to make sure that the residues at the remaining poles and the integrals along the slit make contributions which die out as $\mathscr{R}e\, z \to +\infty$. As for the residues at the poles at the points

$\zeta_n = r_{1,2} \exp \{\pm\, i(2n\omega + \pi/2)\}$, $n = 1, 2, \ldots$, we observe that these contributions are of the form $A e^{z \zeta_n}$, but since $-\omega \leqq \arg z \leqq 0$ it is clear that these contributions die out exponentially when z tends to infinity in the sector $-\omega \leqq \arg z \leqq 0$. As for the integrals along the slit, they are known to die out like $1/z$, as we have seen in similar cases before, or as one can verify by integration by parts. Thus all of the conditions imposed on $f(z)$ and $f_1(z)$ are seen to be satisfied. We observe, however, that the integrals in (5.4.25) over the paths P and P_1 converge only if $\mathscr{Re}\, z \geqq 0$, and hence this representation of our solution is valid only if the bottom slopes down at an angle $\leqq \pi/2$. For an overhanging cliff, when $\omega > \pi/2$, the solution can be obtained by first swinging the path of integration clockwise through $90°$ (and swinging the slit also, of course); the resulting integrals would then be valid for all z such that $\mathscr{Im}\, z \leqq 0$ and the solutions would hold for $0 < \omega \leqq \pi$.

It is perhaps of interest to bring the final formulas together for the simplest special case, i.e. the dock problem for two-dimensional motion (first solved by Friedrichs and Lewy [F.12]), in which the

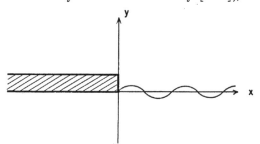

Fig. 5.4.7. The dock problem

angle ω has the value π, as indicated in Figure 5.4.7.* In this case the function $l(\zeta)$ is given by

$$(5.4.26) \qquad l(\zeta) = \frac{1}{2\pi i} \int_{-\infty}^{0} \frac{\log \left(\frac{\xi+i}{\xi-i}\right)}{\xi - \zeta}\, d\xi,$$

and the integral defines it at once in the entire slit ζ-plane. The standing wave solutions $\varphi(x, y) = \mathscr{Re}\, f(z)$ and $\varphi_1(x, y) = \mathscr{Im}\, f_1(z)$ are determined through

* As was mentioned in section 5.1, the dock problem in water of uniform finite depth and for the three-dimensional case was first solved by Heins [H.13] with the aid of the Wiener-Hopf technique.

(5.4.27)
$$f(z) = \frac{1}{2\pi i} \int_P e^{\zeta z} \frac{h(\zeta)}{\zeta + i} \, d\zeta,$$

(5.4.28)
$$f_1(z) = \frac{1}{2\pi i} \int_{P_1} e^{\zeta z} \frac{h(\zeta)}{\zeta + i} \, d\zeta,$$

with $h(\zeta)$ defined by

(5.4.29)
$$h(\zeta) = e^{l(\zeta)}.$$

As was remarked above, the integrals in (5.4.27) and (5.4.28) converge only if $\mathscr{Re}\, z \geqq 0$. However, the analytic continuation into the entire lower half-plane is achieved simply by swinging the paths P and P_1 into the positive imaginary axis (which can be done since $h(\zeta)$ is bounded at ∞), while staying on the Riemann surface of $h(\zeta)$, and these integrals are then valid for all z in the lower half-plane.

Finally, it is also of interest to remark that the functions $f(z)$ and $f_1(z)$ do not behave in the same way at the origin: the first is bounded there, and the second is not, and this behavior holds not only for the special case of the dock problem, but also in all cases under consideration here.

5.5. Diffraction of waves around a vertical wedge.
Sommerfeld's diffraction problem

In this section we are primarily concerned with the problem of determining the effect of a barrier in the form of a vertical rigid wedge, as indicated in Fig. 5.5.1, on a plane simple harmonic wave

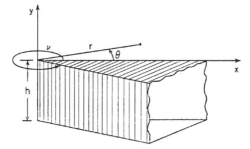

Fig. 5.5.1. Diffraction of a plane wave by a vertical wedge

coming from infinity. In this case it is convenient to make use of cylindrical coordinates (r, θ, y). We seek a harmonic function

$\Phi(r, \theta, y; t)$ in the region $0 < \theta < \nu$, $-h < y < 0$, i.e. in the region exterior to the wedge of angle $2\pi - \nu$ and in water of finite depth h when at rest. The problem is reduced to one in the two independent variables (r, θ) by setting

(5.5.1) $$\Phi(r, \theta, y; t) = f(r, \theta) \cosh m(y + h)e^{i\sigma t}.$$

The boundary conditions $\Phi_\theta = 0$ for $\theta = 0$, $\theta = \nu$ corresponding to the rigid walls of the wedge yield for $f(r, \theta)$ the boundary conditions

(5.5.2) $$f_\theta = 0, \qquad \theta = 0, \qquad \theta = \nu.$$

The free surface condition $g\Phi_y + \Phi_{tt} = 0$ at $y = 0$ yields the condition

(5.5.3) $$m \tanh mh = \sigma^2/g,$$

while the condition $\Phi_y = 0$ at the bottom $y = -h$ is satisfied automatically. Once any real value for the frequency σ is prescribed, equation (5.5.3) is used to determine the real constant m — which will turn out to be the wave number of the waves at ∞ —, and we note that (5.5.3) has exactly one real solution of m except for sign; if the water is infinitely deep we have $m = \sigma^2/g$, and the function $\cosh m(y + h)$ in (5.5.1) is replaced by e^{my}.

Thus the function $f(r, \theta)$ is to be determined as a solution of the reduced wave equation

(5.5.4) $$\nabla^2_{(r, \theta)}f + m^2 f = 0, \qquad 0 < r < \infty, \qquad 0 < \theta < \nu,$$

subject to the boundary conditions (5.5.2). Actually, we shall in the end carry out the solution in detail only for the case of a reflecting rigid plane strip (i.e. for the special case $\nu = 2\pi$), but it will be seen that the same method would furnish the result for any wedge. It is convenient to introduce a new independent variable ϱ, replacing r, by the equation $r = \varrho/m$; in this variable equation (5.5.4) has the form

(5.5.5) $$\nabla^2_{(\varrho, \theta)}f + f = 0, \qquad 0 < \varrho < \infty, \qquad 0 < \theta < \nu,$$

and we assume this equation as the basis for the discussion to follow.

So far we have not formulated conditions at ∞, except for the vague statement that we want to consider the effect of our wedge-shaped barrier on an incoming plane wave from infinity. Of course, we then expect a reflected wave from the barrier and also diffraction effects from the sharp corner at the origin. In conformity with our general practice we wish to formulate these conditions at ∞ in such a way that the solution of the problem will be uniquely determined. It has some point to consider the question of reasonable conditions at ∞ which determine unique solutions of the reduced wave equation under more general circumstances than those considered in the physical problem formulated above. For general domains it is not known how to formulate these conditions at ∞, and, in fact, it would seem to be a very difficult task to do so since such a formulation would almost certainly require consideration of many special cases. In one special case, however, the appropriate condition to be imposed at infinity has been known for a long time. This is the case in which any reflecting or refracting obstacles lie in a bounded domain of the plane, or, stated otherwise, it is the case in which a full neighborhood of the point at infinity is made up entirely of the homogeneous medium in which the waves propagate. In this case, the condition at ∞ which determines the "secondary" waves uniquely is Sommerfeld's radiation condition, which states, roughly speaking, that these waves behave like a cylindrical outgoing progressing wave at ∞. However, if the reflecting or refracting obstacles extend to infinity, the Sommerfeld condition may not be appropriate at all. Consider, for example, the case in which the entire x-axis is a reflecting barrier (i.e. the case $\nu = \pi$), and the primary wave is an incoming plane wave from infinity. It is clear on physical grounds that the secondary wave will be the reflected plane wave, which certainly does not behave at ∞ like a cylindrical wave since, for example, its amplitude does not even tend to zero at ∞. Another case is that of Sommerfeld's classical diffraction problem in which an incoming plane wave is reflected from a barrier consisting of the positive half of the x-axis. In this case, the secondary wave has both a reflected component which has a non-zero amplitude at ∞, and a diffracted part which dies out at ∞. A uniqueness theorem has been derived by Peters and Stoker [P.19] which includes these special cases; we proceed to give this proof both for its own sake and also because it points the way to a straightforward and elementary solution of the special problem formulated above. In Chapters 6 and 7 a different way of looking at the problem of determining appropriate radiation conditions is proposed; it involves considering simple harmonic waves (Chapter 6), or steady waves (Chapter 7) as limits when $t \to \infty$ in appropriately formulated initial value problems which correspond to unsteady motions.

The uniqueness theorem, which is general enough to include the problem above, is formulated in the following way: We assume that $f(x, y)$ is a complex-valued solution* of the equation

$$(5.5.6) \qquad \nabla^2 f + f = 0$$

in a domain D with boundary Γ, part of which may extend to infinity. It is supposed that any circle C in the x, y-plane cuts out of D a domain in which the application of Green's formula is legitimate, and, in addition, that the boundary curve Γ outside a sufficiently

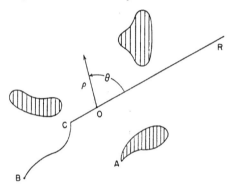

Fig. 5.5.2. The domain D

large circle consists of a single half-ray R going to ∞ (cf. Figure 5.5.2).** On the boundary Γ the condition

$$(5.5.7) \qquad f_n = 0$$

is imposed, i.e. the normal derivative of f vanishes, corresponding to a reflecting barrier. (We could also replace this condition on part, or all, of Γ by the condition $f = 0$.) We now write the solutions of f in D which satisfy (5.5.6) in the form

$$(5.5.8) \qquad f = g + h,$$

in order to formulate the conditions at ∞ in a convenient way. What we have in mind is to separate the solution into a part h which satisfies a radiation condition and a part g which contains,

* It is natural to consider such complex solutions, since, for example, a plane wave is obtained by taking $f(\varrho, \theta) = \exp \{ i\varrho \cos (\theta + \alpha) \}$.
** Our theorem also holds if D is the more general domain in which the ray R is replaced at ∞ by a sector, and the uniqueness proof given below holds with insignificant modifications for this case also.

roughly speaking, the prescribed incoming wave together with any secondary reflected or refracted waves which also do not satisfy a radiation condition. More precisely, we require h to satisfy the following radiation condition:

$$(5.5.9) \qquad \lim_{\varrho \to \infty} \int_C \left| \frac{\partial h}{\partial \varrho} + ih \right|^2 ds = 0.$$

Here C is taken to be a circle, with its center O (cf. Figure 5.5.2) on the ray R going to infinity, and with radius ϱ so large that all obstacle curves except a part of R lie in its interior. This condition clearly follows from the well-known Sommerfeld radiation condition, which requires that

$$(5.5.9)_1 \qquad \lim_{\varrho \to \infty} \varrho^{\frac{1}{2}} \left(\frac{\partial h}{\partial \varrho} + ih \right) \to 0$$

uniformly in θ, and, incidentally, this is a condition independent of the particular point from which ϱ is measured; we observe that if h behaves at ∞ like $e^{-i\varrho}/\sqrt{\varrho}$, i.e. like an outgoing cylindrical wave, then condition $(5.5.9)_1$ is satisfied. We shall make use of the radiation condition in the form (5.5.9) in much the same way as F. John [J.5] who used it to obtain uniqueness theorems for (5.5.6) in cases other than those treated here; his methods were in turn modeled on those of Rellich [R.7].

The behavior of the function g at infinity is prescribed as follows:

$$(5.5.10) \qquad g \sim g_1 + g_2 \text{ at } \infty,$$

with g_1 a function that is *once for all prescribed*,* while g_2 is a function satisfying the same radiation condition as h, i.e. the condition (5.5.9). (That the behavior of g at ∞ is fixed only within an additive function satisfying the radiation condition is natural and inevitable.)

Finally, we prescribe regularity conditions at re-entrant points (such as A, B, C in Figure 5.5.2) of the boundary of D; these conditions are that

$$(5.5.11) \qquad f(\varrho, \theta) \sim c_1, \qquad f_\varrho(\varrho, \theta) \sim \frac{c_2}{\varrho^k}, \qquad k < 1,$$

with (ϱ, θ) polar coordinates centered at the particular singular point, and c_1 and c_2 constants. (These conditions on f mean physically that the radial velocity component may be infinite at a corner, but not

---

* How the function g_1 should be chosen is a matter for later discussion.

as strongly as it would be for a source or sink.) At other boundary points we require continuity of f and its normal derivative.

We can now state our theorem as follows:

Uniqueness theorem: A solution f of (5.5.6) in D is uniquely determined if it 1) satisfies the boundary condition (5.5.9); 2) admits of a decomposition of the form (5.5.8) with h a function satisfying (5.5.9), g a function behaving as prescribed by (5.5.10) at ∞; and 3) satisfies the regularity conditions at the boundary of D.

The proof of this theorem will be given shortly, but we proceed to discuss its implications here. The theorem is at first sight somewhat unsatisfactory since it involves the assumption that every solution considered can be decomposed according to (5.5.8), with $g(\varrho, \theta)$ a certain function the behavior of which at ∞, in so far as the leading term g_1 (cf. (5.5.10)) in its asymptotic development is concerned, is not given a priori. However, it is not difficult in some instances at least to guess, on the basis of physical arguments, how the function $g_1(\varrho, \theta)$ should be defined. For example, suppose the domain D consisted of the exterior of bounded obstacles only. In such a case it seems clear that $g_1(\varrho, \theta)$ should be defined as the function describing the incoming wave—either as a plane wave from infinity, say, or a wave originating from an oscillatory source—since bounded obstacles give rise only to reflected and diffracted components which die out at ∞ and which could be expected to satisfy the radiation condition. Even if there is a ray in the boundary that goes to ∞ (as was postulated above), it still would seem appropriate to take $g_1(\varrho, \theta)$ as the function describing the incoming wave, provided that it arises from an oscillatory point source,* since such a source would hardly lead to reflected or refracted secondary waves that would violate the radiation condition. However, if the incoming wave is a plane wave and an obstacle extends to ∞, one expects an outgoing reflected wave to occur which would in general not satisfy the radiation condition; in this case the function $g_1(\varrho, \theta)$ should be taken as the sum of the incoming plane wave and an outgoing reflected wave. For example, one might consider the case in which the entire x-axis is a reflecting barrier, as in Figure 5.5.3. In this case one would in an altogether natural way define $g_1(\varrho, \theta)$ as the sum of the incoming and of the reflected wave as follows:

* The same statement would doubtlessly hold if the disturbance originated in a bounded region, since this case could be treated by making use of a distribution of oscillatory point sources.

$$(5.5.12) \qquad g_1(\varrho, \theta) = e^{i\varrho \ \cos \ (\theta - \alpha)} + e^{i\varrho \ \cos \ (\theta + \alpha)},$$

with α the angle of incidence of the incoming plane wave. If we were then to set $f \equiv g_1 + h$ (i.e. we set $g \equiv g_1$ everywhere) and prescribe that h should satisfy the radiation condition, it is clear that we would

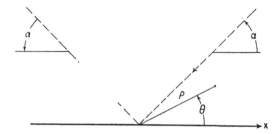

Fig. 5.5.3. Infinite straight line barrier

have a unique solution by taking $h \equiv 0$. Our uniqueness theorem does not apply directly here since there are two infinite reflecting rays going to ∞, but it could be easily modified so that it would apply to this case. Thus we have — for the first time, it seems — a

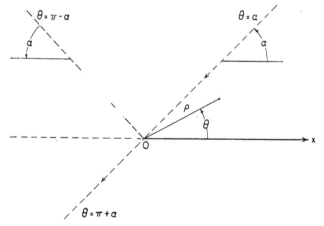

Fig. 5.5.4. Sommerfeld's diffraction problem

uniqueness theorem for this particularly simple problem of the reflection of a plane wave by a rigid plane. A less trivial example is the classical Sommerfeld diffraction problem — in effect, a special case of the problem with which our present discussion began—in which a plane wave coming from infinity at angle α to the x-axis is reflected and diffracted by a rigid half-plane barrier along the positive x-axis, as indicated in Figure 5.5.4. In this case it seems plausible to define the function $g_1(\varrho, \theta)$ as follows:

$$(5.5.13) \quad g_1(\varrho, \theta) = \begin{cases} e^{i\varrho \cos (\theta - \alpha)} + e^{i\varrho \cos (\theta + \alpha)}, & 0 < \theta < \pi - \alpha \\ e^{i\varrho \cos (\theta - \alpha)}, & \pi - \alpha < \theta < \pi + \alpha \\ 0, & \pi + \alpha < \theta < 2\pi. \end{cases}$$

This function is, of course, discontinuous, corresponding to the division of the plane into the regions in which a) the incoming wave and its reflection from the barrier coexist, b) the region in which only the wave transmitted past the edge of the barrier exists, and c) the region in the shadow created by the barrier. Again we would be inclined to take $g \equiv g_1$ (cf. (5.5.8) and (5.5.10)) and set $f \equiv g_1 + h$, with h satisfying the radiation condition. Of course, the function $h(\varrho, \theta)$ in (5.5.8) representing the diffracted wave would then also be discontinuous in that case since the sum $g_1 + h$ is everywhere continuous. It will be seen that the well-known solution given by Sommerfeld can be decomposed in this way and that h then satisfies the radiation condition. Our uniqueness theorem will thus be shown to be applicable in at least the important special case of particular interest in this section.

One might hazard a guess regarding the right way to determine the function g in all cases involving unbounded domains: it seems highly plausible that it would always be correctly given by the methods of geometrical optics. By this we mean, from the mathematical point of view, that g would be the lowest order term in an asymptotic expansion of the solution f with respect to the frequency of the motion that is valid for large frequencies; the methods of geometrical optics would thus be available for determining g. However, to prove a theorem of such generality would seem to be a very difficult task since it would probably require some sort of representation for the solution of wave propagation problems when more or less arbitrary domains and boundary data are prescribed.

Once having proved that the solution of Sommerfeld's diffraction problem could be decomposed in the way indicated above into the sum of two discontinuous functions, one of which satisfies the radiation condition, it was observed that the latter fact opens the way to a new solution of the diffraction problem which is entirely elementary, straightforward, and which can be written down in a few lines. In other words, once the reluctance to work with discontinuous functions is overcome, the solution of the problem is reduced to something quite elementary by comparison with other methods of solution. The problem was solved long ago by Sommerfeld [S.12], and afterwards by many others, including Macdonald [M.1], Bateman [B.5], Copson [C.4], Schwinger [S.5], and Karp [K.3].

We shall first prove the uniqueness theorem. Afterwards, the simple solution of Sommerfeld's problem just referred to will be derived; this solution is in the form of a Fourier series. The Fourier series solution is next transformed to furnish a variety of solutions given by integral representations, including the familiar representation given by Sommerfeld. The new representations are particularly convenient for the purpose of discussing a number of properties of the solution. In particular, two such representations can be used to show that the function h in the decomposition $f = g_1 + h$ (cf. (5.5.13)) satisfies the radiation condition, and that our solution f satisfies the regularity conditions at the origin; thus the solution is shown, by virtue of our uniqueness theorem, to be the only one which behaves at ∞ like g_1 plus a function satisfying the radiation condition. The Stokes' phenomenon encountered in crossing the lines of discontinuity of the functions g_1 and h is also discussed.

The uniqueness theorem formulated above is proved in the following way. Suppose there were two solutions f and f^* (cf. (5.5.8)) with f^* given by

$$(5.5.14) \qquad f^*(\varrho, \theta) = g^*(\varrho, \theta) + h^*(\varrho, \theta).$$

We introduce the difference $\chi(\varrho, \theta)$ of these solutions:

$$(5.5.15) \quad \chi(\varrho, \theta) = f(\varrho, \theta) - f^*(\varrho, \theta)$$
$$= g(\varrho, \theta) - g^*(\varrho, \theta) + h(\varrho, \theta) - h^*(\varrho, \theta)$$

and observe that $\chi(\varrho, \theta)$ satisfies the radiation condition (5.5.9), by virtue of the Schwarz inequality, since h, h^*, and the difference $g - g^*$ all satisfy it by hypothesis; thus we have

$$(5.5.16) \qquad \lim_{\varrho \to \infty} \int_C \left| \frac{\partial \chi}{\partial \varrho} + i\chi \right|^2 ds = 0.$$

The complex-valued function χ is decomposed into its real and imaginary parts:

$$(5.5.17) \qquad \chi = \chi_1 + i\chi_2,$$

and Green's formula

(5.5.18) $$\iint_{D^*} (\chi_2 \nabla^2 \chi_1 - \chi_1 \nabla^2 \chi_2) dx\, dy = \int_{\Gamma^*} \left(\chi_2 \frac{\partial \chi_1}{\partial n} - \chi_1 \frac{\partial \chi_2}{\partial n} \right) ds$$

is applied to χ_1 and χ_2 in the domain D^* indicated in Figure 5.5.5. The domain D^* is bounded by a circle C so large as to include all of the obstacles in its interior except R, by curves which exclude the

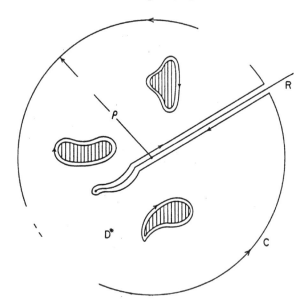

Fig. 5.5.5. The domain D^*

prolongation of R into the interior of C, and by curves excluding the other bounded obstacles. By (5.5.15), χ is a solution of (5.5.6) which also clearly satisfies the boundary condition (5.5.7). Since $\nabla^2 \chi_1 = - \chi_1$ and $\nabla^2 \chi_2 = - \chi_2$, it follows that the integrand of the left-hand side of (5.5.18) vanishes. Because of the regularity conditions at boundary points we are permitted to deform the boundary curve Γ^* into the obstacle curves, and it then follows from the boundary condition (5.5.7) that

(5.5.19) $$\int_C (\chi_2 \chi_{1_n} - \chi_1 \chi_{2_n})\, ds = 0$$

since the contributions at the obstacles all vanish. We now make use of the easily verified identity

(5.5.20) $2(\chi_2 \chi_{1n} - \chi_1 \chi_{2n}) = |\chi_n|^2 + |\chi|^2 - |\chi_n + i\chi|^2$

to deduce from (5.5.19) the condition

(5.5.21) $\int_C (|\chi_n|^2 + |\chi|^2) ds - \int_C |\chi_n + i\chi|^2 ds = 0,$

from which we obtain, in view of (5.5.16) and $\chi_n = \partial\chi/\partial\varrho$ on C:

(5.5.22) $\lim\limits_{\varrho \to \infty} \int_C |\chi|^2 ds = \lim\limits_{\varrho \to \infty} \int_C |\chi_n|^2 ds = 0.$

From the boundary condition (5.5.7), as applied on R, we see that $\chi(\varrho, \theta)$ can be continued as a periodic function of period 4π in θ on C; hence χ can be represented for all sufficiently large values of ϱ by the Fourier series

(5.5.23) $\chi = \sum\limits_{0}^{\infty} A_{n/2}(\varrho) \cos \dfrac{n\theta}{2},$

with $A_{n/2}(\varrho)$, the Fourier coefficient, a certain linear combination of the Bessel functions $J_{n/2}(\varrho)$ and $Y_{n/2}(\varrho)$, since χ is a solution of (5.5.6). The Fourier coefficients are given by

(5.5.24) $A_{n/2}(\varrho) = \dfrac{1}{\pi} \int_0^{2\pi} \chi(\varrho, \theta) \cos \dfrac{n\theta}{2} d\theta,$

and consequently we have

(5.5.25) $|\varrho^{\frac{1}{2}} A_{n/2}(\varrho)| \leq \dfrac{\varrho^{\frac{1}{2}}}{\pi} \int_0^{2\pi} |\chi| d\theta$

$\leq \sqrt{\dfrac{2\varrho}{\pi} \int_0^{2\pi} |\chi|^2 d\theta}$

$\leq \sqrt{\dfrac{2}{\pi} \oint_C |\chi|^2 ds}.$

It follows at once from (5.5.22) that the Fourier coefficients behave for large ϱ as follows:

(5.5.26) $\lim\limits_{\varrho \to \infty} \varrho^{\frac{1}{2}} A_{n/2}(\varrho) = 0.$

Since the Bessel functions $J_{n/2}(\varrho)$ and $Y_{n/2}(\varrho)$ all behave at ∞ like $1/\sqrt{\varrho}$, it follows that all of the coefficients $A_{n/2}(\varrho)$ must vanish. Consequently χ vanishes identically outside a sufficiently large circle, hence it vanishes* throughout its domain of definition, and the

* This could be proved in standard fashion since χ is now seen to satisfy homogeneous boundary conditions in the domain D^* of Fig. 5.3.5.

uniqueness theorem is proved. As was stated above, this uniqueness proof is much like that of Rellich [R.7].

The above proof can be modified easily in such a way as to apply to a region with a sector, rather than a ray, cut out at ∞. The only difference is that the Fourier series for $\chi(\varrho, \theta)$ would then not have the period 4π and that the Bessel functions involved would not be of index $n/2$.

Once it has become clear that the decomposition of the solution into the sum of the two discontinuous functions $g(\varrho, \theta)$ and $h(\varrho, \theta)$ defined earlier is a procedure that is really natural and suitable for this problem, one is then led to the idea that such a decomposition might be explicitly used in such a way as to determine the solution of the original problem ((cf. (5.5.8)) in a direct and straightforward way. Our next purpose is to carry out such a procedure.

We set (cf. Figure (5.5.4) and equation (5.5.13)):

$$(5.5.27) \qquad f(\varrho, \theta) = g(\varrho, \theta) + h(\varrho, \theta)$$

with $g(\varrho, \theta)$ defined by

$$(5.5.28) \qquad g(\varrho, \theta) = \begin{cases} e^{i\varrho \cos (\theta-\alpha)} + e^{i\varrho \cos (\theta+\alpha)}, & 0 < \theta < \pi - \alpha \\ e^{i\varrho \cos (\theta-\alpha)}, & \pi - \alpha < \theta < \pi + \alpha \\ 0, & \pi + \alpha < \theta < 2\pi. \end{cases}$$

In addition we have

$$(5.5.29) \qquad f_\theta = 0 \text{ for } \theta = 0, \qquad \theta = 2\pi$$

and we also require

$$(5.5.30) \qquad \lim_{\varrho \to \infty} \sqrt{\varrho} \left(\frac{\partial h}{\partial \varrho} + ih \right) = 0 \text{ uniformly in } \theta,$$

since the validity of the radiation condition in this strong form can be verified in the end.

The desired solution will be found by developing $f(\varrho, \theta)$ into a Fourier series in θ for fixed ϱ, and determining the coefficients of the series through use of the radiation condition in the strong form (5.5.30); afterwards, the series can easily be summed to yield a convenient integral representation of the solution. That such a process will be successful can be seen very easily: The Fourier series for $f(\varrho, \theta)$ will, on account of the boundary condition (5.5.29) and the fact that f is a solution of the reduced wave equation, be of the form

$\sum c_n J_{n/2}(\varrho) \cos n\theta/2$; the Fourier coefficients for $g(\varrho, \theta)$ as defined by (5.5.28) are given in terms of integrals of the form

$$I_n = \int_0^{2\pi} e^{i\varrho \cos (\theta \pm \alpha)} \cos \frac{n\theta}{2} \, d\theta$$

since this function also satisfies the condition $g_\theta = 0$ for $\theta = 0, 2\pi$. Since $J_{n/2}(\varrho)$, and its derivatives as well, behave like $1/\sqrt{\varrho}$ for large values of ϱ and the integrals I_n — by a straightforward application of the method of stationary phase, for example, — also behave in this way, it is clear that the limit relation (5.5.30) when used in connection with (5.5.27) will serve to determine the coefficients c_n.

We proceed to carry out this program. The finite Fourier transform $\bar{f}$ of f is introduced by the formula

(5.5.31) $$\bar{f}(\varrho, n) = \int_0^{2\pi} f(\varrho, \theta) \cos \frac{n\theta}{2} \, d\theta.$$

Since $f_\theta = 0$ for $\theta = 0, 2\pi$ we find for $f_{\theta\theta}$ the transform

(5.5.32) $$\bar{f}_{\theta\theta} = - \frac{n^2}{4} \bar{f},$$

by using two integrations by parts. Since f is a solution of

(5.5.33) $$\varrho^2 f_{\varrho\varrho} + \varrho f_\varrho + f_{\theta\theta} + \varrho^2 f = 0,$$

it follows that $\bar{f}$ is a solution of

(5.5.34) $$\varrho^2 \bar{f}_{\varrho\varrho} + \varrho \bar{f}_\varrho + \left(\varrho^2 - \frac{n^2}{4} \right) \bar{f} = 0,$$

and solutions of this equation are given by

(5.5.35) $$\bar{f}(\varrho, n) = a_n J_{n/2}(\varrho).$$

(The Bessel functions $Y_{n/2}(\varrho)$ of the second kind are not introduced because they are singular at the origin; the solution we want is in any case obtained without their use.)

The transform of $g(\varrho, \theta)$ is, of course, given by

(5.5.36) $$\bar{g}(\varrho, n) = \int_0^{2\pi} g(\varrho, \theta) \cos \frac{n\theta}{2} \, d\theta,$$

and we have, in view of (5.5.8), the relation:

(5.5.37) $$\int_0^{2\pi} h(\varrho, \theta) \cos \frac{n\theta}{2} \, d\theta = a_n J_{n/2}(\varrho) - \int_0^{2\pi} g(\varrho, \theta) \cos \frac{n\theta}{2} \, d\theta$$

or, also:

(5.5.38) $\qquad \bar{h}(\varrho, n) = a_n J_{n/2}(\varrho) - \bar{g}(\varrho, n).$

We must next apply the operation $\sqrt{\varrho}\,(\partial/\partial\varrho + i)$ to both sides of (5.5.37) and then make the passage to the limit, with the result*

$$(5.5.39) \quad 0 = \lim_{\varrho \to \infty} \sqrt{\varrho} \left(\frac{\partial}{\partial\varrho} + i \right) \left[a_n J_{n/2}(\varrho) - \int_0^{2\pi} g(\varrho, \theta) \cos \frac{n\theta}{2}\, d\theta \right].$$

Since the functions $J_{n/2}(\varrho)$ behave asymptotically as follows:

$$J_{n/2}(\varrho) \sim \sqrt{\frac{2}{\pi\varrho}} \cos\left(\varrho - \frac{\pi}{4} - \frac{n\pi}{4} \right)$$

and since these asymptotic expansions can be differentiated, we have

$$(5.5.40) \qquad \left(\frac{\partial}{\partial\varrho} + i \right) J_{n/2}(\varrho) \sim \sqrt{\frac{2}{\pi\varrho}}\, e^{i(\varrho + \pi/4 - n\pi/4)}$$

as an easy calculation shows. The behavior of the integral over g can be found easily by the well-known method of stationary phase, which (cf. Ch. 6.8) states that

$$\int_a^b \psi(\theta) e^{i\varrho\varphi(\theta)}\, d\theta \sim \sqrt{\frac{2\pi}{\varrho\,|\varphi''(\alpha)|}}\, \psi(\alpha) e^{i\left(\varrho\varphi(\alpha) \pm \frac{\pi}{4} \right)}$$

in which α is a simple zero of the derivative $\varphi'(\theta)$ in the range $a < \theta < b$, and the ambiguous sign in the exponential is to be taken the same as the sign of $\varphi''(\alpha)$. In the present case, in which $g(\varrho, \theta)$ is defined by (5.5.28) one sees at once that there are three points of stationary phase, i.e. at $\theta = \alpha$, $\theta = \pi - \alpha$, and $\theta = \pi + \alpha$. Of the three contributions only the first, i.e. the contribution at $\theta = \alpha$,** furnishes a non-vanishing contribution for $\varrho \to \infty$ when the operator $\sqrt{\varrho}(\partial/\partial\varrho + i)$ is applied to it; one finds, in fact:

$$(5.5.41) \quad \left(\frac{\partial}{\partial\varrho} + i \right) \int_0^{2\pi} g(\varrho, \theta) \cos \frac{n\theta}{2}\, d\theta \sim 2 \sqrt{\frac{2\pi}{\varrho}} \cos \frac{n\alpha}{2}\, e^{i\left(\varrho + \frac{\pi}{4} \right)}.$$

Use of (5.5.40) and (5.5.41) in (5.5.39) furnishes, finally, the coefficients a_n:

$$(5.5.42) \qquad a_n = 2\pi \cos \frac{n\alpha}{2}\, e^{i\frac{n\pi}{4}}.$$

The Fourier series for $f(\varrho, \theta)$ is

---

* It should be noted that the argument goes through if the radiation condition is used in the weak form.

** This has physical significance, since it says that only the incoming wave is effective in determining the Fourier coefficients of the solution.

$$f(\varrho, \theta) = \frac{1}{2\pi} \bar{f}(\varrho, 0) + \frac{1}{\pi} \sum_{n=1}^{\infty} \bar{f}(\varrho, n) \cos \frac{n\theta}{2}$$

or, from (5.5.35) and (5.5.42),

$$f(\varrho, \theta) = J_0(\varrho) + 2 \sum_{n=1}^{\infty} e^{\frac{in\pi}{4}} J_{n/2}(\varrho) \cos \frac{n\alpha}{2} \cos \frac{n\theta}{2}.$$

It is not difficult to sum the series for $f(\varrho, \theta)$. If we use the representation (for a derivation, see Courant-Hilbert [C.10, p. 413])

(5.5.43) $$J_\nu(\varrho) = \frac{e^{i\nu\pi}}{2\pi i} \int_P e^{-\frac{\varrho}{2}\left(\zeta - \frac{1}{\zeta}\right)} \zeta^{-\nu-1} \, d\zeta,$$

where P is the path in the complex ζ-plane shown in Figure 5.5.6, we find that $f(\varrho, \theta)$ can be expressed as the integral of the sum of

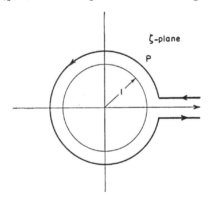

Fig. 5.5.6. The path P in the ζ-plane

a constant plus four geometric series. The summation of the geometric series and a little algebra yields, finally, a solution in the form

(5.5.44)

$$f(\varrho, \theta) = \frac{1}{8\pi i} \int_P \frac{e^{-\frac{\varrho}{2}\left(\zeta - \frac{1}{\zeta}\right)}}{\zeta} \left\{ \begin{array}{l} \dfrac{\zeta^{\frac{1}{2}} + e^{\frac{i}{2}\left(\alpha+\theta+\frac{3\pi}{2}\right)}}{\zeta^{\frac{1}{2}} - e^{\frac{i}{2}\left(\alpha+\theta+\frac{3\pi}{2}\right)}} + \dfrac{\zeta^{\frac{1}{2}} + e^{-\frac{i}{2}\left(\alpha+\theta-\frac{3\pi}{2}\right)}}{\zeta^{\frac{1}{2}} - e^{-\frac{i}{2}\left(\alpha+\theta-\frac{3\pi}{2}\right)}} \\[3mm] + \dfrac{\zeta^{\frac{1}{2}} + e^{\frac{i}{2}\left(\alpha-\frac{\theta+\frac{3\pi}{2}}{}\right)}}{\zeta^{\frac{1}{2}} - e^{\frac{i}{2}\left(\alpha-\theta+\frac{3\pi}{2}\right)}} + \dfrac{\zeta^{\frac{1}{2}} + e^{-\frac{i}{2}\left(\alpha-\theta-\frac{3\pi}{2}\right)}}{\zeta^{\frac{1}{2}} - e^{-\frac{i}{2}\left(\alpha-\theta-\frac{3\pi}{2}\right)}} \end{array} \right\} d\zeta.$$

We proceed to analyze the solution (5.5.44) of our problem with respect to its behavior at ∞ and the origin, and we will show that the conditions needed for the validity of the uniqueness theorem proved above are satisfied. We will also transform it into the solution given by Sommerfeld (cf. equation (5.5.47)). Not all of the details of these calculations will be given: they can be found in the paper by Peters and Stoker [P.19].

If we set

$$(5.5.45) \qquad f(\varrho, \theta) = I(\varrho, \theta + \alpha) + I(\varrho, \theta - \alpha)$$

and define $I(\varrho, \varkappa)$ by

$$(5.5.46) \quad I(\varrho, \varkappa) = \frac{1}{8\pi i} \int_P \frac{e^{-\frac{\varrho}{2}\left(\zeta - \frac{1}{\zeta}\right)}}{\zeta} \left[\frac{\zeta^{\frac{1}{2}} + e^{\frac{i}{2}\left(\varkappa + \frac{3\pi}{2}\right)}}{\zeta^{\frac{1}{2}} - e^{\frac{i}{2}\left(\varkappa + \frac{3\pi}{2}\right)}} + \frac{\zeta^{\frac{1}{2}} + e^{\frac{i}{2}\left(-\varkappa + \frac{3\pi}{2}\right)}}{\zeta^{\frac{1}{2}} - e^{\frac{i}{2}\left(-\varkappa + \frac{3\pi}{2}\right)}} \right] d\zeta$$

we see on comparison with (5.5.44) that (5.5.45) defines $f(\varrho, \theta)$ correctly as the solution we wish to investigate.

Let us first obtain the solution in the form given by Sommerfeld. To this end, the denominators of the fractions in square brackets in (5.5.46) are rationalized, and the fractions combined to yield

$$I(\varrho, \varkappa) = \frac{1}{4\pi i} \int_P \frac{e^{-\frac{\varrho}{2}\left(\zeta - \frac{1}{\zeta}\right)}}{\zeta} \left[\frac{\zeta^2 + 2\zeta^{\frac{3}{2}} e^{i\frac{3\pi}{4}} \cos\frac{\varkappa}{2} + 2i\zeta^{\frac{1}{2}} e^{i\frac{3\pi}{4}} \cos\frac{\varkappa}{2} + 1}{\zeta^2 + 2i\zeta \cos \varkappa - 1} \right] d\zeta.$$

One can then verify readily that I satisfies the differential equation

$$-2\frac{dI}{d\varrho} + (2i\cos\varkappa)I = \frac{1}{4\pi i} \int_P e^{-\frac{\varrho}{2}\left(\zeta - \frac{1}{\zeta}\right)} \left[1 + 2\zeta^{-\frac{1}{2}} e^{i\frac{3\pi}{4}} \cos\frac{\varkappa}{2} + 2i\zeta^{-\frac{3}{2}} e^{i\frac{3\pi}{4}} \cos\frac{\varkappa}{2} + \zeta^{-2} \right] d\zeta$$

$$= \frac{e^{i\frac{3\pi}{4}} \cos\frac{\varkappa}{2}}{2\pi i} \int_P e^{-\frac{\varrho}{2}\left(\zeta - \frac{1}{\zeta}\right)} \left(\zeta^{-\frac{1}{2}} + i\zeta^{-\frac{3}{2}} \right) d\zeta.$$

If we use (5.5.43) and the well-known trigonometric formulas for $J_{1/2}(\varrho)$, $J_{-1/2}(\varrho)$ we see that the last equation is equivalent to

$$-2\frac{dI}{d\varrho} + (2i\cos\varkappa)I = -\sqrt{\frac{2}{\pi\varrho}} e^{i\frac{\pi}{4}} e^{-i\varrho} \cos\frac{\varkappa}{2}.$$

A solution I_N of the non-homogeneous equation which in general vanishes as $\varrho \to \infty$ is readily found:

$$I_N = -\frac{e^{i\frac{\pi}{4}}}{\sqrt{2\pi}} e^{i\varrho \cos\varkappa} \cos\frac{\varkappa}{2} \int_\varrho^\infty \frac{e^{-i\lambda(1 + \cos\varkappa)}}{\sqrt{\lambda}} d\lambda.$$

Thus for I the appropriate solution of the differential equation must be

$$I = e^{i\varrho \cos \varkappa} + I_N.$$

Introduction of a new variable of integration z in the expression for I_N through the relation $2\lambda \cos^2 \varkappa/2 = z^2$, and use of the formula

$$\int_{-\infty}^{\infty} e^{-iz^2}\, dz = \sqrt{\pi}\, e^{-\frac{i\pi}{4}}$$

leads with no difficulty to the expression

$$(5.5.47) \qquad I(\varrho, \varkappa) = \frac{e^{\frac{i\pi}{4}}}{\sqrt{\pi}}\, e^{i\varrho \cos \varkappa} \int_{-\infty}^{\sqrt{2\varrho}\,\cos\frac{\varkappa}{2}} e^{-iz^2}\, dz$$

and this leads, in conjunction with (5.5.45), to Sommerfeld's solution.

To derive the asymptotic behavior of $I(\varrho, \varkappa)$ as $\varrho \to \infty$ we proceed a little differently. The fractions in the square brackets in (5.5.46) are combined, and some algebraic manipulation is applied, to yield

$$I(\varrho, \varkappa) = \frac{e^{-i\varrho}}{4\pi i} \int_P \frac{\dfrac{1}{\sqrt{2}}(\zeta^{1/2} + i\zeta^{-1/2})\, e^{-\varrho\left(\frac{\zeta^{1/2}-i\zeta^{-1/2}}{\sqrt{2}}\right)^2}}{\dfrac{1}{\sqrt{2}}(\zeta^{1/2} - i\zeta^{-1/2}) - \sqrt{2}\, e^{\frac{3\pi i}{4}} \cos\dfrac{\varkappa}{2}}\, \frac{d\zeta}{\zeta}.$$

A new integration variable λ is now introduced by the equation

$$\lambda = \frac{1}{\sqrt{2}}(\zeta^{1/2} - i\zeta^{-1/2}),\quad d\lambda = \frac{1}{2\sqrt{2}}\frac{d\zeta}{\zeta}(\zeta^{1/2} + i\zeta^{-1/2})$$

with the result

$$(5.5.48) \qquad I(\varrho, \varkappa) = \frac{e^{-i\varrho}}{2\pi i} \int_L \frac{e^{-\varrho\lambda^2}\, d\lambda}{\lambda - \sqrt{2}\, e^{\frac{3\pi i}{4}} \cos\dfrac{\varkappa}{2}}.$$

The path P (cf. Fig. 5.5.6) is transformed into the path L shown in Fig. 5.5.7, as one readily can see. The path L leaves the circle of radius $\sqrt{2}$ centered at the origin on its left. This representation of the function $I(\varrho, \varkappa)$ is obviously a good deal simpler than that furnished by (5.5.46), and it is quite advantageous in studying the properties of the solution: for one thing, the plane waves at ∞ can be obtained as the residues at the poles

$$(5.5.49) \qquad \lambda_\pm = \sqrt{2}\, e^{\frac{3\pi i}{4}} \cos\frac{\theta \pm \alpha}{2}.$$

In fact, if there is a pole in the upper half of the λ-plane (and there may or may not be, depending on the values of both θ and α) one

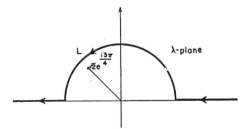

Fig. 5.5.7. The path L in the λ-plane

has, after deformation of the path L over it and into the real axis, for $I(\varrho, \varkappa)$ the result:

$$(5.5.50) \quad I(\varrho, \varkappa) = e^{-i\varrho} e^{2i\varrho \, \cos^2\left(\frac{\varkappa}{2}\right)} - \frac{e^{-i\varrho}}{2\pi i} \int_{-\infty}^{\infty} \frac{e^{-\varrho\lambda^2} \, d\lambda}{\lambda - \sqrt{2} \, e^{\frac{3\pi i}{4}} \, \cos\frac{\varkappa}{2}}$$

$$= e^{i\varrho \, \cos \varkappa} - \frac{e^{-i\varrho}}{2\pi i} \int_{-\infty}^{\infty} \frac{e^{-\varrho\lambda^2} \, d\lambda}{\lambda - \sqrt{2} \, e^{\frac{3\pi i}{4}} \, \cos\frac{\varkappa}{2}} \, .$$

If $\varkappa = \pi$ — the only case in which there is a singularity on the real

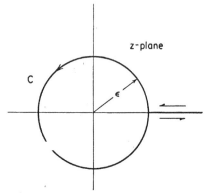

Fig. 5.5.8. The path C in the z-plane

axis, i.e. a pole at $\lambda = 0$ — we assume that the path of integration is deformed near the origin into the upper half-plane. It is convenient to introduce the variable $z = \varrho \lambda^2$ in the integral, with the result

$$(5.5.51) \quad I(\varrho, \varkappa) = e^{i\varrho \cos \varkappa} + \frac{e^{-i\varrho}}{4\pi i \sqrt{\varrho}} \int_C \frac{e^{-z}\, dz}{z^{\frac{1}{2}}\left[\left(\dfrac{z}{\varrho}\right)^{\frac{1}{2}} - \lambda_\varkappa\right]}$$

with $\lambda_\varkappa = \sqrt{2}\, e^{i3\pi/4} \cos \varkappa/2$, and C the path of integration shown in Fig. 5.5.8 For large values of ϱ, and assuming $\lambda_\varkappa \neq 0$, the square bracket in the integrand can be developed in powers of $(z/\varrho)^{1/2}$, and we may write

$$\int_C \frac{e^{-z}\, dz}{z^{\frac{1}{2}}\left[\left(\dfrac{z}{\varrho}\right)^{\frac{1}{2}} - \lambda_\varkappa\right]} = \frac{1}{\lambda_\varkappa} \int_C \frac{e^{-z}\, dz}{z^{\frac{1}{2}}\left[\dfrac{1}{\lambda_\varkappa}\left(\dfrac{z}{\varrho}\right)^{\frac{1}{2}} - 1\right]}$$

$$= \frac{-1}{\lambda_\varkappa} \int_C \frac{e^{-z}}{z^{\frac{1}{2}}}\left[1 + \frac{1}{\lambda_\varkappa}\left(\frac{z}{\varrho}\right)^{\frac{1}{2}} + \frac{1}{\lambda_\varkappa^2}\frac{z}{\varrho} + \ldots\right] dz.$$

It is clear that we may allow $\varepsilon \to 0$ (see Fig. 5.5.8) and hence the path C can be deformed into the two banks of the slit along the real axis; each of the terms in the square brackets then can be evaluated in terms of the Γ-function (cf., for example, MacRobert [M.2], p. 143). It is thus clear that for $\lambda_\varkappa \neq 0$, the leading term in the asymptotic expansion of the integral in (5.5.51) behaves like $1/\sqrt{\varrho}$; in fact, we have for $I(\varrho, \varkappa)$:

$$(5.5.52) \qquad I(\varrho, \varkappa) \sim e^{i\varrho \cos \varkappa} - \frac{e^{-i\varrho}}{4\pi i \lambda_\varkappa} \cdot \frac{-2\Gamma(\frac{1}{2})}{\sqrt{\varrho}}.$$

Since $\Gamma(\frac{1}{2}) = \sqrt{\pi}$ and $\lambda_\varkappa = \sqrt{2}\, e^{\frac{i3\pi}{4}} \cos \dfrac{\varkappa}{2}$ we have

$$(5.5.53) \qquad I(\varrho, \varkappa) \sim e^{i\varrho \cos \varkappa} - \frac{e^{-i\varrho - i\frac{\pi}{4}}}{2\sqrt{2\pi\varrho}\, \cos \dfrac{\varkappa}{2}}.$$

Of course, this holds only if $\varkappa$ lies in the range $0 \leq \varkappa < \pi$ since a pole occurs in the upper half of the λ-plane only when $\cos \varkappa/2$ is positive (cf. (5.5.50)). We must also exclude the value $\varkappa = \pi$, corresponding to $\lambda_\varkappa = 0$. Since $\varkappa = \theta \mp \alpha$, we see that the values $\theta = \pi \pm \alpha$ correspond to the exceptional value $\varkappa = \pi$, and these values of θ, in turn, are those which yield the lines in the physical plane across which our solution f behaves discontinuously at ∞. (Cf. Fig. 5.5.4).

The discussion of the last paragraph yields the result, in conjunction with equation (5.5.45) which defines our solution in terms of $I(\varrho, \varkappa)$:

(5.5.54) $f(\varrho, \theta) \sim e^{i\varrho \cos (\theta-\alpha)} + e^{i\varrho \cos (\theta+\alpha)}$

$$- \frac{e^{-i\varrho-i\frac{\pi}{4}}}{2\sqrt{2\pi\varrho} \, \cos \dfrac{\theta - \alpha}{2}} - \frac{e^{-i\varrho-i\frac{\pi}{4}}}{2\sqrt{2\pi\varrho} \, \cos \dfrac{\theta + \alpha}{2}}$$

for large ϱ and for angles θ such that $0 < \theta < \pi - \alpha$, and α in the range $0 < \alpha < \pi$: only in this case are there poles of both of the integrals in (5.5.45) in the upper halfplane.

The discussion of the behavior of the solution in other sectors of the physical plane and along the exceptional lines can be carried out in the same way as above. For example, if $\lambda_{+} = \sqrt{2} \, e^{i3\pi/4} \cos (\theta+\alpha/2) = 0$, and hence $\theta = \pi - \alpha$, it follows that there is only one pole in the upper halfplane and our solution $f(\varrho, \pi-\alpha)$ is given by (cf. 5.5.48), (5.5.49)):

$$f(\varrho, \pi - \alpha) = e^{i\varrho \cos (\pi-2\alpha)} - \frac{e^{-i\varrho}}{2\pi i} \int_{-\infty}^{\infty} \frac{e^{-\varrho\lambda^2}}{\lambda - \lambda_{-}}$$

$$+ \frac{e^{-i\varrho}}{2\pi i} \int_{L} \frac{e^{-\varrho\lambda^2}}{\lambda} \, d\lambda$$

or also (cf. (5.5.51) and Fig. 5.5.8) by:

$$f(\varrho, \pi - \alpha) = e^{i\varrho \cos (\pi-2\alpha)} + \frac{e^{-i\varrho}}{4\pi i \varrho^{1/2}} \int_{C} \frac{e^{-z} \, dz}{z^{\frac{1}{2}} \left[\left(\dfrac{z}{\varrho} \right)^{\frac{1}{2}} - \lambda_{-} \right]}$$

$$+ \frac{e^{-i\varrho}}{4\pi i} \int_{C} \frac{e^{-z}}{z} \, dz.$$

The asymptotic behavior of f can now be determined in the same way as above; the result is

(5.5.55) $f(\varrho, \pi - \alpha) \sim e^{i\varrho \cos (\pi-2\alpha)} + \dfrac{1}{2} e^{-i\varrho} - \dfrac{e^{-i\varrho-\frac{i\pi}{4}}}{2\sqrt{2\pi\varrho} \, \cos \dfrac{\pi - 2\alpha}{2}},$

the second term resulting from the pole at the origin.

In this fashion the behavior of $f(\varrho, \theta)$ for large values of ϱ is determined, and leads to

$$(5.5.56)$$

$$f(\varrho, \theta) \sim g(\varrho, \theta) = \begin{cases} e^{i\varrho \cos (\theta - \alpha)} + e^{i\varrho \cos (\theta + \alpha)}, & 0 < \theta < \pi - \alpha \\ e^{i\varrho \cos (\pi - 2\alpha)} + \tfrac{1}{2} e^{-i\varrho} & , \quad \theta = \pi - \alpha \\ e^{i\varrho \cos (\theta - \alpha)} & , \quad \pi - \alpha < \theta < \pi + \alpha \\ \tfrac{1}{2} e^{-i\varrho} & , \quad \theta = \pi + \alpha \\ 0 & , \quad \pi + \alpha < \theta < 2\pi. \end{cases}$$

This is, of course, a verification of one of the conditions imposed at ∞. In addition, the next terms in the asymptotic expansion, of order $1/\sqrt{\varrho}$, are also determined, as follows:

$$(5.5.56)'$$

$$f(\varrho, \theta) - g(\varrho, \theta) \sim \begin{cases} -\dfrac{e^{-i\varrho - \frac{i\pi}{4}}}{2\sqrt{2\pi\varrho}\,\cos\dfrac{\theta - \alpha}{2}} - \dfrac{e^{-i\varrho - \frac{i\pi}{4}}}{2\sqrt{2\pi\varrho}\,\cos\dfrac{\theta + \alpha}{2}}, & 0 < \theta < \pi - \alpha \\[3em] -\dfrac{e^{-i\varrho - \frac{i\pi}{4}}}{2\sqrt{2\pi\varrho}\,\cos\dfrac{\pi - 2\alpha}{2}}. & , \quad \theta = \pi - \alpha \\[3em] -\dfrac{e^{-i\varrho - \frac{i\pi}{4}}}{2\sqrt{2\pi\varrho}\,\cos\dfrac{\theta - \alpha}{2}} - \dfrac{e^{-i\varrho - \frac{i\pi}{4}}}{2\sqrt{2\pi\varrho}\,\cos\dfrac{\theta + \alpha}{2}}, & \pi - \alpha < \theta < \pi + \alpha \\[3em] -\dfrac{e^{-i\varrho - \frac{i\pi}{4}}}{2\sqrt{2\pi\varrho}\,\cos\dfrac{\pi + 2\alpha}{2}} & , \quad \theta = \pi + \alpha \\[3em] -\dfrac{e^{-i\varrho - \frac{i\pi}{4}}}{2\sqrt{2\pi\varrho}\,\cos\dfrac{\theta - \alpha}{2}} - \dfrac{e^{-i\varrho - \frac{i\pi}{4}}}{2\sqrt{2\pi\varrho}\,\cos\dfrac{\theta + \alpha}{2}}, & \pi + \alpha < \theta < 2\pi \end{cases}$$

We observe that these expansions do not hold uniformly in θ because of zeros in the denominators for $\theta = \pi \pm \alpha$, i.e. at the lines of discontinuity of the function $g(\varrho, \theta)$.

With the aid of the function $g(\varrho, \theta)$ defined in (5.5.56) we define a function $h(\varrho, \theta)$ by the equation

$$(5.5.57) \qquad f(\varrho, \theta) = g(\varrho, \theta) + h(\varrho, \theta).$$

Thus h is of necessity a discontinuous function since f is continuous while g has jump discontinuities along the lines $\theta = \pi \pm \alpha$. The function h is given by (cf. (5.5.50)):

$$(5.5.58) \qquad h(\varrho, \theta) = \frac{-e^{-i\varrho}}{2\pi i} \int_{-\infty}^{\infty} \frac{e^{-\varrho\lambda^2} d\lambda}{\lambda - \lambda_-} + \int_{-\infty}^{\infty} \frac{e^{-\varrho\lambda^2} d\lambda}{\lambda - \lambda_+},$$

with the proviso that the integrals should be deformed into the upper half-plane in the vicinity of the origin in case either λ_- or λ_+ vanishes: i.e., in case θ has one of its two critical values $\pi \pm \alpha$. That the sum $g + h$ really is our solution f is rather clear in the light of our discussion above; and that it has jump discontinuities which just compensate those of g in order to make f continuous can also be easily verified. We shall not carry out the calculation here. The function $h(\varrho, \theta)$, in view of (5.5.56) and (5.5.57) thus yields what might be called the "scattered" part of the wave.

In order to show that our solution f satisfies the conditions of the uniqueness theorem proved above, we proceed to show that h as defined by (5.5.58) satisfies the radiation condition (5.5.9); afterwards we will prove that f behaves at the origin as prescribed by (5.5.11). Our solution f will thus be proved to be unique.

That the function $h(\varrho, \theta)$ defined by (5.5.58) satisfies the radiation condition is not at all obvious: one sees, for example (cf. (5.5.56)'), that its behavior at ∞ is far from being uniform in the angle θ. In fact, the transformation of h to be introduced below is motivated by the desire to obtain an estimate for the quantity $| \partial h/\partial \varrho + ih |$, which figures in the radiation condition, that is independent of θ; and this in turn means an estimate independent of the quantities λ_- and λ_+ defined by (5.5.49). The function $h(\varrho, \theta)$ can first of all be put in the form

$$h(\varrho, \theta) = \frac{-e^{-i\varrho}}{\pi i} \left\{ \lambda_- \cdot \int_0^{\infty} \frac{e^{-\varrho\lambda^2} d\lambda}{\lambda^2 - \lambda_-^2} + \lambda_+ \cdot \int_0^{\infty} \frac{e^{-\varrho\lambda^2} d\lambda}{\lambda^2 - \lambda_+^2} \right\},$$

as one readily verifies. We proceed as follows: First we write

$$h(\varrho, \theta) = -\frac{e^{-i\varrho}}{\pi i} \left[\lambda_- \cdot \int_0^{\infty} e^{-\varrho\lambda^2} \int_0^{\infty} e^{-(\lambda^2 - \lambda_-^2)t} \, dt d\lambda + \lambda_+ \cdot \int_0^{\infty} e^{-\varrho\lambda^2} \int_0^{\infty} e^{-(\lambda^2 - \lambda_+^2)t} \, dt d\lambda \right]$$

then carry out the integrations with respect to λ to obtain

$$h(\varrho, \theta) = -\frac{e^{-i\varrho}}{2\sqrt{\pi} i} \left[\lambda_- \cdot \int_0^{\infty} \frac{e^{\lambda_-^2 t} \, dt}{(\varrho + t)^{\frac{1}{2}}} + \lambda_+ \cdot \int_0^{\infty} \frac{e^{\lambda_+^2 t} \, dt}{(\varrho + t)^{\frac{1}{2}}} \right].$$

From this representation of h we obtain

$$\frac{\partial h}{\partial \varrho} + ih = \frac{e^{-i\varrho}}{4\sqrt{\pi} i} \left[\lambda_- \cdot \int_0^{\infty} \frac{e^{\lambda_-^2 t} \, dt}{(\varrho + t)^{\frac{3}{2}}} + \lambda_+ \cdot \int_0^{\infty} \frac{e^{\lambda_+^2 t} \, dt}{(\varrho + t)^{\frac{3}{2}}} \right].$$

It is important to observe that λ^2_- and λ^2_+ have pure imaginary values, as we know from (5.5.49). We also observe that the exceptional lines $\theta = \pi \pm \alpha$, which correspond to $\lambda_\mp = 0$, simply have the effect that one of the two terms in the brackets in the last equation vanishes. From the Schwarz inequality we have

$$\left| \frac{\partial h}{\partial \varrho} + ih \right|^2 \leq \frac{|\lambda_-|^2}{8\pi} \left| \int_0^\infty \frac{e^{\lambda^2_- t}\, dt}{(\varrho + t)^{\frac{3}{2}}} \right|^2 + \frac{|\lambda_+|^2}{8\pi} \left| \int_0^\infty \frac{e^{\lambda^2_+ t}\, dt}{(\varrho + t)^{\frac{3}{2}}} \right|^2.$$

Consider the first term on the right; we find:

$$|\lambda_-|^2 \left| \int_0^\infty \frac{e^{\lambda^2_- t}\, dt}{(\varrho + t)^{\frac{3}{2}}} \right|^2 \leq |\lambda_-|^2 \cdot \left| \int_0^\infty \frac{dt}{(\varrho + t)^{\frac{3}{2}}} \right| \cdot \left\{ \left| \frac{e^{\lambda^2_- t}}{\lambda^2_-(\varrho + t)^{\frac{3}{2}}} \right|_0^\infty + \frac{3}{2\lambda^2_-} \left| \int_0^\infty \frac{e^{\lambda^2_- t}\, dt}{(\varrho + t)^{\frac{5}{2}}} \right| \right\}$$

$$\leq \left(-\frac{2}{(\varrho + t)^{\frac{1}{2}}} \bigg|_0^\infty \right) \cdot \left(\frac{1}{\varrho^{\frac{3}{2}}} + \frac{3}{2} \int_0^\infty \frac{dt}{(\varrho + t)^{\frac{5}{2}}} \right)$$

$$\leq \frac{4}{\varrho^2}.$$

Since the same estimate holds for the second term, it follows that

$$\left| \frac{\partial h}{\partial \varrho} + ih \right|^2 \leq \frac{1}{\pi \varrho^2},$$

and this estimate holds for all values of θ, since it holds for the two exceptional values $\theta = \pi \pm \alpha$ as well as for all other values in the range $0 \leq \theta \leq 2\pi$. We have thus verified that the radiation condition holds — in fact, we have shown that it holds in the strong form.

We proceed to show that $f(\varrho, \theta)$ behaves properly at the origin. To this end, we start with the solution in the form (cf. (5.5.48) and (5.5.45)):

$$f(\varrho, \theta) = \frac{e^{-i\varrho}}{2\pi i} \left\{ \int_L \frac{e^{-\varrho \lambda^2}\, d\lambda}{\lambda - \lambda_-} + \int_L \frac{e^{-\varrho \lambda^2}\, d\lambda}{\lambda - \lambda_+} \right\},$$

with L the path of Fig. (5.5.7). The transformation $\lambda = \sqrt{z}$ is then made, so that the new path of integration D is like the path C in Fig. 5.5.8 except that the circular part now has a radius large enough to include the singularities of the integrands in its interior. We may take the radius of the circular part of D to have the value $1/\varrho$, since we care only for small values of ϱ in the present consideration. The transformation $\varrho z = u$ then leads to the following formula for $f(\varrho, \theta)$:

$$f(\varrho, \theta) = \frac{e^{-i\varrho}}{4\pi i} \left\{ \int_{D_1} \frac{e^{-u}\, du}{u^{\frac{1}{2}}(u^{\frac{1}{2}} - \lambda_- \varrho^{\frac{1}{2}})} + \int_D \frac{e^{-u}\, du}{u^{\frac{1}{2}}(u^{\frac{1}{2}} - \lambda_+ \varrho^{\frac{1}{2}})} \right\}$$

with D_1 a path of the same type as D except that the circular part of D_1 is now the circle of unit radius. For small values of ϱ the integrals in the last expression can be expressed in the form

$$\int_{D_1} \frac{e^{-u}}{u} \left(1 + \frac{\varrho^{1/2}\lambda_\pm}{u^{\frac{1}{2}}} + \frac{\varrho\lambda_\pm^2}{u} + \ldots \right) du$$

$$= 2\pi i \left[1 + \varrho^{\frac{1}{2}}\lambda_\pm \int_{D_1} \frac{e^{-u}}{u^{\frac{3}{2}}} \, du + \varrho\lambda_\pm^2 \int_{D_1} \frac{e^{-u}}{u^2} \, du + \ldots \right].$$

From this expansion we see clearly that

$$\begin{cases} f(\varrho, \theta) \sim 1 \\ f_\varrho(\varrho, \theta) \sim \dfrac{c}{\varrho^{\frac{1}{2}}} \end{cases} \quad \text{as} \quad \varrho \to 0.$$

This completes the verification of the conditions needed for the application of the uniqueness theorem to our solution f.

It has been shown by Putnam and Arthur [P.18] (see also Carr and Stelzriede [C.1]) that the theory of diffraction of water waves

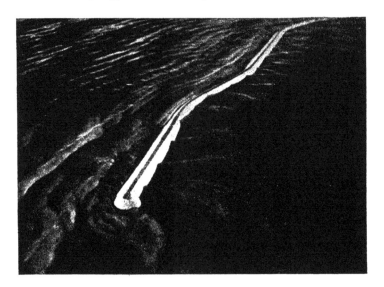

Fig. 5.5.9. Waves behind a breakwater

around a vertical barrier is in good accord with the physical facts, the accuracy being particularly high in the shadow created by the breakwater. Figure 5.5.9 is a photograph (given to the author by J. H. Carr of the Hydrodynamics Laboratory at the California Institute of Technology) of a model of a breakwater which gives some indication of the wave pattern which results.

5.6. Brief discussions of additional applications and of other methods of solution

The object of the present section is to point out a few further problems and methods of dealing with problems concerned, for the most part, with simple harmonic waves of small amplitude.

The first group of problems to be mentioned belongs, generally speaking, to the field of oceanography. For general treatments of this subject the book of Sverdrup, Johnson, and Fleming [S.32] should be consulted. One type of problem of this category which was investigated vigorously during World War II is the problem of wave refraction along a coast, or, in other terms, the problem of the modification in the shape of the wave crests and in the amplitude of ocean waves as they move from deep water into shallow water. We have seen in the preceding sections that it is not entirely easy to give exact solutions in terms of the theory of waves of small amplitude even in relatively simple cases, such, for example, as the case of a uniformly sloping bottom. As a consequence, approximate methods modeled after those of geometrical optics were devised, beginning with the work of Sverdrup and Munk [S.35]. Basically, these methods boil down to the assumption that the local propagation speed of a wave of given length is known at any point from the formulas derived in Chapter 4 for water of constant depth once the depth of the water at that point is known; and that Huygens' principle, or variants of it, can be used to locate wave fronts or to construct the rays orthogonal to them. The errors resulting from such an assumption should not be very great in practice since the depth variations are usually rather gradual. Various schemes of a graphical character have been devised to exploit this idea, for example by Johnson, O'Brien, and Isaacs [J.7], Arthur [A.3], Munk and Traylor [M.16], Suquet [S.30], and Pierson [P.8]. Figure 5.6.1 is a refraction diagram for waves passing over a shoal in an otherwise level bottom in the form of a flat circular hump, and Fig. 5.6.2 is a picture of the actual waves. Both figures were taken from a paper by Pierson [P.8], and they refer to waves in an experimental tank. As one sees, there

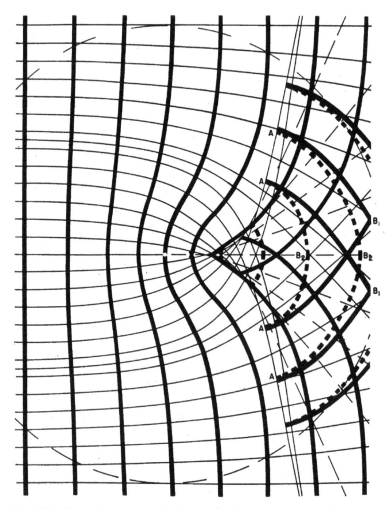

Fig. 5.6.1. Theoretical wave crest-orthogonal pattern for waves passing over a
clock glass. No phase shift

is fair general agreement in the wave patterns—even good agreement
in detail over a good part of the area. However, near the center of
the figures there are considerable discrepancies, since the theoretical
diagram shows, for instance, a sharp point in one of the wave crests

which is lacking in the photograph. The fact is that there is a caustic in the rays constructed by geometrical optics (i.e. the orthogonals to the wave crests have an envelope), and in the vicinity of such a region the approximation by geometrical optics is not good. One of

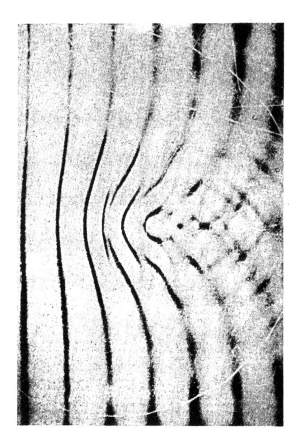

Fig. 5.6.2. Shadowgraph for waves of moderate length passing over a clock glass

the interesting features of Fig. 5.6.2 is that the shoal in the bottom results in wave crests which cross each other on the lee side of the shoal, although the oncoming waves form a single train of plane waves. Figure 5.6.3 is an aerial photograph (again taken from the paper by Pierson) showing the same effect in the ocean at a point off the coast of New Jersey; the arrow points to a region where there would appear to be three wave trains intersecting, but all of them appear to arise from a single train coming in from deep water.

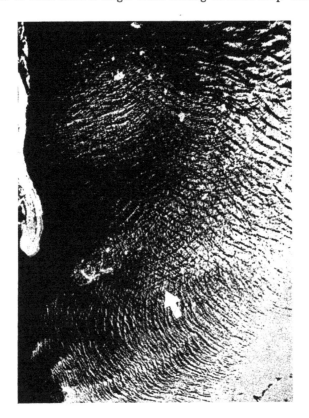

Fig. 5.6.3. Aerial photograph at Great Egg Inlet, New Jersey

In the case of sufficiently shallow water Lowell [L.16] has studied the conditions under which the approximation by geometrical optics is valid; his starting point is the linear shallow water theory (for which see Ch. 10.13) in which the propagation speed of waves is $\sqrt{gh}$, with h the depth of the water, and it is thus independent of the wave length. Eckart [E.2, 3] has devised an approximate theory which makes it possible to deal with waves in both deep and shallow water, as well as in the transition region between the two.

There is an interesting application of the theory of water waves to a problem in seismology which will be explained here even though it is necessary to go somewhat beyond the linear theory on which this part of the book is based. We have seen in Chapter 3 above that the displacements, velocities, and pressure variations in a simple harmonic standing wave die out exponentially in water of infinite depth. However, it was pointed out by Miche [M. 8] that this is true only of the first order terms in the development of the basic nonlinear theory with respect to the wave amplitude; if the development is carried out formally to second order it turns out that the pressure fluctuates with an amplitude that does not die out with the depth, but depends on the square of the amplitude. (For progressing waves, this is not true.) In addition, the second order pressure variation has a frequency which is double the frequency of the linear standing wave. (It is not hard to see in a general way how this latter nonlinear effect arises mathematically. In the Bernoulli law, the nonlinear term of the form $\Phi_x^2 + \Phi_y^2$ would lead, through an iteration process starting with $\Phi = Ae^{my} \cos mx \cos \sigma t$, to terms involving $\cos^2 \sigma t$ and thus to harmonics with the double frequency.) It happens that seismic waves in the earth of very small amplitudes—called microseisms—and of periods of from 3 to 10 seconds are observed by sensitive seismographs; these waves seem unlikely to be the result of earthquakes or local causes; rather, a close connection between microseisms and disturbed weather conditions over the ocean was noticed. However, since it was thought that surface waves in the ocean lead to pressure variations which die out so rapidly in the depth that they could not be expected to generate observable waves in the earth, it was thought unlikely that storms at sea could be a cause for microseisms. The result of Miche stated above was invoked by Longuet-Higgins and Ursell [L. 14] in 1948 to revive the idea that storms at sea can be the origin of microseisms. (See also the paper of 1950 by Longuet-Higgins [L.13].) In addition, Bernard [B.8] had collected evidence in 1941 indicating that the frequency of microseisms near Casablanca was just double that of sea waves reaching the coast nearby; the same ratio of frequencies was noticed by Deacon [D. 6] with respect to microseisms recorded at Kew and waves recorded on the north coast of Cornwall. Further confirmation of the correlation between sea waves and microseisms is given in the paper of Darbyshire [D. 4].

A reasonable explanation for the origin of microseisms thus seems to be available. Of course, this explanation presupposes that standing waves are generated, but Longuet-Higgins has shown that the needed effects are present any time that two trains of progressing waves moving in opposite directions are superimposed, and it is not hard to imagine that such things would occur in a storm area—for example, through the superposition of waves generated in different portions of a given storm area. It might be added that Cooper and Longuet-Higgins [C.3] have carried out experiments which confirm quantitatively the validity of the Miche theory of nonlinear standing waves. It is perhaps also of interest to refer to a paper by Danel [D.2] in which standing waves of large amplitude with sharp crests are discussed.

In Chapter 6 some references will be made to interesting studies concerning the location of storms at sea as determined by observations on shore of the long waves which travel at relatively high speeds outward from the storm area (cf. the paper by Deacon [D.6]).

The general problem of predicting the character of wave conditions along a given shore is, of course, interesting for a variety of reasons, including military reasons (see Bates [B.6], for example). Methods for the forecasting of waves and swell, and of breakers and surf are treated in two pamphlets [U.1, 2] issued by the U.S. Hydrographic Office.

A necessary preliminary to forecasting studies, in general, is an investigation of ways and means of recording, analyzing, and representing mathematically the surface of the ocean as it actually occurs in nature. Among those who have studied such questions we mention here Seiwell [S.9, 10] and Pierson [P.10]. The latter author concerns himself particularly with the problem of obtaining mathematical representations of the sea surface which are on the one hand sufficiently accurate, and on the other hand not so complicated as to be practically unusable. The surface of the open sea is, in fact, usually extraordinarily complicated. Figure 5.6.4 is a photograph of the sea (taken from the paper by Pierson) which bears this out. Pierson first tries representations employing the Fourier integral and comes to the conclusion that such representations would be so awkward as to preclude their use. (In Chapter 6 we shall have an opportunity to see that it is indeed not easy to discuss the results of such representations even for motions generated in the simplest conceivable fashion—by applying an impulse at a point of the surface when the water is initially at rest, for example.) Pierson then goes on to advocate a statistical approach to the problem in which various of the important parameters are assumed to be distributed according to a Gaussian law. These developments are far too extensive for inclusion in this book—besides, the

Fig. 5.6.4. Surface waves on the open sea

author is, by temperament, more interested in deterministic theories in mechanics than in those employing arguments from probability and statistics, while knowing at the same time that such methods are very often the best and most appropriate for dealing with the complex problems which arise concretely in practice. It would, however, seem to the author to be likely that any mathematical representations of the surface of the sea—whether by the Fourier integral or any other integrals—would of necessity be complex and cumbersome in proportion to the complexity of that surface and the degree to which details are desired.

Before leaving this subject, it is of interest to examine another photograph of waves given by Pierson [P.10], and shown in Fig. 5.6.5. Near the right hand edge of the picture the wave crests of the predominant system are turned at about 45° to the coast line, and they are broken rather than continuous; such wave systems are said to be short-crested. About half-way toward shore it is seen that these waves have arranged themselves more nearly parallel to the coast (indicating, of course, that the water has become shallower) and at the same time the crests are longer and less broken in appearance,

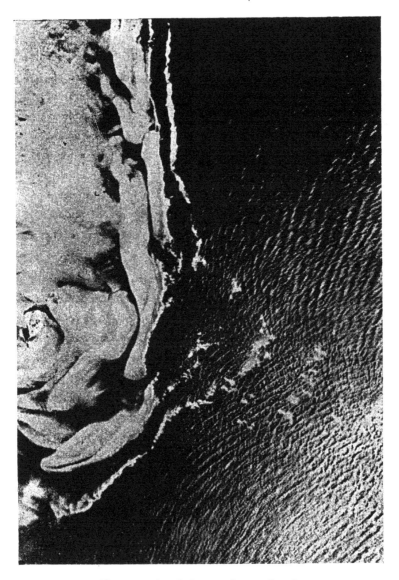

Fig. 5.6.5. Aerial photograph over Oracoke

though no single one of them can be identified for any great distance. Near the shore, the wave crests are relatively long and nearly parallel to it. On the photograph a second train of waves having a shorter wave length and smaller amplitude can be detected; these waves are traveling almost at right angles to the shore (they are probably caused by a breeze blowing along the shore) and they are practically not diffracted. Each of the two wave trains appears to move as though the other were not present: the case of a linear superposition would thus seem to be realized here. One observes also that there is a shoal, as evidenced by the crossed wave trains and the white-water due to breaking over the shoals.

We pass next to a brief discussion of a few problems in which our emphasis is on the *methods* of solution, which are different from those employed in the preceding chapters of Part II. The first such problem to be discussed employs what is called the Wiener-Hopf method of solving certain types of boundary problems by means of an ingenious, though somewhat complicated, procedure which utilizes an integral equation of a special form. This method has been used, as was mentioned in the introduction to this chapter, by Heins [H.12, 13] and by Keller and Weitz [K.9] to solve the dock problem and other problems having a similar character with respect to the geometry of the domains in which the solution is sought. However, it is simpler to explain the underlying ideas of the method by treating a different problem, i.e. the problem of diffraction of waves around a vertical half-plane — in other words, Sommerfeld's diffraction problem, which was treated by a different method in the preceding section. We outline the method, following the presentation of Karp [K.3]. The mathematical formulation of the problem is as follows. A solution $\varphi(x, y)$ of the reduced wave equation

$$(5.6.1) \qquad\qquad \nabla^2\varphi + k^2\varphi = 0$$

is to be found subject to the boundary condition

$$(5.6.2) \qquad\qquad \varphi_y = 0 \quad \text{for } y = 0, \quad x > 0$$

and regular in the domain excluding this ray (cf. Fig. 5.6.6). In addition, a solution in the form

$$(5.6.3) \qquad\qquad \varphi = \varphi_0 + \varphi_1$$

with φ_0 defined by

$$(5.6.4) \qquad \varphi_0 = e^{ik(x \cos \theta_0 + y \sin \theta_0)}, \qquad 0 < \theta_0 < 2\pi,$$

and with φ_1 prescribed to die out at ∞ is wanted. In other words, a plane wave comes from infinity in a direction determined by the angle θ_0, and the scattered wave caused by the presence of the screen, and

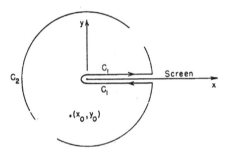

Fig. 5.6.6. Diffraction around a screen

given by φ_1, is to be found. It is a peculiarity of the Wiener-Hopf method—not only in the present problem but in other applications to diffraction problems as well—that the constant k is assumed to be a complex number (rather than a real number, as in the preceding section) given, say, by $k = k_1 + ik_2$, with k_2 small and positive. With this stipulation it is possible to dispense with conditions on φ_1 of the radiation type at ∞, and to replace them by boundedness conditions.

We employ a Green's function in order to obtain a representation of the solution in the form of an integral equation of the type to which the Wiener-Hopf technique applies. In the present case the Green's function $G(x, y; x_0, y_0)$ is defined as that solution of (5.6.1) in the whole plane which has a logarithmic singularity at the point (x_0, y_0) and dies out at ∞ (here the fact that k is complex plays a role). This function is well-known; it is, in fact, the Hankel function $H_0^{(1)}(kr)$ of the first kind:

$$(5.6.5) \quad G(x, y; x_0, y_0) = \frac{i}{4} H_0^{(1)} \left(k[(x - x_0)^2 + (y - y_0)^2]^{\frac{1}{2}} \right).$$

The next step is to apply Green's formula to the functions φ and G in the domain bounded by the circle C_2 and the curves marked C_1 in Fig. 5.6.6. Because of the fact that G is symmetric, has a logarithmic singularity, and that φ and G both satisfy (5.6.1), it follows by arguments that proceed exactly as in potential theory in similar cases that $\varphi(x, y)$ can be represented in the form

$$(5.6.6) \quad \varphi(x, y) = \int_0^\infty [\varphi] \left. \frac{\partial G}{\partial y_0} \right|_{y_0=0} dx_0 + \exp\{ik(x \cos\theta_0 + y \sin\theta_0)\}$$

when the radius of the circle C_2 is allowed to tend to ∞, and the boundary condition (5.6.2), the regularity conditions, and conditions at ∞ are used. (A mild singularity at the edge $x = 0$, $y = 0$ of the screen must also be permitted.) The symbol $[\varphi]$ under the integral sign represents the jump in φ across the screen, which is of course not known in advance. The object of the Wiener-Hopf technique is to determine $[\varphi]$ by using the integral equation (5.6.6); once this is done (5.6.6) yields the solution $\varphi(x, y)$. The first step in this direction is to differentiate both sides of (5.6.6) with respect to y, then set $y = 0$ and confine attention to positive values of x; in view of the boundary condition (5.6.2) we obtain in this way the integral equation

$$(5.6.7) \quad 0 = ik \sin\theta_0\, e^{ikx \cos\theta_0} + \int_0^\infty [\varphi(x_0)]K(x - x_0)dx_0, \qquad x > 0.$$

The kernel $K(x - x_0)$ of the integral equation is given by

$$(5.6.8) \qquad\qquad K(x - x_0) = \left. \frac{\partial^2 G}{\partial y \partial y_0} \right|_{y = y_0 = 0}.$$

Equation (5.6.7) is a typical example of an integral equation solvable by the Wiener-Hopf technique; its earmarks are that the kernel is a function of $(x - x_0)$ and the range of integration is the positive real axis.

The starting point of the method is the observation that the integral in (5.6.7) is strongly reminiscent of the convolution type of integral in the theory of the Fourier transform. In fact, if the limits of integration in (5.6.7) were from $-\infty$ to $+\infty$ and the equation were valid for all values of x, it could be solved at once by making use of the convolution theorem. This theorem states that if

$$\bar{f}(\alpha) = \int_{-\infty}^\infty f(x_0) \exp\{-i\alpha x_0\}dx_0 \text{ and } \overline{K}(\alpha) = \int_{-\infty}^\infty K(x_0) \exp\{-i\alpha x_0\}dx_0$$

—i.e. if $\bar{f}$ and $\overline{K}$ are the Fourier transforms of f and k (cf. Chapter 6)—

then $\overline{\bar{f}(\alpha)\overline{K}(\alpha)} = \int_{-\infty}^\infty f(x_0)K(x - x_0)dx_0$, in other words, the transform of the integral on the right is the product of the Fourier transforms of the function $f(x_0)$ and $K(x_0)$ (cf. Sneddon [S.11], p. 24). Consequently if (5.6.7) held in the wider domain indicated, it could be used to yield

$$0 = \overline{h}(\alpha) + \overline{[\varphi(\alpha)]}\overline{K}(\alpha),$$

with $\overline{h}(\alpha)$ the transform of the nonhomogeneous term in the integral equation. This relation in turn defines the transform $\overline{[\varphi(\alpha)]}$ of $[\varphi(x_0)]$ since $\overline{h}(\alpha)$ and $\overline{K}(\alpha)$ are the transforms of known functions, and hence $[\varphi(x_0)]$ itself. We are, of course, not in a position to proceed at once in this fashion; but the idea of the Wiener-Hopf method is to extend the definitions of the functions involved in such a way that one can do so. To this end the following definitions are made

$$(5.6.9) \quad \begin{cases} g(x) = 0, & x > 0; \quad f(x_0) = [\varphi], \quad x_0 > 0 \\ h(x) = 0, & x < 0; \quad f(x_0) = 0, \quad x_0 < 0 \\ h(x) = ik \sin \theta_0 e^{ikx \cos \theta_0}, & x > 0. \end{cases}$$

Equation (5.6.7) can now be replaced by the equivalent equation

$$(5.6.10) \quad g(x) = h(x) + \int_{-\infty}^{\infty} f(x_0)K(x - x_0)dx_0, \quad -\infty < x < \infty.$$

Here $g(x)$ is unknown for $x < 0$ and $f(x_0)$—the function we seek—is, of course, unknown for $x_0 > 0$; thus we have only one equation for two unknown functions. Nevertheless, both functions can be determined by making use of complex variable methods applied to the Fourier transform of (5.6.10); we proceed to outline the method. We have, to begin with, from (5.6.10):

$$(5.6.11) \qquad\qquad \overline{g}(\alpha) = \overline{h}(\alpha) + \overline{f}(\alpha)\overline{K}(\alpha),$$

with $\overline{h}(\alpha)$ and $\overline{K}(\alpha)$ known functions given by

$$(5.6.12) \qquad\qquad \overline{h}(\alpha) = \frac{k \sin \theta_0}{\alpha - k \cos \theta_0},$$

$$(5.6.13) \qquad\qquad \overline{K}(\alpha) = \frac{i}{2} (k^2 - \alpha^2)^{\frac{1}{2}}.$$

The equation (5.6.11) is next shown to be valid in a strip of the complex α-plane which contains the real axis in its interior. We omit the details of the discussion required to establish this fact; it follows in an elementary way from the assumption that the constant k has a positive imaginary part, and from the conditions of regularity and boundedness imposed on the solution φ of the basic problem. $\overline{K}(\alpha)$ is factored* in the form $(i/2)\overline{K}_-(\alpha) \cdot \overline{K}_+(\alpha)$ with $\overline{K}_-(\alpha) = (k - \alpha)^{1/2}$, $\overline{K}_+(\alpha) = (k + \alpha)^{1/2}$ with $\overline{K}_-$ and $\overline{K}_+$ regular in lower and upper half-

---

* Such a manipulation occurs in general in using this technique; usually a continued product expansion of the transform of the kernel is required.

planes, respectively. The equation (5.6.11) can then be expressed, after some manipulation, in the form

$$(5.6.14) \quad \frac{\bar{g}_+(\alpha)}{(k+\alpha)^{1/2}} - \frac{k \sin \theta_0}{\alpha - k \cos \theta_0} \left[\frac{1}{(k+\alpha)^{1/2}} - \frac{1}{(k+k\cos \theta_0)^{1/2}} \right]$$

$$= \frac{k \sin \theta_0}{(k+k\cos \theta_0)^{1/2}(\alpha - k \cos \theta_0)} + \frac{i}{2} \bar{f}_-(\alpha)(k-\alpha)^{1/2}$$

where the symbols $\bar{g}_+$ and $\bar{f}_-$ refer to the fact that $\bar{g}(\alpha)$ and $\bar{f}(\alpha)$ can be shown to be regular in upper and lower half-planes of the complex α-plane, respectively, each of which overlaps the real axis. In fact, the entire left side of (5.6.14) is regular in such an upper half-plane, and similarly for the right hand side in a lower half-plane. Thus the two sides of the equation define a function which is regular in the entire plane, or, in other words, each side of the equation furnishes the analytic continuation of the function defined by the other side. Finally, it is rather easy to show, by studying the behavior of $\bar{g}(\alpha)$ and $\bar{f}(\alpha)$ at ∞, that the entire function thus defined tends uniformly to zero at ∞; it is therefore identically zero. Thus (5.6.14) defines both $\bar{g}(\alpha)$ and $\bar{f}(\alpha)$ since they can be obtained by equating both sides separately to zero. Thus $g(x)$ and $f(x)$ are determined, and the problem is, in principle, solved.

The Wiener-Hopf method is, evidently, a most amusing and ingenious procedure. However, it also has somewhat the air of a tour de force which uses a good many tools from function theory (while the problem itself can be solved very nicely without going into the complex domain at all, as we have seen in the preceding section) and it also employs the artificial device of assuming a positive imaginary part for the wave number k. (This brings with it, we observe from (5.6.4), that while the primary wave dies out as $x \to + \infty$, it becomes exponentially infinite as $x \to - \infty$.) In addition, the problem of diffraction by a wedge, rather than by a plane barrier, can not be solved by the Wiener-Hopf method, but yields readily to solution by the simple method presented in the preceding section. The author hazards the opinion that problems solvable by the Wiener-Hopf technique will in general prove to be solvable more easily by other methods—for example, by more direct applications of complex integral representations, perhaps along the lines used to solve the difficult mixed boundary problem treated in section 5.4 above.

We mention next two other papers in which integral equations are employed to solve interesting water wave problems. The first of these is the paper by Kreisel [K.19] in which two-dimensional simple harmonic progressing waves in a channel of finite depth containing rigid reflecting obstacles are treated. Integral equations are obtained by using an appropriate Green's function; Kreisel then shows that they can be solved by an iteration method provided that the domain occupied by the water does not differ too much from an infinite strip with parallel sides. (Roseau [R.9] has solved similar problems for certain domains which are not restricted in this way.) It is remarkable that Kreisel is able to obtain in some important cases good and useable upper and lower bounds for the reflection and transmission coefficients. References have already been made to the papers by John [J.5] on the motion of floating bodies. In the second of these papers the problem of the creation of waves by a prescribed simple harmonic motion of a floating body is formulated as an integral equation. This integral equation does not fall immediately into the category of those which can be treated by the Fredholm theory; in fact, its theory has a number of interesting and unusual features since it turns out that the homogeneous integral equation has non-trivial solutions which, however, are of such a nature that the nonhomogeneous problem nevertheless always possesses solutions.

Various problems concerning the effect of obstacles on waves, and of the wave motions created by immersed oscillating bodies, have been treated in a series of notable papers by Ursell [U.3, 4, 5 and U.8, 9, 10]. Ursell usually employs the method of expansions in terms of orthogonal functions, or representations by integrals of the Fourier type, as tools for the solution of the problems.

Finally, it should be mentioned that the approximate variational methods devised by Schwinger [S.5] to treat difficult problems in the theory of electromagnetic waves can also be used to treat problems in water waves (cf. Keller [K.7]). A notable feature of Schwinger's method is that it is a technique which concentrates attention on the quantities which are often of the greatest practical importance, i.e. the reflection and transmission coefficients, and determines them, moreover, without solving the entire problem. Rubin [R.13] has formulated the problem of the finite dock—which has so far defied all efforts to obtain an explicit integral representation for its solution— as a variational problem of a somewhat unconventional type, and proved the existence, on the basis of this formulation, of solutions behaving at ∞ like progressing waves.

An interesting type of problem which might well have been discussed at length in this book is the problem of internal waves. This refers to the occurrence of gravity waves at an interface between two liquids of different density. Such problems are discussed in Lamb [L.3], p. 370. The case of internal waves in media with a continuous variation in density has considerable importance also for tidal motions in both the atmosphere (cf. Wilkes [W.2]) and the oceans (cf. Fjeldstad [F.4]).

MOTIONS STARTING FROM REST. TRANSIENTS.

CHAPTER 6

Unsteady Motions

6.1. General formulation of the problem of unsteady motions

In the region occupied by the water we seek, as usual, a harmonic function $\Phi(x, y, z; t)$ which satisfies appropriate boundary conditions and, in addition, appropriate conditions prescribed at the initial instant $t = 0$. At the free surface we have the boundary conditions

$$(6.1.1) \qquad\qquad -\Phi_y + \eta_t = 0$$
$$(6.1.2) \qquad\qquad \Phi_t + g\eta = -\frac{1}{\varrho}\,p \quad\left.\right\} \text{ for } y = 0, \qquad t > 0$$

in terms of the vertical elevation $\eta(x, z; t)$ of the free surface and the pressure $p(x, z; t)$ prescribed on the surface. As always in mechanics, a specific motion is determined only when initial conditions at the time $t = 0$ are given which furnish the position and velocity of all particles in the system. This would mean prescribing appropriate conditions on Φ throughout the fluid at the time $t = 0$, but since we shall assume Φ to be a harmonic function at $t = 0$ as well as for $t > 0$ it is fairly clear that conditions prescribed at the *boundaries* of the fluid only will suffice since Φ is then determined uniquely throughout its domain of definition in terms of appropriate boundary conditions.* As initial conditions at the free surface, for example, we might therefore take

$$(6.1.3) \qquad\qquad \eta(x, z; 0) = f_1(x, z)$$
$$(6.1.4) \qquad\qquad \eta_t(x, z; 0) = f_2(x, z) \quad\left.\right\} \text{ at } y = 0,$$

with f_1 and f_2 arbitrary functions characterizing the initial elevation and vertical velocity of the free surface.

In water wave problems it is of particular interest to consider cases

---

* We shall see later on (sections 6.2 and 6.9) that the solutions are indeed uniquely determined when the initial conditions are prescribed only for the particles at the boundary of the fluid.

in which the motion of the water is generated by applying an impulsive pressure to the surface when the water is initially at rest. To obtain the condition appropriate for an initial impulse we start from (6.1.2) and integrate it over the small time interval $0 \leqq t \leqq \tau$. The result is

$$(6.1.5) \qquad \int_0^\tau p \, dt = - \varrho \Phi(x, 0, z; \tau) - \varrho g \int_0^\tau \eta \, dt,$$

since $\Phi(x, y, z; 0)$ can be assumed to vanish. One now imagines that $\tau \to 0_+$ while $p \to \infty$ in such a way that the integral on the left tends to a finite value—the impulse I per unit area. Since it is natural to assume that η is finite it follows that the integral on the right vanishes as $\tau \to 0_+$, and we have the formula

$$(6.1.6) \qquad\qquad I = - \varrho \Phi(x, 0, z; 0_+)$$

for the initial impulse per unit area at the free surface in terms of the value of Φ there. If I is prescribed on the free surface (together with appropriate conditions at other boundaries), it follows that $\Phi(x, y, z; 0_+)$ can be determined, or, in other words, the initial velocity of all particles is known.

It is also useful to formulate the initial condition on Φ at the free surface appropriate to the case in which the water is initially at rest under zero pressure, but has an initial elevation $\eta(x, z; 0)$. The condition is obtained at once from (6.1.2); it is

$$(6.1.7) \qquad\qquad \Phi_t(x, 0, z; 0_+) = - g \eta(x, z; 0_+),$$

since $p = 0$ for $t > 0$. Prescribing the initial position and velocity of the free surface is thus equivalent to prescribing the initial values of Φ and its first time derivative Φ_t. From now on the notation 0_+ will not be used in formulating initial conditions—instead we shall simply write 0 instead of 0_+.

6.2. Uniqueness of the unsteady motions in bounded domains

It is of some interest to consider the uniqueness of the unsteady motions, for one thing because of the unusual feature pointed out in the preceding section: it is sufficient to prescribe the initial position and velocity, not of all particles, but only of those on the boundary. A uniqueness proof based on the law of conservation of energy will be given.

To this end, consider the motion of a bounded volume of water confined to a vessel with fixed sides but having a free surface (cf. Fig.

6.2.1). In Chapter 1 we have already discussed the notion of energy and its time rate of change with the following results. For the energy E itself we have, obviously:

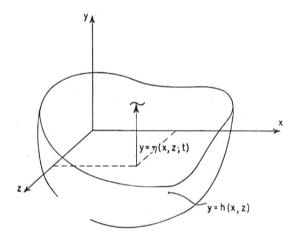

Fig. 6.2.1. Water contained in a vessel

(6.2.1) $E(t) = \varrho \iiint\limits_{R} \left[\tfrac{1}{2}(\Phi_x^2 + \Phi_y^2 + \Phi_z^2) + gy \right] dx \, dy \, dz.$

Here R refers to the volume occupied by the water at any instant. The x, z-plane is, as usual, taken in the plane of equilibrium of the free surface, and $y = \eta(x, z; t)$ and $y = h(x, z)$ are assumed to be the equations of the free surface and of the containing vessel, respectively. The expression for E can now be written in the form

(6.2.2) $E(t) = \dfrac{\varrho}{2} \left\{ \iiint\limits_{R} (\Phi_x^2 + \Phi_y^2 + \Phi_z^2) \, dx \, dy \, dz \right.$

$$\left. + g \iint\limits_{S} (\eta^2 - h^2) \, dx \, dz \right\} \cdot$$

By S is meant the projection on the x, z-plane of the free surface and the containing vessel. In Chapter 1 the following expression for the rate of change of the energy E was derived:

(6.2.3) $\dfrac{dE}{dt} = \iint\limits_{\overline{R}} [\varrho \Phi_t(\Phi_n - v_n) - p v_n] \, dS.$

By $\overline{R}$ is meant the boundary surface of R, while v_n means the normal component of the velocity of $\overline{R}$. It is essential for our uniqueness proof to observe that in the special case in which $p = 0$ on the free surface we have

$$(6.2.4) \qquad \frac{dE}{dt} = 0, \qquad E = \text{const.}$$

This follows at once from the fact that $v_n = \Phi_n = 0$ on the fixed part of the boundary, while $v_n = \Phi_n$ and $p = 0$ on the free surface.

So far no use has been made of the fact that we consider only a linear theory based on the assumption of small oscillations about the equilibrium position. Suppose now that the initial position and velocity of the water particles has been prescribed, or, as we have seen in the preceding section that $\eta(x, z; 0)$ and $\Phi(x, y, z; 0)$ are given functions:

$$(6.2.5) \qquad \begin{cases} \eta(x, z; 0) = f_1(x, z) \\ \Phi(x, y, z; 0) = f_2(x, y, z). \end{cases}$$

We proceed next in the customary way that one uses to prove uniqueness theorems in linear problems. Suppose that η_1, Φ_1, and η_2, Φ_2 are two solutions of the initial value problem. Then $\Phi = \Phi_1 - \Phi_2$ and $\eta = \eta_1 - \eta_2$ are functions which satisfy all of the conditions originally imposed on Φ_i and η_i except that f_1 and f_2 in (6.2.5) would now both vanish, and the free surface pressure would also vanish (cf. (6.1.2) and (6.1.7)). (Here the linearity of our problem is used in an essential way.) It follows therefore that $dE/dt = 0$, and $E = \text{const.}$ when applied to Φ and η, as we have seen. But at the initial instant $\eta = 0$ and $\Phi = 0$, so that

$$(6.2.6) \qquad E = -\frac{\varrho g}{2} \iint_S h^2 \, dx \, dz,$$

from (6.2.2) as applied to $\Phi = \Phi_1 - \Phi_2$ and $\eta = \eta_1 - \eta_2$. Consequently we have the result

$$(6.2.7) \qquad \iiint_R (\Phi_x^2 + \Phi_y^2 + \Phi_z^2) \, dx \, dy \, dz + g \iint_S \eta^2 \, dx \, dz = 0,$$

and this sum obviously vanishes only if grad $\Phi \equiv 0$ and $\eta \equiv 0$—in other words it follows that $\Phi_1 \equiv \Phi_2$ (except for an unessential additive constant), $\eta_1 \equiv \eta_2$, and the uniqueness of the solution of the initial value problem is proved.

The proof given here applies only to a mass of water occupying a bounded region. Nevertheless, it seems clear that the uniqueness of the solution of the initial value problem is to be expected if the water fills an unbounded region, provided that appropriate assumptions concerning the behavior of the solution at ∞ are made. In the following, a variety of such cases will be treated by making use of the technique of the Fourier transform and, although no explicit discussion of the uniqueness question will be carried out, it is well-known that uniqueness theorems (of a somewhat restricted character, it is true) hold in such cases provided only that appropriate conditions at ∞ are prescribed. Recently these uniqueness questions have been treated by Kotik [K.17] and Finkelstein [F. 3]. The latter, for example, proves the uniqueness of unsteady motions in unbounded domains in which rigid obstacles occur, and both writers obtain their uniqueness theorems by imposing relatively weak conditions at infinity. In sec. 9 of this chapter the theory devised by Finkelstein will be discussed.

6.3. Outline of the Fourier transform technique

As indicated above, the solutions of a series of problems of unsteady motions in unbounded regions as determined through appropriate initial conditions will be carried out by using the method of the Fourier transform. The basis for the use of this method is the fact that special solutions Φ of our free surface problems are given—in the case of two-dimensional motion in water of infinite depth, for example —by

$$(6.3.1) \qquad \Phi(x, y; t) = e^{my} \sin(\sigma t + \alpha) \cos m(x - \tau)$$

with

$$(6.3.2) \qquad \sigma^2 = gm$$

and for arbitrary values of α and τ. From these solutions it is possible to build up others by superposition, for example, in the form

$$(6.3.3) \qquad \Phi(x, y; t) = \int_0^\infty h(\sigma) e^{my} \sin(\sigma t + \alpha) dm$$
$$\times \int_{-\infty}^\infty f(\tau) \cos m(x - \tau)\, d\tau,$$

in which $h(\sigma)$ and $f(\tau)$ are arbitrary functions. This in turn suggests that the Fourier integral theorem could be used in order to satisfy given initial conditions, since this theorem states that an arbitrary function $f(x)$ defined for $-\infty < x < \infty$ can be represented in the form

(6.3.4) $$f(x) = \frac{1}{\pi} \int_0^\infty d\alpha \int_{-\infty}^\infty f(\eta) \cos \alpha(\eta - x)\, d\eta$$

provided only that $f(x)$ is sufficiently regular (for example, that $f(x)$ is piecewise continuous with a piecewise continous derivative is more than sufficient) and that $f(x)$ is absolutely integrable:

(6.3.5) $$\int_{-\infty}^\infty | f(x) |\, dx < \infty.$$

Indeed, we see that if we set $h(\sigma) = 1/\pi$ and $\alpha = \pi/2$ in (6.3.3) we would have exactly the integral in (6.3.4) for $t = 0$ and $y = 0$, and hence $\Phi(x, 0; 0)$ would reduce to the arbitrarily given function $f(x)$. Thus a solution would be obtained for an arbitrarily prescribed initial condition on Φ.

It would be perfectly possible to solve the problems treated below by a direct application of (6.3.4), and this is the course followed by Lamb [L.3] in his Chapter IX. Actually the problems were solved first by Cauchy and Poisson (in the early part of the nineteenth century), who derived solutions given by integral representations before the technique of the Fourier integral was known. It might be added that these problems were considered so difficult that they formed the subject of a prize problem of the Académie in Paris.

We prefer, in treating these problems, to make use of the technique of the Fourier transform (following somewhat the presentation given by Sneddon [S.11], Chapter 7) since the building up of the solution to fulfill the prescribed conditions then takes place quite automatically. However, the method is based entirely upon (6.3.4) and thus also requires for its validity that the functions $f(x)$ to which the technique is applied should be representable by the Fourier integral. This is a restriction of a non-trivial character: for example, the basically important solutions given by (6.3.1) are not representable by the Fourier integral.

It is useful to express the Fourier integral in a form different from (6.3.4). We write

$$f(x) = \frac{1}{\pi} \lim_{\xi \to \infty} \int_{-\infty}^\infty f(\eta)\, d\eta \int_0^\xi \cos s(\eta - x)\, ds.$$

But since $\int_{-\xi}^{\xi} \cos s(\eta - x)ds = 2\int_0^{\xi} \cos s(\eta - x)ds$ and

$\int_{-\xi}^{\xi} \sin s(\eta - x)ds = 0$, we may write $\int_0^{\xi} \cos s(\eta - x)ds =$

$\frac{1}{2}\int_{-\xi}^{\xi} \exp \{is(x - \eta)\}\, ds$, and hence

$$(6.3.6) \qquad f(x) = \frac{1}{2\pi}\int_{-\infty}^{\infty} e^{isx}\, ds\int_{-\infty}^{\infty} f(\eta)e^{-i\eta s}\, d\eta.$$

We now set

$$(6.3.7) \qquad \bar{f}(s) = \frac{1}{\sqrt{2\pi}}\int_{-\infty}^{\infty} f(x)e^{-isx}\, dx$$

and call $\bar{f}(s)$ the *Fourier transform* of $f(x)$. It follows at once from (6.3.6) that the original function $f(x)$ is obtained from its transform $\bar{f}(s)$ by the inversion formula

$$(6.3.8) \qquad \bar{f}(x) = \frac{1}{\sqrt{2\pi}}\int_{-\infty}^{\infty} f(s)\ e^{isx}\, ds.$$

In our differential equation problems it will be essential to express the Fourier transform of the derivatives of a function in terms of the transform of the function itself. Consider for this purpose the transform of $d^n f/dx^n$ and integrate by parts (which requires that $d^n f/dx^n$ be continuous):

$$\frac{1}{\sqrt{2\pi}}\int_{-\infty}^{\infty} \frac{d^n f}{dx^n} e^{-isx}\, dx$$

$$= \frac{1}{\sqrt{2\pi}}\left[\frac{d^{n-1}f}{dx^{n-1}} e^{-isx}\ \Big|_{-\infty}^{\infty} + is\int_{-\infty}^{\infty} \frac{d^{n-1}f}{dx^{n-1}} e^{-isx}\, dx\right].$$

If the $(n-1)$-st derivative is to possess a transform it must tend to zero at $\pm \infty$ and hence we have

$$(6.3.9) \qquad \overline{\frac{d^n f}{dx^n}} = is\, \overline{\frac{d^{n-1}f}{dx^{n-1}}},$$

that is, the transform of the n-th derivative is (is) times the transform of the $(n-1)$-st derivative. By repeated application of this formula we obtain the result

(6.3.10)
$$\overline{\frac{d^n f}{dx^n}} = (is)^n \overline{f}$$

provided that $f(x)$ and its first n derivatives are continuous and that all of these functions possess transforms.

A rigorous justification of the transform technique used in the following for solving problems involving partial differential equations is not an entirely trivial affair (see, for example, Courant-Hilbert [C.10], vol. 2, p. 202 ff.). Such a justification could be given, but we shall not carry it out here. Indeed, it would be reasonable to take the attitude that one may proceed quite formally provided that one verifies a posteriori that the solutions obtained in this way really satisfy all conditions of the problem. This is usually not too difficult to do, and, since the relevant uniqueness theorems are available, this course is perfectly satisfactory.

6.4. Motions due to disturbances originating at the surface

We wish to determine first the motion in two dimensions due to the application of an impulse over a segment of the surface $-a \leqq x \leqq a$ at $t = 0$ when the water is at rest in the equilibrium position. We suppose the depth h of the water to be constant and that it extends to infinity in the horizontal direction. The velocity potential $\Phi(x, y; t)$ must satisfy the following conditions. It must be a solution of the Laplace equation:

(6.4.1) $\Phi_{xx} + \Phi_{yy} = 0,$ $- \infty < x < \infty,$ $-h \leqq y \leqq 0,$ $t \geqq 0,$

satisfying the boundary conditions

(6.4.2) $\Phi_{tt} + g\Phi_y = 0;$ $y = 0,$ $t > 0$

and

(6.4.3) $\Phi_y = 0,$ $y = -h,$ $t \geqq 0.$

The first of these conditions states that the pressure on the free surface is zero for $t > 0$. As initial conditions we have, in view of (6.1.6), (6.1.7), and the assumed physical situation:

(6.4.4) $\Phi(x, 0; 0) = -\frac{1}{\varrho} I(x),$

(6.4.5) $\Phi_t(x, 0; 0) = 0,$

with $I(x)$ the impulse per unit area applied to the free surface. In addition, we must impose conditions at ∞. These are that Φ and its first two derivatives with respect to x, y, and t should tend to zero at ∞ in such a way that all of these functions possess Fourier transforms with respect to x. This, in particular, requires that $I(x)$ in (6.4.4) should vanish at ∞. Actually we consider only the special case in which

$$(6.4.6) \qquad I(x) = \begin{cases} P = \text{const.,} & |x| < a \\ 0, & |x| > a, \end{cases}$$

i.e. the case in which a uniform impulse is applied to the segment $|x| < a$, the remainder of the surface being left undisturbed.

The solution $\Phi(x, y; t)$ will now be determined by applying the Fourier transform in x to the relations $(6.4.1)$—$(6.4.5)$ with the object (as always in such problems) of obtaining a simpler problem for the transform $\bar{\Phi}(s, y; t) \equiv \varphi(s, y; t)$. Once the transform φ has been found by solving the latter problem the inversion formula yields the solution Φ. We begin by applying the transform to (6.4.1), i.e. by multiplying by e^{-isx} and integrating over the interval $-\infty < x < \infty$; the result is

$$(6.4.7) \qquad - s^2\varphi(s, y; t) + \varphi_{yy}(s, y; t) = 0$$

in view of (6.3.10) and the assumed behavior of Φ at ∞. (Clearly, it is also necessary to suppose that the operation of differentiating Φ twice with respect to y can be interchanged with the operation of integrating Φ over the infinite interval.) This step already achieves one of the prime objects of the approach using a transform: the transform φ satisfies an ordinary differential equation instead of the partial differential equation satisfied by Φ. The general solution of (6.4.7) is

$$(6.4.8) \qquad \varphi(s, y, t) = A(s; t)e^{|s|y} + B(s; t)e^{-|s|y}$$

in terms of the arbitrary "constants" $A(s; t)$ and $B(s; t)$. It is a simple matter to find the appropriate special solution that also satisfies the bottom condition (6.4.3), and from it to continue (just as is done in what follows) in such a way as to find the solution for water of uniform depth. However, we prefer to take the case of infinite depth and to replace (6.4.3) by the condition that $\Phi_y \to 0$ when $y \to -\infty$. The transform φ then also must have this property so that we obtain for $\varphi(s, y; t)$ in this case the solutions

$$(6.4.9) \qquad \varphi(s; y; t) = A(s; t)e^{|s|y}.$$

The transform is next applied to the free surface condition (6.4.2) to obtain

(6.4.10) $\varphi_{tt} + g\varphi_y = 0, \qquad y = 0, \qquad t > 0$

and upon insertion of $\varphi(s, 0; t)$ from (6.4.9) we find for $A(s; t)$ the differential equation

(6.4.11) $A_{tt} + g|s|A = 0, \qquad t > 0.$

Finally, the initial conditions must be taken into account. The transform of (6.4.5) leads, evidently, to the condition $A_t(s; 0) = 0$, and the solution of (6.4.11) satisfying this condition is

(6.4.12) $A(s; t) = a(s) \cos(\sqrt{g|s|}\, t)$

with $a(s)$ still to be determined by using (6.4.4). From (6.4.4) we have $\varphi(s, 0; 0) = -(1/\varrho)\overline{I}(s)$ in which $\overline{I}(s)$ is, of course, the transform of $I(x)$ as given by (6.4.4); hence $a(s) \equiv -(1/\varrho)\overline{I}(s)$ and we have for $\overline{\Phi}(s, y; t) \equiv \varphi(s, y; t)$ the result

(6.4.13) $\overline{\Phi}(s, y; t) = -\dfrac{1}{\varrho}\overline{I}(s)e^{|s|y} \cos(\sqrt{g|s|}\, t).$

The inversion formula (6.3.8) then leads immediately to the solution

(6.4.14) $\Phi(x, y; t) = -\dfrac{1}{\varrho\sqrt{2\pi}} \displaystyle\int_{-\infty}^{\infty} \overline{I}(s)e^{|s|y}e^{isx} \cos(\sqrt{g|s|}\, t)\, ds.$

In our special case (cf. (6.4.6)) we have for $\overline{I}(s)$:

$$\overline{I}(s) = \frac{P}{\sqrt{2\pi}}\int_{-a}^{a} e^{-isx}\, dx = \frac{2P}{\sqrt{2\pi}}\int_{0}^{a} \cos sx\, dx = \frac{2Pa}{\sqrt{2\pi}} \cdot \frac{\sin sa}{sa},$$

and hence finally for $\Phi(x, y; t)$ the solution

(6.4.15) $\Phi(x, y; t) = -\dfrac{2Pa}{\pi\varrho} \displaystyle\int_{0}^{\infty} \dfrac{\sin sa}{sa} e^{sy} \cos sx \cos(\sqrt{gs}\, t)\, ds,$

as one can readily verify. For the free surface elevation we have (from (6.1.2)):

(6.4.16) $\eta(x; t) = -\dfrac{1}{g}\Phi_t = \dfrac{-2Pa}{\pi\varrho\sqrt{g}} \lim_{y\to 0} \displaystyle\int_{0}^{\infty} \dfrac{\sin sa}{sa} e^{sy} \cos sx$

$$\sin(\sqrt{gs}\, t)\sqrt{s}\, ds.$$

One observes that the integrals converge well for all $y < 0$ because of the exponential factor e^{sy}, i.e. everywhere except possibly on the free surface. These formulas can now be used to obtain the solution for the case of an impulse concentrated on the surface at $x = 0$; one need only suppose that $a \to 0$ while $P \to \infty$ in such a way that the total impulse $2Pa$ tends to a finite limit. For a unit total impulse we would then obviously obtain for Φ and η the formulas:

(6.4.17) $\Phi(x, y; t) = -\dfrac{1}{\pi\varrho} \displaystyle\int_0^\infty e^{sy} \cos sx \cos (\sqrt{gs}\, t)\, ds,$

(6.4.18) $\eta(x; t) = -\dfrac{1}{\pi\varrho\sqrt{g}} \lim_{y\to 0} \displaystyle\int_0^\infty e^{sy} \cos sx \sin (\sqrt{gs}\, t)\sqrt{s}\, ds.$

(We define $\eta(x; t)$ as a limit for $y \to 0$ since the integral obviously diverges for $y = 0$. This would, however, not be necessary in (6.4.16).)

By operating in the same way, one can easily obtain the solutions corresponding to the case of an initial elevation of the free surface at time $t = 0$, with no impulse applied. The only difference would be that Φ in (6.4.4) would be assumed to vanish while Φ_t in (6.4.5) would be different from zero. We simply give the result of such a calculation, but only for the limit case in which the initial elevation is concentrated at the origin. For Φ and η the formulas are:

(6.4.19) $\Phi(x, y; t) = -\dfrac{\sqrt{g}}{\pi} \displaystyle\int_0^\infty e^{sy} \cos sx \sin (\sqrt{gs}\, t)\dfrac{ds}{\sqrt{s}},$

(6.4.20) $\eta(x; t) = \dfrac{1}{\pi} \lim_{y\to 0} \displaystyle\int_0^\infty e^{sy} \cos sx \cos (\sqrt{gs}\, t)\, ds.$

There is no difficulty in treating problems having *cylindrical symmetry* that are exactly analogous to the above two-dimensional cases. In these cases also one could begin with the solutions having symmetry of this type that are simple harmonic in the time (cf. Chapter 3):

(6.4.21) $\Phi(r, y; t) = e^{i\sigma t} e^{my} J_0(mr)$

with $\sigma^2 = gm$ (for water of infinite depth). Here the quantity r is the distance $\sqrt{x^2 + z^2}$ from the y-axis, and $J_0(mr)$ is the Bessel function of order zero that is regular at the origin. One could now build up more complicated solutions by superposition of these solutions and satisfy given initial conditions by using the Fourier-Bessel integral. This is the method followed by Lamb [L.3], p. 429. Instead of this procedure,

one could make use of the Hankel transform in a fashion exactly analogous to the Fourier transform procedure used above (cf. Sneddon
[S.11], p. 290, and Hinze [H.15]). We content ourselves here with
giving the result for the velocity potential $\Phi(r, y; t)$ and the surface
elevation $\eta(r; t)$ due to the application of a concentrated unit impulse
at the origin at $t = 0$:

$$(6.4.22) \quad \Phi(r,y;t) = -\frac{1}{2\pi\varrho} \int_0^\infty e^{sy} J_0(sr) \cos{(\sqrt{gs}\, t)}s \; ds,$$

$$(6.4.23) \quad \eta(r; t) = \frac{-1}{2\pi\varrho\sqrt{g}} \lim_{y\to 0} \int_0^\infty e^{sy} J_0(sr) \sin{(\sqrt{gs}\, t)}s^{3/2} \; ds.$$

Naturally we want to discuss the character of the motions furnished
by the above relations, and in doing so we come upon a fact that holds
good in all problems of this type: it is a comparatively straightforward
matter to obtain an integral representation for the solution, but not
always an easy matter to carry out the details of the discussion of its
properties. The reason for this is not far to seek—it is due to the fact
that the solutions are given in terms of an integral over an integrand
which is oscillatory in character and which changes rather rapidly
over even small intervals of the integration variable for important
ranges in the values of the independent variables. Hence even a numerical integration would not be easy to carry out. The fact is that the
motions are really of a complicated nature, as we shall see, and hence
a mathematical description of them can be expected to present some
difficulties. Indeed, the phenomena under consideration here are
analogous to the refraction and diffraction phenomena of physical
optics and thus depend on intricate interference effects, which are
further complicated in the present instances by the fact that the wave
motions are subject to dispersion, as we have seen in Chapter 3.

Some insight into the nature of the solutions furnished by our formulas can be obtained by expanding the integrands in power series
and integrating term by term (cf. Lamb. [L.3], p. 385).* The result
for $\eta(x; t)$ as given by (6.4.20), for example, is found to be (for $x > 0$):

$$(6.4.24) \quad \eta(x; t) = \frac{1}{\pi x}\left\{ \frac{gt^2}{2x} - \frac{1}{1\cdot3\cdot5}\left(\frac{gt^2}{2x}\right)^3 + \frac{1}{1\cdot3\cdot5\cdot7\cdot9}\left(\frac{gt^2}{2x}\right)^5 - \cdots \right\}.$$

It is clear that there is a singularity for $x = 0$, as one would expect.

---

* The subsequent discussion in this section follows closely the presentation
given by Lamb.

The series converges for all values of the dimensionless quantity $gt^2/2x$, but practically the series is useful only for small values of $gt^2/2x$, i.e. for small values of t, or large values of x. One observes also that any particular "phase" of the disturbance—such as a zero of η, for example—must propagate with a constant *acceleration*, since any such phase is clearly associated with a specific constant value of the quantity $gt^2/2x$.

It is in many respects more useful to find an asymptotic representation for the motion valid in the present case for large values of the quantity $gt^2/2x$, for which the power series are not very useful because of their slow convergence. Indeed, the asymptotic representation yields all of the qualitative features contained in the exact solution (6.4.24), and is also accurate even for rather small values of the quantity $gt^2/2x$ (cf. Sneddon [S.11], p. 287). For this purpose it happens to be rather easy to work out an asymptotic development of the solution that is valid for large values of $gt^2/2x$, and this we proceed to do, following Lamb. We write (6.4.19) in the form

$$(6.4.25) \qquad \Phi(x, y; t) = \frac{-1}{\pi} \left\{ \int_0^\infty e^{\frac{\sigma^2}{g} y} \sin\left(\frac{\sigma^2 x}{g} + \sigma t\right) d\sigma \right.$$
$$\left. - \int_0^\infty e^{\frac{\sigma^2}{g} y} \sin\left(\frac{\sigma^2 x}{g} - \sigma t\right) d\sigma \right\},$$

making use of $\sigma = \sqrt{gs}$, $2\sigma d\sigma = g ds$. New quantities ξ and ω are introduced in (6.4.25) by the relations

$$\xi = \frac{x^{1/2}}{g^{1/2}}\left(\sigma \pm \frac{gt}{2x}\right), \qquad \omega = \left(\frac{gt^2}{4x}\right)^{1/2},$$

from which

$$\xi^2 - \omega^2 = \frac{\sigma^2 x}{g} \pm \sigma t.$$

The expression (6.4.25) is thus readily found to take the form

$$(6.4.26) \qquad \Phi(x, 0; t) = \frac{2g^{1/2}}{\pi x^{1/2}} \int_0^\omega \sin\left(\xi^2 - \omega^2\right) d\xi$$

where ξ is introduced as new variable of integration and y is assumed to vanish. The corresponding free surface elevation is given by

$$(6.4.27) \qquad \eta = -\frac{1}{g}\,\Phi_t\bigg|_{y=0} = \frac{g^{\frac{1}{2}}t}{\pi x^{3/2}}\int_0^{\omega}\cos\,(\xi^2 - \omega^2)\,d\xi$$

as one readily verifies. In order to study the last expression we consider the integral

$$(6.4.28) \qquad \int_0^{\omega} e^{i\,(\xi^2 - \omega^2)}\,d\xi = \int_0^{\infty} e^{i\,(\xi^2 - \omega^2)}\,d\xi - \int_{\omega}^{\infty} e^{i\,(\xi^2 - \omega^2)}\,d\xi.$$

It is well known that

$$(6.4.29) \qquad \int_0^{\infty} e^{i\,(\xi^2 - \omega^2)}\,d\xi = \tfrac{1}{2}\,\sqrt{\pi}\,e^{-i\,(\omega^2 - \pi/4)};$$

while the second contribution can be treated as follows:

$$(6.4.30) \qquad \int_{\omega}^{\infty} e^{i\,(\xi^2 - \omega^2)}\,d\xi = \frac{1}{2}\int_{\omega^2}^{\infty}\frac{1}{\sqrt{t}}\,e^{i\,(t - \omega^2)}\,dt$$

$$= \frac{1}{2i}\left[t^{-\frac{1}{2}}\,e^{i\,(t - \omega^2)}\,\bigg|_{\omega^2}^{\infty} + \frac{1}{2}\int_{\omega^2}^{\infty} t^{-\frac{3}{2}}\,e^{i\,(t - \xi^2)}\,dt \right]$$

through introduction of $t = \xi^2$ as new variable, and an integration by parts. We show next that the final integral is of the order ω^{-1}, as follows:

$$\left| \int_{\omega^2}^{\infty} t^{-\frac{3}{2}}e^{i\,(t - \omega^2)}\,dt \right| \leqq \int_{\omega^2}^{\infty} t^{-\frac{3}{2}}\left| e^{i\,(t - \omega^2)} \right|\,dt$$

$$= \int_{\omega^2}^{\infty} t^{-\frac{3}{2}}\,dt = 2\omega^{-1}.$$

Upon considering the real parts of (6.4.28), (6.4.29), (6.4.30), and inserting in (6.4.27) one finds

$$(6.4.31) \qquad \eta(x;\,t) = \frac{1}{x\,\sqrt{\pi}}\left(\frac{gt^2}{4x}\right)^{\frac{1}{2}}\left[\cos\left(\frac{gt^2}{4x} - \frac{\pi}{4}\right) + O(\omega^{-1})\right]$$

in which the function $O(\omega^{-1})$ refers, as one readily verifies, to a term which behaves like $(gt^2/4x)^{-1/2}$. Consequently, if $\omega^2 = gt^2/4x$ is sufficiently large, we may assume for the free surface elevation due to a concentrated surface elevation at $x = 0$ and $t = 0$ the approximate expression

(6.4.32) $$\eta(x; t) \simeq \frac{1}{x\sqrt{\pi}} \left(\frac{gt^2}{4x}\right)^{\frac{1}{2}} \cos \left(\frac{gt^2}{4x} - \frac{\pi}{4}\right).$$

By continuing the integration by parts, as in (6.4.30), it would be possible to obtain approximations valid up to any order in the quantity $\omega^{-1} = (gt^2/4x)^{-1/2}$, but such an expansion would not be convergent; it is rather an asymptotic expansion correct within a certain order in ω^{-1} when an appropriate finite number of terms in the expansion is taken. Expansions of this type are—as in other branches of mathematical physics—very useful in many of our problems and we shall have many other occasions to employ them.

The case of a concentrated point *impulse* applied at $x = 0$ at the time $t = 0$ can be treated in exactly the same manner as the case just considered: one has only to begin with the solution (6.4.17) instead of (6.4.19), and proceed along similar lines. In particular, the approximate solution valid (to the same order in ω^{-1}) for large values of $gt^2/4x$ can be obtained; the result for the free surface elevation is

(6.4.33) $$\eta(x; t) \simeq \frac{-2}{\varrho g\, xt\sqrt{\pi}} \left(\frac{gt^2}{4x}\right)^{3/2} \sin \left(\frac{gt^2}{4x} - \frac{\pi}{4}\right).$$

The method used to derive these asymptotic formulas is rather special: it cannot be very easily used to study the cylindrical waves given by (6.4.23), for example. We turn, therefore, in the next section to the derivation of asymptotic approximations in all of these cases by the application of Kelvin's method of stationary phase. Afterwards, the motions themselves will be discussed in section 6.6 on the basis of the approximate formulas.

6.5. Application of Kelvin's method of stationary phase.

The integrals of section 6.4 can all be put into the form

(6.5.1) $$I(k) = \int_a^b \psi(\xi, k) e^{ik\varphi(\xi)} \, d\xi$$

without much difficulty, and this is a form peculiarly suited to an approximate treatment valid for large values of the real constant k. In fact, Kelvin seems to have been led to the approximate method known as the method of stationary phase through his interest in problems concerning gravity waves, in particular the ship wave problem. The general idea of the method of approximation is as follows. When k is large the function $\exp\{ik\varphi(\xi)\}$ oscillates rapidly as ξ changes, unless $\varphi(\xi)$ is nearly constant, so that the positive and negative contributions to the value of $I(k)$ largely cancel out, provided that $\psi(\xi, k)$ is not a rapidly oscillating function of ξ when k is large. Hence one might expect the largest contributions to the integral to arise from the neighborhoods of those points in the interval from a to b at which $\varphi(\xi)$, the phase of the oscillatory part of the integral, varies most slowly, i.e., from neighborhoods of the points where $\varphi'(\xi) = 0$. This indeed turns out to be the case. In section 6.8 it will be shown that

$$(6.5.2) \quad I(k) = \sum_r \psi(\alpha_r, k) \left(\frac{2\pi}{k\,|\,\varphi''(\alpha_r)|}\right)^{\frac{1}{2}} \exp\left\{i\left(k\varphi(\alpha_r) \pm \frac{\pi}{4}\right)\right\}$$

$$+ \sum_s \psi(\alpha_s, k)\frac{\Gamma(\frac{1}{3})}{\sqrt{3}}\left(\frac{6}{k\,|\,\varphi'''(\alpha_s)|}\right)^{\frac{1}{3}} \exp\{ik\varphi(\alpha_s)\} + O\left(\frac{1}{k^{2/3}}\right).$$

By $O(1/k^{2/3})$ we mean a function which tends to zero like $1/k^{2/3}$ as $k \to \infty$. In these expressions the sums are taken over all the zeros α_r of $\varphi'(\xi)$ in the interior of the interval $a \leq \xi \leq b$ at which $\varphi''(\alpha_r) \neq 0$ and over the zeros α_s of $\varphi'(\xi)$ at which $\varphi''(\alpha_s) = 0$ but $\varphi'''(\alpha_s) \neq 0$. The sign of the term $\pm \pi/4$ in the first sum should be taken to agree with the sign of $\varphi''(\alpha_r)$. The relation (6.5.2) is valid if $\psi(\xi, k)$ and $\varphi(\xi)$ are analytic functions of ξ in $a \leq \xi \leq b$, and if the only stationary points of $\varphi(\xi)$ are such that $\varphi''(\xi)$ and $\varphi'''(\xi)$ do not vanish simultaneously.*

We proceed to obtain the approximate solution (6.4.32) obtained in the previous section once more by this method. The motion of the water was to be determined for the case of an elevation of the water surface concentrated at a point; the formula for the velocity potential was put in the form (cf. (6.4.25)):

$$(6.5.3) \quad \Phi(x, y; t) = -\frac{1}{\pi}\left\{\int_0^\infty e^{\frac{\sigma^2 y}{g}} \sin\left(\frac{\sigma^2 x}{g} + \sigma t\right) d\sigma\right.$$

$$\left. - \int_0^\infty e^{\frac{\sigma^2 y}{g}} \sin\left(\frac{\sigma^2 x}{g} - \sigma t\right) d\sigma\right\}.$$

---

* If a zero of $\varphi'(\xi)$ of still higher order should occur, then terms of other types would appear, and the error would die out less rapidly in k. It should also be noted that the coefficient function ψ of section 6.8 is assumed to be independent of k, which is not true in some of the examples to follow. However, it is not difficult to see that the proof of section 6.8 can be modified quite easily in such a way as to include all of our cases.

This can in turn be put in the form

$$(6.5.4) \quad \Phi(x, y; t) = -\frac{1}{\pi} \left\{ \int_0^\infty e^{my} e^{i(mx+\sigma t)} \, d\sigma - \int_0^\infty e^{my} e^{i(mx-\sigma t)} \, d\sigma \right\}$$

with $m = \sigma^2/g$. It is understood that the imaginary part only is to be taken at the end. It is convenient to introduce a new dimensionless variable of integration as follows:

$$(6.5.5) \quad \xi = \frac{2x}{gt} \sigma,$$

in terms of which (6.5.4) is readily found to take the form

$$(6.5.6) \quad \Phi(x, y; t) = -\frac{gt}{2\pi x} \left\{ \int_0^\infty e^{my} e^{ik (\xi^2+2\xi)} \, d\xi - \int_0^\infty e^{my} e^{ik(\xi^2-2\xi)} \, d\xi \right\}$$

with

$$(6.5.7) \quad k = \frac{gt^2}{4x}$$

as a dimensionless parameter. The quantity m is of course also a function of ξ, and $\exp\{m(\xi)y\}$ plays the role of the function $\psi(\xi)$ in (6.5.1). When the parameter k is large, we may approximate the solution by using (6.5.2). For the phases $\varphi(\xi)$ we have

$$(6.5.8) \quad \varphi(\xi) = \xi^2 \pm 2\xi$$

with stationary points given by

$$(6.5.9) \quad \varphi'(\xi) = 2\xi \pm 2 = 0,$$

and we see that $\xi = 1$ is the only such point in the interval $0 < \xi < \infty$ over which the integrals are taken. Consequently only the second integral in (6.5.6) possesses a point of stationary phase, and at this point we have

$$(6.5.10) \quad \varphi''(1) = 2, \quad \varphi(1) = -1.$$

We obtain therefore from (6.5.2) the approximate formula

$$(6.5.11) \quad \Phi(x, y; t) \simeq \sqrt{\frac{g}{\pi x}} \, e^{m(1)y} \, e^{i\left(-k+\frac{\pi}{4}\right)},$$

as one readily verifies, and this formula is a good approximation for large values of $k = gt^2/4x$. We can also calculate the free surface elevation η in the same way from $\eta = -(1/g)\,\Phi_t\,|_{y=0}$; the result is easily found to be

$$(6.5.12) \qquad \eta(x;\,t) \simeq \frac{g^{1/2}\,t}{2\sqrt{\pi}x^{3/2}}\cos\left(\frac{gt^2}{4x} - \frac{\pi}{4}\right),$$

just as before (cf. (6.4.32)).

For the case of a concentrated impulse the method of stationary phase as applied to (6.4.17) or (6.4.18) leads to the following approximation valid once again for large values of $gt^2/4x$:

$$(6.5.13) \qquad \eta(x;\,t) \simeq -\frac{g^{1/2}\,t^2}{4\sqrt{\pi}\,\varrho x^{5/2}}\sin\left(\frac{gt^2}{4x} - \frac{\pi}{4}\right),$$

and this coincides with the result given in (6.4.33).

In the case of an impulse distributed over a segment one obtains from (6.4.16) the result

$$(6.5.14) \qquad \eta(x;\,t) = -\frac{2P}{\varrho(\pi g)^{1/2}x^{1/2}}\sin\frac{gt^2a}{4x^2}\sin\left(\frac{gt^2}{4x} - \frac{\pi}{4}\right),$$

valid for large values of $gt^2/4x$.*

For the ring waves furnished by (6.4.23) the asymptotic formula is

$$(6.5.15) \qquad \eta(r;\,t) = -\frac{gt^3}{2^{7/2}\pi\varrho r^4}\sin\frac{gt^2}{4r}.$$

To obtain this formula it is necessary to replace the Bessel function $J_0(sr)$ in (6.4.23) by its integral representation

$$J_0(sr) = \frac{2}{\pi}\int_0^{\pi/2}\cos\,(sr\,\cos\,\beta)\,d\beta$$

and then apply the method of stationary phase twice in succession. Since such a procedure is discussed later on in dealing with the simplified ship wave problem (cf. Chapter 8.1), we omit a discussion of it here, except to remark that the approximate formula (6.5.15) is valid for any $r \neq 0$ and $gt^2/4r$ sufficiently large.

---

* It may seem strange that this formula indicates that $x = 0$ is a singular point for η, while $x = 0$ is not singular in the exact formula (6.4.16). This comes about through the introduction of the new variable (6.5.5) and the parameter k in (6.5.7) which were used to convert the original integral to the form (6.5.1). However, the validity of the formula (6.5.2) is assured, as one can see from section 6.8, only if $x \neq 0$.

6.6. Discussion of the motion of the free surface due to disturbances initiated when the water is at rest

We proceed to discuss the motions of the water surface in accordance with the results given in the preceding section. The general character of the motion is well given by the approximate formulas, and we shall therefore confine our discussion to them. We observe first that the oscillatory factors in the four approximate formulas (6.5.12)—(6.5.15) do not differ essentially, but the slowly varying nonoscillatory factors are different in the various cases: (a) at a fixed point on the water surface the disturbance increases in amplitude linearly in t in the case of an initial elevation concentrated at a point (cf. (6.5.12)), while for a fixed time the amplitude becomes large for small x like $x^{-3/2}$; (b) in the case of an initial impulse concentrated at a point the amplitude increases quadratically in t at a fixed point, while for a fixed time the amplitude increases like $x^{-5/2}$ for small x. (In these limit cases the approximate formulas are valid for $x \neq 0$, since the only other requirement is that the quantity $gt^2/4x$ should be large.) The behavior of these solutions near $x = 0$ is not very surprising since there is a singularity there. The behavior at any fixed point x as $t \to \infty$ is, however, somewhat startling: the amplitude is seen to grow large without limit as the time increases in both of these cases. This rather unrealistic result is a consequence of the fact that the singularity at the origin is very strong. If the initial disturbance were finite and spread over an area, the amplitude of the resulting motion would always remain bounded with increasing time, as one could show by an appeal to the general behavior of Fourier transforms.* This fact is well shown in the special case of a distributed impulse, as we see from (6.5.14), which is valid for all $x \neq 0$ and large t: the amplitude remains bounded as $t \to \infty$.

The general character of the waves generated by a point disturbance is indicated schematically in the accompanying figures which show the variation in surface elevation at a fixed point x when the time increases, and at a fixed time for all x. These figures are based on the formula (6.5.12) for the case of an initial elevation; the results for the case of initial impulse would be of the same general nature.

---

* It is also a curious fact that the motion given by (6.5.14) for the case of an impulse over a segment requires infinite energy input, since the amplitude at any fixed point does not tend to zero. For the case of an initial elevation confined to a segment, however, the wave amplitude would die out with increasing time.

It is worth while to discuss the character of the motion furnished
by (6.5.12) in still more detail. It has already been remarked that any
particular phase—such as a zero, or a maximum or minimum of η—is
of necessity propagated with an *acceleration* since each such phase is
associated with a particular constant value of the quantity $gt^2/4x$: if
the phase is fixed by setting $gt^2/4x = c$, then this phase moves in
accordance with the relation $x = gt^2/4c$. The formula (6.5.12) holds
only where the quantity $gt^2/4x$ is large, and hence the individual phases are accelerated slowly in the region of validity of this formula; or,

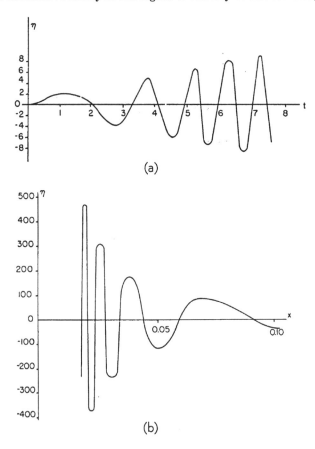

(a)

(b)

Fig. 6.6.1a,b Propagation of waves due to an initial elevation

in other words, the phases move in such regions at nearly constant velocity. Also, for not too great changes in x or t the waves behave very nearly like simple harmonic waves of a certain fixed period and wave length. This can be seen as follows. Suppose that we vary t alone from $t = t_0$ to $t = t_0 + \Delta t$. We may write for the phase φ:

$$\varphi = \frac{gt_0^2}{4x_0}\left[1 + \frac{2\Delta t}{t^0} + \left(\frac{\Delta t}{t_0}\right)^2\right],$$

as one readily verifies. Thus if $\Delta t/t_0$ is small, i.e. if the change Δt in the time is small compared with the total lapse of time since the motion was initiated, we have for the change in phase:

$$\Delta \varphi = \varphi - \varphi_0 \simeq \frac{2\Delta t}{t_0}\left(\frac{gt_0^2}{4x_0}\right).$$

Consequently the period $T = \Delta t$ of the motion corresponding to the change $\Delta\varphi = 2\pi$ in the phase is given approximately by the formula

$$(6.6.1) \qquad\qquad T \simeq \frac{4\pi x_0}{gt_0}.$$

The accuracy of this formula is good, as we know, if $T/t_0 \simeq 4\pi x_0/gt_0^2$ is small, and this is the case since $gt_0^2/4x_0$ is always assumed to be large. Thus the period at any fixed point varies slowly in the time. In the same way one finds for the local wave length λ the approximate formula

$$(6.6.2) \qquad\qquad \lambda \simeq \frac{8\pi x_0^2}{gt_0^2}$$

by varying with respect to x alone, and this is also easily seen to be accurate if $gt_0^2/4x_0$ is large. Thus for a fixed position x the period and wave length both vary slowly, and they decrease as the time increases, while for a fixed time the same quantities increase with x, as is borne out by the figures shown above.

It is of considerable interest next to compute the local phase velocity—the velocity of a zero of η, for example—from $gt^2/4x = c$ when x and t vary independently; the result is

$$(6.6.3) \qquad\qquad \frac{dx}{dt} = \frac{2x}{t}$$

for the velocity of any phase; thus for fixed x the phases move more

slowly as the time increases, but for fixed t more rapidly as x increases
—that is, the waves farther away from the source of the disturbance
move more rapidly, and they are also longer, as we know from (6.6.2).
The wave pattern is thus drawn out continually, and the waves as
they travel outward become longer and move faster. The last fact is
not too surprising since the waves in the vicinity of a particular point
have essentially the simple character of the sine or cosine waves of
fixed period that we have studied earlier, and such waves, as we have
seen in Chapter 3, propagate with speeds that increase with the wave
length. All of the above phenomena can be observed as the result of
throwing a stone into a pond; though the motion in this case is three-
dimensional it is qualitatively the same, as one can see by comparing
(6.5.15) with (6.5.12).

There is another way of looking at the whole matter which is
prompted by the last observations. Apparently, the disturbance at
the origin acts like a source which emits waves of all wave lengths and
frequencies. But since our medium is a dispersive medium in which
the propagation speed of a particular phase increases with its wave
length, it follows that the disturbance as a whole tends with increasing
time to break up into separate trains of waves each of which has approximately the same wave length, since waves whose lengths differ
move with different velocities. However, it would be a mistake to
think that such wave trains or groups of waves themselves move with
the phase speed corresponding to the wave length associated with the
group. If one fixes attention on the group as a whole rather than on
an individual wave of the group, the velocity of the group will be seen
to differ from that of its component waves. The phase velocity for the
present case can be obtained in terms of the local wave length readily
from the equation (6.6.3) by expressing its right hand side in terms
of the local wave length through use of (6.6.2); the result is

$$(6.6.4) \qquad \frac{dx}{dt} = \frac{2x}{t} = \sqrt{\frac{g\lambda}{2\pi}}.$$

On the other hand, the position x of a group of waves of fixed wave
length λ at time t is given closely by the formula

$$(6.6.5) \qquad x = \frac{1}{2}\sqrt{\frac{g\lambda}{2\pi}}\, t,$$

as we see directly from (6.6.2), so that the velocity of the group is

$\frac{1}{2}\sqrt{g\lambda/2\pi}$, which is, evidently, just half the phase speed of its component waves. In other words, the component waves in a particular group move forward through the group with a speed twice that of the group.

Finally, we observe that these results are in perfect accord with the discussion in Chapter 3 concerning the notions of phase and group velocity. The phase speed c for a simple harmonic wave of wave length λ in water of infinite depth is given (cf. $(3.2.3)_1$), by $c = \sqrt{g\lambda/2\pi}$, and this is also the phase speed of the waves whose wave length is λ—as we see from $(6.6.4)$. We have also defined in section 3.4 the notion of group velocity for simple harmonic waves in water of infinite depth, and found it to be just half the phase velocity. The kinematic definition of the group velocity given in section 3.4 was obtained by the superposition of trains of simple harmonic waves of slightly different wave length and amplitude, while in the present case the waves are the result of a superposition of waves of all wave lengths and periods. However, the principle of stationary phase, which furnishes the approximate solution studied here, in effect says that the main motion in certain regions is the result of the superposition of waves whose wave lengths and amplitudes differ arbitrarily little from a certain given value. The results of the analysis in the present case are thus entirely consistent with the analysis of section 3.4.

At any time, therefore, the surface of the water is covered by groups of waves arranged so that the groups having waves of greater length are farther away from the source. These groups, therefore, tend to separate, as one sees from $(6.6.4)$. The waves in a given group do not maintain their amplitude, however, as the group proceeds: one sees readily from $(6.5.12)$ in combination with $(6.6.2)$ that their amplitude is proportional to $1/\sqrt{x}$ for waves of fixed length λ.

The above interpretations of the results of the basic theory are all borne out by experience. Figure 6.6.2 shows a time sequence of photographs of waves (given to the author by Prof. J. W. Johnson of the University of California at Berkeley) created by a disturbance concentrated in a small area: the decrease in wave length at a fixed point with increasing time, the increase in the wave lengths near the front of the outgoing disturbance as the time increases, the general drawing out of the wave pattern with time, the occurrence of well-defined groups, etc. are well depicted.

An interesting development in oceanography has been based on the theory developed in the present section. Deacon [D. 6, 7] and his

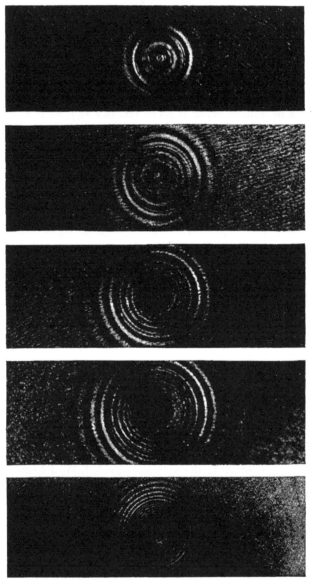

Fig. 6.6.2. Waves due to a concentrated disturbance

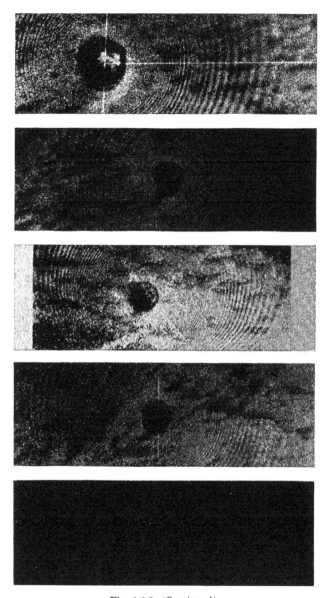

Fig. 6.6.2. (Continued)

associates have carried out studies which correlate the occurrence of storms in the Atlantic with the long waves which move out from the storm areas and reach the coast of Cornwall in a relatively short time. By analyzing the periods of the swell, as determined from actual wave records, it has been possible to identify the swell as having been caused by storms whose location is known from meteorological observations. Aside from the interest of researches of this kind from the purely scientific point of view, it is clear that such hindcasts could, in principle, be turned into methods of forecasting the course of storms at sea in areas lacking meteorological observations.

6.7. Waves due to a periodic impulse applied to the water when initially at rest. Derivation of the radiation condition for purely periodic waves

In section 3 of Chapter 4 we have solved the problem of two-dimensional waves in an infinite ocean when the motion was a simple harmonic motion in the time that was maintained by an application of a pressure at the surface which was also simple harmonic in the time. In doing so, we were forced to prescribe radiation conditions at ∞ — effectively, conditions requiring the waves to behave like outgoing progressing waves at ∞ — in order to have a complete formulation of the problem with a uniquely determined solution. It was remarked at the time that a different approach to the problem would be discussed later on which would require the imposition of boundedness conditions alone at ∞, rather than the much more specific radiation condition. In this section we shall obtain the solution worked out in 4.3 without imposing a radiation condition by considering it as the limit of an unsteady motion as the time tends to infinity. However, it has a certain interest to make a few remarks about the question of radiation conditions in unbounded domains from a more general point of view (cf. [S. 21]).

In wave propagation problems for what will be called here, exceptionally, the steady state, i.e., a motion that is simple harmonic in the time, it is in general not possible to characterize uniquely the solutions having the desired physical characteristics by imposing only boundedness conditions at infinity. It is, in fact, as we have seen in special cases, necessary to impose sharper conditions. In the simplest case in which the medium is such as to include a full neighborhood of the point at infinity that is in addition made up of homogeneous matter, the correct radiation condition is not difficult to guess. It is simply that the wave at infinity behaves like an outgoing spherical wave from an oscillatory point source, and such a condition is what is commonly called the radiation, or Sommerfeld, condition. Among other things this condition precludes the possibility that there might be an incoming wave generated at infinity—which, if not ruled out, would manifestly make a unique solution of the problem impossible.

If the refracting or reflecting obstacles to the propagation of waves happen to extend to infinity—for example, if a rigid reflecting wall should happen to go to infinity—it is by no means clear a priori what conditions should be imposed at infinity in order to ensure the uniqueness of a simple harmonic solution having appropriate properties otherwise.* A point of view which seems to the author reasonable is that *the difficulty arises because the problem of determining simple harmonic motions is an unnatural problem in mechanics.* One should in principle rather formulate and solve an initial value problem by assuming the medium to be originally at rest everywhere outside a sufficiently large sphere, say, and also assume that the periodic disturbances are applied at the initial instant and then maintained with a fixed frequency. As the time goes to infinity the solution of the initial value problem will tend to the desired steady state solution without the necessity to impose any but boundedness conditions at infinity.**

The steady state problem is unnatural—in the author's view, at least—because a hypothesis is made about the motion that holds for all time, while Newtonian mechanics is basically concerned with the prediction—in a unique way, furthermore—of the motion of a mechanical system from given initial conditions. Of course, in mechanics of continua that are unbounded it is necessary to impose conditions at ∞ not derivable directly from Newton's laws, but for the initial value problem it should suffice to impose only boundedness conditions at infinity. In sec. 6.9. the relevant uniqueness theorem for the special case to be considered later is proved.

* For a treatment of the radiation condition in such cases see Rellich [R.7], John [J.5], and Chapter 5.5.

** The formulation of the usual radiation condition is doubtlessly motivated by an instinctive consideration of the same sort of hypothesis combined with the feeling that a homogeneous medium at infinity will have no power to reflect anything back to the finite region. Evidently, we also have in mind here only cases in which no free oscillations having finite energy occur — if such modes of oscillation existed, clearly no uniqueness theorems of the type we have in mind could be derived.

If one wished to be daring one might, on the basis of these remarks, formulate the following general method of obtaining the appropriate radiation condition: Consider any convenient problem in which the part of the domain outside a large sphere is maintained intact and initially at rest. (In other words, one might feel free to modify in any convenient way any bounded part of the medium.) Next solve the initial value problem for an oscillatory point source placed at any convenient point. Afterwards a passage to the limit should be made in allowing the time t to approach ∞, and after that the space variables should be allowed to approach infinity. The behavior at the far distant portions of the domain should then furnish the appropriate radiation conditions independent of the constitution of the finite part of the domain. It might be worth pointing out specifically that this is a case in which the order of the two limit processes cannot be interchanged: obviously, if the time t is first held fixed while the space variables tend to infinity the result would be that the motion would vanish at ∞, and no radiation conditions could be obtained.

The writer would not have set down these remarks—which are of a character so obvious that they must also have occurred to many others—if it were not for two considerations. Every reader will doubt-lessly have said to himself: "That is all very well in principle, but will it not be prohibitively difficult to carry out the solution of the initial value problem and to make the subsequent passages to the limit?" In general, such misgivings are probably all too well founded. How-ever, the problem concerning water waves to be treated here happens to be an interesting special case in which (1) the indicated program can be carried out in all detail, and (2) it is slightly easier to solve the initial value problem than it is to solve the steady state problem with the Sommerfeld condition imposed.

We restrict ourselves to two-dimensional motion in an x, y-plane, with the y-axis taken vertically upward and the x-axis in the originally undisturbed horizontal free surface. The velocity potential $\varphi(x, y; t)$ is a harmonic function in the lower half-plane:

$$(6.7.1) \qquad \varphi_{xx} + \varphi_{yy} = 0, \qquad y < 0, \qquad t > 0.$$

The free surface boundary conditions are (cf. (6.1.1), (6.1.2)):

$$(6.7.2) \qquad -\varphi_y + \eta_t = 0$$

$$(6.7.3) \qquad \varphi_t + g\eta = -\frac{1}{\varrho}p$$

$\left.\begin{array}{c} \\ \\ \end{array}\right\}$ for $y = 0$, $t > 0$.

As usual, $\eta = \eta(x; t)$ represents the vertical displacement of the free surface measured from the x-axis, and $p = p(x; t)$ represents the pressure applied on the free surface. We suppose that φ and its first and second derivatives tend to zero at ∞ for any given time t—in fact that they tend to zero in such a way that Fourier transforms exist— but we do not, in accordance with our discussion above, make any more specific assumptions about the behavior of our functions as $t \to \infty$. At the time $t = 0$ we prescribe the following initial conditions

$$(6.7.4) \qquad \varphi(x, 0; 0) = \varphi_t(x, 0; 0) = 0,$$

which state (cf. (6.1.6), (6.1.7)) that the free surface is initially at rest in its horizontal equilibrium position.

In what follows we consider only the special case in which the surface pressure $p(x; t)$ is given by

$$(6.7.5) \qquad p(x; t) = \delta(x)e^{i\omega t}, \quad t > 0$$

in which $\delta(x)$ is the Dirac δ-function. We have not made explicit use of the δ-function until now, but we have used it implicitly in section 6.4 in dealing with concentrated impulses. It is to be interpreted in the same way here, i.e. as a symbol for a limit process in which the pressure is first distributed over a segment the length of which is considered to grow small while the total pressure is maintained at the constant value one. By inserting this expression for p in (6.7.3) and eliminating the quantity η by making use of (6.7.2) the free surface condition is obtained in the form

$$(6.7.6) \qquad g\varphi_y + \varphi_{tt} = -\frac{i\omega}{\varrho}\delta(x)e^{i\omega t}, \; t > 0.$$

Our problem now consists in finding a solution $\varphi(x, y; t)$ of (6.7.1) which behaves properly at ∞, and which satisfies the free surface condition (6.7.6) and the initial conditions (6.7.4).

We proceed to solve the initial value problem by making use of the Fourier transform applied to the variable x. The result of transforming (6.7.1) is

$$(6.7.7) \qquad -s^2\bar{\varphi} + \bar{\varphi}_{yy} = 0,$$

in which $\bar{\varphi}(s, y; t)$ is the transform of $\varphi(x, y; t)$ and use has been made of the conditions at ∞. The bounded solutions of (6.7.7) for $y < 0$, $s > 0$ are all of the form

(6.7.8) $\bar{\varphi}(s, y; t) = A(s; t)e^s y.$

The transform is now applied to the boundary condition (6.7.6), with the result:

(6.7.9) $g\bar{\varphi}_y + \bar{\varphi}_{tt} = -\dfrac{1}{\sqrt{2\pi}} \dfrac{i\omega}{\varrho} e^{i\omega t},$ for $y = 0,$

and on substitution of $\bar{\varphi}(s, 0; t)$ from (6.7.8) we find

(6.7.10) $A_{tt} + gsA = -\dfrac{1}{\sqrt{2\pi}} \dfrac{i\omega}{\varrho} e^{i\omega t}.$

The initial conditions (6.7.4) now furnish for $A(s; t)$ the conditions

(6.7.11) $A(s; 0) = A_t(s; 0) = 0.$

The solution of (6.7.10) subject to the initial conditions (6.7.11) is

(6.7.12) $A(s; t) = -\dfrac{1}{\sqrt{2\pi}} \dfrac{i\omega}{\varrho} \displaystyle\int_0^t \dfrac{e^{i\omega(t-\tau)}}{\sqrt{gs}} \sin \sqrt{gs}\ \tau d\tau.$

Finally, we insert the last expression for $A(s; t)$ in (6.7.8) and apply the inverse transform to obtain the following integral representation for our solution $\varphi(x, y; t)$:

(6.7.13) $\varphi(x, y; t) = -\dfrac{i\omega}{\varrho\pi} \displaystyle\int_0^\infty e^{sy} \cos sx \displaystyle\int_0^t \dfrac{e^{i\omega(t-\tau)}}{\sqrt{gs}} \sin \sqrt{gs}\ \tau d\tau ds.$

The fact that the solution is an even function of x has been used here. Our object now is to study the behavior of this solution as $t \to \infty$.

Since y is negative (we do not discuss here the limit as $y \to 0$, i.e. the behavior on the free surface) the integral with respect to s converges well and there is no singularity on the positive real axis of the complex s-plane. However, the passage to the limit $t \to \infty$ is more readily carried out by writing the solution in a different form in which a singularity—a pole, in fact—then appears on the real axis of the s-plane. (It seems, indeed, likely that such an occurrence would be the rule in any considerations of the present kind since the limit function as $t \to \infty$ would not usually be a function having a Fourier transform, and one could expect that the limit function would some-how appear as a contribution in the form of a residue at a pole.) It is convenient to deform the path of integration in the s-plane into the path L indicated in the accompanying figure. The path L lies on the

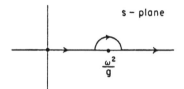

Fig. 6.7.1. Path of integration in s-plane

positive real axis except for a semicircle in the upper half-plane centered at the point $s = \omega^2/g$. By Cauchy's integral theorem this leaves the function φ given in (6.7.13) unchanged.

We now replace $\sin \sqrt{gs}\, \tau$ in (6.7.13) by exponentials and carry out the integration on τ to obtain

$$(6.7.14) \quad \varphi(x, y; t) = -\frac{i\omega e^{i\omega t}}{\pi\varrho} \int_L e^{sy} \cos sx \left\{ \begin{array}{l} -\dfrac{1}{2\sqrt{gs}}\, \dfrac{e^{i(\sqrt{gs}-\omega)t}}{\sqrt{gs}-\omega} \\[2ex] -\dfrac{1}{2\sqrt{gs}}\, \dfrac{e^{-i(\sqrt{gs}+\omega)t}}{\sqrt{gs}+\omega} \\[2ex] +\dfrac{1}{gs-\omega^2} \end{array} \right\} ds.$$

We wish now to consider the three items in the bracket separately, and, as we see, two of them do indeed have a singularity at $s = \omega^2/g$ which is by-passed through our choice of the path L. The first two items are rather obviously the result of the initial conditions and hence could be expected to provide transients which die out as $t \to \infty$. This is in fact the case, as can be seen easily in the following way: That branch of $\sqrt{s}$ is taken which is positive on the positive real axis, and we operate always in the right half-plane. If, in addition, s is in the upper half-plane it follows that $i(\sqrt{gs} \pm \omega)$ has its real part negative (ω being real). Consider now the contribution furnished by the uppermost item in the square brackets. Since the exponential has a negative real part on the semi-circular portion of the path L it is clear that as $t \to +\infty$ this part of the path makes a contribution that tends to zero. The remaining portions of L, which lie on the real axis, are then readily seen to make contributions which die out like $1/t$: this can be seen easily by integration by parts, for example, or by application of known results about Fourier transforms. The middle item in the square brackets has no singularity on the real axis, so that the path L can be taken entirely on the real axis; thus, in accordance with the remarks just made concerning the similar situation for the first item, it is clear that this contribution also dies out like $1/t$. Thus for large t we obtain the following asymptotic representation for φ:

$$(6.7.15) \qquad \varphi(x, y; t) \simeq -\frac{i\omega}{\pi\varrho}\, e^{i\omega t} \int_L \frac{e^{sy} \cos sx}{gs - \omega^2}\, ds.$$

Actually, the right hand side is the solution of the steady state problem—as obtained, for example, in the paper of Lamb [L.2] and by a different method by us in section 4.3 (although in a different form) —when the condition at ∞ is the radiation condition stating that φ behaves like an out-going progressing wave. The steady state solution as obtained in section 4.3 actually was a little more awkward to obtain directly through use of the radiation condition than it was to obtain the solution (6.7.13) of the initial value problem. In particular, the asymptotic behavior of an integral representation had to be investigated in the former case also before the radiation condition could be used. Thus we have seen in this special case that the radiation condition can be replaced by boundedness conditions (in the space variables, that is) if one treats an appropriate initial value problem instead of the steady state problem.

Even though not strictly necessary—since (6.7.15) is known to furnish the desired steady state solution—it is perhaps of interest to show directly that the right hand side of (6.7.15) has the behavior one expects for an out-going progressing wave when $x \to +\infty$. The procedure is the same as that used in discussing (6.7.14): The factor $\cos sx$ is replaced by exponentials to obtain

$$(6.7.16) \quad \varphi(x, y; t) \simeq \frac{\omega}{\varrho}\, e^{i\omega t} \left[\frac{1}{2\pi i} \int_L \frac{e^{sy}\, e^{isx}}{gs - \omega^2}\, ds + \frac{1}{2\pi i} \int_L \frac{e^{sy}\, e^{-isx}}{gs - \omega^2}\, ds\right].$$

By the same argument as above one sees that the first integral makes a contribution that tends to zero as $x \to +\infty$. The second integral is treated by deforming the path L over the pole $s = \omega^2/g$ into a path M which consists of the positive real axis except for a semi-circle *in the lower half-plane*. The contribution of the second integral then consists of the residue at the pole plus the integral over the path M. But the contribution of the latter integral is, once more, seen to tend to zero as $x \to +\infty$ because of the factor e^{-isx}. Thus $\varphi(x, y; t)$ behaves for large x as follows:

$$(6.7.17) \qquad \varphi(x, y) \simeq -\frac{\omega}{g\varrho} e^{\frac{\omega^2}{g} y} e^{-i\left(\frac{\omega^2}{g} x - \omega t\right)},$$

and this does in fact represent a progressing wave in the positive x-direction which, in addition, has the wavelength $2\pi g/\omega^2$ appropriate to a progressing sine wave with the frequency ω in water of infinite depth.

6.8. Justification of the Method of Stationary Phase

In section 6.5 the method of stationary phase was used (and it will be used again later on) to obtain approximations of an asymptotic character for the solutions of a variety of problems when these solutions are given by means of integrals of the form

$$(6.8.1) \qquad I(k) = \int_a^b \psi(x) e^{ik\varphi(x)} \, dx,$$

and the object is to obtain an approximation valid when the real constant k is large. Since we make use of such approximate formulas in so many important cases, it seems worth while to give a mathematical justification of the method of stationary phase, following a procedure due to Poincaré. The presentation given here is based upon the presentation given by Copson [C.5].

Poincaré's proof requires the assumption that $\varphi(z)$ and $\psi(z)$ are regular analytic functions of the complex variable z in a domain containing the segment S: $a \le x \le b$ of the real axis in its interior. (In what follows, we assume a and b to be finite, but an extension to the case of infinite limits would not be difficult.) In addition $\varphi(z)$ is assumed to be real when z is real. These conditions are more restrictive than is necessary for the validity of the final result. For example, the function ψ might also depend on k, provided that $\psi(x, k)$ is not strongly oscillatory, or singular, for large values of k. The assumption of analyticity is also not indispensable. However, these generalizations would complicate both the formulation and proof of our theorem without changing their essentials; consequently we do not consider them here.

It will be shown that *the main contributions to $I(k)$ arise from the points of S near those values of x for which $\varphi'(x) = 0$* —that is, near the points of stationary phase. The term of lowest order in the asymptotic development of $I(k)$ with respect to k will then be obtained on the basis of this observation. Kelvin himself offered a heuristic argument (cf. sec. 5 above) indicating why such a procedure should yield the desired result.

Since $\varphi'(z)$ is regular in the domain containing S, it follows that its zeros are isolated. Hence S can be divided into a finite number of segments on which $\varphi(z)$ has either one stationary point or no stationary point. We shall show first that the contribution to $I(k)$ from a segment containing no stationary point is of order $1/k$. Next it will be shown that a segment containing any given point of stationary phase can be found such that the contribution to the integral furnished by the segment is of lower order than $1/k$, and a formula for this contribution will be derived. It turns out that this contribution of lowest order is independent of the length of the segment containing the point of stationary phase, provided only that the segment has been chosen short enough. Once these facts have been proved, it is clear that the lowest order contributions to the integral are to be found by adding the contributions arising at each of the points of stationary phase.

Suppose, then, that $\varphi(x)$ has no stationary point on a segment $c \leqq x \leqq d$ of S. We may write

$$I_1 = \int_c^d \psi(x) e^{ik\varphi(x)}\, dx = \int_c^d \frac{\psi(x)}{ik\varphi'(x)} \frac{d}{dx} \left(e^{ik\varphi(x)} \right)\, dx,$$

since $\varphi'(x) \neq 0$ in $c \leqq x \leqq d$ by hypothesis. Integration by parts then leads to the result

$$I_1 = \frac{\psi(d)}{ik\varphi'(d)} e^{ik\varphi(d)} - \frac{\psi(c)}{ik\varphi'(c)} e^{ik\varphi(c)} - \frac{1}{ik} \int_c^d e^{ik\varphi} \psi_1\, dx,$$

$$\text{with } \psi_1(x) = \frac{d}{dx} (\psi/\varphi'). \quad \text{Since } \left| \int_c^d e^{ik\varphi} \psi_1\, dx \right| \leqq \int_c^d |\psi_1|\, dx$$

because of the fact that $k\varphi(x)$ is real, it follows that the integral in the above expression is bounded. Thus I_1 is indeed of order $1/k$, as stated above. It might be noted that this argument really does not require the analyticity of φ and ψ, but only that the integrands be integrable and that integration by parts may be performed. Infinite limits for the integrals could also be permitted if $\varphi(x)$ and $\psi(x)$ behave appropriately at ∞.

Suppose now that $\varphi(x)$ has one stationary value at $x = \alpha$ in the segment $\alpha - \varepsilon_1 \leq x \leq \alpha + \varepsilon_1$, $\varepsilon_1 > 0$, i.e., $\varphi'(x)$ vanishes only at $x = \alpha$ in this interval. Suppose, in addition, that the second derivative $\varphi''(x)$ does not vanish at $x = \alpha$, and indeed is positive there: $\varphi''(\alpha) > 0$. (The case in which $\varphi''(\alpha)$ is negative and the more critical case in which $\varphi''(\alpha) = 0$ will be discussed later.) We shall show that a positive number $\varepsilon \leq \varepsilon_1$ exists such that

$$(6.8.2) \quad I_2(k) = \int_{\alpha - \varepsilon}^{\alpha + \varepsilon} \psi(x) e^{ik\varphi(x)}\, dx = \left(\frac{2\pi}{k\varphi''(\alpha)}\right)^{\frac{1}{2}} \psi(\alpha) e^{i\left(k\varphi(\alpha) + \frac{\pi}{4}\right)} + O\left(\frac{1}{k}\right).$$

In other words, we shall show that a fixed segment of length 2ε containing α exists such that its contribution to I is independent of ε and is of order $1/\sqrt{k}$, with an error of order $1/k$.

To prove these statements we begin by introducing new variables as follows:

$$(6.8.3) \qquad x = \alpha + u, \qquad \varphi(x) = \varphi(\alpha) + w(u).$$

Consider first the integral $I_2(k, \varepsilon_1)$:

$$(6.8.4) \quad I_2(k, \varepsilon_1) = e^{ik\varphi(\alpha)} \int_{-\varepsilon_1}^{\varepsilon_1} e^{ikw(u)}\, \psi(\alpha + u)\, du = e^{ik\varphi(\alpha)}\, J.$$

It is convenient to write the integral J as the sum of two terms:

$$(6.8.5) \qquad J = \int_{-\varepsilon_1}^{0} e^{ikw(u_1)}\, \psi(\alpha + u_1)\, du_1 + \int_{0}^{\varepsilon_1} e^{ikw(u_2)}\, \psi(\alpha + u_2)\, du_2$$
$$= J_1 + J_2.$$

Since $\varphi(x)$ has a minimum at $x = \alpha$, it follows that $w(u_1)$ is a positive monotonic function in the interval $-\varepsilon_1 \leq u_1 \leq 0$, and likewise $w(u_2)$ in the interval $0 \leq u_2 \leq \varepsilon_1$. Hence we may introduce a new integration variable t, which is furthermore real, in each of the integrals, defined as follows:

$$(6.8.6) \qquad \begin{cases} t^2 = w(u_1) \text{ in } -\varepsilon_1 \leq u_1 \leq 0, \text{ and} \\ t^2 = w(u_2) \text{ in } 0 \leq u_2 \leq \varepsilon_1. \end{cases}$$

In each interval t is taken as the positive square root. The integrals J_1 and J_2 become, as one readily sees:

$$(6.8.7) \qquad \begin{cases} J_1 = -\int_{0}^{t_1} e^{ikt^2}\, \psi(\alpha + u_1)\, \dfrac{du_1}{dt}\, dt, \\ J_2 = \int_{0}^{t_2} e^{ikt^2}\, \psi(\alpha + u_2)\, \dfrac{du_2}{dt}\, dt \end{cases}$$

with $t_1 = \sqrt{w(-\varepsilon_1)}$, and $t_2 = \sqrt{w(\varepsilon_1)}$. The functions $u_1(t)$, $u_2(t)$ are solutions of $w(u_i) = t^2$. For $w(u)$ we have the power series development

$$(6.8.8) \qquad w(u) = \varphi(\alpha + u) - \varphi(\alpha) = a_2 u^2 + a_3 u^3 + \dots$$

since $w(0) = w'(0) = 0$ (cf. (6.8.3)); in addition $2a_2 = \varphi''(\alpha) > 0$, by assumption. We suppose that this series converges in a circle which contains the entire interval $-\varepsilon_2 \leqq u \leqq \varepsilon_2$ in its interior, with $\varepsilon_2 < \varepsilon_1$. Since $t^2 = w(u_2)$ we may write

$$(6.8.9) \qquad \begin{cases} t = u_2 \sqrt{a_2 + a_3 u_2 + \dots} & \text{for } 0 \leqq u^2 \leqq \varepsilon_2 \quad \text{and} \\ t = -u_1 \sqrt{a_2 + a_3 u_1 + \dots} & \text{for } -\varepsilon_2 \leqq u_1 \leqq 0. \end{cases}$$

Since $a_2 \neq 0$ we may express the square roots as power series in u_i and then invert the series to obtain u_1 and u_2 as power series in t, as follows:

$$(6.8.10) \qquad \begin{cases} u_1 = -c_1 t + c_2 t^2 + \dots, \\ u_2 = c_1 t + c_2 t^2 + \dots. \end{cases}$$

with $c_1 = +\sqrt{2/\varphi''(\alpha)}$. Hence we may write

$$-\psi(\alpha + u_1)\frac{du_1}{dt} = c_1 \psi(\alpha) + t P_1(t),$$

$$\psi(\alpha + u_2)\frac{du_2}{dt} = c_1 \psi(\alpha) + t P_2(t),$$

in which $P_1(t)$ and $P_2(t)$ are convergent power series. It may be that these series do not converge up to the values t_1 and t_2 of the upper limits of the above integrals J_1 and J_2 in (6.8.7). In that case we simply assume the length of the segment is taken to be still less than $2\varepsilon_2$ so that the inversion of the series (6.8.8) is permissible and the series $P_1(t)$ and $P_2(t)$ converge up to appropriate values $\bar{t}_1$ and $\bar{t}_2$. It is clear that numbers $\bar{t}_1$ and $\bar{t}_2$ with these properties exist. The integrals J_1 and J_2 may now be written in the form*

$$(6.8.11) \qquad \begin{cases} J_1 = \displaystyle\int_0^{\bar{t}_1} e^{ikt^2}\{c_1\psi(\alpha) + t P_1(t)\}\, dt, \quad \text{and} \\ J_2 = c_1\psi(\alpha) \displaystyle\int_0^{\bar{t}_2} e^{ikt^2}\, dt + \int_0^{\bar{t}_2} e^{ikt^2} t P_2(t)\, dt = J_3 + J_4. \end{cases}$$

---

* The requirement of analyticity for φ and ψ is used to permit this simple introduction of t as variable of integration. However, the existence of a finite number of derivatives would clearly have sufficed.

We proceed to study the integrals J_3 and J_4. Upon introducing $\theta = kt^2$ as new variable in J_3 we obtain

$$J_3 = \frac{c_1\psi(\alpha)}{2\sqrt{k}} \int_0^{k\bar{t_2^2}} \frac{e^{i\theta}}{\sqrt{\theta}}\, d\theta.$$

But we may write

(6.8.12)
$$\int_0^{k\bar{t_2^2}} \frac{e^{i\theta}}{\sqrt{\theta}}\, d\theta = \int_0^\infty \frac{e^{i\theta}}{\sqrt{\theta}}\, d\theta - \int_{k\bar{t_2^2}}^\infty \frac{e^{i\theta}}{\sqrt{\theta}}\, d\theta$$

$$= \sqrt{\pi}\, e^{\pi i/4} - \int_{k\bar{t_2^2}}^\infty \frac{e^{i\theta}}{\sqrt{\theta}}\, d\theta,$$

by a known formula. The last integral can now be shown to be of order $1/\sqrt{k}$ by integration by parts, as follows:

$$\int_{k\bar{t_2^2}}^\infty \frac{e^{i\theta}}{\sqrt{\theta}}\, d\theta = \frac{1}{i}\left\{ \frac{e^{i\theta}}{\sqrt{\theta}}\bigg|_{k\bar{t_2^2}}^\infty + \frac{1}{2}\int_{k\bar{t_2^2}}^\infty \frac{e^{i\theta}}{\theta^{3/2}}\, d\theta \right\}.$$

The first contribution on the right hand side is obviously of order $1/\sqrt{k}$ since $\bar{t_2}$ is a fixed number; as for the second, we have

$$\left| \int_{k\bar{t_2^2}}^\infty \frac{e^{i\theta}}{\theta^{3/2}}\, d\theta \right| \leq \int_{k\bar{t_2^2}}^\infty \theta^{-3/2}\, d\theta = \frac{2}{\bar{t_2}\sqrt{k}}$$

and hence the second contribution is also of order $1/\sqrt{k}$. Thus for J_3 we have the result

(6.8.13)
$$J_3 = \tfrac{1}{2}c_1\psi(\alpha)\left(\frac{\pi}{k}\right)^{\frac{1}{2}} e^{\frac{1}{4}\pi i}\left[1 + O\left(\frac{1}{k^{1/2}}\right)\right].$$

The integral J_4 is first integrated by parts to obtain

$$J_4 = \int_0^{\bar{t_2}} e^{ikt^2} t P_2(t)\, dt$$

$$= \frac{1}{2ik}\left\{ e^{ik\bar{t_2^2}} P_2(\bar{t_2}) - P_2(0) - \int_0^{\bar{t_2}} e^{ikt^2} P_2'(t)\, dt \right\}$$

and hence

(6.8.14)
$$|J_4| \leq \frac{1}{2k}\left\{ |P_2(\bar{t_2})| + |P_2(0)| + \int_0^{\bar{t_2}} |P_2'(t)|\, dt \right\}$$

and the right hand side is thus of order $1/k$. The integral J_1 can obviously be treated in the same way as J_2 and with an exactly analogous result; consequently we have from (6.8.13), and (6.8.14) for the integral given in (6.8.4) the result

$$(6.8.15) \qquad I_2(k, \varepsilon_3) = I_2(k) = \psi(\alpha) \left(\frac{2\pi}{k\varphi''(\alpha)} \right)^{\frac{1}{2}} e^{i\left(k\varphi(\alpha) + \frac{\pi}{4}\right)} + O\left(\frac{1}{k}\right),$$

once ε_3 has been chosen small enough. One observes how it comes about that the lowest order term is independent of the values of l_1 and l_2, and hence of the length of the segment: the entire argument requires only that l_1 and l_2 be any fixed positive numbers since one needs only the fact that the products kl_1^2 and kl_2^2 grow large with k.

If $\varphi(x)$ had been assumed to have a maximum at $x = \alpha$, with $\varphi''(\alpha) < 0$, the only difference would be that $- k\varphi''(\alpha)$ and $- \pi/4$ would appear in the final formula instead of $+ k\varphi''(\alpha)$ and $+ \pi/4$. Consequently, in all cases in which $\varphi''(\alpha) \neq 0$ we have

$$(6.8.16) \qquad I_2(k) = \psi(\alpha) \left(\frac{2\pi}{k\,|\varphi''(\alpha)|} \right)^{\frac{1}{2}} e^{i\left(k\varphi(\alpha) \pm \frac{\pi}{4}\right)} + O\left(\frac{1}{k}\right)$$

and the sign of the term $i\pi/4$ should agree with the sign of $\varphi''(\alpha)$.

Finally, in case $\varphi''(\alpha) = 0$, but $\varphi'''(\alpha) \neq 0$ it is not difficult to derive the appropriate asymptotic formula for $I(k)$. In fact, the steps are nearly identical with those taken just now for the case $\varphi''(\alpha) \neq 0$. One introduces $x = \alpha + u$, $\varphi(x) = \varphi(\alpha) + w(u)$ as before and then makes use of power series in the variable t defined by $t^3 = w(u)$ in the same way as above. The result is, for ε sufficiently small:

$$(6.8.17) \qquad I_2(k) = \int_{\alpha-\varepsilon}^{\alpha+\varepsilon} e^{ik\varphi(x)}\, \psi(x)\, dx$$

$$= \psi(\alpha) \frac{\Gamma(\frac{1}{3})}{\sqrt{3}} \left(\frac{6}{k\,|\varphi'''(\alpha)|} \right)^{\frac{1}{3}} \cdot e^{ik\varphi(\alpha)} + O\left(\frac{1}{k^{2/3}}\right)$$

where $\Gamma(\frac{1}{3})$ refers to the gamma function. Hence the contribution arising from the stationary point is now of a different order of magnitude, i.e., of order $1/k^{1/3}$ instead of $1/k^{1/2}$. This fact is of significance in the case of the ship wave problem which will be treated later.

Naturally the lowest order terms in $I(k)$ consist of a sum of terms furnished by the contributions of all of the points of stationary phase in the interval S. It is important enough to bear repetition that if no such points exist, then $I(k)$ is in general of order $1/k$.

In case a stationary point falls at an end point $x = a$ or $x = b$ of the interval of integration, one sees readily that the contribution furnished by such a point to $I(k)$ is the same as that given above in case $\varphi'' \neq 0$ except that a factor $1/2$ would appear in the final result. On the other hand, if $\varphi'' = 0$ but $\varphi''' \neq 0$ at an end point, then the contribution differs in phase as well as in the numerical factor from the contribution given above in (6.8.17).

6.9. A time-dependent Green's function. Uniqueness of unsteady motions in unbounded domains when obstacles are present

In sec. 6.2 above the uniqueness of unsteady wave motions for water confined to a vessel of finite dimensions was proved. More general results have been obtained by Kotik [K.17], Kampé de Feriet and Kotik [K.1], and Finkelstein [F.3] with regard to such uniqueness questions. In the present section a rather general uniqueness theorem will be proved, following the methods of Finkelstein, who, unlike the other authors mentioned, obtains uniqueness theorems when obstacles are present in the water. The essential tool for this purpose is a time-dependent Green's function, which is in itself of interest and worth while discussing for its own sake quite apart from its use in deriving uniqueness theorems. With the aid of such a function, for example, all of the problems solved in the preceding sections can be solved once more in a different fashion, and still other and more complicated unsolved problems can be reduced to solving an integral equation, as we shall see.

We shall derive the time-dependent Green's function in question for the case of three-dimensional motion in water of infinite depth, although there would be no difficulty to obtain it in other cases as well. The Green's function G in question is required to be a harmonic function in the variables (x, y, z) with a singularity of appropriate character at a certain point (ξ, η, ζ) which is introduced at the time $t = \tau$ and maintained thereafter; thus G depends upon $\xi, \eta, \zeta; \tau$ and $x, y, z; t$: $G \equiv G(\xi, \eta, \zeta; \tau \mid x, y, z; t)$. In fact, G is the velocity potential which yields the solution of the following water wave problem: A certain disturbance is initiated at the point (ξ, η, ζ) at the time $t = \tau$. The pressure on the free surface of the water is assumed to be zero always, and at the time $t = \tau$ the water is assumed to have been at rest in its equilibrium position. Since G is a harmonic function in x, y, z it is reasonable to expect that the correct singularity to impose at the point (ξ, η, ζ) in order that it should have the properties one likes a Green's function to have is that it behaves there like $1/R$, with $R = \sqrt{(\xi - x)^2 + (\eta - y)^2 + (\zeta - z)^2}$. Thus G should satisfy the following conditions: It should be a solution of the Laplace equation

(6.9.1) $G_{xx} + G_{yy} + G_{zz} = 0$ for $-\infty < y < 0$, $t \geq \tau$,

satisfying the free surface condition

(6.9.2) $G_{tt} + gG_y = 0$, $y = 0$.

At ∞ we require G, G_t and their first derivatives to be uniformly bounded at any given time t. (Actually, they will be seen to tend to zero at ∞.) At the point ξ, η, ζ we require

(6.9.3) $G - \dfrac{1}{R}$ to be bounded.

As initial conditions at the time $t = \tau$ we have (cf. sec. 6.1)

(6.9.4) $G = G_t = 0$ for $t = \tau$, $y = 0$.

As we shall see later on, these conditions determine G uniquely.

We proceed to construct the function G explicitly. As a first step we set

(6.9.5) $G(\xi, \eta, \zeta; \tau \mid x, y, z; t) = A(\xi, \eta, \zeta \mid x, y, z) +$
$$B(\xi, \eta, \zeta; \tau \mid x, y, z; t)$$

with A defined by

(6.9.6) $A \equiv \dfrac{1}{R} - \dfrac{1}{R'}$ with $R' = \sqrt{(\xi - x)^2 + (\eta + y)^2 + (\zeta - z)^2}$.

Thus A contains the prescribed singularity, and we may require B to be regular. Since A is a harmonic function, it follows that B is harmonic; in addition, B satisfies the free surface condition

(6.9.7) $B_{tt} + gB_y = \dfrac{-2g\eta}{[(\xi - x)^2 + \eta^2 + (\zeta - z)^2]^{3/2}}$

$$= 2g \dfrac{\partial}{\partial \eta} \dfrac{1}{[(\xi - x)^2 + \eta^2 + (\zeta - z)^2]^{1/2}} \text{ at } y = 0,$$

as one can readily verify. To determine B from this and the other conditions arising from those imposed on G it would be possible to employ the Hankel transform in exactly the same way as the Fourier transform was used in preceding sections. However, it seems better in the present case to proceed directly by using the special, but well-known, Hankel transform for the function e^{-bs}/s (cf., for example, Sneddon [S.11], p. 528); this yields the formula

(6.9.8)
$$\frac{1}{\sqrt{a^2 + b^2}} = \int_0^\infty e^{-bs} J_0(as) \, ds,$$

valid for $b > 0$. By means of this formula the right hand side of (6.9.7) can be written in a different form to yield

(6.9.9) $\quad B_{tt} + gB_y = 2g \dfrac{\partial}{\partial \eta} \displaystyle\int_0^\infty e^{\eta s} J_0(sr) \, ds = 2g \int_0^\infty s e^{\eta s} J_0(sr) \, ds$

$$\text{at } y = 0$$

valid for $\eta < 0$ and with

(6.9.10)
$$r = \sqrt{(\xi - x)^2 + (\zeta - z)^2}.$$

Since B is a harmonic function in x, y, z, it would seem reasonable to seek it among functions of the form

(6.9.11)
$$B = \int_0^\infty s T(t, s) e^{(y+\eta)s} J_0(sr) \, ds,$$

which are harmonic functions. The free surface condition (6.9.9) will now be satisfied, as one can easily see, if $T(t)$ satisfies the differential equation

(6.9.12)
$$T_{tt} + gsT = 2g.$$

The function T is now uniquely determined from (6.9.12) and the initial conditions $T = T_t = 0$ for $t = \tau$ derived from (6.9.4); the result is

(6.9.13)
$$T(t, s) = 2 \frac{1 - \cos \sqrt{gs} \, (\tau - t)}{s}.$$

Thus we have for G the function

(6.9.14) $\quad G(\xi, \eta, \zeta; \tau \mid x, y, z; t) = \dfrac{1}{R} - \dfrac{1}{R'}$

$$+ 2 \int_0^\infty e^{s(y+\eta)} \, [1 - \cos \sqrt{gs} \, (\tau - t) \,] J_0(sr) \, ds,$$

and it clearly satisfies all of the conditions prescribed above, except possibly the conditions at ∞, which we shall presently investigate in some detail because of later requirements. Before doing so, however, we observe the important fact that G is symmetrical not only in the space variables ξ, η, ζ and x, y, z, but also in the time variables τ and t, i.e. that

(6.9.15) $G(\xi, \eta, \zeta; \tau \mid x, y, z; t) \equiv G(x, y, z; t \mid \xi, \eta, \zeta; \tau)$ and
$$\equiv G(\xi, \eta, \zeta; t \mid x, y, z; \tau).$$

We turn next to the discussion of the behavior of G at ∞. Consider first the function $A = 1/R - 1/R'$. This function evidently will behave at ∞ like a dipole; hence if σ represents distance from the origin it follows that A and its radial derivative A_σ behave as follows for large σ:

(6.9.16)
$$\begin{cases} A \sim 1/\sigma^2 \\ A_\sigma \sim 1/\sigma^3 . \end{cases}$$

On the free surface where $y = 0$ we have

(6.9.17)
$$\begin{cases} A = 0 & \text{for } y = 0, \\ A_y \sim 1/\sigma^3 & \text{for } y = 0 \text{ and large } \sigma. \end{cases}$$

To determine the behavior of B—i.e. of the integral in (6.9.14)—we expand $[1 - \cos \sqrt{gs}\,(\tau - t)]$ in a power series in $\tau - t$ and write

(6.9.18) $B = 2 \displaystyle\int_0^\infty e^{s(y+\eta)}\, J_0(sr) \left[\dfrac{gs(\tau - t)^2}{2!} - \dfrac{g^2 s^2 (\tau - t)^4}{4!} \cdots \right]\, ds.$

It is clearly legitimate to integrate term-wise for y negative. The formula (6.9.8) can be expressed in the form

(6.9.19)
$$\frac{1}{R'} = \int_0^\infty e^{s(y+\eta)}\, J_0(sr)\, ds,$$

and from it we obtain

(6.9.20) $\dfrac{\partial^n}{\partial y^n}\left[\dfrac{1}{R'} \right] = \dfrac{n!\, P_n(\mu)}{R'^{n+1}} = \displaystyle\int_0^\infty s^n\, e^{s(y+\eta)}\, J_0(sr)\, ds,$

with $\mu = \cos \theta$, by a well-known formula for spherical harmonics. It follows, since $P_n(\mu)$ are bounded functions, that the leading term in the asymptotic expansion of B arises from the first term in the square bracket. Hence the behavior of B is seen from (6.9.20) for the case $n = 1$ to be given by

(6.9.21) $B \sim 1/\sigma^2,$

for σ large and any fixed values of τ and t. The derivative B_y is seen, also from (6.9.20), to behave like $1/\sigma^3$ and the derivative B_r also can be seen to behave like $1/\sigma^3$; thus the radial derivative B_σ behaves in the same way and we have

(6.9.22) $B_\sigma \sim 1/\sigma^3, \qquad B_y \sim 1/\sigma^3.$

Summing up, we have for the Green's function G the following behavior at ∞:

$$(6.9.23) \qquad \begin{cases} G \sim 1/\sigma^2 \\ G_\sigma \sim 1/\sigma^3 \\ G_y \sim 1/\sigma^3. \end{cases}$$

All of these conditions hold uniformly for any fixed finite ranges in the values of τ and t.

We turn next to the consideration of a water wave problem of very general character, as follows. The space $y < 0$ is filled with water and in addition there are immersed surfaces S_i of finite dimensions having a prescribed motion (which, of course, must of necessity be a motion of small amplitude near to a rest position of equilibrium). The pressure on the free surface S_f is prescribed for all time, and the initial position and velocity of the particles on the free surface and the immersed surfaces are given at the time $t = 0$. At infinity the displacement and velocity of all particles are assumed to be bounded. The resulting motion can be described for all times $t > 0$ in terms of a velocity potential $\Phi(x, y, z; t)$ which satisfies conditions of the kind studied in the first section of this chapter; these conditions are:

$$(6.9.24) \qquad \nabla^2_{x, y, z} \Phi = 0$$

in the region R consisting of the half space $y < 0$ exterior to the immersed surfaces S_i. On the free surface the condition

$$(6.9.25) \qquad \Phi_{tt} + g\Phi_y = -\frac{1}{\varrho} p_t = P(x, 0, z; t), \qquad t > 0, \qquad y = 0$$

is prescribed, with p the given surface pressure (cf. (6.1.1) and (6.1.2)). At the equilibrium position of the immersed surfaces the condition

$$(6.9.26) \qquad \Phi_n = V \text{ on } S_i, \qquad t \geq 0,$$

with V the normal velocity of S_i, is prescribed. The initial position of S_i at $t = 0$ is, of course, assumed known, and for the initial conditions otherwise we know (cf. 6.1) that it suffices to prescribe Φ and Φ_t on the free surface at $t = 0$:

$$(6.9.27) \qquad \begin{cases} \Phi(x, 0, z; 0) = f_1(x, z) \\ \Phi_t(x, 0, z; 0) = f_2(x, z). \end{cases}$$

At ∞ we assume that Φ, Φ_t and their first derivatives are uniformly bounded.

We proceed now to set up a representation for the function Φ by using the Green's function obtained above. In case there are no immersed surfaces this representation furnishes an explicit solution of the problem, and in the other cases it leads to an integral equation for it. In all cases, however, a uniqueness theorem can be obtained.

To carry out this program we begin, in the usual fashion, by applying Green's formula to the Green's function G and to Φ_t (rather than Φ) in a sphere centered at the origin of radius a large enough to include the immersed surfaces and the singular point (ξ, η, ζ) of the Green's function minus a small sphere of radius ε centered at the singular point. Since G and Φ_t are both harmonic functions and G behaves like $1/R$ at the singular point, it follows by the usual arguments in potential theory that $\Phi_t(x, y, z; t)$ is obtained in the form of a surface integral, as follows:

$$(6.9.28) \qquad \Phi_t(x, y, z; t) = \frac{1}{4\pi} \iint\limits_{\xi, \eta, \zeta} (G\Phi_{tn} - \Phi_t G_n) \, dS.$$

The symmetry of G has been used at this point. The integration variables are ξ, η, ζ. Even though G depends on the difference $t - \tau$ the integral in (6.9.28) depends only on t; that is, only the singular part of the behavior of G matters in applying Green's formula, and the resulting expression for Φ_t depends only on the time at which Φ_t and Φ_{tn} are measured. The surface integral is taken over the boundary of the region just described (cf. Fig. 6.9.1), and n is the normal taken

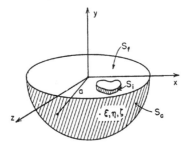

Fig.6.9.1. Domain for application of Green's formula

outward from the region. The boundary is composed of three different parts: the portion of the sphere S_a of radius a lying below the plane $y = 0$, the part S_f of the plane $y = 0$ cut out by the sphere S_a, and the immersed surfaces S_i (which might possibly cut out portions of the plane $y = 0$).

It is important to show first of all that the contribution to the surface integral provided by S_a tends to zero as $a \to \infty$, and that the integral over S_f exists as $a \to \infty$. The second part is readily shown: The integrand to be studied is $G\Phi_{ty} - \Phi_t G_y$. From the symmetry of G and (6.9.23) we see that the above integrand behaves like $1/a^2$ for large a since Φ_{ty} and Φ_t are assumed to be uniformly bounded at ∞; hence the integral converges uniformly in t and τ for any fixed ranges of these variables. To show that the integral of $G\Phi_{t\sigma} - \Phi_t G_\sigma$ over S_a tends to zero for $a \to \infty$ requires a lengthier argument. Consider first the term $\Phi_t G_\sigma$. Since G_σ behaves like $1/a^3$ for large a while Φ_t is bounded, it is clear that the integral of this term behaves like $1/a$ and hence tends to zero as $a \to \infty$. The integral over the remaining term is broken up into two parts, as follows:

(6.9.29)
$$\iint_{S_a} \Phi_{t\sigma} G \, dS = \int_0^{2\pi} \int_{(\pi/2)+\delta}^{\pi} \Phi_{t\sigma} G a^2 \sin\theta \, d\theta \, d\omega$$
$$+ \int_0^{2\pi} \int_{\pi/2}^{(\pi/2)+\delta} \Phi_{t\sigma} G a^2 \sin\theta \, d\theta \, d\omega.$$

The integrations are carried out in polar coordinates, and δ is a small angle (cf. Fig. 6.9.2); the second integral represents the contribution from a thin strip of the sphere S_a adjacent to the free surface. Since

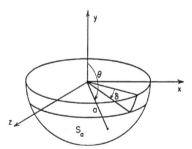

Fig. 6.9.2. The sphere S_a

$\Phi_{t\sigma}$ is bounded and G behaves like $1/a^2$ for large a, it is clear that the absolute value of the second contribution (i.e. that from the thin strip) can be made less than $\varepsilon/2$, say, if δ is chosen small enough. Once δ has been fixed, it can be seen that the contribution of the remaining part of S_a can also be made less than $\varepsilon/2$ in absolute value if a is taken large enough. If this is once shown it is then clear that the integral in question vanishes in the limit as $a \to \infty$. The proof of this fact is, however, not difficult: we need only observe that Φ_t is by assumption bounded at ∞ and it is a well-known fact* that $\Phi_{t\sigma}$ then tends to zero uniformly like $1/a$ along any ray from the origin which makes an angle $\geq \delta$ with the plane $y = 0$. Thus the integrand in the first term of (6.9.29) behaves like $1/a$ and it therefore can be made arbitrarily small by taking a sufficiently large. Thus for Φ_t we now have the representation

$$(6.9.30) \qquad \Phi_t(x, y, z; t) = \frac{1}{4\pi} \iint\limits_{\eta=0} (G\Phi_{t\eta} - \Phi_t G_\eta)\, d\xi\, d\zeta$$

$$+ \frac{1}{4\pi} \iint\limits_{S_t} (G\Phi_{tn} - \Phi_t G_n)\, dS,$$

in which it is, of course, understood that any parts of the plane $y = 0$ cut out by S_i are omitted in the first integral. The next step is to integrate both sides of (6.9.30) with respect to t from 0 to τ. The result is

$$(6.9.31) \qquad \Phi(x, y, z; \tau) - \Phi(x, y, z; 0)$$

$$= \frac{1}{4\pi} \iint\limits_{\eta=0} \left[\int_0^\tau (G\Phi_{t\eta} - \Phi_t G_\eta)\, dt\right] d\xi\, d\zeta + \frac{1}{4\pi} \iint\limits_{S_t} \left[\int_0^\tau (G\Phi_{tn} - \Phi_t G_n)\, dt\right] dS$$

$$= \frac{1}{4\pi} \iint\limits_{\eta=0} \left[\left.\left(G\Phi_\eta + \frac{1}{g}\Phi_t G_t\right)\right|_0^\tau - \int_0^\tau \left(\Phi_\eta G_t + \frac{1}{g}\Phi_{tt}G_t\right) dt \right] d\xi\, d\zeta + I$$

when $G_{tt} + gG_y = 0$ for $y = 0$ is used (cf. (6.9.2)) and I is intended as a symbol for the integral over S_t. We have $G = G_t = 0$ for $t = \tau$; while for $t = 0$ we have $\Phi_t = f_2$, and $\Phi_y|_{t=0}$ uniquely determined by f_1** from the conditions (6.9.27). In addition, we know that $\Phi_y + (1/g)\Phi_{tt} = (1/g)P$ for $t > 0$ from (6.9.25). It follows that (6.9.31) can be written in the form

---

* One way to prove it is to use the Poisson integral formula expressing Φ_t at any point in terms of its values on the surface of a sphere centered at the point in question. Differentiation of this formula yields for any first derivative of Φ_t an estimate of the form M/b where M depends only on the bound for Φ_t on the sphere and b is the radius of the sphere. Finally, since our domain for $\theta > (\pi/2)+\delta$ contains spheres of arbitrarily large radius at points arbitrarily far from the origin, the result we need follows.

** Since $\Phi(x, y, z; 0)$ is harmonic, it is uniquely determined by its boundary values on $y = 0$ and the boundedness conditions at ∞.

(6.9.32) $\Phi(x, y, z; \tau) - \Phi(x, y, z; 0)$

$$= -\frac{1}{4\pi} \iint_{\eta=0} \left[\left(G\Phi_\eta + \frac{1}{g} f_2 G_t\right)\Big|_{t=0} + \int_0^\tau \frac{1}{g} G_t P dt \right] d\xi \, d\zeta$$

$$+ \frac{1}{4\pi} \iint_{S_i} \left[\int_0^\tau G\Phi_{tn} dt \right] dS - \frac{1}{4\pi} \iint_{S_i} \left[\int_0^\tau \Phi_t G_n dt \right] dS.$$

We now see that if there are no immersed surfaces S_i an explicit solution $\Phi(x, y, z; \tau)$ is given at once by (6.9.32) in terms of the initial conditions, which fix $\Phi_y|_{t=0}$ and f_2, and the condition on the free surface pressure fixing P—in fact, our general argument shows that every solution having the required properties is representable in this form. Consequently, the uniqueness theorem is proved for these cases. In particular, the Green's function constructed above is therefore uniquely determined since its regular part, B, satisfies the conditions imposed above on Φ.

In case there are immersed surfaces present the equation (6.9.32) does not yield the solution Φ, but it does yield an integral equation for it in the following way (which is the standard way of obtaining an integral equation for a harmonic function satisfying various boundary conditions): One goes back to the derivation of (6.9.30), but considers that the singularity is at a point (x, y, z) of S_i. If S_i is sufficiently smooth (and we assume that it is) the equation (6.9.30) still holds, except that the factor $1/4\pi$ is replaced by $1/2\pi$, and Φ is then of course given only on S_i. The integration on t from 0 to τ is once more performed, and an equation analogous to (6.9.32) is obtained; it can be written in the form

(6.9.33) $\Phi(x, y, z; \tau) = F(x, y, z; \tau) - \dfrac{1}{2\pi} \iint_{S_i} \left[\int_0^\tau \Phi_t G_n dt \right] dS$

with F a known function obtained by adding together what corresponds to the first two integrals on the right hand side of (6.9.32). As we see, this is an integral equation for the determination of $\Phi(x, y, z; \tau)$ on S_i. If it were once solved, the value of Φ on S_i could be used in (6.9.32) to furnish the values of Φ everywhere.

We may make use of (6.9.32) to obtain our uniqueness theorem in the following fashion. Suppose there were two solutions Φ_1 and Φ_2. Set $\Phi = \Phi_1 - \Phi_2$. Then Φ satisfies all of the conditions imposed on Φ_1 and Φ_2 except that the nonhomogeneous boundary conditions and

initial conditions are now replaced by homogeneous conditions, i.e.
$f_2 = P = 0$; $f_1 = 0$ and hence $\Phi_y \big|_{t=0} = 0$ since $\Phi(x, y, z; 0)$ is a
harmonic function which vanishes for $y = 0$, and $\Phi_{tn} = 0$ since
$\Phi_n \equiv 0$ on S_i. Thus for Φ we would have the integral representation:

$$(6.9.34) \qquad \Phi(x, y, z; \tau) = -\frac{1}{4\pi} \iint_{S_i} \left[\int_0^\tau \Phi_t G_n \, dt \right] dS.$$

Since G_n behaves at ∞ like $1/\sigma^3$ (cf. (6.9.23)) and values of Φ_t on the
bounded surfaces S_i are alone in question, it follows that Φ also behaves like $1/\sigma^3$ at ∞ for any fixed τ since the surfaces S_i are bounded.
The derivatives of Φ could also be shown to die out at ∞ at least as
rapidly as $1/\sigma^3$ since the derivatives of G could be shown to have this
property — for example, by proceeding in the fashion used to obtain
(6.9.23).

As a consequence the following function of t (essentially the energy
integral) exists:*

$$(6.9.35) \quad E(t) = \frac{1}{2} \iiint_R [\Phi_x^2 + \Phi_y^2 + \Phi_z^2] \, dxdydz + \frac{1}{2g} \iint_{S_f} \Phi_t^2 \, dxdz.$$

Differentiation of both sides with respect to t yields

$$(6.9.36) \quad E'(t) = \iiint_R \left[(\Phi_t)_x \Phi_x + (\Phi_t)_y (\Phi)_y + (\Phi_t)_z (\Phi)_z \right] dxdydz$$

$$+ \frac{1}{g} \iint_{S_f} \Phi_t \Phi_{tt} \, dxdz$$

$$= \iint_B \Phi_t \Phi_n dS + \frac{1}{g} \iint_{S_f} \Phi_t \Phi_{tt} \, dxdz,$$

by application of Green's first formula, with $B = S_i + S_f$ the
boundary of R. But $\Phi_n = 0$ on S_i and $\Phi_n = \Phi_y = -(1/g)\Phi_{tt}$ on S_f.
It follows therefore that $E'(t) = 0$ and $E = \text{const}$. But $\Phi \equiv 0$ at
$t = 0$ and hence $E \equiv 0$ from (6.9.35). It follows that Φ_x, Φ_y, Φ_z are
identically zero, and Φ thus also vanishes identically. Hence $\Phi_1 \equiv \Phi_2$
and our uniqueness theorem is proved.

---

* It should perhaps be noted that the energy integral for the original motions
need not, and in general will not exist, since the velocity potential and its derivatives are required only to be bounded at ∞.

Waves on a Running Stream. Ship Waves

In this concluding section of Part II made up of Chapters 7, 8, and 9, we treat problems which involve small disturbances on a running stream with a free surface; that is, motions which take place in the neighborhood of a uniform flow, rather than in the neighborhood of the state of rest, as has been the case in all of the preceding chapters of Part II. In Chapter 7 the classical problems concerning *steady two-dimensional motions* in water of uniform (finite or infinite) depth are treated first. It is of considerable interest, however, to consider also *unsteady motions* (which seem to have been neglected hitherto) both because of their intrinsic interest and because such a study throws some light on various aspects of the problems concerning steady motions. In Chapter 8 the classical ship wave problem, in which the ship is idealized as a disturbance concentrated at a point on the surface of a running stream, is studied in considerable detail. In particular, a method of justifying the asymptotic treatment of the solution through the repeated use of the method of stationary phase is given, and the description of the character of the waves for both straight and curved courses is carried out at length. Finally, in Chapter 9 the problem of the motion of a ship of given hull shape is treated under very general conditions: the ship is assumed to be a rigid body having six degrees of freedom and to move in the water subject only to the propeller thrust, gravity, and the pressure of the water, while the motion of the water is not restricted in any way.

CHAPTER 7

Two-dimensional Waves on a Running Stream in Water of Uniform Depth

As indicated in Fig. 7.0.1 we consider waves created in a channel

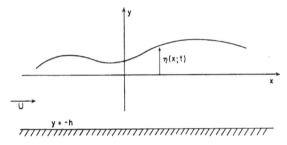

Fig. 7.0.1. Waves on a running stream

of constant depth h, when the stream has uniform velocity U in the positive x-direction in the undisturbed state. Such a uniform flow can readily be seen to fulfill the conditions derived in Chapter 1 for a potential flow with $y = 0$ as a free surface under constant pressure. We assume that the motions arising from disturbances created in the uniform stream have a velocity potential $\Phi(x, y; t)$, and we set

$$(7.0.1) \quad \Phi(x, y; t) = Ux + \varphi(x, y; t), \quad -\infty < x < \infty, \, -h < y < \eta.$$

Since $\Phi(x, y; t)$ is a harmonic function of x and y it follows that $\varphi(x, y; t)$ is also harmonic:

$$(7.0.2) \qquad \qquad \nabla^2 \varphi = 0.$$

The function $\varphi(x, y; t)$ is assumed to yield a small disturbance on the running stream, and we interpret this to mean that φ and its derivatives are all small quantities and that quadratic and higher order terms in them can be neglected in comparison with linear terms. We assume also that the vertical displacement $y = \eta(x; t)$ of the free

198

surface, as measured from the undisturbed position $y = 0$, is also a small quantity of the same order as $\varphi(x, y; t)$. Under these circumstances the dynamic free surface condition as given by Bernoulli's law (cf. (1.4.6)) and the kinematic free surface condition (cf.(1.4.5)) take the forms

$$(7.0.3) \quad \frac{p}{\varrho} + g\eta + \varphi_t + U\varphi_x + \frac{1}{2} U^2 = 0 \qquad \left.\begin{array}{c} \\ \\ \end{array}\right\} \quad \text{at } y = 0,$$

$$(7.0.4) \quad \eta_t + U\eta_x - \varphi_y = 0$$

when quadratic terms in φ and η are neglected and an unessential additive constant is ignored in (7.0.3).* At the same time, it is proper and consistent in such an approximation to satisfy the free surface conditions at $y = 0$ instead of at the displaced position $y = \eta$. (The reason for this is explained in Chapter 2—actually only for the case $U = 0$, but the discussion would be the same in the present case.) At the bottom $y = -h$ we have the condition

$$(7.0.5) \qquad \varphi_y = 0 \qquad \text{at } y = -h.$$

In case the channel has infinite depth we replace (7.0.5) with

$(7.0.5)'$ φ and its derivatives up to second order are bounded at $y = -\infty$.

In addition to the conditions (7.0.2) to (7.0.5) it is necessary also to postulate conditions at $x = \pm \infty$ and, unless the motion to be studied is a steady** motion with φ independent of t, it is also necessary to impose initial conditions at the time $t = 0$. The cases to be treated in the remainder of this chapter differ with respect to these various types of conditions, and we shall formulate them as they are needed.

7.1. Steady motions in water of infinite depth with $p = 0$ on the free surface

If the disturbance potential φ is independent of t, and if $p = 0$ on the free surface it follows that $\varphi(x, y)$ satisfies the conditions

---

* It is perhaps worth noting explicitly that it would be inappropriate to assume that U, the velocity of the stream, is a small quantity of the same order as η and φ: to do so would lead to the elimination of the terms in U and the resulting theory would not differ from that of the preceding chapters.

** In this chapter the term "steady motion" is used in the customary way to describe a flow which is the same at each point in space for all times. In the preceding chapters we have sometimes used this term (in conformity with established custom in the literature dealing with wave propagation) in a different sense.

$$(7.1.1) \qquad \nabla^2 \varphi = 0, \qquad -\infty < y \leqq 0,$$

$$(7.1.2) \qquad \varphi_y + \frac{U^2}{g} \varphi_{xx} = 0, \qquad y = 0.$$

In addition, we require that

(7.1.3) φ and its derivatives up to second order are bounded at ∞,

though this condition is more restrictive than is necessary. The second of these conditions was obtained from (7.0.3) and (7.0.4) by differentiating (7.0.3) and eliminating η.

It is interesting to find *all* functions $\varphi(x, y)$ satisfying these conditions, and it is easy to do so following the same arguments as were used in Chapter 3.1. Using (7.1.1) we may re-write (7.1.2) in the form

$$(7.1.4) \qquad \varphi_y - \frac{U^2}{g} \varphi_{yy} = 0, \qquad y = 0.$$

(This of course makes use of the fact that φ is harmonic for $y = 0$, which we assume to be true. One could easily show, in fact, that the free surface condition (7.1.2) permits an analytic continuation of φ over $y = 0$, so that φ is actually harmonic in a domain including $y = 0$ in its interior.) We observe that (7.1.4) is the same condition on φ_y as was imposed on the function called φ in Chapter 3, and we proceed as we did there by introducing a harmonic function $\psi(x, y)$ through

$$(7.1.5) \qquad \psi = \varphi_y - \frac{U^2}{g} \varphi_{yy} \quad , \quad y \leqq 0.$$

This function vanishes on $y = 0$, and can therefore be continued analytically by reflection into the upper half plane. Since φ and its derivatives were assumed to be bounded in the lower half plane, it follows that ψ is bounded in the entire plane and hence by Liouville's theorem it is a constant; hence ψ vanishes identically since $\psi = 0$ for $y = 0$. Thus we have for φ_y a differential equation given by (7.1.5) with $\psi \equiv 0$, and it has as its only solutions the functions

$$(7.1.6) \qquad \varphi_y = c(x)e^{\frac{g}{U^2}y} \ .$$

Since φ_y is also a harmonic function, it follows that $c(x)$ is a solution of the differential equation

$$(7.1.7) \qquad \frac{d^2 c}{dx^2} + \left(\frac{g}{U^2}\right)^2 c = 0.$$

Hence φ is given by

(7.1.8) $\varphi(x, y) = A e^{\frac{g}{U^2} y} \cos \left(\frac{g}{U^2} x + \alpha \right) + c_1(x)$

with A and α constants and $c_1(x)$ an arbitrary function of x. By making use of (7.1.2), however, one finds that $\frac{d^2 c_1}{dx^2} = 0$, and hence that $c_1 \equiv$ const. since φ is bounded at ∞. There is no loss of generality in taking $c_1 = 0$. The only solutions of our problem are therefore given by

(7.1.9) $\varphi(x, y) = A e^{\frac{g}{U^2} y} \cos \left(\frac{gx}{U^2} + \alpha \right).$

Thus the only steady motions satisfying our conditions, aside from a uniform flow, are periodic in x with the fixed wave length λ given by

(7.1.10) $\lambda = 2\pi \frac{U^2}{g}.$

The amplitude and phase of the motions are arbitrary. If we were to observe these waves from a system of coordinates moving in the x-direction with the constant velocity U, we would see a train of progressing waves given by

$$\varphi = A e^{my} \cos m \, (x + Ut)$$

with

$$m = \frac{g}{U^2} = \frac{2\pi}{\lambda}.$$

These waves are identical with those already studied in Chapter 3 (cf. sec. 3.2). The phase speed of these waves would of course be the velocity U and the wave length λ would, as it should, satisfy the relation (3.2.8) for waves having this propagation speed. In other words, the only waves we find are identical (when observed from a coordinate system moving with velocity U) with the progressing waves that are simple harmonic in the time and which have such a wave length that they would travel at velocity U in still water.

7.2. Steady motions in water of infinite depth with a disturbing pressure on the free surface

The same hypotheses are made as in the previous section, except that we assume the pressure on the free surface to be a function

$p_0(x) \not\equiv 0$ over the segment $-a \leq x \leq a$ and zero otherwise, as indicated in Fig. 7.2.1. The free surface condition, as obtained from (7.0.8)

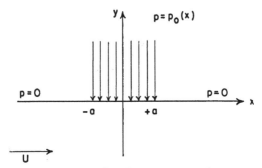

Fig. 7.2.1. Pressure disturbance on a running stream

and (7.0.4) by eliminating η and assuming η and φ to be independent of t, is now given by

$$(7.2.1) \qquad \varphi_{xx} + \frac{g}{U^2}\varphi_y = -\frac{p_x}{U\varrho}, \qquad \text{on } y = 0,$$

as one readily verifies. We prescribe in addition that φ and its first two derivatives are bounded at ∞.

The solutions φ of our problems are conveniently derived by introducing the analytic function $f(z)$ of the complex variable $z = x + iy$ whose real part is φ:

$$(7.2.2) \qquad f(z) = \varphi(x, y) + i\psi(x, y).$$

Since $\varphi_y = -\psi_x$, the condition (7.2.1) can be put in the form

$$(7.2.3) \qquad \varphi_x - \frac{g}{U^2}\psi = -\frac{p}{U\varrho} + \text{const.}, \qquad \text{on } y = 0,$$

and the constant can be taken as zero without loss of generality, since adding a constant to p can not affect the motion.

We consider now only the case in which the surface pressure p is a constant $p = p_0$ over the segment $|x| \leq a$, and zero otherwise. Since this surface pressure is discontinuous at $x = \pm a$, it is necessary to admit a singularity at these points; we shall see that a unique solution of our problem is obtained if we require that φ is bounded at these points while φ_x and φ_y behave like $1/r^{1-\varepsilon}$, $\varepsilon > 0$, with r the distance from the points $x = \pm a$ on the free surface. (This singularity is weaker than the logarithmic singularity of φ appropriate at a source or sink.)

In terms of $f(z)$, the free surface condition (7.2.3) clearly can be put in the form

$$(7.2.4) \quad \mathscr{I}m \left(if_z - \frac{g}{U^2} \right) f = \begin{cases} -\dfrac{p_0}{U\varrho} = \text{const.}, \ |x| \leqq a \\ 0 \qquad\qquad\quad , \ |x| > a \end{cases} \quad \text{for } \mathscr{I}m \, z = 0.$$

The device of applying the boundary condition in this form seems to have been used first by Keldysh [K.21]. We now introduce the analytic function $F(z)$ defined in the lower half plane by the equation

$$(7.2.5) \qquad\qquad F(z) = if_z - \frac{g}{U^2} f.$$

This function has the following properties: 1) Its imaginary part is prescribed on the real axis. 2) The first derivatives of its imaginary part are bounded at ∞, since the first two derivatives of φ are assumed to have this property and hence f_{zz} and f_z are bounded in view of the Cauchy-Riemann equations. 3) Near $z = \pm a$ its imaginary part behaves like $1/|z \mp a|^{1-\varepsilon}$, $\varepsilon > 0$, as one readily sees. It is now easy to show that $F(z)$ is uniquely determined,* within an additive real constant, as follows: Let $G = F_1 - F_2$ be the difference of two functions satisfying these three conditions. $\mathscr{I}m \, G$ then vanishes on the entire real axis, except possibly at the points $(\pm a, 0)$, and G can therefore be continued as a single-valued function into the whole plane except at the points $(\pm a, 0)$. However, the singularity prescribed at the points $(\pm a, 0)$ is weaker than that of a pole of first order, and hence the singularities at these points are removable. Since the first derivatives of $\mathscr{I}m \, G$ are bounded at ∞, it follows from the Cauchy-Riemann equations that G_z is bounded at ∞. Hence G_z is constant, by Liouville's theorem, and G is the linear function: $G = cz + d$. Since $\mathscr{I}m \, G = 0$ on the real axis, it follows that c and d are real constants. However, a term of the form $cz + d$ on the left hand side of (7.2.5) leads to a term of the form $\alpha z + \beta$, with $-\dfrac{g}{U^2} \alpha = c$, in the solution of this

---

* In Chapter 4, the function $F(z)$ given by (4.3.10) had a real part which satisfied identical conditions except that the condition 2) is slightly more restrictive in the present case.

equation for $f(z)$, and since $f(z)$ is assumed to be bounded at ∞. it follows that $c = 0$.

We have here the identical situation that has been dealt with in sec. 3 of Chapter 4, except that it was the real part of the function $F(z)$, rather than the imaginary part, that was prescribed on the real axis, and we can take over for our present purposes a number of the results obtained there. The function $F(z)$, now known to be uniquely determined within an additive real constant, is given by

$$(7.2.6) \qquad\qquad F(z) = \frac{p_0}{U\varrho\pi} \log \frac{z - a}{z + a},$$

which differs from $F(z)$ as given by (4.3.12) essentially only in the factor i—as it should. In any case, one can readily verify that $F(z)$ satisfies the conditions imposed above. We take that branch of the logarithm that is real for z real and $|z| > a$, and specify a branch cut starting at $z = -a$ and going to ∞ along the positive real axis. The equation (7.2.5) is now an ordinary differential equation for the function $f(z)$ which we are seeking.

The differential equation (7.2.5) has, of course, many solutions, and this means that the free surface condition and the boundedness conditions at ∞ and at the points $(\pm\,a, 0)$ are not sufficient to ensure that a unique solution exists. In fact, it is clear that the non-vanishing solution of the homogeneous problem found in the preceding section could always be added to the solution of the problem formulated up to now. A condition at ∞ is needed similar to the radiation condition imposed in the analogous circumstances in Chapter 4. In the present case, the solution can be made unique by requiring that the disturbance created by the pressure over the segment $|x| \leqq a$ should die out on the *upstream* side of the channel, i.e. at $x = -\infty$. The only justification for such an assumption—aside from the fact that it makes the solution unique—is based on the observation that one never sees anything else in nature.* In sec. 7.4 we shall give a more satisfactory discussion of this point which is based on studying the unsteady flow that arises when the motion is created by a disturbance initiated at the time $t = 0$, and the steady state is obtained in the limit as $t \to \infty$. In this formulation, the condition that the motion

---

* Lamb [L.3], p. 399, makes use, once more, of the device of introducing dissipative forces of a very artificial character which then lead to a steady state problem with a unique solution when only boundedness conditions are prescribed at ∞.

should die out on the upstream side is not imposed; instead, it turns out to be satisfied automatically.

A solution of the differential equation (7.2.5) (in dimensionless form) has been obtained in Chapter 4 (cf. (4.3.13)) which has exactly the properties desired in the present case; it is:

$$(7.2.7) \quad f(z) = -\frac{p_0 i}{U \varrho \pi} e^{-\frac{ig}{U^2} z} \int_{+i\infty}^{z} e^{\frac{ig}{U^2} t} \log \frac{t-a}{t+a} \, dt, \qquad \mathscr{I}m \, z \lessgtr 0.$$

The path of integration (cf. Fig. 4.3.1) comes from $i\infty$ along the positive imaginary axis and encircles the origin in such a way as to leave it and the point $(-a, 0)$ to the left. That (7.2.7) yields a solution of (7.2.5) is easily checked. One can also verify easily that $\varphi = \mathscr{R}e \, f(z)$ satisfied all of the boundary and regularity conditions, except perhaps the condition at ∞ on the upstream side. In Chapter 4, however, it was found (cf. (4.3.15)) that $f(z)$ behaves at ∞ as follows:

$$(7.2.8) \quad f(z) = \begin{cases} O\left(\dfrac{1}{z}\right) & \text{for } \mathscr{R}e \, z < 0, \\[2ex] -\dfrac{4p_0 U}{g\varrho} \sin \dfrac{ga}{U^2} \cdot e^{-\frac{ig}{U^2} z} + O\left(\dfrac{1}{z}\right) & \text{for } \mathscr{R}e \, z > 0. \end{cases}$$

Thus $f(z)$ dies out as $x \to -\infty$, but there are in general waves of nonzero amplitude far downstream, i.e. at $x = +\infty$. The uniquely-determined harmonic function $\varphi = \mathscr{R}e \, f(z)$ is now seen to satisfy all conditions that were imposed.

The waves at $x = +\infty$ are identical (within a term of order $1/z$) with the steady waves that we have found in the preceding section to be possible when the stream is subject to no disturbance (cf. (7.1.9)), and the wave far downstream has the wave length $\lambda = 2\pi U^2/g$. However, we observe the curious and interesting fact (pointed out by Lamb [L.3], p. 404) that this wave may also vanish: clearly if $ga/U^2 = n\pi$, $n = 1, 2, \ldots$, $\varphi = \mathscr{R}e \, f(z)$ vanishes downstream as well as upstream, and this occurs whenever $2a/\lambda$ is an integer, i.e. whenever the length of the segment over which the disturbing pressure is applied is an integral multiple of the wave length of a steady wave in water of velocity U (with no disturbance anywhere). This in turn gives rise to the observation that there exist rigid bodies of such a shape that they create only a local disturbance when immersed in a running stream: one need only calculate the shape of the free surface—which is, of course, a streamline—for $ga/U^2 = n\pi$, take a rigid body having the shape of a segment of this surface and put it into the water. (Involved here is, as one sees, a uniqueness theorem for problems in which the shape of the upper surface of the liquid, rather than the pressure, is prescribed over a segment, but such a theorem could be proved along the lines of the uniqueness proof of the analogous theorem for simple harmonic waves given by F. John [J.5].) This fact has an interesting physical consequence, i.e., that such bodies are not subject to any wave resistance (by which we mean that the resultant of the pressure forces on the body has no horizontal component) while in general a resistance would be felt. This can be seen as follows: Observe the motion from a coordinate system moving with velocity U in the x-direction. All forces remain the same relative to this system, but the wave at $+ \infty$ would now be a progressing wave simple harmonic in the time and having the propagation speed $- U$, while at $- \infty$ the wave amplitude is zero. Thus if we consider two vertical planes extending from the free surface down into the water, one far upstream, the other far downstream we know from the discussion in Chapter 3.3 that there is a net flow of energy into the water through these planes since energy streams in at the right, but no energy streams out at the left since the wave amplitude at the left is zero. Consequently, work must be done on the water by the disturbance pressure and this work is done at the rate $RU = F$, where R represents the horizontal resistance and F the net energy flux into the water through two planes containing the disturbing body between them. Thus if $F = 0$—which is the case if the wave amplitude dies out downstream as well as upstream— then $R = 0$. This result might have practical applications. For example, pontoon bridges lead to motions which are approximately two-dimensional, and hence it might pay to shape the bottoms of the pontoons in such a way as to decrease the wave resistance and hence the required strength of the moorings. However, such a design would yield an optimum result, as we have seen, only at a definite velocity of the stream; in addition, the wave resistance is probably small compared with the resistance due to friction, etc., except in a stream flowing with high velocity.

We conclude this section by giving the solution of the problem of determining the waves created in a stream when the disturbance is concentrated at a point, i.e. in the case in which the length $2a$ of the segment over which the pressure p_0 is applied tends to zero but $\lim_{a \to 0} 2p_0 a = P_0$. The desired solution is obtained at once from (7.2.7); it is:

$$(7.2.9) \qquad f(z) = \frac{P_0 i}{U \varrho \pi} e^{-\frac{ig}{U^2}z} \int_{i\infty}^{z} \frac{1}{t} e^{\frac{ig}{U^2}t} dt.$$

This solution behaves like $1/z$ far upstream and like $(-2P_0/U\varrho)$ exp $\{-igz/U^2\}$ far downstream. Note that the amplitude downstream does not vanish for any special values of U in this case. It is perhaps also of interest to observe that $f(z)$ behaves near the origin like $i \log z$, and hence the singularity at the point of disturbance has the character of a vortex point; we recall that the singularity in the analogous case of the waves created by an oscillatory point source that were studied in Chapter 4 had the character of a source point, since $f(z)$ behaved like $\log z$ rather than like $i \log z$ (cf. 4.3.28)), with a strength factor oscillatory in the time. When one thinks of the physical circumstances in these two different cases one sees that the present result fits the physical intuition.

7.3. Steady waves in water of constant finite depth

In water of constant finite depth the circumstances are more complicated, and in several respects more interesting, than in water of infinite depth. This is already indicated in the simplest case, in which the free surface pressure is assumed to be everywhere zero and the motion is assumed to be steady. In this case we seek a function $\varphi(x, y)$ satisfying the conditions (7.0.2) to (7.0.5), with φ_t and η_t both identically zero. The boundary conditions are thus

$$(7.3.1) \qquad \varphi_y + \frac{U^2}{g} \varphi_{xx} = 0, \qquad y = 0,$$

and

$$(7.3.2) \qquad \varphi_y = 0, \qquad y = -h.$$

A harmonic function which satisfies these conditions is given by:

$$(7.3.3) \qquad \varphi(x, y) = A \cosh m(y + h) \cos (mx + \alpha)$$

with A and α arbitrary constants, and m a root of the equation

$$(7.3.4) \qquad \frac{U^2}{gh} = \frac{\tanh mh}{mh}.$$

The condition (7.3.4) ensures that the boundary condition on the free surface is satisfied, as one can easily verify. It is very important for the discussion in this and the following section to study the roots of the equation (7.3.4). The curves $\zeta = \tanh \xi$ and $\zeta = (U^2/gh)\,\xi$ are plotted in Fig. (7.3.1). The roots of (7.3.4) are of course furnished by the intersections $\xi = mh$ of these curves. One observes: 1) $m = 0$ is always a root; 2) there are two real roots different from zero if $U^2/gh < 1$;

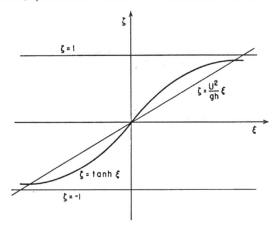

Fig. 7.3.1. Roots of the transcendental equation ($U^2/gh < 1$)

3) there are no real roots other than zero if $U^2/gh \geq 1$; 4) if $U^2/gh = 1$ the function $U^2 m - g \tanh mh$ vanishes at $m = 0$ like m^3; 5) since $\tan i\xi = i \tanh \xi$, it follows that (7.3.4) has infinitely many pure imaginary roots no matter what value is assigned to U^2/gh.

On the basis of this discussion of the roots of (7.3.4) we therefore expect that no motions other than the steady flow with no surface disturbance (for which $\varphi \equiv$ const.) will exist unless $U^2/gh < 1$. These waves are then seen to have the wave length appropriate for simple harmonic waves of propagation speed $c = U$ in water of depth h, as can be seen from (3.2.1), (3.2.2), and (3.2.3). It is possible to give a rigorous proof of this uniqueness theorem — which holds when no conditions at ∞ other than boundedness conditions are imposed — by making use of an appropriate Green's function, or by making use of the method devised by Weinstein [W.7] for simple harmonic waves in water of finite depth, but we will not do so here.

More interesting problems arise when we suppose that steady waves are created by disturbances on the free surface, or perhaps also on the bottom. Mathematically this means that a nonhomogeneous boundary condition would replace one, or perhaps both, of the homogeneous boundary conditions (7.3.1) and (7.3.2). In addition, as we infer from the discussion of the preceding section, it is also necessary in general to prescribe a condition of "radiation" type at ∞ in addition to boundedness conditions, and an appropriate such condition is that the disturbance should die out upstream. In the present problem, however, the additional parameter furnished by the depth of the water leads to some peculiarities that are conditioned in part by the difference in behavior of the solutions of the homogeneous problem in their dependence on the parameter U^2/gh: Since the only solution of the homogeneous problem in the case $U^2/gh \geq 1$ is $\varphi \equiv 0$, one expects that the solution of the nonhomogeneous problem will be uniquely determined in this case without the necessity of prescribing a radiation condition at ∞. However, if $U^2/gh < 1$ it is clear that the nonhomogeneous problem can not have a unique solution unless a condition — such as that requiring the disturbance to die out upstream — is imposed that will rule out the otherwise possible addition of the non-vanishing solution of the homogeneous problem. These cases have been worked out (cf. Lamb [L.3], p. 407) with the expected results, as outlined above, for $U^2/gh > 1$ and $U^2/gh < 1$, but the known representations of these solutions for the steady state make the wave amplitudes large for $U^2/gh = 1$ and $|x|$ large.

We shall not solve these steady state problems directly here because the peculiarities — not to say obscurities — indicated above can all be clarified and understood by re-casting the formulation of the problem in a way that has already been employed in the previous chapter (cf. sec. 6.7)). The basic idea (cf. Stoker [S.22]) is to abandon the formulation of the problem in terms of a *steady motion* in favor of a formulation involving appropriate initial conditions at the time $t = 0$, and afterwards to make a passage to the limit in the solutions for the unsteady motion by allowing the time to tend to ∞. As was indicated in sec. 6.7, the advantage of such a procedure is that the initial value problem, being the natural dynamical problem in Newtonian mechanics (while the steady state is an artificial problem), has a unique solution when no conditions other than boundedness conditions are imposed at ∞. If a steady state exists at all, it should then result upon letting $t \to \infty$, and the limit state would then automatically have those properties at ∞ which satisfy what one calls radiation conditions, and which one has to guess at if the steady state problem is taken as the starting point of the investigation.

We shall proceed along these lines in the next section in attacking the problem of the waves created in a stream of uniform depth when a disturbance is created in the undisturbed uniform stream at the time $t = 0$. The subsequent unsteady motion will be determined when only boundedness conditions are imposed at ∞. It will then be seen that the behavior of the solutions as $t \to \infty$ is indeed as indicated above, i.e. the waves die out at infinity both upstream and downstream when $U^2/gh > 1$, that they die out upstream but not downstream when $U^2/gh < 1$. One might be inclined to say: "Well, what of it, since one guessed the correct condition on the upstream side anyway?" However, we now get a further insight, which we did not possess before, i.e. that for $U^2/gh = 1$ there just simply is no steady state when $t \to \infty$ although a uniquely determined unsteady motion exists for every given value of the time t. In fact it will be shown that the disturbance potential becomes infinite like $t^{2/3}$ at all points of the fluid when $t \to \infty$ and $U^2/gh = 1$, and that the velocity also becomes infinite everywhere when $t \to \infty$.

7.4. Unsteady waves created by a disturbance on the surface of a running stream

The boundary conditions on the disturbance potential $\varphi(x, y; t)$ at the free surface (cf. Fig. 7.0.1 and equations (7.0.3) and (7.0.4)) are

$$(7.4.1) \qquad \frac{p}{\varrho} + g\eta + \varphi_t + U\varphi_x + \frac{U^2}{2} = 0,$$

$$(7.4.2) \qquad \eta_t + U\eta_x - \varphi_y = 0,$$

to be satisfied at $y = 0$ for all times $t > 0$. The quantity $p = p(x; t)$ is the pressure prescribed on the free surface. At the bottom $y = -h$ we have, of course, the condition

$$(7.4.3) \qquad \varphi_y = 0, \qquad t \geqq 0.$$

At the initial instant $t = 0$ we suppose the flow to be the undisturbed uniform flow, and hence we prescribe the initial conditions:

$$(7.4.4) \qquad \varphi(x, y; 0) = \eta(x; 0) = p(x; 0) = 0.$$

From (7.4.1), which we assume to hold at $t = 0$, we thus have the condition

$$(7.4.5) \qquad \varphi_t(x, y; 0) = 0.$$

Finally, we prescribe the surface pressure p for $t > 0$:

$$(7.4.6) \qquad p = p(x), \qquad t > 0.$$

(The surface pressure is thus constant in time.) At ∞ we make no assumptions other than bouhdedness assumptions. We shall not formulate these boundedness conditions explicitly: instead, they are used implicitly in what follows because of the fact that Fourier transforms in x for $-\infty < x < \infty$ are applied to φ and p and their derivatives. Of course, this means that these quantities must not only be bounded but also must tend to zero at ∞, and this seems reasonable since the initial conditions leave the water undisturbed at ∞.

We have, therefore, the problem of finding the surface elevation $\eta(x; t)$ and the velocity potential $\varphi(x, y; t)$ in the strip $-h \leq y \leq 0$, $-\infty < x < \infty$, which satisfy the conditions (7.4.1) to (7.4.6). We begin the solution of our problem by eliminating the surface elevation η from the first two boundary conditions to obtain:

$$(7.4.7) \quad \varphi_{tt} + U^2 \varphi_{xx} + 2U\varphi_{xt} + g\varphi_y = -\frac{U}{\varrho} p_x, \quad \text{at } y = 0.$$

The Fourier transform with respect to x is now applied to $\varphi_{xx} + \varphi_{yy} = 0$ to yield (cf. sec. 6.3):

$$(7.4.8) \qquad\qquad \bar{\varphi}_{yy} - s^2 \bar{\varphi} = 0,$$

where the bar over $\bar{\varphi}$ refers to the transform $\bar{\varphi} = \bar{\varphi}(s, y; t)$ of φ. From (7.4.3) we have $\bar{\varphi}_y = 0$ for $y = -h$; hence $\bar{\varphi}$, in view of (7.4.8) must be of the form

$$(7.4.9) \qquad\qquad \bar{\varphi}(s, y; t) = A(s; t) \cosh s(y + h),$$

with $A(s; t)$ a function to be determined. The transform is next applied to (7.4.7) with the result:

$$(7.4.10) \quad \bar{\varphi}_{tt} + 2isU\bar{\varphi}_t + g\bar{\varphi}_y - U^2 s^2 \bar{\varphi} = -\frac{isU}{\varrho} \bar{p}, \quad \text{at } y = 0,$$

and this yields, from (7.4.9) for $y = 0$, the differential equation

$$(7.4.11) \quad A_{tt} + 2isUA_t + [gs \tanh sh - s^2 U^2]A = -\frac{isU\bar{p}}{\varrho \cosh sh}.$$

Here $\bar{p}(s)$ is of course the transform of $p(x)$. As initial conditions at $t = 0$ for $A(s; t)$ we have from (7.4.4) and (7.4.5) the conditions (again in conjunction with (7.4.9)):

$$(7.4.12) \qquad\qquad A(s; 0) = A_t(s; 0) = 0.$$

The function $A(s; t)$ is then easily found; it is

$$(7.4.13) \quad A(s;t) = \frac{isU\bar{p}}{\varrho \cosh sh} \cdot \left\{ \begin{array}{l} \dfrac{1}{s^2 U^2 - gs \tanh sh} \\[2ex] + \dfrac{1}{2\sqrt{gs \tanh sh}} \cdot \dfrac{e^{-it\,(sU + \sqrt{gs \tanh sh})}}{sU + \sqrt{gs \tanh sh}} \\[2ex] - \dfrac{1}{2\sqrt{gs \tanh sh}} \cdot \dfrac{e^{-it\,(sU - \sqrt{gs \tanh sh})}}{sU - \sqrt{gs \tanh sh}} \end{array} \right\}$$

The solution $\varphi(x, y; t)$ of our problem is of course now obtained by inverting $\bar{\varphi}(s, y; t)$:

$$(7.4.14) \quad \varphi(x, y; t) = \frac{1}{\sqrt{2\pi}} \int_{-\infty}^{\infty} A(s;t) \cosh s(y + h)\, e^{isx}\, ds.$$

The path of integration is the real axis. One finds easily that the integrand behaves for large s like $e^{|s|y}/s$, since the denominators of the terms in the square brackets in (7.4.13) behave like s^2, the ratio $\cosh s(y + h)/\cosh sh$ behaves like $e^{|s|y}$ for large s, and $\bar{p}(s)$ tends to zero at ∞ in general. Since y is negative (cf. Fig. 7.0.1) it is clear that the integral converges uniformly. (We omit a discussion of the behavior on the free surface corresponding to $y = 0$, although such a discussion would not present any real difficulties.) Upon examining the function $A(s; t)$ in (7.4.13) it might seem that it has singularities at zeros of the denominators (and such zeros can occur, as we shall see) but in reality one can easily verify that the function has no singularities when the three terms in the square brackets are taken together— or, as one might also put it, any singularities in the individual terms cancel each other. Thus the solution given by (7.4.14) is a regular harmonic function in the strip $-h \leqq y < 0$ for all time t, or, in other words, a motion exists no matter what values are given to the parameters. In addition, the fact that the integral exists ensures that φ (and also its derivatives) tends to zero for any given time when $|x| \to \infty$—this is the content of the so-called Riemann-Lebesgue theorem. This means that the amplitude of the disturbance dies out at infinity at any given time t—a not unexpected result since a certain time must elapse before any appreciable effects of a disturbance are felt at a distance from the seat of the disturbance.*

However, we know from our earlier discussion (and from everyday

---

* It should be pointed out once more that disturbances propagate at infinite speed since our medium is incompressible. Each Fourier component, however, propagates with a finite speed.

observation of streams, for that matter) that as $t \to \infty$ it may happen that a disturbance also propagates downstream as a wave with non-vanishing amplitude. Our main interest here is to study such a passage to the limit. It is clear that one cannot accomplish such a purpose simply by letting $t \to \infty$ in (7.4.14), since, for one thing, the transform $\bar{\varphi}$ of φ cannot exist if φ does not tend to zero at ∞. What we wish to do is to consider the contributions of the separate items in the brackets in (7.4.13), and to avoid any singularities caused by zeros in their denominators by regarding $A(s; t)$ as an analytic function in the neighborhood of the real axis of a complex s-plane and deforming the path of integration in (7.4.14) by Cauchy's integral theorem in such a way as to avoid such singularities. One can then study the limit situation as $t \to \infty$.

In carrying out this program it is essential to study the separate terms defining the function $A(s; t)$ given by (7.4.13). To begin with, we observe that the function $\sqrt{gs \tanh sh}$ can be defined as an analytic and single-valued function in a neighborhood of the real axis since the function $s \tanh sh$ has a power series development at $s = 0$ that is valid for all s and begins with a term in s^2, and, in addition, the function has no real zero except $s = 0$. Once the function $\sqrt{gs \tanh sh}$ has been so defined, it follows that each of the terms in (7.4.13) is an analytic function in a strip containing the real axis except at real zeros of the denominators. It is important to take account of these zeros, as we have already done in sec. 7.3. For our present purposes it is useful to consider the function

$$(7.4.15) \quad W(s) = gs \left(\frac{U^2}{gh} \cdot sh - \tanh sh \right) = s^2 U^2 - gs \tanh sh$$
$$= (sU + \sqrt{gs \tanh sh})(sU - \sqrt{gs \tanh sh})$$
$$= f_+(s)f_-(s).$$

With reference to Fig. 7.3.1 above and the accompanying discussion, one sees that there are at most three real zeros of the function $W(s)$: $s = 0$ is in all cases a root, and there exist in addition two other real roots if the dimensionless parameter gh/U^2 is greater than unity. Also, it is clear that if $gh/U^2 \neq 1$ the origin is a double root of $W(s)$, but is a quadruple root if $gh/U^2 = 1$. In case $gh/U^2 > 1$ the real roots $\pm \beta$ of $W(s)$ are simple roots. (It might be noted in passing that $W(s)$ has infinitely many pure imaginary zeros $\pm i\beta_n$, $n = 1, 2, \ldots$.)

It follows at once that if we deform the path of integration in

(7.4.14) from the real axis to the path P shown in Figure 7.4.1 we can consider separately the contributions to the integral furnished by each of the three items in the square brackets in (7.4.13), since the separate

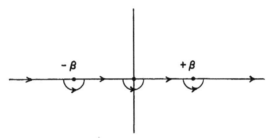

Fig. 7.4.1. The path P in the s-plane

integrals would then exist. This we proceed to do, except that we prefer to consider the velocity components φ_x and φ_y of the disturbance rather than φ itself. For φ_x we write*

(7.4.16) $$\varphi_x = \varphi_x^{(s)} + \varphi_x^{(t)},$$

with $\varphi_x^{(s)}$ and $\varphi_x^{(t)}$ defined (in accordance with (7.4.13) and (7.4.14)) as follows:

(7.4.17) $$\varphi_x^{(s)} = \frac{-U}{\varrho\sqrt{2\pi}} \int_P \frac{\overline{p}(s)s^2 \cosh s(y+h)}{W(s) \cosh sh} e^{isx}\, ds,$$

7.4.18) $$\varphi_x^{(t)} = \frac{-U}{2\varrho\sqrt{2\pi}} \int_P \frac{\overline{p}(s)s^2 \cosh s(y+h)}{\cosh sh \sqrt{gs \tanh sh}} \left[\frac{e^{-itf_+(s)}}{f_+(s)} - \frac{e^{-itf_-(s)}}{f_-(s)} \right] e^{isx}\, ds.$$

The functions $W(s)$, $f_-(s)$, and $f_+(s)$ have been defined in (7.4.15). Evidently the notation $\varphi_x^{(s)}$, $\varphi_x^{(t)}$ has been chosen to point to the fact that $\varphi_x^{(s)}$ should yield the steady part of the motion while $\varphi_x^{(t)}$ should furnish "transients" which die out as $t \to \infty$. This is indeed the case, as we now show, at least when the parameter gh/U^2 is not equal to unity, its critical value.

Consider first the case $gh/U^2 < 1$. In this case there are no singularities on the real axis, even at the origin (cf. (7.4.18)), since f_+ and f_- vanish to the first power and $\sqrt{gs \tanh sh}$ vanishes to the first power also at $s = 0$. Since $\overline{p}(s)$ is regular at $s = 0$ and s^2 occurs in the numerator of the integrand our statement follows. Consequently the path P can be deformed back again into the real axis. In this case the

---

* The discussion would differ in no essential way for φ_y instead of φ_x.

behavior of $\varphi_x^{(t)}$ for large t can be obtained by the principle of stationary phase (cf. sec. 6.8). In the present case the functions $f_+(s)$ and $f_-(s)$ have non-vanishing first derivatives for all s, and consequently $\varphi_x^{(t)} \to 0$ at least like $1/t$ since there are no points where the phase is stationary. (Here and in what follows no attempt is made to give the asymptotic behavior with any more precision than is necessary for the purposes in view.) As $t \to \infty$ therefore we obtain the steady state solution $\varphi_x^{(s)}$. The behavior of $\varphi_x^{(s)}$ for $|x|$ large is also obtained at once: one sees that the integrand in (7.4.17) has no singularities in this case also, and it follows at once from the Riemann-Lebesgue theorem that $\varphi_x^{(s)} \to 0$ as $|x| \to \infty$. Thus a steady state exists, and it has the property that the disturbances die out both upstream and downstream.

We turn next to the more complicated case in which $gh/U^2 > 1$. The integrand for $\varphi_x^{(t)}$ has no singularity at the origin, but it has simple poles at $s = \pm \beta$ (cf. Figure 7.4.1) furnished by simple zeros of $f_-(s)$ at these points. Again we show that $\varphi_x^{(t)} \to 0$ as $t \to \infty$. Consider first the contribution of the semicircles at $s = \pm \beta$. (Since $s = 0$ is not a singularity, we deform the path back into the real axis there.) In the lower half-plane near $s = \pm \beta$ one sees readily that $f_-(s)$ has a negative imaginary part, and thus the exponent in $\exp\{-itf_-(s)\}$ has a negative real part, since $f_-(s)$ is real on the real axis and its first derivative $f'_-(s)$ is positive there (so that $f_-(s)$ behaves like $c(s \mp \beta)$ with c a positive constant). Thus for any closed portion of the semicircles which excludes the end-points the contribution to the integral tends to zero as $t \to \infty$, and hence also for the whole of the semicircles. On the straight parts of the path the principle of stationary phase can be used again to show that $\varphi_x^{(t)} \to 0$ as $t \to \infty$. In fact, this function behaves like $1/\sqrt{t}$ since one can easily verify that $f_-(s)$ has exactly two points of stationary phase, i.e. two points $\pm \beta_0$ where $f'_-(\pm \beta_0) = 0$ and $f''_-(\pm \beta_0) \neq 0$. (The point $s = \beta_0$ lies between the origin and the point $s = \beta$ where $f_-(s)$ vanishes.) Thus the steady state is again given by $\varphi_x^{(s)}$. However, unlike the preceding case, the steady state does not furnish a motion which dies out both upstream and downstream. This can be seen as follows. Consider first the behavior upstream, i.e. for $x < 0$. On the semicircular parts of the path P in the lower half-plane we see that the exponent in e^{isx} in (7.4.17) has a negative real part, and therefore by the same argument as above, these parts of P make contributions which vanish as $x \to -\infty$. The straight parts of P also make contributions which vanish for large x (either positive or negative), by the Riemann-Lebesgue theorem. Thus the disturbance vanishes upstream. On the downstream side, i.e. for $x > 0$, we cannot conclude that the semi-circular parts of P make vanishing contributions for large x since the exponent in e^{isx} now has a positive real part. We therefore make use of the standard procedure of deforming the path P through the poles at $s = \pm \beta$ and subtracting the residues at these poles. It is clear that the semicircles in the upper half-plane yield vanishing contributions to $\varphi_x^{(s)}$ when $x \to + \infty$: the argument is the same as was used above. This leads to the following asymptotic representation (obtained from the contributions at the poles), valid for x large and positive:

$$(7.4.19) \quad \varphi_x(x, y; \infty) \simeq \frac{- \sqrt{2\pi} \, i\beta^2 U \cosh \beta(y + h)}{\varrho \cosh \beta h W'(\beta)} (\bar{p}(\beta)e^{i\beta x} - \bar{p}(-\beta)e^{-i\beta x}).$$

Here $W'(\beta) \neq 0$ is the value of the derivative of W (cf. (7.4.15)) at $s = \beta$, and the fact that $W'(\beta)$ is an odd function has been used. In particular, if the surface pressure $p(x)$ were given by the delta function $p(x) = \delta(x)$, i.e. if the disturbance were caused by a concentrated pressure point at the origin, (7.4.19) would yield

$$(7.4.19)_1 \qquad \varphi_x(x, y; \infty) \simeq \frac{2\beta^2 U \cosh \beta(y + h)}{\varrho \cosh \beta h W'(\beta)} \sin \beta x$$

since the transform of $\delta(x)$ is $1/\sqrt{2\pi}$. Another interesting special case is that in which $p(x)$ is a constant p_0 over the interval $- a \leq x \leq a$ and zero over the rest of the free surface. In this case $\bar{p} = (2p_0/\sqrt{2\pi})(\sin sa)/s$ and φ_x behaves for large positive x and t as follows:

$$(7.4.19)_2 \quad \varphi_x(x, y; \infty) \simeq \frac{4p_0\beta U \cosh \beta(y + h)}{\varrho \cosh \beta h W'(\beta)} \sin \beta a \sin \beta x.$$

This yields the curious result (mentioned above) that under the proper circumstances the disturbance may die out downstream as well as upstream; it will in fact do so if $\beta a = n\pi$, i.e. if the length $2a$ of the segment over which the disturbing pressure is applied is an integral multiple of the wave length at ∞—which is, in turn, fixed by the velocity U and the depth h.

Finally we consider the critical case $gh/U^2 = 1$, and begin by discussing the behavior of the time dependent terms in φ as $t \to \infty$. For this purpose it is convenient to deal first with the time derivative of this function:

(7.4.20) $\quad \varphi_t = \dfrac{U}{2\varrho\sqrt{2\pi}} \displaystyle\int_P \dfrac{\overline{p}(s)s\cosh s(y+h)}{\cosh sh\sqrt{gs\tanh sh}} \left(e^{-itf_+(s)} - e^{-itf_-(s)}\right) e^{isx} ds.$

The integrand has no singularities on the real axis and consequently the path P can be deformed into the real axis. Thus the principle of stationary phase can be employed once more. Since the derivative of $f_+(s) = sU + \sqrt{gs\tanh sh}$ evidently does not vanish for any real s while the derivative of $f_-(s)$ has one zero at $s = 0$, it follows that the leading term in the asymptotic development of $\varphi_t^{(t)}$ for large t arises from the term $\exp\{-itf_-(s)\}$. Since, in addition, $f_-''(0) = 0$ but $f_-'''(0) \neq 0$ we have (cf. sec. 6.8):

(7.4.21) $\quad \varphi_t^{(t)} \simeq A\overline{p}(0) \cdot \dfrac{1}{(f_-'''(0) \cdot t)^{1/3}}, \qquad A = \text{const.} \neq 0.$

Since $\overline{p}(0)$ is in general different from zero, it follows that $\varphi_t^{(t)}$ behaves like $t^{-1/3}$ and hence that $\varphi^{(t)}$ becomes infinite everywhere (for all x and y, that is) like $t^{2/3}$ as $t \to \infty$.* Thus a steady state does not exist if one considers it to be the limit as $t \to \infty$. It might be thought that the existence in practice of dissipative forces could lead to the vanishing of the transients and thus still leave the steady state $\varphi_x^{(s)}$ as given by (7.4.17) as a representation of the final motion. That is, however, also not satisfactory since $\varphi_x^{(s)}$ becomes unbounded for x large when $gh/U^2 = 1$: at the origin there is a pole of order two since $W(s)$ behaves like s^4 and consequently the term isx in the power series for e^{isx} leads to a contribution from this pole which is linear in x. In linear theories based on assuming small disturbances one is reconciled to singularities and infinities at isolated points, but hardly to arbitrarily large disturbances in whole regions. All of this suggests that the reasonable attitude to take in these circumstances is that the linear theory, which assumes small disturbances, fails altogether for flows at the critical speed $U^2/gh = 1$ and that one should go over to a non-

---

* It might seem odd that we have chosen to discuss the function $\varphi_t^{(t)}$ rather than the function $\varphi_{xt}^{(t)}$ (as we did in the other cases). The reason is that the asymptotic behavior of $\varphi_{xt}^{(t)}$ is not easily obtained directly by the method of stationary phase in the present case since the coefficient of the leading term in this development would be zero. However, one could show (by using Watson's lemma, for example, which yields the complete asymptotic expansion of the integral) that $\varphi_{xt}^{(t)}$ behaves like $t^{-2/3}$, and hence that $\varphi_x^{(t)}$ behaves like $t^{1/3}$.

linear theory in order to obtain reasonable results from the physical point of view. In Chapter 10.9, which deals with the solitary wave (an essentially nonlinear phenomenon), we shall see that such a steady wave exists for flows with velocities in the neighborhood of the critical value.

CHAPTER 8

Waves Caused by a Moving Pressure Point. Kelvin's Theory of the Wave Pattern Created by a Moving Ship

8.1. An idealized version of the ship wave problem. Treatment by the method of stationary phase

The peculiar pattern of the waves created by objects moving over the surface of the water on a straight course has been noticed by everyone: the disturbance follows the moving object unchanged in form and it is confined to a region behind the object that has the same v-shape whether the moving object is a duck or a battleship. An explanation and treatment of the phenomenon was first given by Kelvin [K.11], and this work deserves high rank among the many imaginative things created by him. As was mentioned earlier, Kelvin invented his method of stationary phase as a tool for approximating the solution of this particular problem, and it is indeed a beautiful and strikingly successful example of its usefulness.

It should be stated at once that there is no notion in this and the next following section of solving the problem of the waves created by an actual ship in the sense that the shape of the ship's hull is to be taken into account; such problems will be considered in the next chapter. For practical purposes an analysis of the waves in such cases is very much desired, since a fraction—even a large fraction if the speed of the ship is large—of the resistance to the forward motion of a ship is due to the energy used up in maintaining the system of gravity waves which accompanies it. The problem has of course been studied, in particular, in a long series of notable papers by Havelock,* but the difficulties in carrying out the discussion in terms of parameters which fix the shape of the ship are very great. Indeed, a more or less complete discussion of the solution to all orders of approximation even in the very much idealized case to be studied in the present chapter, is by no means an easy task—in fact, such a complete discussion, along

---

* References to some of these papers will be given in the next chapter.

lines quite different from those of Kelvin, has been carried out only rather recently by A. S. Peters [P. 4] (cf. also the earlier paper by Hogner [H.16]). However, we shall follow Kelvin's procedure here in a general way, but with many differences in detail.

The problem we have in mind to discuss as a primitive substitute for the case of an actual ship is the problem of the surface waves created by a point impulse which moves over the surface of the water (assumed to be infinite in depth). We shall take the solution of section 6.5 for the wave motion due to a point impulse and integrate it along the course of the "ship"—in effect, the surface waves caused by the ship are considered to be the cumulative result of impulses delivered at each point along its course. The result will be an integral representation for the solution, in the form of a triple integral, which can be discussed by the method of stationary phase. However, it is necessary to apply the method of stationary phase three times in succession, and if this is not done with some care it is not clear that the approximation is valid at all; or what is perhaps equally bad from the physical point of view, it may not be clear *where* the approximation can be expected to be accurate. Thus it seems worth while to consider the problem with some attention to the mathematical details; this will be done in the present section, and the interpretation of the results of the approximation will be carried out in the next section (which, it should be said, can be read pretty much independently of the present section).

From section 6.4 the vertical displacement* $\eta(x, y, z; t)$ of the water particles due to a point impulse applied on the surface at the origin and at the time $t = 0$ can be put in the form

$$(8.1.1) \quad \eta(x, y, z; t) = - \frac{1}{2\pi g \varrho} \int_0^\infty \sigma \sin \sigma t \cdot e^{m y} m \, dm \int_0^{\pi/2} \cos(mr \cos \beta) \, d\beta$$

in which $\sigma^2 = gm$ and $r^2 = x^2 + z^2$. We have replaced the Bessel function $J_0(mr)$ by its integral representation

$$J_0(mr) = \frac{2}{\pi} \int_0^{\pi/2} \cos(mr \cos \beta) \, d\beta$$

for reasons which will become clear in a moment. As we have indicated, our intention is to sum up the effect of such impulses as the "ship" moves along its course C. The notations to be used for

---

* Actually, we have considered only the displacement of the free surface in that section, but it is readily seen that (8.1.1) furnishes the vertical displacement of any points in the water.

this purpose are indicated in Figure 8.1.1, which is to be considered as a vertical projection of the free surface on any plane $y = \text{const}$. The course of the ship is given in terms of a parameter t by the relations

(8.1.2) $\begin{cases} x_1 = x_1(t) \\ z_1 = z_1(t) \end{cases}$ $0 \leq t \leq T$,

and t is assumed to mean the time required for the ship to travel from any point $Q(x_1, z_1)$ on its course to its present position at the

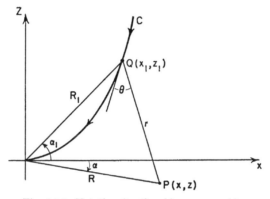

Fig. 8.1.1. Notation for the ship wave problem

origin. We seek the displacement of the water at (x, y, z) when the ship is at the origin; it is therefore determined by the integral

(8.1.3) $\eta(x, y, z)$

$$= -\frac{1}{2\pi g\varrho} \int_0^T k(t)\, dt \int_0^\infty \sigma \sin \sigma t\, e^{my} m\, dm \int_0^{\pi/2} \cos\,(mr \cos \beta)d\beta$$

In this formula $k(t)$ represents the strength of the impulse, which we might reasonably assume to be constant if the speed of the ship is constant; this constant is therefore the only parameter at our disposal which might serve to represent the effect of the volume, shape, etc. of a ship's hull. We write the last relation in the form

(8.1.4) $\eta(x, y, z) = K \int\limits_0^T\int\limits_0^\infty\int\limits_0^{\pi/2} \sigma m e^{my}[e^{i(\sigma t - mr \cos \beta)} + e^{i(\sigma t + mr \cos \beta)}]\, d\beta\, dm\, dt$

with the understanding that the imaginary part of the integral is to be taken. (K is a constant the value of which is not important for the discussion to follow.) It should be noted that $r^2 = (x - x_1)^2 + (z - z_1)^2$. Since $y < 0$, the integral converges strongly because of the exponential factor.

One of the puzzling features (to the author, at least) of existing treatments of the problem by the method of stationary phase is that it is not made clear what parameter is large in the exponentials as the method is applied to each of the three integrals in turn, so that one is not quite sure whether there might not be an inconsistency. The matter is easily clarified by introduction of appropriate dimensionless quantities, as follows (cf. Figure 8.1.1):

(8.1.5)

$$
\begin{cases}
x = R \cos \alpha, \ x_1 = R_1 \cos \alpha_1, \qquad z = R \sin \alpha, \ z_1 = R_1 \sin \alpha_1, \\[2mm]
r = R \sqrt{(\lambda \cos \alpha_1 - \cos \alpha)^2 + (\lambda \sin \alpha_1 - \sin \alpha)^2} = R \cdot l, \\[2mm]
\tau = ct/R, \qquad R_1/R = \lambda, \qquad \varkappa = \dfrac{gR}{4c^2}, \qquad m = \dfrac{gt^2}{4r^2} \xi^2.
\end{cases}
$$

Here the quantity c represents the speed of the ship in its course. It should be noted that x, y, and z are held fixed—they represent the point at which the displacement is to be observed—, but that x_1, z_1 (and hence R_1 and α_1), and r all depend on t. We have also introduced a new variable of integration ξ, replacing m, which depends on t. The Jacobian $\partial(m, t)/\partial(\xi, \tau)$ has the value $gt^2 R\xi/(2cr^2)$ and hence vanishes only for $t = 0$. In terms of the new quantities the integral (8.1.4) is found to take the form:

(8.1.6) $\quad \eta(x, y, z)$

$$
= 4K \int_0^{\tau_0} \int_0^{\infty} \int_0^{\pi/2} \frac{\varkappa^3 \tau^5 \xi^4}{R^2 l^5} e^{\frac{\varkappa \tau^2 \xi^2 y}{R l^2}} \left\{ e^{i\varkappa \frac{(2\xi - \xi^2 \cos \beta)\tau^2}{l}} + e^{i\varkappa \frac{(2\xi + \xi^2 \cos \beta)\tau^2}{l}} \right\} d\beta \, d\xi \, d\tau,
$$

where $\tau_0 = cT/R$.

Again we remark that the integral converges uniformly for $y < 0$. However, the integrand has a singularity if the point (x, y, z) happens to be vertically under a point on the course of the ship: in such a case we have $R = R_1$ (i.e. $\lambda = 1$), and $\alpha = \alpha_1$, so that $l = 0$ for a certain value $\tau \neq 0$ in the interval $0 \leq \tau \leq \tau_0$. Because of the exponential factor, the integral continues to exist, however. Indeed, one sees from (8.1.4) that taking $r = 0$ does not make the integrand singular;

the fact that a singularity crops up in (8.1.6) arises from our choice of the variable ξ which replaces m. This disadvantage caused by introduction of the new variables is much more than outweighed by the fact that we now can see that the approximation by the method of stationary phase depends only on one parameter, i.e. the parameter $\varkappa = gR/4c^2$ in the exponentials. We can expect the use of the method of stationary phase to yield an accurate result if this parameter is large, and that in turn is certainly the case if R is large, i.e. for points not too near the vertical axis through the present location of the ship.

The application of the method of stationary phase to the integral in (8.1.6) can now be justified by an appeal to the arguments used in section 6.8. In doing so, the multiple integral is evaluated by integrating with respect to each variable in turn; at the same time, the integrands are replaced by their asymptotic representations as furnished by the method of stationary phase. One need only observe, in verifying the correctness of such a procedure, that the integrands remain, after each integration, in a form such that the arguments of that section apply—in particular that they remain analytic functions of their arguments provided only that points (x, y, z) on or under the ship's course are avoided*—and that an asymptotic series can be integrated termwise. It is not difficult to see that the contributions to $\eta(x, y, z)$ of lowest order in $1/\varkappa$ are made by arbitrarily small domains containing in their interiors a point where the derivatives $\varphi_\beta, \varphi_\xi, \varphi_\tau$ of the phase $\varphi = (2\xi - \xi^2 \cos \beta)\tau^2/l(\tau)$ vanish simultaneously.

Even for points on the ship's course the argument of section 6.8 will still hold provided that no stationary point of the phase φ occurs for a value of τ such that $l(\tau) = 0$: the reason for this is that the assumption of analyticity was used in section 6.8 only to treat a neighborhood of a point of stationary phase, while for other segments of the field of integration only the assumptions of integrability and the possibility of integration by parts are needed. It happens that the cases to be treated later on are such that $l(\tau)$ does not vanish at any points of stationary phase, and hence for them the asymptotic approximation is valid also for points on the ship's course.

There is one further mathematical point to be mentioned. The

---

* In section 6.8 the integrals studied were of the form $\int_a^b \psi(x) \exp \{ik\varphi(x)\}\, dx$, while here the integral is of the form $\int_a^b \psi(x, k) \exp \{ik\varphi(x)\}\, dx$. However, one can verify that the argument used in section 6.8 can easily be generalized to include the present case.

above discussion requires that we take $y < 0$, and it is not entirely
clear that the passage to the limit $y \to 0$ is legitimate in the approximate formulas, so that the validity of the discussion might be thought
open to question for points on the free surface. Indeed, it would appear
to be difficult to justify such a limit procedure for the integral in
(8.1.1), for instance, since it certainly does not converge if we set
$y = 0$ since the integrand then does not even approach zero as $m \to \infty$.
However, this is a consequence of dealing with a point impulse. If we
had assumed as model for our ship a moving circular disk of radius a
over which a constant distribution of impulse is taken, the result for
the vertical displacement due to such a distributed impulse applied
at $t = 0$ could be shown to be given by

$$\eta(x, y, z; t) = K_1 \int_0^\infty \sigma \sin \sigma t \cdot e^{mv} J_0(mr) J_1(ma) \, dm$$

with $J_1(ma)$ the Bessel function of order one and K_1 a certain constant.
This integral converges uniformly for $y \leqq 0$, as one can see from the
asymptotic behavior of $J_0(mr)$ and $J_1(ma)$. Consequently $\eta(x, y, z; t)$
is continuous for $y = 0$. On the other hand, if the radius a of the disk
is small the result cannot be much different from that for the point
impulse. Thus we might think of the results obtained in the next
section, which start with the formula (8.1.1) for a point impulse, as an
approximation on the free surface to the case of an impulse distributed
over a disk of small radius.

It has already been mentioned that the problem under discussion
here has been treated by A. S. Peters [P.4] by a different method.
Peters obtains a representation for the solution based on contour
integrals in the complex plane, which can then be treated by the
saddle point method to obtain the complete asymptotic development
of the solution with respect to the parameter $\varkappa$ defined above, while
we obtain here only the term of lowest order in such a development.
However, the methods used by Peters lead to rather intricate developments.

8.2. The classical ship wave problem. Details of the solution

In the preceding section we have justified the repeated application
of the method of stationary phase to obtain an approximate solution
for the problem of the waves created when a point impulse moves over
the surface of water of infinite depth. In particular, it was seen that
the approximation obtained in that way is valid at all points on the

surface of the water not too near to the position of the "ship" at the instant when the motion is to be determined (provided only that a certain condition is satisfied at points on the ship's course). In this section we carry out the calculations and discuss the results, returning however to the original variables since no gain in simplicity would be achieved from the use of the dimensionless variables of the preceding section.

Kelvin carried out his solution of the ship wave problem for the case of a straight line course traversed at constant speed. Up to a certain point there is no difficulty in considering more general courses

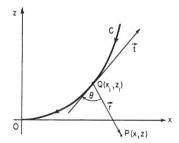

Fig. 8.2.1. Notation for the ship wave problem

for the ship. In Figure 8.2.1 we indicate the course C as any curve given in terms of a parameter t by the equations

(8.2.1) $\qquad \begin{cases} x_1 = x_1(t) \\ z_1 = z_1(t) \end{cases} \qquad$ for $0 \leq t \leq T.$

The parameter t is taken to represent the time required for the ship to pass from any point (x_1, z_1) to its present position at the origin O, but it is convenient to take $t = 0$ to correspond to the origin so that the point (x_1, y_1) moves backward along the ship's course as t increases. The shape of the waves on the free surface is to be determined at the moment when the ship is at the origin. The x-axis is taken along the tangent to the course C, but is taken positive in the direction opposite to the direction of travel of the ship, Since we have taken $t = 0$ at the origin the parameter t in (8.2.1) is really the negative of the time; as a consequence the tangent vector t to C at a point $Q(x_1, y_1)$ as given by

(8.2.2) $\qquad \qquad \mathbf{t} = \left(\dfrac{dx_1}{dt}, \dfrac{dz_1}{dt} \right)$

is in the direction opposite to that of the ship in traversing the course

C. The speed $c(t)$ of the ship is the length of the vector **t** and is given by

$(8.2.2)_1$ $$c(t) = \sqrt{\left(\frac{dx_1}{dt}\right)^2 + \left(\frac{dz_1}{dt}\right)^2}.$$

The point $P(x, z)$ is the point at which the amplitude of the surface waves is to be computed; it is located by means of the vector **r**:

$(8.2.3)$ $$\mathbf{r} = (x - x_1, z - z_1).$$

The angle θ indicated on the figure is the angle $(\leq \pi)$ between the vectors **r** and $-$ **t**.

As we have stated earlier, the surface elevation $\eta(x, z)$ at $P(x, z)$ is to be determined by integrating the elevations due to a point impulse moving along C. The effect of an impulse at the point Q is assumed to be given by the approximate formula (6.5.15), in which, however, we omit a constant multiplier which is unessential for the discussion to follow:

$(8.2.4)$ $$\eta(r; t) \simeq \frac{-t^3}{r^4} \sin \frac{gt^2}{4r}.$$

In other words, we assume that the formula (8.1.1) for the surface elevation η has been approximated by two successive applications of the method of stationary phase. This formula yields the effect at time t and at a point distant r from the point where the impulse was applied at the time $t = 0$; it therefore applies in the present situation with

$(8.2.5)$ $$r^2 = (x - x_1)^2 + (z - z_1)^2,$$

since t does indeed represent the length of time elapsed since the "ship" passed the point Q on its way to its present position at O. The integrated effect of all the point impulses is therefore given by

$(8.2.6)$ $$\eta(x, z) = k_0 \int_0^T \frac{t^3}{r^4} \sin \frac{gt^2}{4r} \, dt,$$

with k_0 a certain constant. For points on the ship's course, where $r = 0$ for some value $t = t_0$ in the interval $0 \leq t \leq T$, this integral evidently does not exist. However, it has been shown in the preceding section that neighborhoods of such points can be ignored in calculating η approximately provided that they are not points of stationary phase. This condition will be met in general, and hence we may imagine that a small interval about a point where $r(t_0) = 0$ has been excluded from the range of integration in case we wish the wave amplitude at a point on the ship's course. We write the integral in the form

$$(8.2.7) \qquad \eta(x,z) = \int_0^T \psi(t) e^{i\varphi(t)} \, dt,$$

and take the imaginary part. The function $\psi(t)$ and the *phase* $\varphi(t)$ are given by

$$(8.2.8) \qquad \psi(t) = k_0 t^3 / r^4$$

$$(8.2.9) \qquad \varphi(t) = g t^2 / 4r.$$

We proceed to make the calculations called for in applying the stationary phase method. In the integral given by (8.2.7) no large parameter multiplying the phase is put explicitly in evidence; however, from the discussion of the preceding section we know that the approximation will be good if the dimensionless quantity $gR/4c^2$, with R the distance from the ship, is large. It could also be verified that (8.2.6) would result if the integrations in (8.1.6) on β and ξ were first approximated by stationary phase followed by a re-introduction of the original variables. We therefore begin by calculating $d\varphi/dt$:

$$(8.2.10) \qquad \frac{d\varphi}{dt} = \frac{g}{4}\left(\frac{2t}{r} - \frac{t^2}{r^2}\frac{dr}{dt}\right).$$

Hence the condition of stationary phase, $d\varphi/dt = 0$, leads to the important relation

$$(8.2.11) \qquad \frac{dr}{dt} = \frac{2r}{t}.$$

The quantity dr/dt is next calculated for the ship's course using (8.2.5); we find (cf. Figure 8.2.1):

$$(8.2.12) \qquad r\frac{dr}{dt} = -\left[(x - x_1)\frac{dx_1}{dt} + (z - z_1)\frac{dz_1}{dt}\right]$$

$$= -\mathbf{r} \cdot \mathbf{t} = cr \cos\theta,$$

in which $c(t)$ is once more the speed of the ship. Thus

$$(8.2.13) \qquad \frac{dr}{dt} = c \cos\theta,$$

which is a rather obvious result geometrically. Combining (8.2.11) and (8.2.13) yields the stationary phase condition in the form

$$(8.2.14) \qquad r = \tfrac{1}{2} ct \cos\theta.$$

We recall once more the significance of this relation: for a fixed point $P(x, y)$ it yields those points Q_i on C which are the sole points effective (within the order of the approximation considered) in creating the disturbance at P—the contributions from all other points being, in effect, cancelled out through mutual interference. It is helpful to introduce the term *influence points* for the points Q_i determined in this way relative to a point P at which the surface elevation of the water is to be calculated.

The last observation makes it possible to draw an interesting conclusion at once from (8.2.14), which can be interpreted in the following way (cf. Figure 8.2.2): At point Q the speed c of the ship and t are

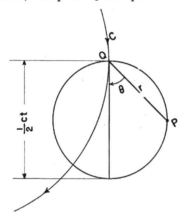

Fig. 8.2.2. Points influenced by a given point Q

known. The relation (8.2.14) then yields the polar coordinates (r, θ), with respect to Q, of *all* points P for which Q is the influence point in the sense of the stationary phase approximation. Such points P evidently lie on a circle with a diameter tangent to the course C of the ship at Q, and Q is at one end of the diameter. The center of the circle is located on the tangent line from Q in the direction toward which the ship moves (i.e. in the direction $-$ t). We repeat that the points P on the circle just described are the *only* points for which Q is a point of stationary phase of the integral (8.2.7), and consequently the contribution of the impulse applied at Q vanishes (within the order considered by us) for all points except those on the circle. It now becomes obvious that the disturbance created by the ship does not affect the whole surface of the water, since only those points are affected which lie on one or more of the circles of influence of all points Q on the ship's course. In other words, the surface waves created by the moving ship will be confined to the region covered by all the influence circles, and thus to the region bounded by the envelope of this one-parameter family of curves. This makes it possible to construct graphically the outline of the disturbed region for any given course traversed at any given speed: one need only draw the circles in the manner indicated at a sufficient number of points Q and then sketch the envelope. Two such cases, one of them a straight course traversed at constant speed, the other a circular course, are shown in Figure 8.2.3. In the case of the straight course it is clear that the envelope

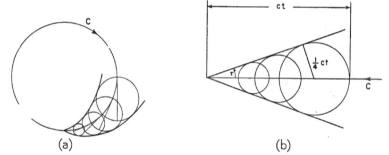

Fig. 8.2.3. Region of disturbance (a) Circular course (b) Straight course

is a pair of straight lines; the disturbance is confined to a sector of semi-angle τ given by $\tau = \text{arc sin } 1/3 = 19°28'$, as one readily sees from Figure 8.2.3. This is already an interesting result: it says that the waves following the ship not only are confined to such a sector but that the angle of the sector is independent of the speed of the ship as long as the speed is constant. If the speed were not constant along a straight course, the region of disturbance would be bounded by curved lines, and its shape would also change with the time. It is, of course, not true that the disturbance is exactly zero outside the region of disturbance as we have defined it here; but rather it is small of a different order from the disturbance inside that region. The observations of actual moving ships bear out this conclusion in a quite startling way, as one sees from Figures 8.2.4 and 8.2.5.

The discussion of the region of disturbance has furnished us with a certain amount of interesting information, but we wish to know a good deal more. In particular, we wish to determine the character of the wave pattern created by the ship and the amplitude of the waves. For these purposes a more thoroughgoing analysis is necessary, and it will be carried out later.

In the special case of a straight course traversed at constant speed it is possible to draw quite a few additional conclusions through further discussion of the condition (8.2.14) of stationary phase. In the above discussion we asked for the points P influenced by a given point Q on the ship's course. We now reverse the question and ask for

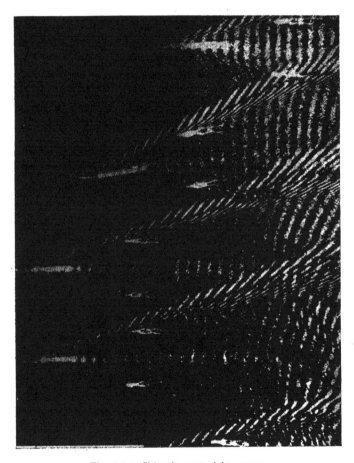

Fig. 8.2.4. Ships in a straight course

the location of all influence points Q_i that correspond to a given point P. This question can be answered in our special case by making another simple geometrical construction (cf. Lamb [L.3], p. 435), as indicated in Figure 8.2.6. In this figure O represents the location of

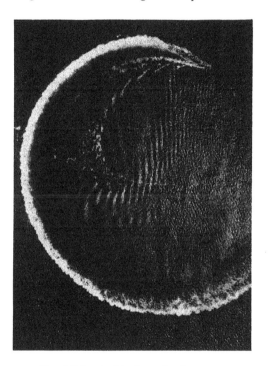

Fig. 8.2.5a. A ship in a circular course

the ship, P the point for which the influence points are to be determined. The construction is made as follows: A circle through P with center on OP and diameter half the length of $\overline{OP}$ is constructed; its intersections with the ship's course are denoted by S_1 and S_2. From the latter points lines are drawn to P and segments orthogonal to them at P are drawn to their intersections Q_1 and Q_2 on the ship's course. The points Q_1 and Q_2 are the desired influence points. To prove that the construction yields the desired result requires only a verification that P does indeed lie on the influence circles determined by the points Q_1 and Q_2 in the manner explained above. Consider the point

Q_1, for example. Since the angle $S_1PQ_1 = 90°$, it follows that a circle with S_1Q_1 as diameter contains the point P. The segments RS_1 and PQ_1 are parallel since both are at right angles to S_1P; by considering the triangle OPQ_1 one now sees that the segment OS_1 is just

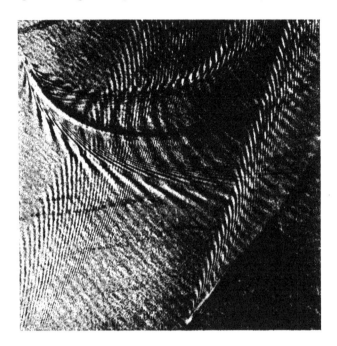

Fig. 8.2.5b. Ships in curved courses

half the length of OQ_1, and that is all that is necessary to show that the circle having S_1Q_1 as diameter is the influence circle for Q_1. Thus there are in general two influence points or no influence points, the latter case corresponding to points P outside the influence region; the transition occurs when P is on the boundary of the region of influence (i.e. when the circle of Figure 8.2.6 having PR as diameter is tangent to the course OQ_2 of the ship), and one sees that in this limit case the two influence points Q_1 and Q_2 coalesce. Consequently one might well expect that the amplitude of the waves at the boundary of the region of disturbance will be higher than at other places, and this phenomenon is indeed one of the prominent features always observed physically. In addition, the direction of the curves of constant phase—a wave crest, or trough, for example—can be determined graphically by the above construction: one expects these curves to be orthogonal to the

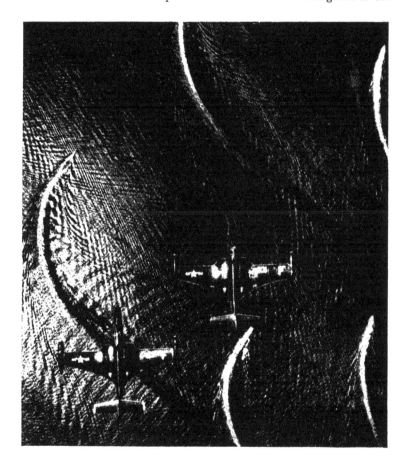

Fig. 8.2.5c. Aircraft carriers maneuvering (from Life Magazine)

lines PQ_1 and PQ_2 drawn back from a point P to each of the points of influence corresponding to P. That this is indeed the case will be seen later, but it is evidently a consequence of the fact that the wave at P is the sum of two circular waves, one generated at Q_1 and the other at

Q_2. Thus we see that the wave pattern behind the ship is made up of two different trains of waves—another fact that is a matter of common observation and which is well shown in Figures 8.2.4 and 8.2.5.

We have been able to draw a considerable number of interesting and

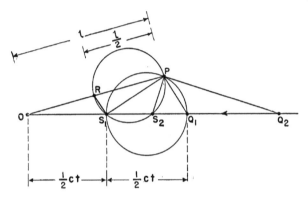

Fig. 8.2.6. Influence points corresponding to a given point

basic conclusions of a qualitative character through use of the condition of stationary phase (8.2.14). We proceed next to study analytically the shape of the disturbed water surface by determining the curves of constant phase, and later on by determining the amplitude of the waves. To calculate the curves of constant phase it is convenient to express the basic condition (8.2.14) of stationary phase in other forms through introduction of the following quantity a, which has the dimension of length:

$$(8.2.15) \qquad a = \frac{2c^2}{g}\,\varphi = \frac{c^2 t^2}{2r}.$$

From (8.2.14) one then finds

$$(8.2.16) \qquad ct = a\cos\theta, \text{ and}$$

$$(8.2.17) \qquad r = \tfrac{1}{2}a\cos^2\theta,$$

as equivalent expressions of the stationary phase condition.

It would be possible to calculate the curves of constant phase for any given course of the ship. We carry this out for the case of a circular course (this case has been treated by L. N. Sretenski [S.15]) and a straight course traversed at constant speed. The notation for the case of the circular course is indicated in Figure 8.2.7, which should be compared with Figure 8.2.1. For the past position (x_1, z_1) of the ship we have

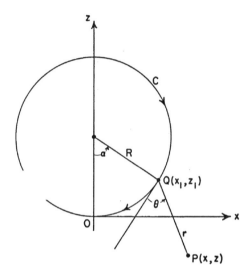

Fig. 8.2.7. Case of a circular course

$$(8.2.18) \qquad \begin{cases} x_1 = R \sin \alpha \\ z_1 = R(1 - \cos \alpha) \end{cases}$$

with

$$(8.2.19) \qquad \alpha = ct/R.$$

Here R is the turning radius of the ship, t the time required for it to travel from Q to O, and c is the constant speed of the ship. The coordinates of the point P, where the disturbance created by the ship is to be found, are given by

$$(8.2.20) \qquad \begin{cases} x = x_1 - r \cos (\alpha + \theta) \\ z = z_1 - r \sin (\alpha + \theta) \end{cases}$$

in which r and θ are the distance and angle noted on the figure. In these equations we replace x_1 and z_1 from (8.2.18) and make use of (8.2.17) to obtain

$$(8.2.21) \qquad \begin{cases} x = R \sin \alpha - \dfrac{a}{2} \cos^2 \theta \, \cos (\alpha + \theta) \\ z = R(1 - \cos \alpha) - \dfrac{a}{2} \cos^2 \theta \, \sin (\alpha + \theta). \end{cases}$$

We wish to find the locus of points (x, z) such that the phase φ remains fixed, i.e. such that the quantity a in (8.2.15) is constant (cf. the remarks following (8.2.9)). It is convenient to introduce the dimensionless parameter $\varkappa$ through

(8.2.22) $\varkappa = a/R.$

One then finds that the angle α (cf. (8.2.19)) is given by

(8.2.23) $\alpha = \varkappa \cos \theta,$

through use of (8.2.16). In terms of these quantities the relations (8.2.21) can be put in the following dimensionless form:

(8.2.24)
$$
\begin{cases}
x/R = \sin (\varkappa \cos \theta) - \dfrac{\varkappa}{2} \cos^2 \theta \, \cos (\theta + \varkappa \cos \theta) \\[2mm]
z/R = 1 - \cos (\varkappa \cos \theta) - \dfrac{\varkappa}{2} \cos^2 \theta \sin (\theta + \varkappa \cos \theta).
\end{cases}
$$

These equations furnish the curves of stationary phase in terms of θ as parameter. Each fixed value of $\varkappa$ furnishes one such curve, since fixing $\varkappa$ (for a fixed turning radius R) is equivalent to fixing the phase φ. In Figure 8.2.8 a few curves of constant phase, as well as the

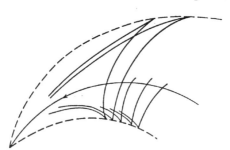

Fig. 8.2.8. Wave crests for a circular course

outline of the region of disturbance, as calculated from (8.2.24), are shown; the successive curves differ by 2π in phase. These curves should be compared with the photographs of actual cases given in Figures 8.2.4 and 8.2.5. One sees that the wave pattern is given correctly by the theory, at least qualitatively. The agreement between theory and observation is particularly striking in view of the manner in which the action of a ship has been idealized as a moving pressure point. In particular there are two distinct sets of waves apparent, in conformity with the fact that we expect each point in the disturbed region to correspond to two influence points: one set which seems to emanate from the ship's bow, and another set which is arranged roughly at right angles to the ship's course. These two systems of waves are called the diverging and the transverse systems, respectively.

From (8.2.24) we can obtain the more important case of the ship waves for a straight course by letting $R \to \infty$ while $\varkappa \to 0$ in such a way that $R\varkappa \to a$ (cf. (8.2.22)). The result is

(8.2.25)
$$
\begin{cases}
x = \dfrac{a}{2} \left(2 \cos \theta - \cos^3 \theta \right) \\[2mm]
z = -\dfrac{a}{2} \cos^2 \theta \sin \theta
\end{cases}
$$

for the curves of constant phase. In Figure 8.2.9 the results of calculations from these equations are shown. These should once more be compared with Figure 8.2.4, which shows an actual case. Again the agreement is striking in a qualitative way. Actually, the agreement

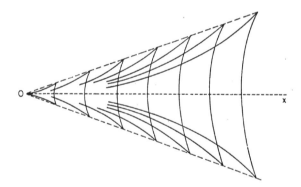

Fig. 8.2.9. Wave crests for a straight course

would be still better if the two systems of waves—the diverging and transverse systems—had been drawn in Figure 8.2.9 with a relative phase difference: the photograph indicates that the crests of the two systems do not join with a common tangent at the boundary of the region of disturbance. We shall see shortly that a closer examination of our approximate solution shows the two systems of waves to have a phase difference there. It is worth while to verify in the present case a general observation made earlier, i.e. that the curves of constant phase are orthogonal to the lines drawn back to the corresponding influence points. One finds from (8.2.25):

$$(8.2.26) \quad \begin{cases} \dfrac{dx}{d\theta} = -\dfrac{a}{2}\,(3\sin^2\theta - 1)\sin\theta \\[2mm] \dfrac{dz}{d\theta} = \dfrac{a}{2}\,(3\sin^2\theta - 1)\cos\theta. \end{cases}$$

Hence $dz/dx = -1/\tan\theta$, which (cf. Figure 8.2.10) means that the curves of constant phase are indeed orthogonal to the lines drawn to

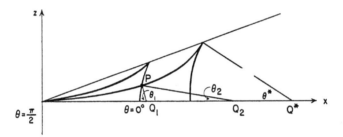

Fig. 8.2.10. Construction of curves of constant phase

the influence points. The values $\theta = \theta^*$ at which $3\sin^2\theta - 1 = 0$ are singular points of the curves; they correspond to points P at the boundary of the influence region where the influence points Q_1 and Q_2 coincide. Evidently there are cusps at these points. One sees also that the diverging set of waves (for $z > 0$, say) is obtained when θ varies in the range $\theta^* \leq \theta \leq \pi/2$, while the transverse waves correspond to values of θ in the range $0 \leq \theta \leq \theta^*$. In addition, we observe that to any point on the ship's course there corresponds (for $\theta = 0^\circ$) only one influence point (of type Q_2) and it does not coincide with the point P. (One sees, in fact, that the diverging wave does not occur on the ship's course.) This is a fact that is needed to justify the application of the method of stationary phase to points on the ship's course, as we have remarked earlier in this section (cf. also the preceding section).

In order to complete our discussion we must consider the amplitude of the surface waves, as given by our approximation, as well as the shape of the curves of constant phase. To this end we must calculate φ and $d^2\varphi/dt^2$ (and even $d^3\varphi/dt^3$) for such values of t as satisfy the stationary phase condition $d\varphi/dt = 0$, as we know from the discussion of section 6.5 and section 6.8. From (8.2.10) we find easily

(8.2.27)
$$\frac{d^2\varphi}{dt^2} = \frac{g}{2r}\left(1 - \frac{t^2}{2r}\frac{d^2r}{dt^2}\right)$$

in view of (8.2.11). We shall also need the value of $d^3\varphi/dt^3$ at points such that $d\varphi/dt = d^2\varphi/dt^2 = 0$; it is readily found to be given by

(8.2.28)
$$\frac{d^3\varphi}{dt^3} = -\frac{gt^2}{4r^2}\frac{d^3r}{dt^3}.$$

We wish to express our results in terms of the parameter θ instead of t. Since $dr/dt = c\cos\theta$ from (8.2.13) we have

(8.2.29)
$$\frac{d^2r}{dt^2} = -c\sin\theta\,\frac{d\theta}{dt}$$

with c, the speed of the ship, now assumed to be constant. In order to calculate $d\theta/dt$ we introduce the angles β and τ indicated in Figure 8.2.11. We have $\theta = \pi - (\beta + \tau)$, and hence

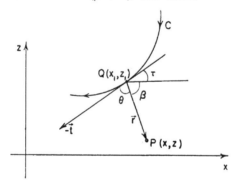

Fig. 8.2.11. The angles β and τ

(8.2.30)
$$\frac{d\theta}{dt} = \left(-\frac{d\beta}{ds} - \frac{d\tau}{ds}\right)\frac{ds}{dt} = -c\left(\frac{d\beta}{ds} + \frac{d\tau}{ds}\right)$$

in which s refers to the arc length of C. But $d\tau/ds = 1/R$, with R the radius of curvature of C; and since $\beta = \arctan(z - z_1)/(x - x_1)$ we find

(8.2.31)
$$\frac{d\beta}{ds} = -\frac{1}{r^2}\left[(x - x_1)\frac{dz_1}{ds} - (z - z_1)\frac{dx_1}{ds}\right] = \frac{\sin\theta}{r}.$$

since the quantity in the square brackets is the vector product of $\mathbf{r}$ and $\mathbf{t}/|\mathbf{t}|$. The expression for $d^2\varphi/dt^2$ given by (8.2.27) can now be expressed in terms of θ and r as follows:

(8.2.32)
$$\frac{d^2\varphi}{dt^2} = \frac{g}{2r}\left[1 - 2\tan^2\theta - \frac{a}{R}\sin\theta\right]$$

$$= \frac{g}{2r}\left[\frac{1 - 3\sin^2\theta - \dfrac{a}{R}\sin\theta\,(1 - \sin^2\theta)}{\cos^2\theta}\right]$$

as one can easily verify. The quantity a is defined by (8.2.15), and the relation (8.2.17), in addition to those immediately above, has been used. The points on the boundary of the region of disturbance could be determined analytically, as follows: the set of all influence points is the one-parameter family of circles given by $d\varphi/dt = \chi(x,z,t) = 0$, and the region of disturbance is bounded by the envelope of these circles, i.e. by the points at which $d^2\varphi/dt^2 = \partial\chi/\partial t = 0$ in addition to $\chi = 0$. In the case of a straight course traversed at constant speed, for example, we see from (8.2.32) for $R = \infty$ that θ then has the value θ^* given by $1 - 3\sin^2\theta = 0$—a result found above, where the value $\theta = \theta^*$ also was seen to characterize cusps on the loci of constant phase. From the form of the relation (8.2.32) one can conclude that the only courses for which the pattern of waves behind the ship follows it without change (i.e. follows it like a rigid body) are those for which $R = $ const.; and thus only the straight and the circular courses have this property.

Finally, we have to consider the amplitude $\eta(x, z)$ of the waves given by our approximate solution. The contribution of a point t_0 of stationary phase to (8.2.7) is given by (cf. (6.5.2)):

(8.2.33) $$\eta(x, z) = \psi(r, \theta)\left(\frac{2\pi}{|\varphi''(r, \theta)|}\right)^{\frac{1}{2}} e^{i\left(\varphi(r,\theta) \pm \frac{\pi}{4}\right)}$$

in which (r, θ) are polar coordinates which locate the point of stationary phase on the course C relative to the point (x, z) (cf. Figure 8.2.1). The sign of the term $\pm \pi/4$ is to be taken the same as that of $\varphi'' = d^2\varphi/dt^2$. In principle, the surface elevation can be calculated for any course, but the results are not very tractable except for the simplest case; i.e. a straight course. We confine our discussion of amplitudes, therefore, to this case in what follows. From (8.2.32) we have

$$(8.2.34) \qquad \frac{d^2\varphi}{dt^2} = \frac{g}{2r}\left(\frac{1-3\sin^2\theta}{\cos^2\theta}\right).$$

We know that there are two values of θ—call them θ_1 and θ_2—at each point in the disturbed region for which $d\varphi/dt = 0$: one belonging for $0 \leq \theta_1 \leq \theta^* = $ arc sin $1/\sqrt{3}$ to the transverse system, the other for $\theta^* \leq \theta_2 < \pi/2$ to the diverging system of waves. In the former case $d^2\varphi/dt^2$ is positive; in the latter case negative. (At the boundary of the region of disturbance, where $\varphi'' = 0$, the formula (8.2.33) is not valid, as we know. This case will be dealt with later.) For points in the interior of the region of disturbance we have, therefore,

$$(8.2.35) \qquad \eta(x,z) \simeq \sqrt{2\pi}\left[\psi(r_1,\theta_1)\frac{1}{\sqrt{|\varphi''(r_1,\theta_1)|}}e^{i\left(\varphi(r_1,\theta_1)+\frac{\pi}{4}\right)}\right.$$
$$\left. + \psi(r_2,\theta_2)\frac{1}{\sqrt{|\varphi''(r_2,\theta_2)|}}e^{i\left(\varphi(r_2,\theta_2)-\frac{\pi}{4}\right)}\right].$$

Since $r_i = \frac{1}{2}ct_i\cos\theta_i$, $r_i = \frac{1}{2}a_i\cos^2\theta_i$, $a_i = 2c^2\varphi_i/g = c^2t_i^2/2r_i$, and $\psi = k_0t_i^3/r_i^4$ (cf. (8.2.15), (8.2.8)) at the points of stationary phase, we may write (8.2.35) in the form

$$(8.2.36) \qquad \eta(x,z) \simeq 2^{9/2}k_0\pi^{1/2}/c^3g^{1/2}\left\{\frac{\sec^3\theta_1}{a_1^{\frac{1}{2}}\sqrt{|1-3\sin^2\theta_1|}}e^{i\left(\frac{ga_1}{2c^2}+\frac{\pi}{4}\right)}\right.$$
$$\left. + \frac{\sec^3\theta_2}{a_2^{\frac{1}{2}}\sqrt{|1-3\sin^2\theta_2|}}e^{i\left(\frac{ga_2}{2c^2}-\frac{\pi}{4}\right)}\right\}.$$

The two systems of waves are thus seen, as was stated above, to have a relative phase difference of $\pi/2$ at any point where $a_1 = a_2$. Their amplitudes die out like $1/\sqrt{a_i}$ on going away from the ship, and that means that they die out like the inverse square root of the distance from the ship. The wave amplitudes of both systems of waves become infinite according to these formulas for $\theta = \theta^*$, i.e. for points at the boundary of the disturbed region, but the asymptotic formula (8.2.33) is not valid at such points since $\varphi'' = 0$ there. We shall consider these points in a moment. The diverging system also has infinite amplitude for $\theta_2 = \pi/2$, but this corresponds to the origin, and the infinite amplitude there results from our assumption of a moving point impulse as a model for our ship.

To determine the amplitude of the waves along the boundary of the disturbed region, we must calculate the value of $d^3\varphi/dt^3$ at such points in order to evaluate the appropriate term in (6.5.2). (The problem of the character of the waves in this region has been treated by Hogner [H.13].) By differentiating (8.2.29) after replacing $d\theta/dt$ by $c \sin \theta / r$ (cf. (8.2.30) and (8.2.31) for $R = \infty$), one finds readily

$$(8.2.37) \qquad \frac{d^3r}{dt^3} = - \frac{c^3 \cos \theta \sin^2 \theta}{r^2},$$

and from (8.2.28) in combination with $r = \frac{1}{2}ct \cos \theta, r = \frac{1}{2}a \cos^2 \theta$:

$$(8.2.38) \qquad \frac{d^3\varphi}{dt^3} = \frac{4gc}{a^2} \frac{\sin^2 \theta}{\cos^5 \theta}.$$

The amplitude of the waves along the boundary of the disturbed region is given by (cf. (6.5.2)):

$$(8.2.39) \qquad \eta(x, z) \simeq \frac{\Gamma(\frac{1}{3})}{\sqrt{3}} \psi(\theta^*) \left(\frac{6}{|\varphi'''(\theta^*)|} \right)^{\frac{1}{3}} e^{i\varphi(\theta^*)},$$

with all functions evaluated for $\theta = \theta^* = \arc \sin 1/\sqrt{3}$. The final result is

$$(8.2.40) \qquad \eta \simeq \frac{k_1}{a^{1/3}} \exp \{iga/2c^2\}.$$

The quantity k_1 is a certain constant. We observe that the wave amplitudes now die out like $1/a^{1/3}$ instead of like $1/a^{1/2}$, as they do in the interior of the disturbed region; i.e. the wave amplitudes are now of a different, and higher, order of magnitude. As we have seen in all of our illustrations of ship waves, the wave amplitudes are quite noticeably higher along the boundary of the disturbed region. The phase also differs now by $\pi/4$ from the former values. On some of the photographs (cf. especially Fig. 8.2.4), there is some evidence of a rather abrupt change of phase in the region of the boundary, though it may be that one should interpret this effect as due rather to the finite dimensions of the actual ship, which then acts as though several moving point sources were acting simultaneously.

In the treatment of the present problem by A. S. Peters [P.4] mentioned in the preceding section, the complete asymptotic development of the solution was obtained.

The above developments hold only for the case of a point impulse moving on the surface of water of infinite depth. It has some interest to point out that there are considerable differences in the results if the depth of the water is finite. Havelock [H.8] has carried out the approximation to the solution by the method of stationary phase for

the case of constant finite depth, with the following general results:
1) If the speed c of the ship and the depth h satisfy the inequality
$c^2/gh < 1$, the general pattern of the waves is much the same as for
water of infinite depth except that the angle of the sector within
which the main part of the disturbance is found is now larger than
for water of infinite depth. 2) If $c^2/gh > 1$ holds, the system of transverse waves no longer occurs, but the diverging system is found.

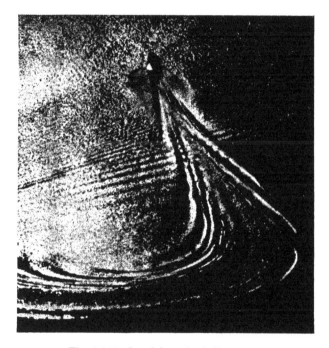

Fig. 8.2.12. Speed boat in shallow water

Figure 8.2.12 is a photograph of a speed boat creating waves, presumably in shallow water, in view of the difference in the wave pattern
when compared with Fig. 8.2.4. Finally, if $c^2/gh = 1$ (i.e. for the case
of the critical speed), the method of stationary phase yields no reasonable results; that this should be so is perhaps to be understood in
the light of the discussion of the corresponding two-dimensional
problem in Chapter 7.4.

CHAPTER 9

The Motion of a Ship, as a Floating Rigid Body, in a Seaway

9.1. Introduction and summary

The purpose of this chapter is to develop a mathematical theory for the motion of a ship, to be treated as a freely floating rigid body under the action of given external forces (a propeller thrust, for example), under the most general conditions compatible with a linear theory and the assumption of an infinite ocean.* This of course requires the amplitude of the surface waves to be small and, in general, that the motion of the water should be a small oscillation near its rest position of equilibrium; it also requires the ship to have the shape of a thin disk so that it can have a translatory motion with finite velocity and still create only small disturbances in the water. In addition, the motion of the ship itself must be assumed to consist of small oscillations relative to a motion of translation with constant velocity. Within these limitations, however, the theory presented is quite general in the sense that no arbitrary assumptions about the interaction of the ship with the water are made, nor about the character of the coupling between the different degrees of freedom of the ship, nor about the waves present on the surface of the sea: the combined system of ship and sea is treated by using the basic mathematical theory of the hydrodynamics of a non-turbulent perfect fluid. For example, the theory presented here would make it possible in principle to determine the motion of a ship under given forces which is started with arbitrary initial conditions on a sea subjected to given surface pressures and initial conditions, or on a sea covered with waves of prescribed character coming from infinity.

It is of course well known that such a linear theory for the non-turbulent motion of a perfect fluid, complicated though it is, still does not contain all of the important elements needed for a thoroughgoing discussion of the practical problems involved. For example, it ignores

---

* The presentation of the theory given here is essentially the same as that given in a report of Peters and Stoker [P.7].

the boundary-layer effects, turbulence effects, the existence in general of a wake, and other important effects of a non-linear character. Good discussions of these matters can be found in papers of Lunde and Wigley [L.18], and Havelock [H.7]. Nevertheless, it seems clear that an approach to the problem of predicting mathematically the motion of ships in a seaway under quite general conditions is a worthwhile enterprise, and that the problem should be attacked even though it is recognized at the outset that all of the important physical factors can not be taken into account. In fact, the theory presented here leads at once to a number of important qualitative statements without the necessity of producing actual solutions—for example, we shall see that certain resonant frequencies appear quite naturally, and in addition that they can be calculated solely with reference to the mass distribution and the given shape of the hull of the ship. Interesting observations about the character of the coupling between the various degrees of freedom, and about the nature of the interaction between the ship and the water, are also obtained simply by examining the equations which the theory yields.

In order to describe the theory and results to be worked out in later sections of this chapter, it is necessary to introduce our notation and to go somewhat into details. In Fig. 9.1.1 the disposition of two of the coordinate systems used is indicated. The system (X, Y, Z) is a

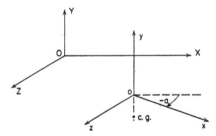

Fig. 9.1.1. Fixed and moving coordinate systems

system fixed in space with the X, Z-plane in the undisturbed free surface of the water and the Y-axis vertically upward. A moving system of coordinates (x, y, z) is introduced; in this system the x, z-plane is assumed to coincide always with the X, Z-plane, and the y-axis is assumed to contain the center of gravity (abbreviated to c.g. in the following) of the ship. The course of the ship is fixed by the motion of the origin of the moving system, and the x-axis is taken along

the tangent to the course. It is then convenient to introduce the
speed $s(t)$ of the ship in its course: the speed $s(t)$ is simply the magnitude of the vector representing the instantaneous velocity of this point.
At the same time we introduce the angular speed $\omega(t)$ of the moving
system relative to the fixed system: one quantity fixes this rotation
because the vertical axes remain always parallel. The angle $\alpha(t)$
indicated in Fig. 9.1.1 is defined by

$$(9.1.1) \qquad\qquad \alpha(t) = \int_0^t \omega(t)\, dt,$$

implying that $t = 0$ corresponds to an instant when the x-axis and
X-axis are parallel. In order to deal with the motion of the ship as a
rigid body it is convenient, as always, to introduce a system of coordinates fixed in the body. Such a system (x', y', z') is indicated in
Fig. 9.1.2. The x', y'-plane is assumed to be in the fore-and-aft plane

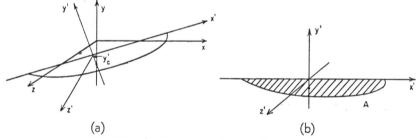

Fig. 9.1.2a, b. Another moving coordinate system

of symmetry of the ship's hull, and the y'-axis is assumed to contain
the c.g. of the ship. The moving system (x', y', z') is assumed to coincide with the (x, y, z) system when the ship and the water are at rest
in their equilibrium positions. The c.g. of the ship will thus coincide
with the origin of the (x', y', z') system only in case it is at the level
of the equilibrium water line on the ship; we therefore introduce the
constant y'_c as the vertical coordinate of the c.g. in the primed coordinate system.

The motion of the water is assumed to be given by a velocity potential $\Phi(X, Y, Z; t)$ which is therefore to be determined as a solution
of Laplace's equation satisfying appropriate boundary conditions at
the free surface of the water, on the hull of the ship, at infinity, and
also initial conditions at the time $t = 0$. The boundary conditions on
the hull of the ship clearly will depend on the motion of the ship,

which in its turn is fixed, through the differential equations for the motion of a rigid body with six degrees of freedom, by the forces acting on it — including the pressure of the water — and its position and velocity at the time $t = 0$. As was already stated, no further restrictive assumptions except those needed to linearize the problem are made.

Before discussing methods of linearization we interpolate a brief discussion of the relation of the theory presented here to that of other writers who have discussed the problem of ship motions by means of the linear theory of irrotational waves. The subject has a lengthy history, beginning with Michell in 1898, and continuing over a long period of years in a sequence of notable papers by Havelock, starting in 1909. This work is, of course, included as a special case in what is presented here. Extensive and up-to-date bibliographies can be found in the papers by Weinblum [W.3] and Lunde [L.19]. Most of this work considers the ship to be held fixed in space while the water streams past; the question of interest is then the calculation of the wave resistance in its dependence on the form of the ship. Of particular interest to us here are papers of Krylov [K.20], St. Denis and Weinblum [S.1], Pierson and St. Denis [P.9] and Haskind [H.4], all of whom deal with less restricted types of motion. Krylov seeks the motion of the ship on the assumption that the pressure on its hull is fixed by the prescribed motion of the water without reference to the back effect on the motion of the water induced by the motion of the ship. St. Denis and Weinblum, and Pierson and St. Denis, employ a combined theoretical and empirical approach to the problem which involves writing down equations of motion of the ship with coefficients which should be in part determined by model experiments; it is assumed in addition that there is no coupling between the different degrees of freedom involved in the general motion of the ship. Haskind attacks the problem in the same degree of generality, and under the same general assumptions, as are made here; in the end, however, Haskind derives his theory completely only in a certain special case. Haskind's theory is also not the same as the theory presented here, and this is caused by a fundamental difference in the procedure used to derive the linear theory from the underlying, basically nonlinear, theory. Haskind develops his theory by assuming that he knows a priori the relative orders of magnitude of the various quantities involved. The problem is attacked in this chapter by a formal development with respect to a small parameter (essentially a thickness-length ratio of the ship); in doing so every quantity is developed systematically in a formal series (for a similar type of discussion see F. John [J.5]). In this way a correct theory should be obtained, assuming the convergence of the series—and there would seem to be no reason to doubt that the series would converge for sufficiently small values of the parameter. Aside from the relative safety of such a method—purchased, it is true, at the price of making rather bulky calculations—it has an additional advantage, i.e., it makes possible a consistent procedure for determining any desired higher order corrections. It is not easy to compare Haskind's theory in detail with the theory presented here. However, it can be stated that certain terms, called damping terms by Haskind, are terms that would be of higher order than any of those retained here. A more precise statement on this point will be made later.

One of the possible procedures for linearizing the problem begins by writing the equation of the hull of the ship relative to the coordinate system fixed in the ship in the form

$$(9.1.2) \qquad z' = \pm\, \beta h(x', y'), \qquad z' \gtrless 0,$$

with β a small dimensionless parameter.* This is the parameter with respect to which all quantities will be developed. In particular, the velocity potential $\Phi(X, Y, Z; t; \beta) \equiv \varphi(x, y, z; t; \beta)$ is assumed to possess the development

$$(9.1.3) \quad \varphi(x, y, z; t; \beta) = \beta\varphi_1(x, y, z; t) + \beta^2\varphi_2(x, y, z; t) + \ldots$$

The free surface elevation $\eta(x, z; t; \beta)$ and the speed $s(t; \beta)$ and angular velocity $\omega(t; \beta)$ (cf. (9.1.1)) are assumed to have the developments

$$(9.1.4) \quad \eta(x, z; t; \beta) = \beta\eta_1(x, z; t) + \beta^2\eta_2(x, z; t) + \ldots,$$

$$(9.1.5) \qquad s(t; \beta) = s_0(t) + \beta s_1(t) + \ldots,$$

$$(9.1.6) \qquad \omega(t; \beta) = \omega_0(t) + \beta\omega_1(t) + \ldots.$$

Finally, the vertical displacement $y_c(t)$ of the center of gravity and the angular displacements** $\theta_1, \theta_2, \theta_3$ of the ship with respect to the $x, y,$ and z axes respectively are assumed given by

---

* It is important to consider other means of linearization, and we shall discuss some of them later. However, it should be said here that the essential point is that a linearization can be made for any body having the form of a thin disk: it is not at all essential that the plane of the disk should be assumed to be vertical, as we have done in writing equation (9.1.2).

** Since we consider only small displacements of the ship relative to a uniform translation, it is convenient to assume at the outset that the angular displacement can be given without ambiguity as a vector with the components $\theta_1, \theta_2, \theta_3$ relative to the x, y, z-coordinate system.

(9.1.7) $\theta_i(t; \beta) = \beta\theta_{i1}(t) + \beta^2\theta_{i2}(t) + \dots$, $i = 1, 2, 3$,

(9.1.8) $y_c(t; \beta) - y_c' = \beta y_1(t) + \beta^2 y_2(t) + \dots$.

These relations imply that the velocity of the water and the elevation of its free surface are small of the same order as the "slenderness parameter" β of the ship. On the other hand, the speed $s(t)$ of the ship is assumed to be of zero order. The other quantities fixing the motion of the ship are assumed to be of first order, except for $\omega(t)$, but it turns out in the end that $\omega_0(t)$ vanishes so that ω is also of first order. The quantity y_c' in (9.1.8) was defined in connection with the description of Fig. 9.1.2; it is to be noted that we have chosen to express all quantities with respect to the moving coordinate system (x, y, z) indicated in that figure. The formulas for changes of coordinates must be used, and they also are to be developed in powers of β; for example, the equation of the hull relative to the (x, y, z) coordinate system is found to be

$$z + \beta\theta_{21}x - \beta\theta_{11}(y - y_c') - \beta h(x, y) + \dots = 0$$

after developing and rejecting second and higher order terms in β.

In marine engineering there is an accepted terminology for describing the motion of a ship; we wish to put it into relation with the notation just introduced. In doing so, the case of small deviations from a straight course is the only one in question. The angular displacements are named as follows: θ_1 is the *rolling*, $\theta_2 + \alpha$ is the *yawing*, and θ_3 is the *pitching* oscillation. The quantity $\beta s_1(t)$ in (9.1.5) is called the *surge* (i.e., it is the small fore-and-aft motion relative to the finite speed $s_0(t)$ of the ship, which turns out to be necessarily a constant), while $y_c - y_c'$ fixes the *heave*. In addition there is the sidewise displacement δz referred to as the *sway*; this quantity, in lowest order, can be calculated in terms of $s_0(t)$ and the angle α defined by (9.1.1) in terms of $\omega(t)$ as follows:

(9.1.9) $\delta z = s_0\alpha = \beta s_0 \int_0^t \omega_1(t)dt$,

since $\omega_0(t)$ turns out to vanish.

In one of the problems of most practical interest, i.e. the problem of a ship that has been moving for a long time (so that all transients have disappeared) under a constant propeller thrust (considered to be simply a force of constant magnitude parallel to the keel of the ship)

into a seaway consisting of a given system of simple harmonic progressing waves of given frequency, one expects that the displacement components would in general be the sum of two terms, one independent of
the time and representing the displacements that would arise from
motion with uniform velocity through a calm sea, the other a term
simple harmonic in the time that has its origin in the forces arising
from the waves coming from infinity. On account of the symmetry of
the hull only two displacements of the first category would differ
from zero: one the vertical displacement, i.e. the heave, the other the
pitching angle, i.e. the angle θ_3. The latter two displacements apparently are referred to as the *trim* of the ship. In all, then, there would be
in this case nine quantities to be fixed as far as the motion of the ship
is concerned: the amplitudes of the oscillations in each of the six
degrees of freedom, the speed s_0, and the two quantities determining
the trim. A procedure to determine all of them will next be outlined.

We proceed to give a summary of the theory obtained when the
series (9.1.2) to (9.1.8) are inserted in all of the equations fixing the
motion of the system, which includes both the differential equations
and the boundary conditions, and any functions involving β are in
turn developed in powers of β. For example, one needs to evaluate φ_x
on the free surface $y = \eta$ in order to express the boundary conditions
there; one calculates it as follows (using (9.1.3) and (9.1.4)):

$$(9.1.10) \quad \varphi_x(x, \eta, z; t; \beta) = \beta[\varphi_{1x}(x, 0, z; t) + \eta\varphi_{1xy}(x, 0, z; t) + \ldots]$$
$$+ \beta^2[\varphi_{2x} + \eta\varphi_{2xy} + \ldots] + \ldots$$
$$= \beta\varphi_{1x}(x, 0, z; t) + \beta^2[\eta_1\varphi_{1xy}(x, 0, z; t) + \varphi_{2x}(x, 0, z; t)] + \ldots .$$

We observe the important fact—to which reference will be made
later—that the coefficients of the powers of β are evaluated at $y = 0$,
i.e. at the undisturbed equilibrium position of the free surface of the
water. In the same way, it turns out that the boundary conditions
for the hull of the ship are automatically to be satisfied on the vertical
longitudinal mid-section of the hull. The end result of such calculations, carried out in such a way as to include all terms of first order in
β is as follows: The differential equation for φ_1 is, of course, the Laplace equation:

$$(9.1.11) \qquad \varphi_{1xx} + \varphi_{1yy} + \varphi_{1zz} = 0$$

in the domain $y < 0$, i.e. the lower half-space, excluding the plane
area A of the x, y-plane which is the orthogonal projection of the

hull (cf. Fig. 9.1.2b), in its equilibrium position, on the x, y-plane. The boundary conditions on φ_1 are

$$(9.1.12) \begin{cases} \varphi_{1z} = -s_0(h_x - \theta_{21}) - (\omega_1 + \dot{\theta}_{21})x + \dot{\theta}_{11}(y - y_c'), \text{ on } A_+ \\ \varphi_{1z} = s_0(h_x + \theta_{21}) - (\omega_1 + \dot{\theta}_{21})x + \dot{\theta}_{11}(y - y_c'), \text{ on } A_- \end{cases}$$

in which A_+ and A_- refer to the two sides $z = 0_+$ and $z = 0_-$ of the plane disk A. The boundary conditions on the free surface are

$$(9.1.13) \qquad \begin{cases} -g\eta_1 + s_0\varphi_{1x} - \varphi_{1t} = 0 \\ -\varphi_{1y} - s_0\eta_{1x} + \eta_{1t} = 0 \end{cases} \qquad \text{at } y = 0.$$

The first of these results from the condition that the pressure vanishes on the free surface, the second arises from the kinematic free surface condition. If s_0, ω_1, θ_{21}, and θ_{11} were known functions of t, these boundary conditions in conjunction with (9.1.11) and appropriate initial conditions would serve to determine the functions φ_1 and η_1 uniquely; i.e. the velocity potential and the free surface elevation would be known. Of course, the really interesting problems for us here are those in which the quantities s_0, ω_1, θ_{21}, and θ_{11}, referring to the motion of the ship, are not given in advance but are rather unknown functions of the time to be determined as part of the solution of the boundary problem. In principle, one method of approach would be to apply the Laplace transform with respect to the time t to (9.1.11), (9.1.12), and (9.1.13)—of course taking account of initial conditions at the time $t = 0$—and then to solve the resulting boundary value problem for the transform $\bar{\varphi}_1(x, y, z; \sigma)$ regarding s_0 and the transforms $\bar{\omega}_1(\sigma)$, $\bar{\theta}_{21}(\sigma)$, and $\bar{\theta}_{11}(\sigma)$ as parameters. However, for the purposes of this introduction it is better to concentrate on the most important special case (already mentioned above) in which the ship has a motion of translation with uniform speed combined with small simple harmonic oscillations of the ship and the sea having the same frequency.* In this case we write the velocity potential $\varphi_1(x, y, z; t)$, the surface elevation η_1, and the other dependent quantities in the form

$$(9.1.14) \begin{cases} \varphi_1(x, y, z; t) = \psi_0(x, y, z) + \psi_1(x, y, z)e^{i\sigma t} \\ \eta_1(x, z; t) = H_0(x, z) + H_1(x, z)e^{i\sigma t} \\ \omega_1 = \Omega_1 e^{i\sigma t}, \ \theta_{11} = \Theta_{11}e^{i\sigma t}, \ \theta_{21} = \Theta_{21}e^{i\sigma t}. \end{cases}$$

The functions ψ_0 and ψ_1 are of course both harmonic functions. We expect the functions φ_1 and η_1 to have time-independent components

---

* It can be seen, however, that the discussion which follows would take much the same course if more general motions were to be assumed.

due to the forward motion of the ship; certainly they would appear in the absence of any oscillatory components due, say, to a wave train in the sea. Upon insertion of these expressions in equations (9.1.12). and (9.1.13) we find for ψ_0 the conditions:

$$(9.1.14)_0 \quad \begin{cases} \psi_{0z} = \mp s_0 h_x, \quad \text{on } A_{\pm}, \\ gH_0 - s_0\psi_{0x} = 0 \\ \psi_{0y} + s_0 H_{0x} = 0 \end{cases} \quad \text{at } y = 0,$$

and for ψ_1 the conditions

$$(9.1.14)_1 \quad \begin{cases} -\psi_{1z} = -s_0\Theta_{21} + (\Omega_1 + i\sigma\Theta_{21})x - i\sigma\Theta_{11}(y - y_c') \text{ on } A_{\pm} \\ -gH_1 + s_0\psi_{1x} - i\sigma\psi_1 = 0 \\ -\psi_{1y} - s_0 H_{1x} + i\sigma H_1 = 0 \end{cases} \quad \text{at } y = 0.$$

We observe, in passing, that ψ_0 satisfies the same boundary conditions as in the classical Michell-Havelock theory. A little later we shall see, in fact, that the wave resistance is indeed independent of all components of the motion of the ship (to lowest order in β, that is) except its uniform forward motion with speed s_0, and that the wave resistance is determined in exactly the same way as in the Michell-Havelock theory. We continue the description of the equations which determine the motion of the ship, and which arise from developing the equations of motion with respect to β and retaining only the terms of order β and β^2. (We observe that it is necessary to consider terms of both orders.) In doing so the mass M of the ship is given by $M = M_1\beta$, with M_1 a constant, since we assume the average density of the ship to be finite and its volume is of course of order β. The moments of inertia are then also of order β. The propeller thrust is assumed to be a force of magnitude T acting in the x'-direction and in the x', y'-plane at a point whose vertical distance from the c.g. is $-l$; the thrust T is assumed to be of order β^2, since the mass is of order β and accelerations are also of order β.* The propeller thrust could also, of course, be called the wave resistance.

The terms of order β yield the following conditions:

* We have in mind problems in which the motion of the ship is a small deviation from a translatory motion with uniform finite speed. If it were desired to study motions with finite accelerations — as would be necessary, for example, if the ship were to be considered as starting from rest — it would clearly be necessary to suppose the development of the propeller thrust T to begin with a term of first order in β, since the mass of the ship is of this order. In that case, the motion of the ship at finite speed and acceleration would be determined independently of the motion of the water: in other words, it would be conditioned solely by the inert mass of the ship and the thrust of order β.

(9.1.15) $\dot{s}_0 = 0,$

(9.1.16) $2\varrho g \displaystyle\int_A \beta h dA = M_1 \beta g,$

(9.1.17) $\displaystyle\int_A x\beta h dA = 0,$

(9.1.18) $\displaystyle\int_A [(\varphi_{1t} - s_0\varphi_{1x})]_-^+ dA = 0,$

(9.1.19) $\displaystyle\int_A [x(\varphi_{1t} - s_0\varphi_{1x})]_-^+ dA = 0,$

(9.1.20) $\displaystyle\int_A [y(\varphi_{1t} - s_0\varphi_{1x})]_-^+ dA = 0.$

The symbol $[\ \ ]_-^+$ occurring here means that the jump in the quantity in brackets on going from the positive to the negative side of the projected area A of the ship's hull is to be taken. The variables of integration are x and y. The equation (9.1.15) states that the term of order zero in the speed is a constant, and hence the motion in the x-direction is a small oscillation relative to a motion with uniform velocity. (This really comes about because we assume the propeller thrust T to be of order β^2.) Equation (9.1.16) is an expression of the law of Archimedes: the rest position of equilibrium must be such that the weight of the ship just equals the weight of the water it displaces. Equation (9.1.17) expresses another law of equilibrium of a floating body, i.e. that the center of buoyancy should be on the same vertical line as the center of gravity of the ship. The remaining three equations (9.1.18), (9.1.19), and (9.1.20) in the group serve to determine the displacements θ_{11}, θ_{21}, and ω_1, which occur in the boundary condition (9.1.12) for the velocity potential φ_1. In the special case we consider (cf. (9.1.14)) we observe that these three equations would determine the values of the constants Ω_1, Θ_{11}, and Θ_{21} (the complex amplitudes of certain displacements of the ship) which occur as parameters in the boundary conditions for the harmonic function $\psi_1(x, y, z)$ given in $(9.1.14)_1$.

We are now able to draw some interesting conclusions. Once the speed s_0 is fixed, it follows that the problem of determining the harmonic function φ_1 is completely formulated through the equations (9.1.14), $(9.1.14)_0$, $(9.1.14)_1$, and (9.1.18) to (9.1.20) inclusive (together with appropriate conditions at ∞). In other words, the motion of the water, which is fixed solely by φ_1, is entirely independent of the pitching displacement $\theta_{31}(t)$, the heave $y_1(t)$, and the surge $s_1(t)$, i.e. of all displacements in the vertical plane except the constant forward speed s_0. A little reflection, however, makes this result quite plausible: Our theory is based on the assumption that the ship is a thin disk disposed vertically in the water, whose thickness is a quantity of first order. Hence only finite displacements of the disk parallel to this vertical plane could create oscillations in the water that are of first order. On the other hand, displacements of first order of the disk at right angles to itself will create motions in the water that are also of first order. One might seek to describe the situation crudely in the following fashion. Imagine a knife blade held vertically in the water. Up-and-down motions of the knife evidently produce motions of the water which are of a quite different order of magnitude from motions produced by displacements of the knife perpendicular to the plane of its blade. Stress is laid on this phenomenon here because it helps to promote understanding of other occurrences to be described later.

The terms of second order in β yield, finally, the following conditions:

(9.1.21)

$$M_1 \dot{s}_1 = - \varrho \int_A h_x [(\varphi_{1t} - s_0\varphi_{1x})^+ + (\varphi_{1t} - s_0\varphi_{1x})^-] dA + T$$

(9.1.22)

$$M_1 \ddot{y}_1 = -2\varrho g \int_L (y_1 + x\theta_{31}) h dx - \varrho \int_A h_y [(\varphi_{1t} - s_0\varphi_{1x})^+ + (\varphi_{1t} - s_0\varphi_{1x})^-] dA$$

(9.1.23)

$$I_{31} \ddot{\theta}_{31} = -2\varrho g \theta_{31} \int_A (y - y_c') h dA - 2\varrho g y_1 \int_L x h dx$$

$$- 2\varrho g \theta_{31} \int_L x^2 h dx + lT$$

$$- \varrho \int_A [x h_y - (y - y_c') h_x] [(\varphi_{1t} - s_0\varphi_{1x})^+ + (\varphi_{1t} - s_0\varphi_{1x})^-] dA.$$

We note that integrals over the projected water-line L of the ship on the vertical plane when in its equilibrium position occur in addition to integrals over the vertical projection A of the entire hull. The quantity I_{31} arises from the relation $I = \beta I_{31}$ for the moment of inertia I of the ship with respect to an axis through its c.g. parallel to the z'-axis. The equation (9.1.21) determines the surge s_1, and also the speed s_0 (or, if one wishes, the thrust T is determined if s_0 is assumed to be given). Furthermore, the speed s_0 is fixed solely by T and the geometry of the ship's hull. This can be seen, with reference to (9.1.14) and the discussion that accompanies it, in the following way: The term $\psi_0(x, y, z)$ in (9.1.14) is the term in φ_1 that is independent of t. It therefore determines T upon insertion of φ_1 in (9.1.21). This term, however, is obtained by finding the harmonic function ψ_0 as a solution of the boundary problem for ψ_0 formulated in $(9.1.14)_0$. In fact, the relation between s_0 and T is now seen to be exactly the same relation as was obtained by Michell. (It will be written down in a later section.) In other words, the wave resistance depends only on the basic translatory motion with uniform speed of the ship, and not at all on its small oscillations relative to that motion. If, then, effects on the wave resistance due to the oscillation of the ship are to be obtained from the theory, it will be necessary to take account of higher order terms. Once the thrust T has been determined the equations (9.1.22) and (9.1.23) form a coupled system for the determination of y_1 and θ_{31}, since φ_1 and θ_{11} have presumably been determined previously. Thus our system is one in which there is a considerable amount of cross-coupling. It might also be noted that the trim, i.e. the constant values of y_1 and θ_{31} about which the oscillations in these degrees of freedom occur are determined from (9.1.22) and (9.1.23) by the time-independent terms in these equations—including, for example, the moment lT of the thrust about the c.g.

We proceed to the discussion of other conclusions arising from our developments and concerning two questions which recur again and again in the literature. These issues center around the question: what is the correct manner of satisfying the boundary conditions on the curved hull of the ship? Michell employed the condition (9.1.12), naturally with $\theta_{11} = \theta_{21} = \omega_1 = 0$, on the basis of the physical argument that $s_0 h_x$ represents the component of the velocity of the water normal to the hull, and since the hull is slender, a good approximation would result by using as boundary condition the jump condition furnished by (9.1.12). Havelock and others have usually followed the same practice. However, one finds constant criticism of the resulting theory in the literature (particularly in the engineering literature) because of the fact that the boundary condition is not satisfied at the actual position of the ship's hull, and various proposals have been made to improve the approximation. This criticism would seem to be beside the point, since the condition (9.1.12) is simply the consequence of a reasonable linearization of the problem. To take account of the boundary condition at the actual position of the hull would, of course, be more accurate — but then, it would be necessary to deal with the full nonlinear problem and make sure that *all* of the essential correction terms of a given order were obtained. In particular, it would be necessary to examine the higher order terms in the conditions at the free surface — after all, the conditions (9.1.13), which are also used by Michell and Havelock (and everyone else, for that matter), are satisfied at $y = 0$ and not on the actual displaced position of the free surface. One way to obtain a more accurate theory would be, of course, to carry out the perturbation scheme outlined here to higher order terms.

Still another point has come up for frequent discussion (cf., for example, Lunde and Wigley [L.18]) with reference to the boundary condition on the hull. It is fairly common in the literature to refer to ships of Michell's type, by which is meant ships which are slender not only in the fore-and-aft direction, but which are also slender in the cross-sections at right angles to this direction (cf. Fig. 9.1.3) so

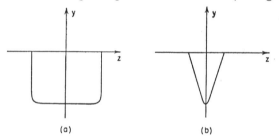

(a)   (b)

Fig. 9.1.3a, b. Ships with full and with narrow mid-sections

that h_y, in our notation, is small. Thus ships with a rather broad bottom (cf. Fig. 9.1.3a), or, as it is also put, with a full mid-section, are often considered as ships to which the present theory does not apply. On the other hand, there are experimental results (cf. Havelock [H.7]) which indicate that the theory is just as accurate for ships with a full mid-section as it is for ships of Michell's type. When the problem is examined from the point of view taken here, i.e. as a problem to be solved by a development with respect to a parameter characterizing the slenderness of the ship, the difference in the two cases would seem to be that ships with a full mid-section should be regarded as slender in both draft and beam, (otherwise no linearization based on assuming small disturbances in the water would be reasonable), while a ship of Michell's type is one in which the draft is finite and the beam is small. In the former case a development different from the one given above would result: the mass and moments of inertia would be of second order, for instance, rather than first order. Later on we shall have occasion to mention other possible ways of introducing the development parameter.

We continue by pointing out a number of conclusions, in addition to those already given, which can be inferred from our equations without solving them. Consider, for example, the equations (9.1.22) and (9.1.23) for the heave y_1 and the pitching oscillation θ_{31}, and make the assumption that

$$(9.1.24) \qquad\qquad \int_L xh\,dx = 0$$

(which means that the horizontal section of the ship at the water line has the c.g. of its area on the same vertical as that of the whole ship). If this condition is satisfied it is immediately seen that the oscillations θ_{31} and y_1 are not coupled. Furthermore, these equations are seen to have the form

$$(9.1.25) \qquad\qquad \ddot{y}_1 + \lambda_1^2 y_1 = p(t)$$

$$(9.1.26) \qquad\qquad \ddot{\theta}_{31} + \lambda_2^2 \theta_{31} = q(t)$$

with

$$(9.1.27) \qquad \lambda_1^2 = \frac{2\varrho g \int_L h\,dx}{M_1},$$

$$(9.1.28) \qquad \lambda_2^2 = \frac{2\varrho g \left[\int_A (y - y_c')h\,dA + \int_L x^2 h\,dx \right]}{I_{31}}.$$

It follows that resonance* is possible if $p(t)$ has a harmonic component of the form $A\cos(\lambda_1 t + B)$ or $q(t)$ a component of the form $A\cos(\lambda_2 t + B)$: in other words, one could expect exceptionally heavy oscillations if the speed of the ship and the seaway were to be such as to lead to forced oscillations having frequencies close to these values. One observes also that these resonant frequencies can be computed without reference to the motion of the sea or the ship: the quantities λ_1, λ_2 depend only on the shape of the hull.**

* The term resonance is used here in the strict sense, i.e. that an infinite amplitude is theoretically possible at the resonant frequency.
** The equation (9.1.27) can be interpreted in the following way: it furnishes the frequency of free vibration of a system with one degree of freedom in which the restoring force is proportional to the weight of water displaced by a cylinder of cross-section area $2\int_L h\,dx$ when it is immersed vertically in water to a depth y_1.

In spite of the fact that the linear theory presented here must be used with caution in relation to the actual practical problems concerning ships in motion, it still seems likely that such resonant frequencies would be significant if they happened to occur as harmonic components in the terms $p(t)$ or $q(t)$ with appreciable amplitudes. Suppose, for instance, that the ship is moving in a sea-way that consists of a single train of simple harmonic progressing plane waves with circular frequency σ which have their crests at right angles to the course of the ship. If the speed of the ship is s_0 one finds that the circular excitation frequency of the disturbances caused by such waves, as viewed from the moving coordinate system (x, y, z) that is used in the discussion here, is $\sigma + s_0 \sigma^2/g$, since σ^2/g is 2π times the reciprocal of the wave length of the wave train. Thus if λ_1 or λ_2 should happen to lie near this value, a heavy oscillation might be expected. One can also see that a change of course to one quartering the waves at angle γ would lead to a circular excitation frequency $\sigma + s_0 \cos \gamma \cdot \sigma^2/g$ and naturally this would have an effect on the amplitude of the response.

It has already been stated that the theory presented here is closely related to the theory published by Haskind [H.4] in 1946, and it was indicated that the two theories differ in some respects. We have not made a comparison of the two theories in the general case, which would not be easy to do, but it is possible to make a comparison rather easily in the special case treated by Haskind in detail. This is the special case dealt with in the second of his two papers in which the ship is assumed to oscillate only in the vertical plane—as would be possible if the seaway consisted of trains of plane waves all having their crests at right angles to the course of the ship. Thus only the quantities $y_1(t)$ and $\theta_{31}(t)$, which are denoted in Haskind's paper by $\zeta(t)$ and $\psi(t)$, are of interest. Haskind finds differential equations of second order for these quantities, but these equations are not the same as the corresponding equations (9.1.22), (9.1.23) above. One observes that (9.1.22) contains as its only derivative the second derivative $\ddot{y}_1$ and (9.1.23) contains as its sole derivative a term with $\ddot{\theta}_{31}$; in other words there are no first derivative terms at all, and the coupling arises solely through the undifferentiated terms. Haskind's equations are quite different since first and second derivatives of both dependent functions occur in both of the two equations; thus Haskind, on the basis of his theory, can speak, for example, of damping terms, while the theory presented here yields no such terms. On the basis of the theory presented so far there should be no damping terms of this order for the following reasons: In the absence of frictional resistances, the only way in which energy can be dissipated is through the transport of energy to infinity by means of out-going progressing waves. However, we have already given valid reasons for the fact that those oscillations of the ship which consist solely of displacements parallel to the vertical plane produce waves in the water with amplitudes that are of higher order than those considered in the first approximation. Thus no such dissipation of energy should occur.* In any case, our theory has this fact as one of its consequences. Haskind [H.4] also says, and we quote from the translation of his paper (see page 59): "Thus, for a ship symmetric with respect to its midship section . . ., only in the absence of translatory motion, i.e., for $s_0 = 0$, are the heaving and pitching oscillations independent." This statement does not hold in our version of the theory. As one sees from (9.1.22) and (9.1.23) coupling occurs if, and only if, $\int_L xh\,dx \neq 0$, whether s_0 vanishes or not. In addition, Haskind obtains no resonant frequencies in these displacements because of the presence of first-derivative terms in his equation; the author feels that such resonant frequencies may well be an important feature of the problem. Thus it seems likely that Haskind's theory differs from that presented here because he includes a number of terms which are of higher order than those retained here. Of course, it does not matter too much if some terms of higher order are included in a perturbation theory, at least if all the terms of lowest order are really present: at worst, one might be deceived in giving too much significance to such higher order terms.

The fact that the theory presented so far leads to the conclusion that no damping of the pitching, surging, and heaving oscillations occurs is naturally an important fact in relation to the practical problems. Unfortunately, actual hulls of ships seem in many cases to be designed in such a way that damping terms in the heaving and pitching oscillations are numerically of the same order as other terms in the equations of motion of a ship. (At least, there seems to be experimental evidence from model studies—see the paper by Korvin-Krukovsky and Lewis [K.16]—which bears out this statement.) Consequently, one must conclude that either actual ships are not

---

* It is, however, important to state explicitly that there would be damping of the rolling, yawing, and swaying oscillations, since these motions create waves having amplitudes of the order retained in the first approximation, and thus energy would be carried off to infinity as a consequence of such motions.

sufficiently slender for the lowest order theory developed here to apply with accuracy, or that important physical factors such as turbulence, viscosity, etc., have effects so large that they cannot be safely neglected. If it is the second factor that is decisive, rather than the loss of energy due to the creation of waves through pitching and heaving, it is clear that only a basic theory different from the one proposed here would serve to include such effects. If, however, the damping has its origin in the creation of gravity waves we need not be entirely helpless in dealing with it in terms of the sort of theory contemplated here. It would not be helpful, though, to try to overcome the difficulty by carrying the development to terms of higher order, for example, even though there would certainly then be damping effects in pitching and heaving: such damping effects of higher order could evidently not introduce damping into the motions of lower order. This is fortunately not the only way in which the difficulty can be attacked. One rather obvious procedure would be to retain the present theory, and simply add damping terms with coefficients to be fixed empirically, in somewhat the same fashion as has been proposed by St. Denis and Weinblum [S.1], for example.

There are still other possibilities for the derivation of theories which would include damping effects without requiring a semi-empirical treatment, but rather a different development with respect to a slenderness parameter. One such possibility has already been hinted at above in the course of the discussion of ships of broad mid-section compared with ships of Michell's type. If the ship is considered to be slender in both draft and beam the waves due to oscillations of the ship would be of the same order with respect to all of the degrees of freedom; a theory utilizing this observation is being investigated. Another possibility would be to regard the draft as small while the beam is finite (thus the ship is thought of as a flat body with a planing motion over the water), i.e. to base the perturbation scheme on the following equation for the hull (instead of (9.1.2)):

$$y' = \beta g(x', z'),$$

and to carry out the development with respect to β. This theory has been worked out in all detail, though it has not yet been published. With respect to damping effects the situation is now in some respects just the reverse of that described above: now it is the oscillations in the vertical plane, together with the rolling oscillation, that are damped to lowest order, while the yawing and swaying oscillations are undamped. It would seem reasonable therefore to investigate the results of such a theory for conventional hulls and make comparisons with model experiments. This still does not exhaust all of the possibilities with respect to various types of perturbation schemes, particularly if hulls of special shape are introduced. Consider, for example, a hull of the kind used for some types of sailing yachts, and shown schematically in Fig. 9.1.4. Such a hull has the property that its beam

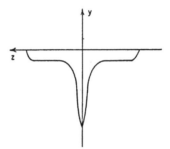

Fig. 9.1.4. Cross section of hull of a yacht

and draft are both finite, but the hull cross section consists of two thin disks joined at right angles like a T. In this case an appropriate development with respect to a slenderness parameter can also be made in regarding both disks as being slender of the same order. The result is a theory in which all oscillations, except the surge, would be damped; this theory has been worked out too but not yet published.

It would take up an inordinate amount of space in this book to deal in detail with all of the various types of possible perturbation schemes mentioned above. In addition, only one of them seems so far to permit explicit solutions even in special cases, and that is the generalization of the Michell theory which was explained at some length above. Consequently, only this theory (in fact, only a special case of it) will be developed in detail in the remainder of the chapter. In all other theories, it seems necessary to solve certain integral equations before the motion of the ship can be determined even under the most restrictive hypotheses—such as a motion of pure translation with no oscillations whatever, for example. Even in the case of the generalized Michell theory (i.e. the case of a ship regarded as a thin disk disposed vertically) an explicit solution of the problem for the lowest order approximation φ_1 to the velocity potential—in terms of an integral representation, say—seems out of the question. In fact, as soon as rolling or yawing motions occur, explicit solutions are unlikely to be found. The best that has been done so far in such cases has been to formulate an integral equation for the values of φ_1 over the vertical projection A of the ship's hull; this method of attack, which looks possible and somewhat hopeful for numerical purposes since the motion of the ship requires the knowledge of φ_1 only over the area A, is under investigation. However, if the motion of the ship is confined to a vertical plane, so that $\omega_1 = \theta_{11} = \theta_{21} = 0$, it is possible to solve the problems explicitly. This can be seen with reference to the boundary conditions (9.1.12) and (9.1.13) which in this case are identical with those of the classical theory of Michell and Havelock, and hence permit an explicit solution for φ_1 which is given later on in section 9.4. After φ_1 is determined, it can be inserted in (9.1.21), (9.1.22), and (9.1.23) to find the forward speed s_0 corresponding to the thrust T, the two quantities fixing the trim, and the surge, pitching, and heaving oscillations.* In all, six quantities fixing the motion of the ship can be determined explicitly. Only this version of the theory will be presented in detail in the remainder of the chapter.

The theory discussed here is very general, and it therefore could be applied to the study of a wide variety of different problems. For example, the stability of the oscillations of a ship could be in principle investigated on a rational dynamical basis, rather than as at present by assuming the water to remain at rest when the ship oscillates. It would be possible to investigate theoretically how a ship would move with a given rudder setting, and find the turning radius, angle of heel, etc. The problem of stabilization of a ship by gyroscopes or other devices could be attacked in a very general way: the dynamical equations for the stabilizers would simply be included in the formulation of the problem together with the forces arising from the interactions of the water with the hull of the ship.

In sec. 9.2 the general formulation of the problem is given; in sec. 9.3 the details of the linearization process are carried out for the case of a ship which is slender in beam (i.e. under the condition implied in the classical Michell-Havelock theory); and in sec. 9.4 a solution of the problem is given for the case of motion confined to the vertical plane, including a verification of the fact that the wave resistance is given by the same formula as was found by Michell.

---

* These free undamped vibrations are uniquely determined only when initial conditions are given.

9.2. General formulation of the problem

We derive here a theory for the most general motion of a rigid body through water of infinite depth which is in its turn also in motion in any manner. As always we assume that a velocity potential exists. Since we deal with a moving rigid body it is convenient to refer the motion to various types of moving coordinate systems as well as to a fixed coordinate system. The fixed coordinate system is denoted by $O - X, Y, Z$ and has the disposition used throughout this book: The X, Z-plane is in the equilibrium position of the free surface of the water, and the Y-axis is positive upwards. The first of the two moving coordinate systems we use (the second will be introduced later) is denoted by $o - x, y, z$ and is specified as follows (cf. Fig. 9.2.1):

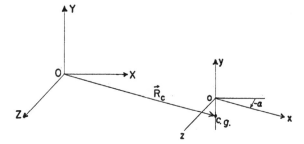

Fig. 9.2.1. Fixed and moving coordinate system

The x, z-plane coincides with the X, Z-plane (i.e. it lies in the undisturbed free surface), the y-axis is vertically upward and contains the center of gravity of the ship. The x-axis has always the direction of the horizontal component of the velocity of the center of gravity of the ship. (If we define the course of the ship as the vertical projection of the path of its center of gravity on the X, Z-plane, then our convention about the x-axis means that this axis is taken tangent to the ship's course.) Thus if $\mathbf{R}_c = (X_c, Y_c, Z_c)$ is the position vector of the center of gravity of the ship relative to the fixed coordinate system and hence $\dot{\mathbf{R}}_c = (\dot{X}_c, \dot{Y}_c, \dot{Z}_c)$ is the velocity of the c.g., it follows that the x-axis has the direction of the vector $\mathbf{u}$ given by

$$(9.2.1) \qquad\qquad \mathbf{u} = \dot{X}_c \mathbf{I} + \dot{Z}_c \mathbf{K}$$

with $\mathbf{I}$ and $\mathbf{K}$ unit vectors along the X-axis and the Z-axis. If $\mathbf{i}$ is a unit vector along the x-axis we may write

$$(9.2.2) \qquad\qquad s(t)\mathbf{i} = \mathbf{u},$$

thus introducing the speed $s(t)$ of the ship relative to a horizontal plane. For later purposes we also introduce the angular velocity vector $\boldsymbol{\omega}$ of the moving coordinate system:

$$(9.2.3) \qquad \boldsymbol{\omega} = \omega(t)\mathbf{J},$$

and the angle α (cf. Fig. 9.2.1) by

$$(9.2.4) \qquad \alpha(t) = \int_0^t \omega(\tau)d\tau.$$

The equations of transformation from one coordinate system to the other are

$$(9.2.5) \quad \begin{cases} X = x \cos \alpha + z \sin \alpha + X_c \ ; \ x = (X - X_c) \cos \alpha - (Z - Z_c) \sin \alpha \\ Y = y \qquad\qquad\qquad\qquad ; \ y = Y \\ Z = -x \sin \alpha + z \cos \alpha + Z_c ; \ z = (X - X_c) \sin \alpha + (Z - Z_c) \cos \alpha. \end{cases}$$

By $\Phi(X, Y, Z; t)$ we denote the velocity potential and write

$$(9.2.6) \quad \Phi(X, Y, Z; t)$$
$$= \Phi(x \cos \alpha + z \sin \alpha + X_c, y, -x \sin \alpha + z \cos \alpha + Z_c; t)$$
$$\equiv \varphi(x, y, z; t).$$

In addition to the transformation formulas for the coordinates, we also need the formulas for the transformation of various derivatives. One finds without difficulty the following formulas:

$$(9.2.7) \quad \begin{cases} \Phi_X = \varphi_x \cos \alpha + \varphi_z \sin \alpha \\ \Phi_Y = \varphi_y \\ \Phi_Z = -\varphi_x \sin \alpha + \varphi_z \cos \alpha. \end{cases}$$

It is clear that $\operatorname{grad}^2 \Phi(X, Y, Z; t) = \operatorname{grad}^2 \varphi(x, y, z; t)$ and that φ is a harmonic function in x, y, z since Φ is harmonic in X, Y, Z. To calculate Φ_t is a little more difficult; the result is

$$(9.2.8) \qquad \Phi_t = -(s + \omega z)\varphi_x + \omega x \varphi_z + \varphi_t.$$

(To verify this formula, one uses $\Phi_t = \varphi_x x_t + \varphi_y y_t + \varphi_z z_t + \varphi_t$ and the relations (9.2.5) together with $s \cos \alpha = \dot{X}_c$, $s \sin \alpha = -\dot{Z}_c$.) The last two sets of equations make it possible to express Bernoulli's law in terms of $\varphi(x, y, z; t)$; one has:

$$(9.2.9) \quad \frac{p}{\varrho} + gy + \frac{1}{2}(\operatorname{grad} \varphi)^2 - (s + \omega z)\varphi_x + \omega x \varphi_z + \varphi_t = 0.$$

Suppose now that $F(X, Y, Z; t) = 0$ is a boundary surface (fixed or moving) and set

$$(9.2.10) \quad F(x \cos \alpha + \ldots, y, -x \sin \alpha + \ldots; t) \equiv f(x, y, z; t),$$

so that $f(x, y, z; t) = 0$ is the equation of the boundary surface relative to the moving coordinate system. The kinematic condition to be
satisfied on such a boundary surface is that the particle derivative
dF/dt vanishes, and this leads to the boundary condition

$$(9.2.11) \quad \varphi_x f_x + \varphi_y f_y + \varphi_z f_z - (s + \omega z) f_x + \omega x f_z + f_t = 0$$

relative to the moving coordinate system upon using the appropriate
transformation formulas. In particular, if $y - \eta(x, z; t) = 0$ is the
equation of the free surface of the water, the appropriate kinematic
condition is

$$(9.2.12) \quad - \varphi_x \eta_x + \varphi_y - \varphi_z \eta_z + (s + \omega z) \eta_x - \omega x \eta_z - \eta_t = 0$$

to be satisfied for $y = \eta$. (The dynamic free surface condition is of
course obtained for $y = \eta$ from (9.2.9) by setting $p = 0$.)

We turn next to the derivation of the appropriate conditions, both
kinematic and dynamic, on the ship's hull. To this end it is convenient
to introduce another moving coordinate system $o' - x', y', z'$ which
is rigidly attached to the ship. It is assumed that the hull of the ship
has a vertical plane of symmetry (which also contains the center of
gravity of the ship); we locate the x', y'-plane in it (cf. Fig. 9.2.2) and
suppose that the y'-axis contains the center of gravity. The $o' - x'$,
y', z' system, like the other moving system, is supposed to coincide

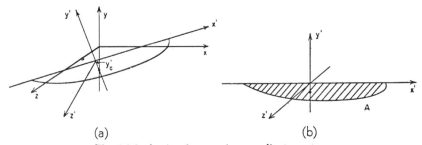

(a)   (b)

Fig. 9.2.2a, b. Another moving coordinate system

with the fixed system in the rest position of equilibrium. The center
of gravity of the ship will thus be located at a definite point on the
y'-axis, say at distance y'_c from the origin o': in other words, the system
of coordinates attached rigidly to the ship is such that the center of
gravity has the coordinate $(0, y'_c, 0)$.

In the present section we do not wish in general to carry out linearizations. However, since we shall in the end deal only with motions

which involve small oscillations of the ship relative to the first moving coordinate system $o - x, y, z$, it is convenient and saves time and space to suppose even at this point that the angular displacement of the ship relative to the $o - x, y, z$ system is so small that it can be treated as a vector $\boldsymbol{\theta}$:

(9.2.13) $$\boldsymbol{\theta} = \theta_1 \mathbf{i} + \theta_2 \mathbf{j} + \theta_3 \mathbf{k}.$$

The transformation formulas, correct up to first order terms in the components θ_i of $\boldsymbol{\theta}$, are then given by:

(9.2.14)
$$\begin{cases} x' = x + \theta_3(y - y_c) - \theta_2 z \\ y' = y - (y_c - y'_c) + \theta_1 z - \theta_3 x \\ z' = z + \theta_2 x - \theta_1(y - y_c) \end{cases}$$

with y_c of course representing the y-coordinate of the center of gravity in the unprimed system. It is assumed that $y_c - y'_c$ is a small quantity of the same order as θ_i and only linear terms in this quantity have been retained. (The verification of (9.2.14) is easily carried out by making use of the vector-product formula $\boldsymbol{\delta} = \boldsymbol{\theta} \times \mathbf{r}$, for the small displacement $\boldsymbol{\delta}$ of a rigid body under a small rotation $\boldsymbol{\theta}$.)

The equation of the hull of the ship (assumed to be symmetrical with respect to the x', y'-plane) is now supposed given relative to the primed system of coordinates in the form:

(9.2.15) $$z' = \pm \zeta(x', y'), \qquad z' \gtrless 0.$$

The equation of the hull relative to the $o - x, y, z$-system can now be written in the form

(9.2.16) $$z + \theta_2 x - \theta_1(y - y'_c) - \zeta(x, y) + [\theta_2 z - \theta_3(y - y'_c)]\zeta_x(x,y)$$
$$+ [(y_c - y'_c) - \theta_1 z + \theta_3 x]\zeta_y(x, y) = 0, \qquad z' > 0,$$

when higher order terms in $(y_c - y'_c)$ and θ_i are neglected. The left hand side of this equation could now be inserted for f in (9.2.11) to yield the kinematic boundary condition on the hull of the ship, but we postpone this step until the next section.

The dynamical conditions on the ship's hull are obtained from the assumption that the ship is a rigid body in motion under the action of the propeller thrust $\mathbf{T}$, its weight Mg, and the pressure p of the water on its hull. The principle of the motion of the center of gravity yields the condition

(9.2.17) $$M \frac{d}{dt}(s\mathbf{i} + \dot{y}_c\mathbf{j}) = \int p\mathbf{n}\, dS + \mathbf{T} - Mg\mathbf{j}.$$

By **n** we mean the inward unit normal on the hull. Our moving coordinate system $o - x, y, z$ is such that $di/dt = -\omega\mathbf{k}$ and $dj/dt = 0$, so that (9.2.17) can be written in the form

$$(9.2.18) \quad M\ddot{s}\mathbf{i} - Ms\omega\mathbf{k} + M\ddot{y}_c\mathbf{j} = \int_S p\mathbf{n}\, dS + \mathbf{T} - Mg\mathbf{j},$$

with p defined by (9.2.9). The law of conservation of angular momentum is taken in the form:

$$(9.2.19) \quad \frac{d}{dt}\int_M (\mathbf{R} - \mathbf{R}_c) \times (\dot{\mathbf{R}} - \dot{\mathbf{R}}_c)dm$$

$$= \int_S p(\mathbf{R} - \mathbf{R}_c)\,\mathbf{n}\, dS + (\mathbf{R}_T - \mathbf{R}_c) \times \mathbf{T}.$$

The crosses all indicate vector products. By **R** is meant the position vector of the element of mass dm relative to the fixed coordinate system. $\mathbf{R}_c$ (cf. Fig. 9.2.1) fixes the position of the c.g. and $\mathbf{R}_T$ locates the point of application of the propeller thrust T. We introduce $\mathbf{r} = (x, y, z)$ as the position vector of any point in the ship in the moving coordinate system and set

$$(9.2.20) \qquad\qquad \mathbf{q} = \mathbf{r} - y_c\mathbf{j},$$

so that **q** is a vector from the c.g. to any point in the ship. The relation

$$(9.2.21) \qquad\qquad \dot{\mathbf{R}} = \dot{\mathbf{R}}_c + (\boldsymbol{\omega} + \dot{\boldsymbol{\theta}}) \times \mathbf{q}$$

holds, since $\boldsymbol{\omega} + \dot{\boldsymbol{\theta}}$ is the angular velocity of the ship; thus (9.2.21) is simply the statement of a basic kinematic property of rigid bodies. By using the last two relations the dynamical condition (9.2.19) can be expressed in terms of quantities measured with respect to the moving coordinate system $o - x, y, z$, as follows:

$$(9.2.22) \quad \frac{d}{dt}\int_M (\mathbf{r} - y_c\mathbf{j}) \times [(\boldsymbol{\omega} + \dot{\boldsymbol{\theta}}) \times (\mathbf{r} - y_c\mathbf{j})]dm$$

$$= \int_S p(\mathbf{r} - y_c\mathbf{j}) \times \mathbf{n}\, dS + (\mathbf{R}_T - \mathbf{R}_c) \times \mathbf{T}.$$

We have now derived the basic equations for the motion of the ship. What would be wanted in general would be a velocity potential $\varphi(x, y, z; t)$ satisfying (9.2.11) on the hull of the ship, conditions (9.2.9) (with $p = 0$) and (9.2.12) on the free surface of the water; and conditions (9.2.17) and (9.2.22), which involve φ under integral signs through the pressure p as given by (9.2.9). Of course, the quantities fixing the motion of the ship must also be determined in such a way that all of the conditions are satisfied. In addition, there would be initial conditions and conditions at ∞ to be satisfied. Detailed consideration of these conditions, and the complete formulation of the problem of determining $\varphi(x, y, z; t)$ under various conditions will be postponed until later on since we wish to carry out a linearization of all of the conditions formulated here.

9.3. Linearization by a formal perturbation procedure

Because of the complicated nature of our conditions, it seems wise (as was indicated in sec. 1 of this chapter) to carry out the linearization by a formal development in order to make sure that all terms of a given order are retained; this is all the more necessary since terms of different orders must be considered. The linearization carried out here is based on the assumption that the motion of the water relative to the fixed coordinate system is a small oscillation about the rest position of equilibrium. It follows, in particular, that the elevation of the free surface of the water should be assumed to be small and, of course, that the motion of the ship relative to the first moving coordinate system $o - x, y, z$ should be treated as a small oscillation. We do not, however, wish to consider the speed of the ship with respect to the fixed coordinate system to be a small quantity: it should rather be considered a finite quantity. This brings with it the necessity to restrict the form of the ship so that its motion through the water does not cause disturbances so large as to violate our basic assumption; in other words, we must assume the ship to have the form of a thin disk. In addition, it is clear that the velocity of such a disk-like ship must of necessity maintain a direction that does not depart too much from the plane of the thin disk if small disturbances only are to be created. Thus we assume that the equation of the ship's hull is given by

$$(9.3.1) \qquad\qquad z' = \beta h(x', y'), \qquad z' > 0,$$

with β a small dimensionless parameter, so that the ship is a thin disk symmetrical with respect to the x', y'-plane, and βh takes the place of ζ in (9.2.15). (It has already been noted in the introduction to this chapter that this is not the most general way to describe the shape of a disk that would be suitable for a linearization of the type carried out here.) We have already assumed that the motion of the ship is a small oscillation relative to the moving coordinate system

$o - x, y, z$. It seems reasonable, therefore, to develop all our basic quantities (taken as functions of $x, y, z; t$) in powers of β, as follows:

(9.3.2) $\varphi(x, y, z; t; \beta) = \beta\varphi_1 + \beta^2\varphi_2 + \cdots,$

(9.3.3) $\eta(x, z; t; \beta) = \beta\eta_1 + \beta^2\eta_2 + \cdots,$

(9.3.4) $s(t; \beta) = s_0 + \beta s_1 + \beta^2 s_2 + \cdots,$

(9.3.5) $\omega(t; \beta) = \omega_0 + \beta\omega_1 + \beta^2\omega_2 + \cdots,$

(9.3.6) $\theta_i(t; \beta) = \beta\theta_{i1} + \beta^2\theta_{i2} + \cdots,$ $i = 1, 2, 3$

(9.3.7) $y_c - y'_o = \beta y_1 + \beta^2 y_2 + \cdots.$

The first and second conditions state that the velocity potential and the surface wave amplitudes, as seen from the moving system, are small of order β. The speed of the ship, on the other hand, and the angular velocity of the moving coordinate system about the vertical axis of the fixed coordinate system, are assumed to be of order zero. (It will turn out, however, that ω_0 must vanish—a not unexpected result.) The relations (9.3.6) and (9.3.7) serve to make precise our assumption that the motion of the ship is a small oscillation relative to the system $o - x, y, z$.

We must now insert these developments in the conditions derived in the previous section. The free surface conditions are treated first. As a preliminary step we observe that

(9.3.8) $\varphi_x(x, \eta, z; t; \beta) = \beta[\varphi_{1x}(x, 0, z; t) + \eta\varphi_{1xy}(x, 0, z; t) + \cdots]$
$$+ \beta^2[\varphi_{2x} + \eta\varphi_{2xy} + \cdots]$$
$$+ \cdots \cdots$$
$$= \beta\varphi_{1x}(x, 0, z; t) + \beta^2[\eta_1\varphi_{1xy}(x, 0, z; t) + \varphi_{2x}(x, 0, z; t)]$$
$$+ \cdots \cdots,$$

with similar formulas for other quantities when they are evaluated on the free surface $y = \eta$. Here we have used the fact that η is small of order β and have developed in Taylor series. Consequently, the dynamic free surface condition for $y = \eta$ arising from (9.2.9) with $p = 0$ can be expressed in the form

(9.3.9) $g[\beta\eta_1 + \beta^2\eta_2 + \cdots] + \tfrac{1}{2}\beta^2[(\mathrm{grad}\ \varphi_1)^2 + \cdots]$
$$- [s_0 + \beta s_1 + \cdots + z(\omega_0 + \beta\omega_1 + \cdots)][\beta\varphi_{1x} +$$
$$\beta^2(\eta_1\varphi_{1xy} + \varphi_{2x}) + \cdots]$$
$$+ x(\omega_0 + \beta\omega_1 + \cdots)[\beta\varphi_{1z} + \beta^2(\eta_1\varphi_{1zy} + \varphi_{2z}) + \cdots]$$
$$+ [\beta\varphi_{1t} + \beta^2(\eta_1\varphi_{1ty} + \varphi_{2t}) + \cdots] = 0$$

and this condition is to be satisfied for $y = 0$. In fact, as always in problems of small oscillations of continuous media, the boundary conditions are satisfied in general at the equilibrium position of the boundaries. Upon equating the coefficient of the lowest order term to zero we obtain the dynamical free surface condition

$$(9.3.10) \quad - g\eta_1 + (s_0 + \omega_0 z)\varphi_{1x} - \omega_0 x \varphi_{1z} - \varphi_{1t} = 0 \qquad \text{for } y = 0,$$

and it is clear that conditions on the higher order terms could also be obtained if desired. In a similar fashion the kinematic free surface condition can be derived from (9.2.12); the lowest order term in β yields this condition in the form:

$$(9.3.11) \quad \varphi_{1y} + (s_0 + \omega_0 z)\eta_{1x} - \omega_0 x \eta_{1z} - \eta_{1t} = 0 \qquad \text{for } y = 0.$$

We turn next to the derivation of the linearized boundary conditions on the ship's hull. In view of (9.3.6) and (9.3.7), the transformation formulas (9.2.14) can be put in the form

$$(9.3.12) \quad \begin{cases} x' = x + \beta\theta_{31}(y - y_c') - \beta\theta_{21}z \\ y' = y - \beta y_1 + \beta\theta_{11}z - \beta\theta_{31}x \\ z' = z + \beta\theta_{21}x - \beta\theta_{11}(y - y_c') \end{cases}$$

when terms involving second and higher powers of β are rejected. Consequently, the equation (9.2.16) of the ship's hull, up to terms in β^2, can be written as follows:

$$z + \beta\theta_{21}x - \beta\theta_{11}(y - y_c') - \beta h[x + \beta\theta_{31}(y - y_c') - \beta\theta_{21}z,$$
$$y - \beta y_1 + \beta\theta_{11}z - \beta\theta_{31}x] = 0,$$

and, upon expanding the function h, the equation becomes

$$(9.3.13) \qquad z + \beta\theta_{21}x - \beta\theta_{11}(y - y_c') - \beta h(x, y) + \ldots = 0,$$

the dots representing higher order terms in β. We can now obtain the kinematic boundary condition for the hull by inserting the left hand side of (9.3.13) for the function f in (9.2.11); the result is

$$(9.3.14) \quad \begin{cases} \omega_0 = 0 \\ \varphi_{1z} = s_0(\theta_{21} - h_x) - x\omega_1 - \dot\theta_{21}x + \dot\theta_{11}(y - y_c') \end{cases}$$

when the terms of zero and first order only are taken into account. It is clear that these conditions are to be satisfied over the domain A of the x, y-plane that is covered by the projection of the hull on the plane when the ship is in the rest position of equilibrium. As was mentioned earlier, it turns out that $\omega_0 = 0$, i.e., that the angular velocity about the y-axis must be small of first order, or, as it could also be put, the curvature of the ship's course must be small since the speed in the course is finite. The quantity $s_1(t)$ in (9.3.4) evidently yields the oscillation of the ship in the direction of the x-axis (the so-called "surge").

It should also be noted that if we use $z' = -\beta h(x', y')$ we find, corresponding to (9.3.14), that

$$\varphi_{1z} = s_0(\theta_{21} + h_x) - (\omega_1 + \dot\theta_{21})x + \dot\theta_{11}(y - y_c').$$

This means that A must be regarded as two sided, and that the last equation is to be satisfied on the side of A which faces the negative z-axis. The last equation and (9.3.14) imply that φ may have discontinuities at the disk A.

The next step in the procedure is to substitute the developments with respect to β, (9.3.2)—(9.3.7), in the conditions for the ship's hull given by (9.2.18) and (9.2.22). Let us begin with the integral $\int_S pn\, dS$ which appears in (9.2.18). In this integral S is the immersed surface of the hull, n is the inward unit normal to this surface and p is the pressure on it which is to be calculated from (9.2.9). With respect to the $o - x$, y, z coordinate system the last equations of the symmetrical halves of the hull are

$$(9.3.15) \quad \begin{cases} S_1: & z = H_1(x, y; t; \beta) = f_1 + f_2 \\ S_2: & z = H_2(x, y; t; \beta) = -f_1 + f_2 \end{cases}$$

where

$$(9.3.16) \quad \begin{cases} f_1 = \beta h + \beta^2[\theta_{31}(y - y_c')h_x - (\theta_{31}x + y_1)h_y] + O(\beta^3) \\ f_2 = -\beta\theta_{21}x + \beta\theta_{11}(y - y_c') + O(\beta^2). \end{cases}$$

We can now write

$$\int_S pn\, dS = \int_{S_1} pn_1\, dS_1 + \int_{S_2} pn_2\, dS_2$$

in which n_1 and n_2 are given by

$$n_1 = \frac{H_{1x}i + H_{1y}j - k}{\sqrt{1 + H_{1x}^2 + H_{1y}^2}}; \qquad n_2 = \frac{-H_{2x}i - H_{2y}j + k}{\sqrt{1 + H_{2x}^2 + H_{2y}^2}}.$$

We can also write

$$\int_S pn\, dS = -\varrho g \int_S yn\, dS + \int_S p_1 n\, dS$$

$$= -\varrho g \int_S yn\, dS + \int_{S_1} p_1 n_1 dS_1 + \int_{S_2} p_1 n_2\, dS_2$$

where p_1, from (9.2.9), is given by

$$(9.3.17) \quad p_1 = -\varrho[\tfrac{1}{2}(\mathrm{grad}\ \varphi)^2 - (s + \omega z)\varphi_x + x\omega\varphi_z + \varphi_t].$$

If S_0 is the hull surface below the x, z-plane, the surface area $S_0 - S$ is of order β and in this area each of the quantities y, H_1, H_2 is of order β. Hence one finds the following to hold:

$$-\int_S y\mathbf{n}\, dS = -\int_{S_0} y\,\mathbf{n}\, dS + (\mathbf{i} + \mathbf{j})O(\beta^3) + \mathbf{k}O(\beta^2).$$

From the divergence theorem we have

$$-\int_{S_0} y\mathbf{n}\, dS = V\mathbf{j},$$

where V is the volume bounded by S_0 and the x, z-plane. With an accuracy of order β^3, V is given by

$$V = 2\beta\int_A h\, dA - \int_B \beta(y_1 + \theta_{31}x)dB = 2\beta\int_A h\, dA - 2\beta^2\int_L (y_1 + \theta_{31}x)h\, dx.$$

Here A is the projection of the hull on the vertical plane when the hull is in the equilibrium position, B is the equilibrium water line area, and L is the projection of the equilibrium water line on the x-axis.

If W_1, W_2 are the respective projections of the immersed surfaces S_1, S_2 on the x, y-plane we have

$$\int_S p_1\mathbf{n}\, dS = \mathbf{i}\left\{\int_{W_1} p_1(x, y, H_1; t)H_{1x}dW_1 - \int_{W_2} p_1(x, y, H_2; t)H_{2x}dW_2\right\}$$

$$+\mathbf{j}\left\{\int_{W_1} p_1(x, y, H_1; t)H_{1y}dW_1 - \int_{W_2} p_1(x, y, H_2; t)H_{2y}dW_2\right\}$$

$$-\mathbf{k}\left\{\int_{W_1} p_1(x, y, H_1; t)dW_1 - \int_{W_2} p_1(x, y, H_2; t)dW_2\right\}.$$

Neither W_1 nor W_2 is identical with A. Each of the differences $W_1 - A$, $W_2 - A$ is, however, an area of order β. From this and the fact that each of the quantities p, H_{1x}, H_{1y}, H_{2x}, H_{2y} is of order β, it follows that

(9.3.18)

$$\int_S p_1\mathbf{n}\, dS = \mathbf{i}\left\{\int_A [p_1(x, y, H_1; t)H_{1x} - p_1(x, y, H_2; t)H_{2x}]dA + O(\beta^3)\right\}$$

$$+\mathbf{j}\left\{\int_A [p_1(x, y, H_1; t)H_{1y} - p_1(x, y, H_2; t)H_{2y}]dA + O(\beta^3)\right\}$$

$$-\mathbf{k}\left\{\int_A [p_1(x, y, H_1; t) - p_1(x, y, H_2; t)]dA + O(\beta^2)\right\}.$$

It was pointed out above that φ may be discontinuous on A. Hence from (9.3.17), (9.3.2), (9.3.4) we write

$$(9.3.19) \quad \begin{cases} p_1(x, y, H_1; t) = \varrho\beta(s_0\varphi_{1x} - \varphi_{1t})^+ + O(\beta^2) \\ p_1(x, y, H_2; t) = \varrho\beta(s_0\varphi_{1x} - \varphi_{1t})^- + O(\beta^2). \end{cases}$$

Here the $+$ and $-$ superscripts denote values at the positive and negative sides of the disk A whose positive side is regarded as the side which faces the positive z-axis. If we substitute the developments of $H_{1x}, H_{1y}, H_{2x}, H_{2y}$, and (9.3.19) in (9.3.18), then collect the previous results, we find

$$\int_S p\mathbf{n}\,dS = \mathbf{i}\left\{\varrho\beta^2\int_A [(h_x - \theta_{21})(s_0\varphi_{1x} - \varphi_{1t})^+ + (h_x + \theta_{21})(s_0\varphi_{1x} - \varphi_{1t})^-]dA + O(\beta^3)\right\}$$

$$(9.3.20) \quad +\mathbf{j}\left\{\begin{matrix} 2\varrho g\beta\int_A h\,dA - 2\varrho g\beta^2\int_L (y_1 + x\theta_{31})h\,dx \\ +\varrho\beta^2\int_A [(h_y + \theta_{11})(s_0\varphi_{1x} - \varphi_{1t})^+ + (h_y - \theta_{11})(s_0\varphi_{1x} - \varphi_{1t})^-]dA + O(\beta^3) \end{matrix}\right\}$$

$$-\mathbf{k}\left\{\varrho\beta\int_A [(s_0\varphi_{1x} - \varphi_{1t})^+ - (s_0\varphi_{1x} - \varphi_{1t})^-]dA + O(\beta^2)\right\}.$$

The integral $\int_S p(\mathbf{r} - y_c\mathbf{j}) \times \mathbf{n}\,dS$ which appears in (9.2.22) can be written

$$\int_S p(\mathbf{r} - y_c\mathbf{j}) \times \mathbf{n}\,dS = -\varrho g\int_S y(\mathbf{r} - y_c\mathbf{j}) \times \mathbf{n}\,dS$$

$$+ \int_{S_1} p_1(\mathbf{r} - y_c\mathbf{j}) \times \mathbf{n}_1\,dS_1$$

$$+ \int_{S_2} p_1(\mathbf{r} - y_c\mathbf{j}) \times \mathbf{n}_2\,dS_2.$$

If we use the same procedure as was used above for the expansion of $\int_S p\mathbf{n}\,dS$ we find

$$p(\mathbf{r} - y_c\mathbf{j}) \times \mathbf{n}\,dS = -\mathbf{i}\left\{\varrho\beta\int_A [(y - y_c)(s_0\varphi_{1x} - \varphi_{1t})^+ - (y - y_c)(s_0\varphi_{1x} - \varphi_{1t})^-]dA + O(\beta^2)\right.$$

$$\left. +\mathbf{j}\left\{\varrho\beta\int_A [x(s_0\varphi_{1x} - \varphi_{1t})^+ - x(s_0\varphi_{1x} - \varphi_{1t})^-]dA + O(\beta^2)\right.\right\}$$

.3.21)

$$\mathbf{k}\begin{cases} 2\varrho g\beta \int_A xhdA - 2\varrho g\beta^2\theta_{31} \int_A (y-y_c')hdA - 2\varrho g\beta^2 y_1 \int_L xhdx - 2\varrho g\beta^2\theta_{31} \int_L x^2 hdx \\[2ex] +\varrho\beta^2 \int_A [x(h_y+\theta_{11})(s_0\varphi_{1x}-\varphi_{1t})^+ + x(h_y-\theta_{11})(s_0\varphi_{1x}-\varphi_{1t})^-]dA \\[2ex] -\varrho\beta^2\int_A [(y-y_c)(h_x-\theta_{21})(s_0\varphi_{1x}-\varphi_{1t})^+ + (y-y_c)(h_x+\theta_{21})(s_0\varphi_{1x}-\varphi_{1t})^-]dA + O(\beta \end{cases}$$

We now assume that the propeller thrust $\mathbf{T}$ is of order β^2 and is directed parallel to the x'-axis: that is

$$\mathbf{T} = \beta^2 T\mathbf{i}'$$

where $\mathbf{i}'$ is the unit vector along the x'-axis. We also assume that $\mathbf{T}$ is applied at a point in the longitudinal plane of symmetry of the ship l units *below* the center of mass. Thus we have the relations

(9.3.22) $$\mathbf{T} = \beta^2 T\mathbf{i} + O(\beta^3),$$

and

(9.3.23) $$\begin{aligned}(\mathbf{R}_T - \mathbf{R}_c) \times \mathbf{T} &= -l\mathbf{j} \times \mathbf{T} \\ &= l\beta^2 T\mathbf{k} + O(\beta^3).\end{aligned}$$

The mass of the ship is of order β. If we write $M = M_1\beta$ and expand the left hand side of (9.2.18) in powers of β it becomes

$$\mathbf{i}[M_1\beta\dot{s}_0 + M_1\beta^2\dot{s}_1 + O(\beta^3)] + \mathbf{j}[M_1\beta^2\ddot{y}_1 + O(\beta^3)] - \mathbf{k}[O(\beta^2)]$$

(9.3.24) $$= \int_S p\mathbf{n}\,dS + \mathbf{T} - M_1\beta g\mathbf{j}.$$

The expansion of the left hand side of (9.2.22) gives

(9.3.25) $$\mathbf{i}[O(\beta^2)] + \mathbf{j}[O(\beta^2)] + \mathbf{k}[I_{31}\beta^2\ddot{\theta}_{31} + O(\beta^3)]$$

$$= \int_S p(\mathbf{r} - y_c\mathbf{j}) \times \mathbf{n}\,dS + (\mathbf{R}_T - \mathbf{R}_c) \times \mathbf{T}$$

where βI_{31} is the moment of inertia of the ship about the axis which is perpendicular to the longitudinal plane of symmetry of the ship and which passes through the center of mass.

If we replace the pressure integrals and thrust terms in the last two equations by (9.3.20), (9.3.21), (9.3.22), (9.3.23), and then equate the coefficients of like powers of β in (9.3.24) and (9.3.25) we obtain the following linearized equations of motion of the ship. From the first order terms we find

(9.3.26)
$$s_0 = 0$$

(9.3.27)
$$2\varrho g \int_A \beta h dA = M_1 \beta g$$

(9.3.28)
$$\int_A x\beta h dA = 0$$

(9.3.29)
$$\int_A [(s_0\varphi_{1x} - \varphi_{1t})^+ - (s_0\varphi_{1x} - \varphi_{1t})^-] dA = 0$$

(9.3.30)
$$\int_A [x(s_0\varphi_{1x} - \varphi_{1t})^+ - x(s_0\varphi_{1x} - \varphi_{1t})^-] dA = 0$$

(9.3.31)
$$\int_A [(y - y_c')(s_0\varphi_{1x} - \varphi_{1t})^+ - (y - y_c')(s_0\varphi_{1x} - \varphi_{1t})^-] dA = 0$$

or by (9.3.29)

(9.3.32)
$$\int_A [y(s_0\varphi_{1x} - \varphi_{1t})^+ - y(s_0\varphi_{1x} - \varphi_{1t})^-] dA = 0.$$

From the second order terms we find

(9.3.33)
$$M_1\dot{s}_1 = \varrho \int_A [(h_x - \theta_{21})(s_0\varphi_{1x} - \varphi_{1t})^+ + (h_x + \theta_{21})(s_0\varphi_{1x} - \varphi_{1t})^-] dA + T$$

$$= \varrho \int_A [h_x(s_0\varphi_{1x} - \varphi_{1t})^+ + h_x(s_0\varphi_{1x} - \varphi_{1t})^-] dA + T'$$

(9.3.34)
$$M_1\ddot{y}_1 = -2\varrho g \int_L (y_1 + x\theta_{31})h dx$$

$$+ \varrho \int_A [(h_y + \theta_{11})(s_0\varphi_{1x} - \varphi_{1t})^+ + (h_y - \theta_{11})(s_0\varphi_{1x} - \varphi_{1t})^-] dA$$

$$= -2\varrho g \int_L (y_1 + x\theta_{31})h dx + \varrho \int_A [h_y(s_0\varphi_{1x} - \varphi_{1t})^+ + h_y(s_0\varphi_{1x} - \varphi_{1t})^-] dA,$$

$$I_{31}\ddot{\theta}_{31} = -2\varrho g \theta_{31} \int_A (y - y_c')h dA - 2\varrho g y_1 \int_L xh dx - 2\varrho g \theta_{31} \int_L x^2 h dx + lT$$

$$+ \varrho \int_A [x(h_y + \theta_{11})(s_0\varphi_{1x} - \varphi_{1t})^+ + x(h_y - \theta_{11})(s_0\varphi_{1x} - \varphi_{1t})^-] dA$$

$$- \varrho \int_A [(y - y_c')(h_x - \theta_{21})(s_0\varphi_{1x} - \varphi_{1t})^+$$
$$+ (y - y_c')(h_x + \theta_{21})(s_0\varphi_{1x} - \varphi_{1t})^-] dA$$

or by (9.3.30), (9.3.31)

$$I_{31}\ddot{\theta}_{31} = -2\varrho g\theta_{31}\int_A (y-y_c')h\,dA - 2\varrho g y_1\int_L xh\,dx - 2\varrho g\theta_{31}\int_L x^2h\,dx + lT$$

(9.3.35)

$$+\varrho\int_A [xh_y - (y-y_c')h_x][(s_0\varphi_{1x}-\varphi_{1t})^+ + (s_0\varphi_{1x}-\varphi_{1t})^-]\,dA.$$

Equation (9.3.26) states that the motion in the x-direction is a small oscillation relative to a motion with uniform speed $s_0 = $ const. Equation (9.3.27) is an expression of Archimedes' law: the rest position of equilibrium must be such that the weight of the water displaced by the ship just equals the weight of the ship. The center of buoyancy of the ship is in the plane of symmetry, and equation (9.3.28) is an expression of the second law of equilibrium of a floating body; namely that the center of buoyancy for the equilibrium position is on the same vertical line, the y'-axis, as the center of gravity of the ship.

The function φ_1 must satisfy

$$\varphi_{1xx} + \varphi_{1yy} + \varphi_{1zz} = 0$$

in the domain $D - A$ where D is the half space $y < 0$, and A is the plane disk defined by the projection of the submerged hull on the x, y-plane when the ship is in the equilibrium position. We assume that A intersects the x, z-plane. The boundary conditions at each side of A are

(9.3.36) $\begin{cases} \varphi_{1z}^+ = -s_0(h_x - \theta_{21}) - (\omega_1 + \dot{\theta}_{21})x + \dot{\theta}_{11}(y - y_c') \\ \varphi_{1z}^- = +s_0(h_x + \theta_{21}) - (\omega_1 + \dot{\theta}_{21})x + \dot{\theta}_{11}(y - y_c'). \end{cases}$

The boundary condition at $y = 0$ is found by eliminating η_1 from (9.3.10) and (9.3.11). Since $\omega_0 = 0$ these equations are

$$-g\eta_1 + s_0\varphi_{1x} - \varphi_{1t} = 0$$
$$-\varphi_{1y} - s_0\eta_{1x} + \eta_{1t} = 0$$

and they yield

(9.3.37) $s_0^2\varphi_{1xx} - 2s_0\varphi_{1xt} + g\varphi_{1y} + \varphi_{1tt} = 0$

for $y = 0$. The boundary conditions (9.3.36) and (9.3.37) show that φ_1 depends on $\omega_1(t)$, $\theta_{11}(t)$ and $\theta_{21}(t)$. The problem in potential theory for φ_1 can in principle be solved in the form

$$\varphi_1 = \varphi_1[x, y, z; t; \omega_1(t), \theta_{11}(t), \theta_{21}(t)]$$

without using (9.3.29), (9.3.30), (9.3.32). The significance of this has already been discussed in sec. 9.1 in relation to equations (9.1.14). The general procedure to be followed in solving all problems was also discussed there.

The remainder of this chapter is concerned with the special case of a ship which moves along a straight course into waves whose crests are at right angles to the course. In this case there are surging, heaving and pitching motions, but we have $\theta_1 = 0$, $\theta_2 = 0$, $\omega = 0$; in addition we note that the potential function φ can be assumed to be an even function of z. Under these conditions the equations of motion are much simpler. They are

$$(9.3.38) \quad M_1 \dot{s}_1 = 2\varrho \int_A h_x (s_0 \varphi_{1x} - \varphi_{1t}) dA + T$$

$$(9.3.39) \quad M_1 \ddot{y}_1 = -2\varrho g y_1 \int_L h dx - 2\varrho g \theta_{31} \int_L x h dx + 2\varrho \int_A h_y (s_0 \varphi_{1x} - \varphi_{1t}) dA$$

$$(9.3.40) \quad I_{31} \ddot{\theta}_{31} = -2\varrho g \theta_{31} \int_A (y - y_c') h dA - 2\varrho g y_1 \int_L x h dx$$

$$-2\varrho g \theta_{31} \int_L x^2 h dx + lT$$

$$+2\varrho \int_A [x h_y - (y - y_c') h_x](s_0 \varphi_{1x} - \varphi_{1t}) dA.$$

It will be shown in the next section that an explicit integral representation can be found for the corresponding potential function and that this leads to integral representations for the surge s_1, the heave y_1 and the pitching oscillation θ_{31}.

9.4. Method of solution of the problem of pitching and heaving of a ship in a seaway having normal incidence

In this section we derive a method of solution of the problem of calculating the pitching, surging, and heaving motions in a seaway consisting of a train of waves with crests at right angles to the course of the ship, which is assumed to be a straight line (i.e., $\omega \equiv 0$). The propeller thrust is assumed to be a constant vector.

The harmonic function φ_1 and the surface elevation η_1 therefore satisfy the following free surface conditions (cf. (9.3.10) and (9.3.11), with $\omega_0 = 0$):

$$(9.4.1) \qquad \begin{cases} - g\eta_1 + s_0\varphi_{1x} - \varphi_{1t} = 0 \\ - \varphi_{1y} - s_0\eta_{1x} + \eta_{1t} = 0 \end{cases} \qquad \text{at } y = 0.$$

The kinematic condition arising from the hull of the ship is (cf. (9.3.14) with $\theta_{21} = \theta_{11} = \omega_1 = 0$):

$$(9.4.2) \qquad \varphi_{1z} = - s_0 h_x.$$

Before writing down other conditions, including conditions at ∞, we express φ_1 as a sum of two harmonic functions, as follows

$$(9.4.3) \qquad \varphi_1(x, y, z; t) = \chi_0(x, y, z) + \chi_1(x, y, z; t).$$

Here χ_0 is a harmonic function independent of t which is also an even function of z. We now suppose that the motion of the ship is a steady simple harmonic motion in the time when observed from the moving coordinate system $o - x, y, z$. (Presumably such a state would result after a long time upon starting from rest under a constant propeller thrust.) Consequently we interpret $\chi_0(x, y, z)$ as the disturbance caused by the ship, which therefore dies out at ∞; while $\chi_1(x, y, z; t)$ represents a train of simple harmonic plane waves covering the whole surface of the water. Thus χ_1 is given, with respect to the fixed coordinate system $O-X, Y, Z$ by the well-known formula (cf. Chapter 3):

$$\chi_1 = C e^{\frac{\sigma^2}{g} Y} \sin\left(\sigma t + \frac{\sigma^2}{g} X + \gamma\right),$$

with σ the frequency of the waves. In the $o - x, y, z$ system we have, therefore:

$$(9.4.4) \quad \chi_1(x, y, z; t) = C e^{\frac{\sigma^2}{g} y} \sin\left[\frac{\sigma^2}{g} x + \left(\sigma + \frac{s_0\sigma^2}{g}\right) t + \gamma\right].$$

We observe that the frequency, relative to the ship, is increased above the value σ if s_0 is positive — i.e. if the ship is heading into the waves —and this is, of course, to be expected. With this choice of χ_1, it is easy to verify that χ_0 satisfies the following conditions:

$$(9.4.5) \qquad s_0^2\chi_{0xx} + g\chi_{0y} = 0 \qquad \text{at } y = 0,$$

obtained after eliminating η_1 from (9.4.1), and

$$(9.4.6) \qquad \chi_{0z} = - s_0 h_x \qquad \text{on } A,$$

with A, as above, the projection of the ship's hull (for $z > 0$) on its vertical mid-section. In addition, we require that $\chi_0 \to 0$ at ∞.

It should be remarked at this point that the classical problem concerning the waves created by the hull of a ship, first treated by Michell [M.9], Havelock [H.7], and many others, is exactly the problem of determining χ_0 from the conditions (9.4.5) and (9.4.6). Afterwards, the insertion of $\varphi_1 \equiv \chi_0$ in (9.3.38), with $\dot{s}_1 = 0$, $\varphi_{1t} \equiv 0$, leads to the formula for the wave resistance of the ship—i.e. the propeller thrust T is determined. Since y_1 and θ_3 are independent of the time in this case, one sees that the other dynamical equations, (9.3.39) and (9.3.40), yield the displacement of the c.g. relative to the rest position of equilibrium (the heave), and the longitudinal tilt angle (the pitching angle). However, in the literature cited, the latter two quantities are taken to be zero, which implies that appropriate constraints would be needed to hold the ship in such a position relative to the water. The main quantity of interest, though, is the wave resistance, and it is not affected (in the first order theory, at least) by the heave and pitch.

We proceed to the determination of χ_0, using a method different from the classical method and following, rather, a course which it is hoped can be generalized in such a way as to yield solutions in more difficult cases.

Suppose that we know the Green's function $G^*(\xi, \eta, \zeta; x, y, z)$ such that G^* is a harmonic function for $\eta < 0$, $\zeta > 0$ except at (x, y, z) where it has the singularity $1/r$; and G^* satisfies the boundary conditions

(9.4.7) $$G_{\xi\xi}^* + kG_\eta^* = 0 \qquad \text{on } \eta = 0$$
$$G_\zeta^* = 0 \qquad \text{on } \zeta = 0$$

where $k = g/s_0^2$. We shall obtain this function explicitly in a moment, and will proceed here to indicate how it is used. Let Σ denote the half plane $\eta = 0$, $\zeta > 0$; and let Ω denote the half plane $\zeta = 0$, $\eta < 0$. From Green's formula and the classical argument involving the singularity $1/r$ we have

$$4\pi\chi_0 = -\iint_\Sigma \chi_0 G^* d\xi d\zeta + \iint_\Sigma \chi_{0\eta} G^* d\xi d\zeta - \iint_\Omega \chi_{0\zeta} G^* d\xi d\eta.$$

Then, since

$$-\iint_\Sigma \chi_0 G_\eta^* d\xi d\zeta + \iint_\Sigma \chi_{0\eta} G^* d\xi d\zeta = \frac{1}{k} \iint_\Sigma (\chi_0 G_{\xi\xi}^* - \chi_{0\xi\xi} G^*) d\xi d\zeta$$

$$= \frac{1}{k} \iint_\Sigma \frac{\partial}{\partial\xi} (\chi_0 G_\xi^* - \chi_{0\xi} G^*) d\xi d\zeta$$

$$= 0,$$

we have an explicit representation of the solution in the form

$$\chi_0(x, y, z) = -\frac{1}{4\pi} \iint_\Omega \chi_{0\xi} G^* d\xi d\eta, \qquad \text{or}$$

(9.4.8) $$\chi_0(x, y, z) = \frac{s_0}{4\pi} \iint_A h_\xi(\xi, \eta) G^*(\xi, \eta, 0; x, y, z) \, d\xi \, d\eta,$$

upon using (9.4.6).

In order to determine G^* consider the Green's function $G(\xi, \eta, \zeta; x, y, z)$ for the half space $\eta < 0$ which satisfies

$$G_{\xi\xi} + kG_\eta = 0$$

on $\eta = 0$. This function can be written as

$$G = \frac{1}{r_1} - \frac{1}{r_2} + g$$

where

$$\frac{1}{r_1} = \frac{1}{\sqrt{(\xi-x)^2+(\eta-y)^2+(\zeta-z)^2}}$$

$$\frac{1}{r_2} = \frac{1}{\sqrt{(\xi-x)^2+(\eta+y)^2+(\zeta-z)^2}}$$

and g is a potential function in $\eta < 0$ which satisfies

$$g_{\xi\xi} + kg_\eta = 2k \frac{\partial}{\partial y} \frac{1}{\sqrt{(\xi - x)^2 + y^2 + (\zeta - z)^2}}$$

on $\eta = 0$. The formula

$$2k \frac{\partial}{\partial y} \frac{1}{\sqrt{(\xi-x)^2+y^2+(\zeta-z)^2}} = 2k \int_0^\infty p e^{py} J_0[p\sqrt{(\xi-x)^2+(\zeta-z)^2}] dp,$$

(obtained from the well-known analogous representation for $1/r$) in which the Bessel function J_0 can be expressed as

$$J_0[p\sqrt{(\xi-x)^2+(\zeta-z)^2}] = \frac{2}{\pi} \int_0^{\pi/2} \cos [p(\xi-x) \cos \theta] \cos [p(\zeta-z) \sin \theta] \, d\theta,$$

allows us to write

$$g_{\xi\xi} + kg_\eta = \frac{4k}{\pi} \int_0^\infty \int_0^{\pi/2} p e^{py} \cos [p(\xi-x) \cos \theta] \cos [p(\zeta-z) \sin \theta] \, d\theta \, dp$$

for $\eta = 0$ and $y < 0$. It is now easy to see that

$$g_{\xi\xi}+kg_\eta = \frac{4k}{\pi}\int_0^\infty\int_0^{\pi/2} pe^{p(\nu+\eta)}\cos\,[p(\xi-x)\cos\theta]\cos\,[p(\zeta-z)\sin\theta]\,d\theta\,dp$$

is a potential function in $\eta < 0$ which satisfies the boundary condition. An interchange of the order of integration gives

$$g_{\xi\xi}+kg_\eta = \frac{4k}{\tau}\int_0^{\pi/2} d\theta\,\mathscr{R}e\int_0^\infty p\cos\,[p(\zeta-z)\sin\theta]e^{p[(\nu+\eta)+i(\xi-x)\cos\theta]}\,dp$$

where $\mathscr{R}e$ denotes the real part. If we think of p as a complex variable, the path from 0 to ∞ in the last result can be replaced by any equivalent path L, to be chosen later:

$$g_{\xi\xi}+kg_\eta = \frac{4k}{\pi}\int_0^{\pi/2} d\theta\,\mathscr{R}e\int_L p\cos\,[p(\zeta-z)\sin\theta]e^{p[(\nu+\eta)+i(\xi-x)\cos\theta]}\,dp.$$

Since the right hand side of this differential equation for g is expressed as a superposition of exponentials in ξ and η it is to be expected that a solution of it can be found in the form

$$g = \frac{4k}{\pi}\int_0^{\pi/2} d\theta\,\mathscr{R}e\int_L \frac{p\cos\,[p(\zeta-z)\sin\theta]}{kp-p^2\cos^2\theta}\,e^{p[(\nu+\eta)+i(\xi-x)\cos\theta]}\,dp$$

provided the path L can be properly chosen. The path L, which will be fixed by a condition given below, must, of course, avoid the pole $p = k/\cos^2\theta$.

It can now be seen that the function $G^*(\xi,\eta,\zeta;x,y,z) = G(\xi,\eta,\zeta;x,y,z) + G(\xi,\eta,\zeta;x,y,-z)$ satisfies all the conditions imposed on the Green's function employed in (9.4.8): the sum on the right has the proper singularity in $\eta < 0$, $\zeta > 0$, it satisfies the boundary condition (9.4.7) and

$$G_\zeta(\xi,\eta,\zeta;x,y,z) + G_\zeta(\xi,\eta,\zeta;x,y,-z)$$

is zero at $\zeta = 0$. Thus we have for G^* the representation:

$$G^*\Big|_{\zeta=0} = 2\left[\frac{1}{\sqrt{(\xi-x)^2+(\eta-y)^2+z^2}} - \frac{1}{\sqrt{(\xi-x)^2+(\eta+y)^2+z^2}}\right]$$
$$+ \frac{8k}{\pi}\int_0^{\pi/2} d\theta\,\mathscr{R}e\int_L \frac{\cos\,(pz\sin\theta)\,e^{p[(\nu+\eta)+i(\xi-x)\cos\theta]}\,dp}{k-p\cos^2\theta}.$$

The substitution of this in (9.4.8) gives finally

$$\chi_0(x,y,z) = \frac{s_0}{2\pi}\iint_A h_\xi(\xi,\eta)\left\{\frac{1}{\sqrt{(\xi-x)^2+(\eta-y)^2+z^2}} - \frac{1}{\sqrt{(\xi-x)^2+(\eta+y)^2+z^2}}\right\}d\xi\,d\eta$$
$$+ \frac{2ks_0}{\pi^2}\iint_A h_\xi(\xi,\eta)\left\{\int_0^{\pi/2} d\theta\,\mathscr{R}e\int_L \frac{\cos\,(pz\sin\theta)e^{p[(\nu+\eta)+i(\xi-x)\cos\theta]}\,dp}{k-p\cos^2\theta}\right\}d\xi\,d\eta.$$

A condition imposed on $\chi_0(x, y, z)$ is that $\chi_0(x, y, z) \to 0$ as $x \to +\infty$. This condition is satisfied if we take L to be the path shown in Fig. 9.4.1.

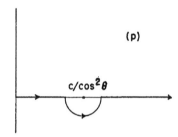

Fig. 9.4.1. The path L in the p-plane, with $c = k$

The function φ_1 is given by

$$\varphi_1 = \chi_1 + \chi_0 = Ce^{\frac{\sigma^2 y}{g}} \sin\left[\frac{\sigma^2 x}{g} + \left(\sigma + \frac{s_0 \sigma^2}{g}\right)t + \gamma\right] + \chi_0$$

and therefore the important quantity $s_0 \varphi_{1x} - \varphi_{1t}$ is given by

$$(9.4.9) \quad s_0 \varphi_{1x} - \varphi_{1t} = -C\sigma e^{\frac{\sigma^2 y}{g}} \cos\left[\frac{\sigma^2 x}{g} + \left(\sigma + \frac{s_0 \sigma^2}{g}\right)t + \gamma\right] + s_0 \chi_{0x}.$$

If this is substituted in the equation (9.3.38) for the surge we have

$$M_1 \dot{s}_1 = -2\varrho C\sigma \iint_A h_x e^{\frac{\sigma^2 y}{g}} \cos\left[\frac{\sigma^2 x}{g} + \left(\sigma + \frac{s_0 \sigma^2}{g}\right)t + \gamma\right] dx\,dy$$

$$+ 2\varrho s_0 \iint_A h_x \chi_{0x}\,dx\,dy + T.$$

The last equation shows that in order to keep s_1 bounded for all t we must take for T the value

$$(9.4.10) \qquad T = -2\varrho s_0 \iint_A h_x \chi_{0x}\,dx\,dy$$

where

$$\chi_{0x}(x, y, 0) = \frac{s_0}{2\pi} \iint_A h_\xi(\xi, \eta)\left\{\frac{(\xi - x)}{[(\xi - x)^2 + (\eta - y)^2]^{3/2}} - \frac{(\xi - x)}{[(\xi - x)^2 + (\eta + y)^2]^{3/2}}\right\} d\xi\,d\eta$$

$$- \frac{2s_0}{\pi^2} \iint_A h_\xi(\xi, \eta)\left\{\int_0^{\pi/2} d\theta\,\mathcal{R}e\int_L \frac{igp\cos\theta\,e^{p[(\nu+\eta)+i(\xi-x)\cos\theta]}}{g - s_0^2 p\cos^2\theta}\,dp\right\} d\xi\,d\eta.$$

In effect, T is determined by the other time-independent term in the equation of motion. Equation (9.4.10) gives the thrust necessary to maintain the speed s_0, or inversely it gives the speed s_0 which corresponds to a given thrust. The integral in (9.4.10) is called the wave resistance integral. As one sees, it does not depend on the seaway. The integral can be expressed in a simpler form as follows.

The function $\chi_{0x}(x, y, 0)$ is a sum of integrals of the type

$$\iint_A h_\xi(\xi, \eta) f(\xi, \eta; x, y) \, d\xi d\eta.$$

If an integral of this type is substituted in the wave resistance integral we have

$$\iint_A \iint_A h_x(x, y) h_\xi(\xi, \eta) f(\xi, \eta; x, y) \, d\xi d\eta dx dy = I$$

say. This is the same as

$$\iint_A \iint_A h_\xi(\xi, \eta) h_x(x, y) f(x, y; \xi, \eta) \, dx dy d\xi d\eta = I$$

and we see that $I = 0$ if

$$f(\xi, \eta; x, y) = -f(x, y; \xi, \eta).$$

Therefore

$$T = \frac{4\varrho s_0^2}{\pi^2} \iint_A \iint_A h_x(x, y) h_\xi(\xi, \eta) f_1 \, d\xi d\eta dx dy$$

where

$$f_1 = \int_0^{\pi/2} d\theta \, \mathscr{R}e \int_L \frac{i g p \cos \theta \, e^{p(y+\eta)} \cos \left[p(\xi - x) \cos \theta \right] dp}{g - s_0^2 p \cos^2 \theta}.$$

Since $\mathscr{R}e \int_L$ is zero except for the residue from the integration along the semi-circular path centered at the point

$$\frac{g}{s_0^2 \cos^2 \theta} = \frac{k}{\cos^2 \theta},$$

we find from the evaluation of this residue that

$$f_1 = \frac{\pi g^2}{s_0^4} \int_0^{\pi/2} \sec^3 \theta \, e^{k(y+\eta) \sec^2 \theta} \cos \left[k(\xi - x) \cos \theta \right] d\theta.$$

We introduce Michell's notation:

$$P(\theta) = \iint_A h_x(x, y) e^{ky \, \sec^2 \theta} \, \cos{(kx \sec \theta)} \, dxdy$$

$$Q(\theta) = \iint_A h_x(x, y) e^{ky \, \sec^2 \theta} \, \sin{(kx \sec \theta)} \, dxdy$$

and can then write

$$T = \frac{4\varrho g^2}{\pi s_0^2} \int_0^{\pi/2} (P^2 + Q^2) \sec^3 \theta \, d\theta.$$

This is the familiar formula of Michell for the wave resistance. The surge is given by

$$s_1 = \frac{-2\varrho C \sigma g}{(g\sigma + s_0\sigma^2) M_1} \iint_A h_x e^{\frac{\sigma^2 y}{g}} \sin\left[\frac{\sigma^2}{g} x + \left(\sigma + \frac{s_0\sigma^2}{g} \right) t + \gamma \right] dxdy.$$

Hereafter we will suppose for simplicity that there is no coupling between (9.3.39) and (9.3.40), so that $\int_L xhdx = 0$. The substitution of (9.4.9) in (9.3.39) therefore gives the following equation for the heave:

$$M_1\ddot{y}_1 + \left[2\varrho g \int_L hdx \right] y_1 = -2\varrho C \sigma \iint_A h_y e^{\frac{\sigma^2 y}{g}} \cos\left[\frac{\sigma^2 x}{g} + \left(\sigma + \frac{s_0\sigma^2}{g} \right) t + \gamma \right] dxdy$$

$$+ 2\varrho s_0 \iint_A h_y \chi_{0x} \, dxdy.$$

The time independent part of y_1, the heave component of the trim, we denote by y_1^*; it is given by

(9.4.11) $$\left(g \int_L h \, dx \right) y_1^* = s_0 \iint_A h_y \chi_{0x} \, dxdy.$$

Here y_1^* is the vertical displacement of the center of gravity of the ship from its rest position when moving in calm water. The integral on the right hand side of (9.4.11) is even more difficult to evaluate than the wave resistance integral.

The response to the seaway in the heave component is given by

$$y_1^{**} = \frac{-2\varrho C\sigma \iint_A h_y e^{\frac{\sigma^2 y}{g}} \cos\left[\frac{\sigma^2 x}{g} + \left(\sigma + s_0 \frac{\sigma^2}{g}\right)t + \gamma\right] dxdy}{2\varrho g \int_L hdx - M_1\left(\sigma + \frac{s_0\sigma^2}{g}\right)^2}.$$

For the case under consideration, the theory predicts that resonance in the heave occurs when

$$\sigma + \frac{s_0\sigma^2}{g} = \left[\frac{2\varrho g}{M_1}\int_L hdx\right]^{1/2}.$$

The equation for the pitching angle is

$$I_{31}\ddot{\theta}_{31} + 2\varrho g \left[\int_A (y-y_c')hdA + \int_L x^2 hdx\right]\theta_{31}$$

$$= -2\varrho C\sigma \iint_A [xh_y - (y-y_c')h_x] \cos\left\{\frac{\sigma^2 x}{g} + \left(\sigma + \frac{s_0\sigma^2}{g}\right)t + \gamma\right\} dxdy$$

$$+ lT + 2\varrho s_0 \int_A [xh_y - (y-y_c')h_x]\chi_{0x}dA.$$

The time independent part of θ_{31}, which we denote by θ_{31}^* is given by

$$2\varrho g \left[\int_A (y-y_c')hdA + \int_L x^2 hdx\right]\theta_{31}^*$$

$$= lT + 2\varrho s_0 \int_A [xh_y - (y-y_c')h_x]\chi_{0x}dA$$

$$= (l-y_c')T + 2\varrho s_0 \int_A [xh_y - yh_x]\chi_{0x}dA.$$

The angle θ_{31}^* is called the angle of trim; it is the angular displacement of a ship which moves with the speed s_0 in calm water.

The oscillatory part of the heave θ_{31} to the sea is

$$\theta_{31}^{**} = \frac{-2\varrho C\sigma \iint_A [xh_y - (y-y_c')h_x] \cos\left\{\frac{\sigma^2 x}{g} + \left(\sigma + \frac{s_0\sigma^2}{g}\right)t + \gamma\right\} dxdy}{2\varrho g \left[\int_A (y-y_c')hdA + \int_L x^2 hdx\right] - I_{31}\left(\sigma + \frac{s_0\sigma^2}{g}\right)^2}.$$

and we see that the theory predicts resonance when

$$\sigma + \frac{s_0\sigma^2}{g} = \left\{\frac{2\varrho g}{I_{31}}\left[\int_A (y-y_c')hdA + \int_L x^2 hdx\right]\right\}^{1/2}.$$

Of course, the differential equations for y_1 and θ_{31} permit also solutions of the type of free undamped oscillations of a definite frequency (in fact, having the resonant frequencies just discussed) but with arbitrary amplitudes which could be fixed by appropriate initial conditions. This point has been discussed at length in the introduction to this chapter.

PART III

CHAPTER 10

Long Waves in Shallow Water

10.1. Introductory Remarks and Recapitulation of the Basic Equations

The basic theory for waves in shallow water has already been derived at length in Chapter 2 in two different ways: one derivation,
along conventional lines, proceeded on the basis of assuming the
pressure to be determined by the hydrostatic pressure law $p =
g\varrho(\eta - y)$ (see Fig. 10.1.1), the other by making a formal develop-
ment in powers of a parameter σ; the two theories are the same in

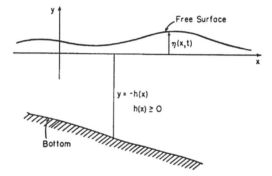

Fig. 10.1.1. Long waves in shallow water

lowest order. With one exception, the present chapter will make use
only of the theory to lowest order and consequently the derivation of
it given in sections 2 and 3 of Chapter 2 suffices for all sections of this
chapter except section 9.

We recapitulate the basic equations. In terms of the horizontal
velocity component $u = u(x, t)$, and the free surface elevation
$\eta = \eta(x, t)$ the differential equations (cf. (2.2.11), (2.2.12)) are

(10.1.1) $$u_t + uu_x = - g\eta_x,$$

(10.1.2) $$[u(\eta + h)]_x = - \eta_t.$$

It is sometimes useful and interesting to make reference to the gas dynamics analogy, by introducing the "density" $\bar{\varrho}$ through

(10.1.3) $\bar{\varrho} = \varrho(\eta + h)$,

and the "pressure" $\bar{p}$ by $\bar{p} = \int_{-h}^{\eta} p\,dy$, which in view of the hydrostatic pressure law yields the relation

(10.1.4) $\bar{p} = \dfrac{g}{2\varrho}\,\bar{\varrho}^2$.

This is an "adiabatic law" with "adiabatic exponent" 2 connecting pressure and density. As one sees, it is the depth of the water, essentially, which plays the role of the density in a gas. In terms of these quantities, the equations (10.1.1) and (10.1.2) take the form

(10.1.5) $\bar{\varrho}(u_t + uu_x) = -\bar{p}_x + g\bar{\varrho}h_x$,

(10.1.6) $(\bar{\varrho}u)_x = -\bar{\varrho}_t$.

These equations, together with (10.1.4), correspond exactly to the equations of compressible gas dynamics for a one-dimensional flow if $h_x = 0$, i.e. if the depth of the undisturbed stream is constant. It follows that a "sound speed" or propagation speed c for the phenomena governed by these equations is defined by $c = \sqrt{d\bar{p}/d\bar{\varrho}}$, as in acoustics, and this quantity in our case has the value

(10.1.7) $c = \sqrt{\dfrac{g\bar{\varrho}}{\varrho}} = \sqrt{g(\eta + h)}$,

as we see from (10.1.4) and (10.1.3). Later on, we shall see that it is indeed justified to call the quantity c the propagation speed since it represents the local speed of propagation of "small disturbances" relative to the moving stream. We observe the important fact that c (which obviously is a function of x and t) is proportional to the square root of the depth of the water.

The propagation speed $c(x, t)$ is a quantity of such importance that it is worthwhile to reformulate the basic equations (10.1.1) and (10.1.2) with c in place of η. Since $c_x = (g\eta_x + gh_x)/2c$ and $c_t = g\eta_t/2c$ one finds readily

(10.1.8) $u_t + uu_x + 2cc_x - H_x = 0$,

(10.1.9) $2c_t + 2uc_x + cu_x = 0$,

with

(10.1.10) $H = gh$.

The verification in the general case that the quantity c represents a wave propagation speed requires a rather thorough study of certain basic properties of the differential equations. However, if we restrict ourselves to motions which depart only slightly from the rest position of equilibrium (i.e. the state with $\eta \equiv 0$, $u \equiv 0$) it is easy to verify that the quantity c then is indeed the propagation speed. From (10.1.7) we would have in this case $c = c_0 + \varepsilon(x, t)$, with $c_0 = \sqrt{gh}$ and ε a small quantity of first order. We assume u and its derivatives also to be small of first order and, in addition, take the case in which the depth h is constant. Under these circumstances the equations (10.1.8) and (10.1.9) yield

(10.1.11)
$$u_t + 2c_0\varepsilon_x = 0,$$

(10.1.12)
$$2\varepsilon_t + c_0 u_x = 0$$

if first order terms only are retained. By eliminating ε we obtain for u the differential equation

(10.1.13)
$$u_{tt} - c_0^2 u_{xx} = 0.$$

This is the classical linear wave equation all solutions of which are functions of the form $u = u(x \pm c_0 t)$ and this means that the motions are superpositions of waves with constant propagation speed $c_0 = \sqrt{gh}$.

The role of the quantity c as a propagation speed (together with many other pertinent facts) can be understood most readily by discussing the underlying integration theory of equations (10.1.8) and (10.1.9) by using what is called the method of characteristics; we turn therefore to a discussion of this method in the next section.

10.2. Integration of the Differential Equations by the Method of Characteristics

The theory of our basic differential equations (10.1.8) and (10.1.9), which are of the same form as those in compressible gas dynamics, has been very extensively developed because of the practical necessity for dealing with the flow of compressible gases. The purpose of the present section is to summarize those features of this theory which can be made useful for discussing the propagation of surface waves in shallow water. In doing so, extensive use has been made of the presentation given in the book by Courant and Friedrichs [C.9]; in fact, a good deal of the material in sections 10.2 to 10.7, inclusive, follows the presentation given there.

The essential point is that the partial differential equations (10.1.8) and (10.1.9) are of such a form that the initial value problems associated with them admit of a rather simple discussion in terms of a pair of ordinary differential equations called the *characteristic differential equations*. We proceed to derive the characteristic equations for the special case in which [cf. (10.1.10)]

(10.2.1) $H_x = m = \text{const.}$

i.e. the case in which the bottom slope is constant. In fact, this is the only case we consider in this chapter. If we add equations (10.1.8) and (10.1.9) it is readily seen that the result can be written in the form:

(10.2.2) $\left\{ \dfrac{\partial}{\partial t} + (u + c) \dfrac{\partial}{\partial x} \right\} \cdot (u + 2c - mt) = 0.$

The expression in brackets is, of course, to be understood as a differential operator. Similarly, a subtraction of (10.1.9) from (10.1.8) yields

(10.2.3) $\left\{ \dfrac{\partial}{\partial t} + (u - c) \dfrac{\partial}{\partial x} \right\} \cdot (u - 2c - mt) = 0.$

But the interpretation of the operations defined in (10.2.2) and (10.2.3) is well known (cf. (1.1.3)): the relation (10.2.2), for example, states that the function $(u + 2c - mt)$ is constant for a point moving through the fluid with the velocity $(u + c)$, or, as we may also put it, for a point whose motion is characterized by the ordinary differential equation $dx/dt = u + c$. Equation (10.2.3) can be similarly interpreted. That is, we have the following situation in the x, t-plane: There are two sets of curves, C_1 and C_2, called characteristics, which are the solution curves of the ordinary differential equations

(10.2.4) $\begin{cases} C_1 : \dfrac{dx}{dt} = u + c, \text{ and} \\[2mm] C_2 : \dfrac{dx}{dt} = u - c, \end{cases}$

and we have the relations

(10.2.5) $\begin{cases} u + 2c - mt = k_1 = \text{const. along a curve } C_1 \text{ and} \\ u - 2c - mt = k_2 = \text{const. along a curve } C_2. \end{cases}$

Of course the constants k_1 and k_2 will be different on different curves in general. It should also be observed that the two families of characteristics determined by (10.2.4) are really distinct because of the fact that $c = \sqrt{g(\eta + h)} \neq 0$ since we suppose that $\eta > -h$, i.e. that the water surface never touches the bottom.

By reversing the above procedure it can be seen rather easily that the system of relations (10.2.4) and (10.2.5) is completely equivalent to the system of equations (10.1.8) and (10.1.9) for the case of constant bottom slope, so that a solution of either system yields a solution of the other. In fact, if we set $f(x, t) = u + 2c - mt$ and observe that $f(x, t) = k_1 = $ const. along any curve $x = x(t)$ for which $dx/dt = u + c$ it follows that along such curves

$$(10.2.6) \qquad f_t + f_x \frac{dx}{dt} = f_t + (u + c)f_x = 0.$$

In the same way the function $g(x, t) = u - 2c - mt$ satisfies the relation

$$(10.2.7) \qquad g_t + (u - c)g_x = 0$$

along the curves for which $dx/dt = u - c$. Thus wherever the curve families C_1 and C_2 cover the x, t-plane in such a way as to form a non-singular curvilinear coordinate system the relations (10.2.6) and (10.2.7) hold. If now equations (10.2.6) and (10.2.7) are added and the definitions of $f(x, t)$ and $g(x, t)$ are recalled it is readily seen that equation (10.1.8) results. By subtracting (10.2.7) from (10.2.6) equation (10.1.9) is obtained. In other words, any functions u and c which satisfy the relations (10.2.4) and (10.2.5) will also satisfy (10.1.8) and (10.1.9) and the two systems of equations are therefore now seen to be completely equivalent.

As we would expect on physical grounds, a solution of the original dynamical equations (10.1.8) and (10.1.9) could be shown to be uniquely determined when appropriate initial conditions (for $t = 0$, say) are prescribed; it follows that a solution of (10.2.4) and (10.2. 5) is also uniquely determined when initial conditions are prescribed since we know that the two systems of equations are equivalent.

At first sight one might be inclined to regard the relations (10.2.4) and (10.2.5) as more complicated to deal with than the original differential equations, particularly since the right hand sides of (10.2.4) are not known in advance and hence the characteristic curves are also not known: they must, in fact, be determined in the course of determining the unknown functions u and c which constitute the desired solution. Nevertheless, the formulation of our problems in terms of the characteristic form is quite useful in studying properties of the solutions and also in studying questions referring to the appropriateness of various boundary and initial conditions. It is useful to begin by describing briefly a method of determining the characteristics and thus the solution of a given problem by a method of successive approximation which at the same time makes possible a number of useful observations and interpretations regarding the role played by the characteristics in general. Let us for this purpose consider a problem in which the values of the velocity u and the surface elevation η (or, what amounts to the same thing, the propagation or wave speed $c = \sqrt{g(\eta + h)}$) are prescribed for all values of x at the initial instant $t = 0$. We wish to calculate the solution for $t > 0$ by determining u and c through use of (10.2.4) and (10.2.5) and the given initial conditions. At $t = 0$ we assume that

$$(10.2.8) \qquad \begin{cases} u(x, 0) = \bar{u}(x) \\ c(x, 0) = \bar{c}(x) \end{cases}$$

in which $\bar{u}(x)$ and $\bar{c}(x)$ are given functions. We can approximate the values of u and c for small values of t as follows: consider a series of points on the x-axis (cf. Fig. 10.2.1) a small distance δx apart. At all of these points the values of u and c are known from (10.2.8). Consequently the slopes of the characteristics C_1 and C_2 at these points are

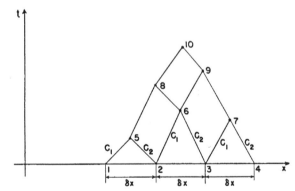

Fig. 10.2.1. Integration by finite differences

known from (10.2.4). From the points 1, 2, 3, 4 straight line segments with these slopes are drawn until they intersect at points 5, 6, and 7, and if δx is chosen sufficiently small it is reasonable to expect that

the positions of these points will be good approximations to the intersections of the characteristics issuing from the points 1, 2, 3, 4 since
we are simply replacing short segments of these curves by their
tangents. The values of both x and t at points 5, 6, and 7 are now
known—they can be determined graphically for example—and
through the use of (10.2.5) and the initial conditions we can also
determine the approximate values of u and c at these points. For this
purpose we observe that along any particular segment issuing from
the points 1, 2, 3 or 4 the values of $u + 2c - mt$ and $u - 2c - mt$
are known constants since the values of u and c are fixed by (10.2.8)
for $t = 0$; hence we have

$$(10.2.9) \quad \begin{cases} \text{along } C_1\colon u + 2c - mt = \bar{u} + 2\bar{c}, \text{ and} \\ \text{along } C_2\colon u - 2c - mt = \bar{u} - 2\bar{c}. \end{cases}$$

At the points 5, 6, and 7 we know the values of t and hence (10.2.9)
furnishes two independent linear equations for the determination of
the values of u and c at each of these points. Once u and c are known
at points 5, 6, and 7 the slopes of the characteristics issuing from these
points can be determined once more from (10.2.4) and the entire process can be carried out again to yield the additional points 8 and 9
and the approximate values of u and c at these points. In this way
we can approximate the values of u and c at the points of a net over
a certain region of the x, t-plane, and can then obtain approximate
values for u and c at any points in the same region either by interpolation or by refining the net inside the region. It is quite plausible
and could be proved mathematically that the above process would
converge as $\delta x \to 0$ to the unique solution of (10.2.4) and (10.2.5)
corresponding to the given initial conditions for sufficiently small
values of t (i.e. for a region of the x, t-plane not too far from the x-axis)
provided that the prescribed initial values of u and c are sufficiently
regular functions of x—for example, if they have piecewise continuous first derivatives.

It should be clear that once the characteristics are known the values
of u and c for all points of the x, t-plane covered by them are also
known, since the constants k_1 and k_2 in (10.2.5) are known on each
characteristic through the initial data and hence the values of u and c
for any point (x, t) can be calculated by solving the linear equation
(10.2.5) for the characteristics through that point. This statement of
course implies that each one of the two families of characteristics
covers a region of the x, t-plane simply and that no two members of

different families are tangent to each other—in other words it is
implied that the two families of characteristics form a regular curvilinear coordinate system over the region of the x, t-plane in question.
One of the points of major interest in the later discussion centers
around the question of determining where and when the characteristics cease to have this property, and of interpreting the physical
meaning of such occurrences.

   The method of finite differences used above to determine the characteristics can be interpreted in such a way as to throw a strong light
on the physical properties of the solution. Consider the point 10 of
Fig. 10.2.1 for example. We recall that the approximate values u_{10}
and c_{10} of u and c at point 10 were obtained through making use of the
initial values of u and c at points 1, 2, 3, 4 on the x-axis only, and
furthermore that the values u_{10} and c_{10} required the use of points confined solely to the region within the approximate characteristics joining point 10 with points 1 and 4. Since the finite difference scheme
outlined above converges as $\delta x \to 0$ to yield the exact characteristics
we are led to make the following important statement: *the values of u
and c at any point $P(x, t)$ within the region of existence of the solution
are determined solely by the initial values prescribed on the segment of
the x-axis which is subtended by the two characteristics issuing from P.*

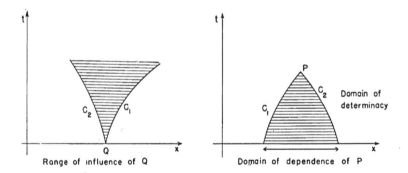

Fig. 10.2.2. Domain of dependence and range of influence

   In addition, the two characteristics issuing from P are also determined
solely by the initial values on the segment subtended by them. Such
a segment of the x-axis is often called the *domain of dependence* of the

point P. Correspondingly we may define the *range of influence* of a point Q on the x-axis as the region of the x, t-plane in which the values of u and c are influenced by the initial values assigned to point Q. In Fig. 10.2.2 we indicate these two regions. It is also useful on occasion to introduce the notion of *domain of determinacy* relative to a given domain of dependence. It is the region in which the motion is determined solely by the data over a certain segment of the x-axis. These regions are outlined by characteristic curves, as indicated in Fig. 10.2.2, in an easily understandable fashion in view of the discussion above.

We are now in a position to understand why it is appropriate to call the quantity c the propagation or wave speed. To this end we suppose that a certain motion of water exists at a definite time, which we take to be $t = 0$. This means, of course, that u and c are known at that time, and, as we have just seen, the motion would be uniquely determined for $t > 0$. However, we raise the question: what difference would there be in the subsequent motion if we created a disturbance in some part of the fluid, say over a segment Q_1Q_2 of the x-axis (cf. Fig. 10.2.3)? This amounts to asking for a comparison of *two solutions* of our equations which differ only because of a difference in the initial

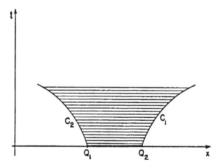

Fig. 10.2.3. Propagation of disturbances

conditions over the segment Q_1Q_2. Our whole discussion shows that the two solutions in question would differ only in the shaded region of Fig. 10.2.3, which comprises all points of the x, t-plane influenced by the data on the segment Q_1Q_2, and which is bounded by characteristics C_2 and C_1 issuing from the endpoints Q_1 and Q_2 of the segment. These curves, however, satisfy the differential equations $dx/dt = u - c$, $dx/dt = u + c$. Since u represents the velocity of the water, it is then clear that c represents the speed relative to the flowing stream at which the disturbance on the segment Q_1Q_2 spreads over the water. This implies that the data in our two problems really differ at points Q_1 and Q_2 and that these differences persist along the characteristics issuing from these points. Actually, only discontinuities in derivatives at Q_1 and Q_2 (and not of the functions themselves) are permitted in the above theory, and it could be shown that such discontinuities would never smooth out entirely along the characteristics C_1 and C_2. We are therefore justified in referring to the quantity $c = \sqrt{g(\eta + h)}$ as the (local) propagation speed of small disturbances—that is, small in the sense that only discontinuities in derivatives occur at the front of a disturbance.

10.3. The Notion of a Simple Wave

There is an important class of problems in which the theory of characteristics as presented in the preceding section becomes particularly simple. These are the problems in which (1) the initial undisturbed depth h of the water is constant so that the quantity m in (10.2.1) (cf. also (10.1.10)) is zero, (2) the water extends from the origin to infinity at least in one direction, say in the positive direction of the x-axis, and (3) the water is either at rest or moves with constant velocity and the elevation of its free surface is zero at the time $t = 0$. In other words, the water is in a uniform state at time $t = 0$ such that $u = u_0 = \text{const.}$ and $c = c_0 = \sqrt{gh} = \text{const.}$ at that instant. Our discussion from here on is modeled closely on the discussion given by Courant and Friedrichs [C.9], Chapter III.

We now suppose that a disturbance is initiated at the origin $x = 0$ so that either the particle velocity u, or the surface elevation η (or the wave velocity $c = \sqrt{g(\eta + h)}$) changes with the time in a prescribed manner.* That is, a disturbance at one point in the water propagates into water of constant depth and uniform velocity. Under these circumstances we show that *one of the two families of characteristics furnished by (10.2.4) consists entirely of straight lines along each of which u and c are constant*. The corresponding motion we call a *simple wave*.

---

* One might accomplish this experimentally in a tank as follows: To obtain a prescribed velocity u at one point it would only be necesary to place a vertical plate in the water extending from the surface of the water to the bottom of the tank and to move it with the prescribed velocity. To change η at one point water might be either poured into the tank or pumped out of it at that point at an appropriate rate.

Our statement is an immediate consequence of the following fundamental fact: *if the values of u and c on any characteristic curve, C_1^0 say
(i.e. a solution curve of the first of the two ordinary differential equations
(10.2.4)), are constant, then C_1^0 is a straight line and furthermore it is
embedded in a family of straight line characteristics along each of which
u and c are constant*, at least in a region of the x, t-plane where $u(x, t)$
and $c(x, t)$ are without singularities and which is covered by the
two distinct families of characteristics. The proof is easily given. To
begin with, the curve C_1^0 is a straight line if u and c are constant along
it, since the slope of the curve is constant in that case from (10.2.4).
Next, let C_1 be another characteristic near to C_1^0. We consider any two
points A_0 and B_0 on C_1^0 together with the characteristics of the family
C_2 through A_0 and B_0 and suppose that the latter characteristics
intersect C_1 at points A and B (cf. Fig. 10.3.1): To prove our statement

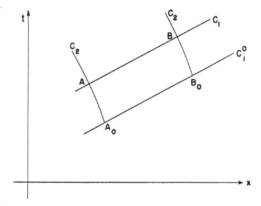

Fig. 10.3.1. Region containing a straight characteristic

we need only show that $u(A) = u(B)$ and $c(A) = c(B)$ since then u and c
would be constant on C_1 (because of the fact that A and B are any
arbitrary points on C_1) and hence the slope of the curve C_1 would be
constant, just as was argued for C_1^0. We have $u(A_0) = u(B_0)$ and
$c(A_0) = c(B_0)$ and consequently we may write

(10.3.1) $\begin{cases} u_A - 2c_A = u_{A_0} - 2c_{A_0}, \\ u_B - 2c_B = u_{B_0} - 2c_{B_0} = u_{A_0} - 2c_{A_0} \end{cases}$

by making use of the second relation of (10.2.5) (which holds along
the characteristics C_2) and observing that $m = 0$ since the original

depth of the water is assumed to be constant. Next we make use of the first relation of (10.2.5) for C_1 to obtain

$$(10.3.2) \qquad u_A + 2c_A = u_B + 2c_B.$$

But from (10.3.1) we have

$$(10.3.3) \qquad u_A - 2c_A = u_B - 2c_B,$$

and (10.3.2) and (10.3.3) are obviously satisfied only if $u_A = u_B$ and $c_A = c_B$. Our statement is therefore proved.

The problems formulated in the first paragraph of this section are at once seen to have solutions (at least in certain regions of the x, t-plane) of the type we have just defined as simple waves since there is a region near the x-axis in the x, t-plane throughout which the particle velocity u and wave speed c are constant, and in which therefore the characteristics are two sets of parallel straight lines. The circumstances are illustrated in Fig. 10.3.2 below: There is a zone I along

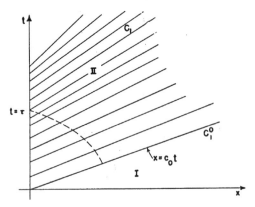

Fig. 10.3.2. A simple wave

the x-axis which might be called the zone of quiet* inside which the characteristics are obviously straight lines $x \pm c_0 t = $ const. (These lines are not drawn in the figure). This region is terminated on the upper side by an "initial characteristic" $x = c_0 t$ which divides the

---

* In a "zone of quiet" we permit the particle velocity u to be a non zero constant, but the free surface elevation η is taken to be zero in such a region. In case $u = u_0 = $ const. $\neq 0$ initially, the motion can be thought of as observed from a coordinate system moving with that velocity; thus there is no real loss of generality in assuming $u_0 = 0$, as we frequently do in the following.

region of quiet from the disturbed region above it. The physical interpretation of this is of course that the disturbance initiated at the
time $t = 0$ propagates into the region of quiet, and the water at any
point remains unaffected until sufficient time has elapsed to allow
the disturbance to reach that point. The exact nature of the motion
in the disturbed region is determined, of course, by the character of
the disturbance created at the point $x = 0$, i.e., by appropriate data
prescribed along the t-axis.* One set of characteristics, i.e., the set
containing the initial characteristic C_1^0, therefore consists of straight
lines. (That the characteristics C_2 in the zone II are necessarily curved
lines and not straight lines can be seen from the fact that they would
otherwise be the continuations of the straight characteristics from the
zone I of quiet and hence the zone II would also be a zone of quiet, as
one sees immediately). Furthermore, the set of straight characteristics
C_1 in zone II is completely determined by appropriate conditions prescribed at $x = 0$ for all t, i.e., along the t-axis. What these conditions
should be can be inferred from the following discussion. Consider any
straight characteristic issuing from a point $t = \tau$ on the t-axis. We
know that the slope dx/dt of this straight line is given in view of
(10.2.4), by

$$(10.3.4) \qquad \frac{dx}{dt} = u(\tau) + c(\tau).$$

Suppose now that there is a curved characteristic C_2 going back
from $t = \tau$ on the t-axis to the initial characteristic C_1^0 (see the dotted
curve in Fig. 10.3.2). We have the following relation from (10.2.5):

$$(10.3.5) \qquad u(\tau) - 2c(\tau) = u_0 - 2c_0,$$

in which u_0 and c_0 are the known values of u and c in the zone of quiet.
Hence the slope of any of the straight characteristics issuing from the
t-axis can be given in either of the two forms:

$$(10.3.6) \qquad \begin{cases} \dfrac{dx}{dt} = \dfrac{1}{2}\left[3u(\tau) - u_0\right] + c_0, \text{ or} \\[2mm] \dfrac{dx}{dt} = 3c(\tau) - 2c_0 + u_0, \end{cases}$$

as one sees from (10.3.4) and (10.3.5). Thus if *either* $u(\tau)$ *or* $c(\tau)$ is

* Our discussion in the preceding section centered about the initial value
problem for the case in which the initial data are prescribed on the x-axis, but one
sees readily that the same discussion would apply with only slight modifications
to the present case, in which what is commonly called a boundary condition (i.e.
at the boundary point $x = 0$), rather than an initial condition, is prescribed.

given, i.e. if either u or c is prescribed along the t-axis, then the slopes of the straight characteristics C_1 and with them the characteristics C_1 themselves are determined. Since we know, from (10.3.5), the values of *both* u and c along the t-axis if either one is given, and since u and c are clearly constant along the straight characteristics, it follows that we know the values of u and c throughout the entire disturbed region—in other words, the motion is completely determined.

So far, we have considered only the case in which the curved characteristics (i.e., those of the type C_2) which issue from the boundary $x = c_0 t$ of the disturbed region actually reach the t-axis. This, however, need not be the case. Suppose, for example, that u_0 is positive and $u_0 > c_0 = \sqrt{gh}$. In this case the slope dx/dt of the curves C_2 is positive, and we cannot expect that they will turn to the left, as in Fig. 10.3.2. Indeed, in such a case one does not expect that a disturbance will propagate upstream (that is, to the left in our case) since the stream velocity is greater than the propagation speed. In gas dynamics one would say that the flow is supersonic, while in hydraulics the flow is said to be *supercritical*. One could also look at the matter in another way: For not too large values of t the velocity u can be expected to remain supersonic and hence for such values of t both sets of characteristics issuing from the t-axis would go into the right half plane (u being again taken positive). Thus we would have the situation indicated in Fig. 10.3.3, in which a segment of the t-axis is subtended by two

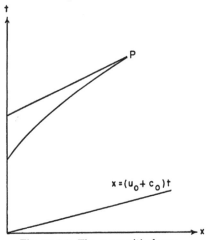

Fig. 10.3.3. The supercritical case

characteristics drawn backward from P. In this case, as in the case of the initial value problem treated in the preceding section, we must prescribe the values of both u and c along the t-axis. If we do so, then the solution is once more determined through (10.3.4) and the fact that $u + 2c$ is constant along one set of characteristics and $u - 2c$ is constant along the other.

In either of our two cases, i.e. of subcritical or supercritical flow, we see therefore that the simple wave can be determined. One sees also how useful the formulation in terms of the characteristics can be in determining appropriate subsidiary conditions such as boundary conditions.

If we wish to know the values of u and c for any particular time $t = t_0$, once the simple wave configuration is determined, we need only draw the line $t = t_0$ and observe its intersections with the straight characteristics since the values of u and c are presumably known on each one of the latter. Thus u and c would be known as functions of x for that particular time. Of course, the surface elevation η would also be known from

$$c = \sqrt{g(h + \eta)}.$$

10.4. Propagation of disturbances into still water of constant depth

In the preceding section we have seen how the method of characteristics leads to the notion of a simple wave in terms of which we can describe with surprising ease the propagation of a disturbance initiated at a point into water of constant depth moving with uniform speed. In the present section we consider in more detail the character of the simple waves which occur in two important special cases. We assume always that the pulse is initiated at $x = 0$ and that it then propagates in the positive x-direction into still water. Thus we are considering cases in which the flow is subcritical at the outset.

One of the most striking and important features of our whole discussion is that there is an essential difference between the propagation of a pulse which is created by steadily *decreasing* the surface elevation η at $x = 0$ and of a pulse which results by steadily *increasing* the elevation at $x = 0$. If the pulse is created by initiating a change in the particle velocity u at $x = 0$ (which might be achieved simply by moving a vertical barrier at $x = 0$ with the prescribed particle velocity) instead of by changing the surface elevation η the same typical differences will result if u is in the first case decreased from zero through negative values, and in the other case is gradually increased so that it becomes positive (i.e. if the particles at $x = 0$ are given in the first case a negative acceleration and in the second case a positive acceleration.) The qualitative difference between the two cases from the physical point of view is of course that in the first case it is a *depression* in the water surface and in the second case an *elevation above the undisturbed surface*—sometimes referred to later on as a *hump*—which propagates into still water.

If we were to consider waves of very small amplitude so that we might linearize our equations (as was done in deriving equation (10.1.13)) there would be no essential qualitative distinction between the motions in the two cases; that there is actually a distinction between the two is a consequence of the nonlinearity of the differential equations.

In the preceding section we have seen that the motions in either of our two cases can be described in the x, t-plane by means of a family of straight characteristics which issue from the t-axis. In Figure 10.4.1 we show these characteristics together with a curve indicating a set of prescribed values for $c = \sqrt{g(h+\eta)} = c(t)$ at $x = 0$, which in turn result from prescribed values of η at that point. We assume that $u = u_0 = 0$ in the zone of quiet I. Hence the slope dx/dt of any straight characteristic issuing from a point $t = \tau$ on the t-axis is given, in accordance with (10.3.6) by

$$(10.4.1) \qquad \frac{dx}{dt} = 3c(\tau) - 2c_0.$$

When τ is varied (10.4.1) yields the complete set of straight characteristics in the zone II. The values of u and c along the same characteristic are constant (as we have seen in the preceding section), so that the value of u along a characteristic is determined, from (10.3.5) by·

$$(10.4.2) \qquad u(\tau) = 2[c(\tau) - c_0],$$

since u_0 is assumed to be zero and $c(\tau)$ is given.

We are now in a position to note a crucial difference between the two cases described above. In the first of the two cases—i.e. that of a depression moving into still water—the elevation $\eta(t)$ at $x = 0$ is assumed to be a decreasing function of t so that $c(t)$ also decreases with increase of t. It follows that the slopes dx/dt of the straight line characteristics as given by (10.4.1) decrease as t increases* so that the family of straight characteristics diverge on moving out from the

---

* One should observe that decreasing values of dx/dt mean that the characteristics make increasing angles with the x-axis, i.e. that they become steeper with respect to the horizontal.

t-axis. (This is the case indicated in Fig. 10.4.1). In the second case, however, the value of η and thus of c is assumed to be an increasing function of t at $x = 0$ so that the straight characteristics must even-

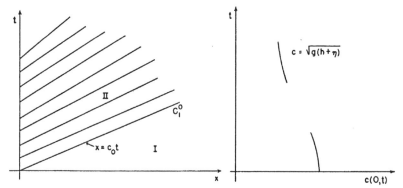

Fig. 10.4.1. Propagation of pulses into still water

tually intersect—in fact, they will have an envelope in general— and this in turn means that our problem can not be expected to have a continuous solution for values of x and t beyond those for which such intersections exist. In the first case the motion is continuous throughout. What happens in the second case beyond the point where the solution is continuous can not be discussed mathematically until we have widened our basic theory, but in terms of the physical behavior of the water we might expect the wave to break, or to develop what is called a bore,* some time after the solution ceases to be continuous. In later sections we propose to discuss the question of the development of breakers and bores in some detail.

The two cases discussed above are the exact analogues of two cases well known in gas dynamics: Consider a long tube filled with gas at rest and closed by a piston at one section. If the piston is moved away from the gas with increasing speed in such a way as to cause a rarefaction wave to move into the quiet gas, then a continuous motion results. However, if the piston is pushed with increasing speed into the gas so as to create a compression wave, then such a wave always

---

* In certain estuaries in various parts of the world the incoming tides from the ocean are sometimes observed to result in the formation of a nearly vertical wall of water, called a bore, which advances more or less unaltered in form over quite large distances. What is called a hydraulic jump is another phenomenon of the same sort. Such phenomena will be discussed in detail later on.

develops eventually into a shock wave. That is, the development of a shock in gas dynamics is analogous to the development of a bore (and also of a hydraulic jump) in water.

10.5. Propagation of depression waves into still water of constant depth

In this section we give a detailed treatment of the first type of motion in which a depression of the water surface propagates into still water. However, it is interesting and instructive to prescribe the disturbance in terms of the velocity of the water rather than in terms of the surface elevation. We assume, in addition, that the velocity is prescribed by giving the displacement $x = x(t)$ of the water particles originally in the vertical plane at $x = 0$,* and this, as we have remarked before, could be achieved experimentally simply by moving a vertical plate at the end of a tank in such a way that its displacement is $x(t)$.** Figure 10.5.1 indicates the straight characteristics which

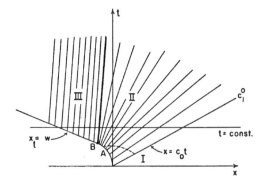

Fig. 10.5.1. A depression wave

initiate on the "piston curve" $x = x(t)$. The piston is assumed to start from rest and move in the negative direction with increasing speed until it reaches a certain speed $w < 0$, after which the speed remains constant. That is, x_t decreases monotonically from zero at $t = 0$ until it attains the value w, after which it stays constant at that value. In Fig. 10.5.1 this point is marked B; clearly the piston curve is

---

* In our theory, it should be recalled, the particles originally in a vertical plane remain always in a vertical plane.

** Moving such a plate at the end of a tank of course corresponds in gas dynamics to moving a piston in a gas-filled tube.

a straight line from there one. At any point A on the piston curve we have $u_A = x_t(t)$, corresponding to the physical assumption that the water particles in contact with the piston remain in contact with it and thus have the same velocity. If we consider the curved characteristic drawn from A back to the initial characteristic C_1^0 which terminates the zone I of rest we obtain from (10.3.5) the relation

$$(10.5.1) \qquad c_A = \tfrac{1}{2}u_A + c_0,$$

since in our case $u_0 = 0$. The slope of the straight characteristic at A is thus given by (cf. (10.4.1)):

$$(10.5.2) \qquad \frac{dx}{dt} = \frac{3}{2}u_A + c_0.$$

Since we have assumed that $u_A = x_t(t)$ always decreases as t increases until $x_t = w$ it follows from (10.5.2) that dx/dt also decreases as t increases in this range of values of t so that the characteristics diverge as they go outward from the piston curve. Beyond the point B the straight characteristics are *parallel* straight lines, since $u_A = w =$ const. on that part of the piston curve, and the state of the water is therefore constant in the zone marked III in Fig. 10.5.1. The zone II is thus a region of non-constant state connecting two regions of different constant states. Since $c_A = \sqrt{g(h+\eta_A)}$, where η_A refers to the elevation of the water surface at the piston, it follows from (10.5.1) that η_A decreases in the zone II as t increases, i.e. the water surface at the "piston" moves downward as the piston moves to the left, since we assume that u_A decreases as A moves out along the piston curve. Since u and c are constant along any straight characteristic it is not difficult to describe the character of the motion corresponding to the disturbed zone II at any time t: Consider any straight line $t =$ const. Its intersection with a characteristic yields the values of u and c at that point — they are the values of u and c which are attached to that characteristic. Since the characteristics diverge from the piston curve one sees that the elevation η steadily increases upon moving from the piston to the right and the particle velocity decreases in magnitude, until the initial characteristic C_1^0 is reached after which the water is undisturbed. On the other hand, if attention is fixed on a definite point $x > 0$ in the water and the motion is observed as the time increases it is clear — once more because the characteristics diverge — that the water remains undisturbed until the time reaches the value determined by $x = c_0 t$, after which the water surface falls steadily while the water particles passing that point move more and more rapidly in the negative x-direction.

In the foregoing discussion of a depression we have made an assumption without saying so explicitly, i.e. that the speed u_A of the piston is such that $c_A = \frac{1}{2}u_A + c_0$ (cf. (10.5.1)) is not negative, and this in turn requires that

(10.5.3) $-u_A \leqq 2c_0.$

Since $-u_A$ increases monotonically to the terminal value $-w$ it follows that $-w$ must be assumed in the above discussion to have at

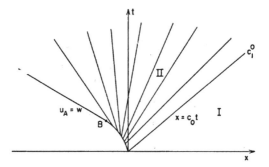

Fig. 10.5.2. A limit case

most the value $2c_0$. The limit case in which $-w$ just equals $2c_0$ is interesting. Since the straight characteristics have the slope $dx/dt = u + c$ and since $c_A = 0$ from (10.5.1) when $u_A = -2c_0$, it follows in this case that $dx/dt = u_A$ on the straight part of the piston curve. But this means that the straight characteristics have all coalesced into the piston curve itself in this region, or in other words that the zone *III* has disappeared in this limit case. The circumstances are indicated in Fig. 10.5.2. At the front of the wave for values of x to the left of B the elevation η_A of the water is equal to $-h$ from $c_A = \sqrt{g(h + \eta_A)} = 0$, which means that the water surface just touches the bottom at the advancing front of the wave.

It is now clear what would happen if the terminal speed $-w$ of the piston were greater than $2c_0$: The zone *II* would terminate on the tangent to the piston curve drawn from the point where the piston speed $-x_t$ just equals $2c_0$. The region between this terminal characteristic and the remainder of the piston curve beyond it might be called the zone of cavitation, since no water would exist for (x, t)

values in such a region. In other words, the piston eventually pulls itself completely free from the water in this case. Quite generally we see that the piston will lose contact with the water (under the circumstances postulated in this section, of course) if, and only if, it finally exceeds the speed $2c_0$. Once this happens it is clear that the piston has no further effect on the motion of the water. These circumstances are indicated in Fig. 10.5.3.

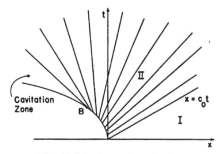

Fig. 10.5.3. Case of cavitation

If the acceleration of the piston is assumed to be infinite so that its speed changes instantly from zero to the constant terminal value $- w$, the motion which results can be described very simply by explicit formulas. The general situation in the x, t-plane is indicated in Fig. 10.5.4. This case might be considered a limit case of the one indicated in Fig. 10.5.1 which results when the portion of the piston curve extending from the origin to point B shrinks to a point. The

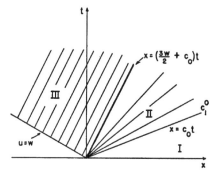

Fig. 10.5.4. Centered simple wave

consequence is that the straight characteristics in zone *II* all pass through the origin. The zone *III* is again one of constant state. In the zone *II* we have obviously for the slopes of the characteristics

$$(10.5.4) \qquad \frac{dx}{dt} = \frac{x}{t}.$$

At the same time we have from (10.5.2) $dx/dt = \frac{3}{2}u + c_0$ so that

$$(10.5.5) \qquad \frac{x}{t} = \frac{3}{2}u + c_0.$$

It follows that the zone *II* is terminated on the upper side by the line

$$(10.5.6) \qquad x = \left(\frac{3}{2}w + c_0\right)t.$$

From (10.5.5) and (10.5.1) we can obtain the values of u and c within zone *II*:

$$(10.5.7) \qquad u = \frac{2}{3}\left(\frac{x}{t} - c_0\right) \quad \text{and}$$

$$(10.5.8) \qquad c = \frac{1}{2}u + c_0 = \frac{1}{3}\left(\frac{x}{t} + 2c_0\right).$$

Since $c \geqq 0$ we must have $-x/t \leqq 2c_0$ so that $-w$ must be $\leqq 2c_0$ from (10.5.6) in conformity with a similar result above. If $w = -2c_0$, the terminal characteristic of zone *II* is given, from (10.5.6), by $x = -2c_0t = wt$ and this line falls on the piston curve since the slope of the piston curve is w. In this limit case, therefore, the zone *III* collapses into the piston curve. If the piston is moved at still higher speed, then cavitation occurs as in the cases discussed above since $c = 0$ at the front of the wave, or in other words, the water surface touches the bottom.

From (10.5.8) we can calculate the elevation η of the water surface since $c = \sqrt{g(h + \eta)}$;

$$(10.5.9) \qquad \eta + h = \frac{1}{9g}\left(\frac{x}{t} + 2c_0\right)^2.$$

In the case of incipient cavitation, i.e. $-w = 2c_0$, we have $\eta = -h$ at the front of the wave. The curve of the water surface at any time t is a parabola from the front of the wave to the point $x = c_0t$ (corresponding to the characteristic which delimits the zone of quiet), after which it is horizontal. In Fig. 10.5.5 the total depth $\eta + h$ of the water is plotted against x for a fixed time t. The surface of the water is tangent to the bottom at the front $x = -2c_0 t$ of the moving water. The region in which the water is in motion extends from this point back to the point $x = c_0 t$. From (10.5.7) we can draw the following somewhat unexpected conclusion in this case: Since t may be given arbitrarily large values it follows that the velocity u of the water at any fixed point x tends to the values $-\frac{2}{3}c_0$ as t grows large.

The case of cavitation may have a certain interest in practice: the motion of the water might be considered as an approximation to the flow which would result from the sudden destruction of a dam built in a valley with very steep sides and not too great bottom slope (cf.

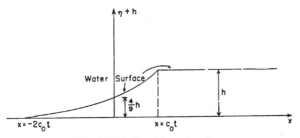

Fig. 10.5.5. Breaking of a dam

the paper of Ré [R.5]). If the water behind the dam were 200 feet high, for example, our results indicate that the front of the wave would move down the valley at a speed of about 110 miles per hour. By setting $x = 0$ in (10.5.9) we observe that the depth of the water at the site of the dam is always constant and has the value $\frac{4}{9}h$, i.e. four-ninths of the original depth of the water behind the dam. The velocity of the water at this point is also constant and has the value $u = -\frac{2}{3}c_0 = -\frac{2}{3}\sqrt{gh}$, as we see from (10.5.7). The volume rate of discharge of water at the original location of the dam is thus constant.

So far we have not considered the motion of the individual water particles. However, that is readily done in all cases once the velocity $u(x, t)$ is known: We have only to integrate the ordinary differential equation

(10.5.10)
$$\frac{dx}{dt} = u.$$

In zone *II* in our present case we have

(10.5.11)
$$\frac{dx}{dt} = \frac{2}{3}\left(\frac{x}{t} - c_0\right).$$

By setting $\xi = x + 2c_0 t$ one finds readily that ξ satisfies the differential equation $d\xi/dt = 2\xi/3t$, from which $\xi = At^{2/3}$ with A an arbitrary constant. Hence we have for the position $x(t)$ of any particle in zone II

(10.5.12) $x = t\{At^{-1/3} - 2c_0\}.$

In the case of cavitation this formula holds for arbitrarily large t so that we have for large t the asymptotic expression for x:

(10.5.13) $x \sim -2c_0 t.$

(This is not in contradiction with our above result that $u \sim -\frac{2}{3}c_0$ for large t and fixed x since in that case different particles pass the point in question at different times, while (10.5.13) refers always to the same particle).

In the first section of Chapter 12 this same problem of the breaking of a dam will be treated by using the exact nonlinear theory in such a manner as to determine the motion during its early stages after the dam has been broken—in other words, at the times when the shallow water theory is most likely to be inaccurate.

10.6. Discontinuity, or shock, conditions

The difference in behavior of a depression which propagates into still water as compared with the behavior of a hump has already been pointed out: in the first case the motion is continuous throughout, but in the second case the motion can not be continuous after a certain time. The general situation is indicated in Fig. 10.6.1, which shows the characteristics in the x, t-plane for the motion which results when a "piston" at the end of a tank is pushed into the water with steadily increased speed. As before, the slope dx/dt of a straight characteristic issuing from the "piston curve" $x = x(t)$ is given (cf. (10.5.2)) by $dx/dt = \frac{3}{2}u_A + c_0$, in which $u_A = x_t(t)$ is the velocity of the piston. Since u_A is assumed to increase with t it is clear that the characteristics will cut each other. In general, they have an envelope as indicated by the heavy line in the figure. The continuous solutions furnished by our theory, which have been the only ones under consideration so far, are thus valid in the region of the x, t-plane between the initial characteristic and the piston curve up to the curved characteristic (indicated by the curve segment ED) through the "first" point E on the envelope of the straight characteristics, but not beyond ED.

What happens "beyond the envelope" can in principle therefore

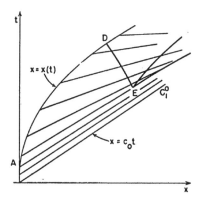

Fig. 10.6.1. Initial point of breaking

not be studied by the theory presented up to now. However, it seems very likely that discontinuous solutions may develop as the time increases beyond the value corresponding to the point E, which are then to be interpreted physically as motions involving the gradual development of bores and breakers in the water.

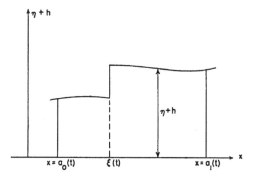

Fig. 10.6.2. Discontinuity conditions

There is a particularly simple limit case of the situation indicated in Fig. 10.6.2 for which a discontinuous solution can be found once we have obtained the discontinuity conditions that result from the fundamental laws of mechanics. That is the case in which the "piston" is accelerated instantaneously from rest to a constant forward velocity so that the piston curve is a straight line issuing from the origin in the x, t-plane. It is the exact counterpart of the case discussed at the end of the preceding section in which the piston was withdrawn from the water at a uniform speed.

To obtain the conditions at a discontinuity we consider a region made up of the water lying between two vertical planes $x = a_0(t)$ and $x = a_1(t)$ with $a_1 > a_0$ and such that these planes contain always the same particles. Such an assumption can be made, we recall from Chapter 2, since in our theory the particles which are in a vertical plane at any instant always remain in a vertical plane. Hence the horizontal particle velocity component u is the same throughout any vertical plane. We now suppose that there is a finite discontinuity in the surface elevation η at a point $x = \xi(t)$ within the column of water between $x = a_0(t)$ and $x = a_1(t)$, as indicated in Fig. 10.6.2.

The laws of conservation of mass and of momentum as applied to our column of water yield the relations

$$(10.6.1) \qquad \frac{d}{dt} \int_{a_0(t)}^{a_1(t)} \varrho(\eta + h)dx = 0$$

and

$$(10.6.2) \quad \frac{d}{dt} \int_{a_0(t)}^{a_1(t)} \varrho(\eta + h)u \, dx = \int_{-h}^{\eta_0} p_0 \, dy - \int_{-h}^{\eta_1} p_1 \, dy$$
$$= \tfrac{1}{2} g\varrho(\eta_0 + h)^2 - \tfrac{1}{2} g\varrho(\eta_1 + h)^2,$$

when the formula $p = g\varrho(\eta - y)$ for the pressure in the water is used. The second relation states that the change in momentum of the water column is equal to the difference of the resultant forces over the end sections of the column.

The integrals in these relations have the form

$$I = \int_{a_0(t)}^{a_1(t)} \psi(x, t) \, dx$$

in which $\psi(x, t)$ has a discontinuity at $x = \xi(t)$. Differentiation of this integral yields the relation

$$\frac{dI}{dt} = \frac{d}{dt} \int_{a_0(t)}^{\xi(t)} \psi \, dx + \frac{d}{dt} \int_{\xi(t)}^{a_1(t)} \psi \, dx$$

$$(10.6.3)$$
$$= \int_{a_0(t)}^{a_1(t)} \frac{\partial \psi}{\partial t} \, dx + \psi \, (\xi_-, t)\dot{\xi}(t) - \psi(a_0(t), t)u_0$$
$$+ \psi(a_1(t), t)u_1 - \psi(\xi_+, t)\dot{\xi}(t).$$

The quantities u_0 and u_1 are the velocities $\dot{a}_0(t)$ and $\dot{a}_1(t)$ at the ends of the column, $\dot{\xi}$ is the velocity of the discontinuity, and $\psi(\xi_-, t)$ and $\psi(\xi_+, t)$ mean that the limit values of ψ to the left and to the right of $x = \xi$ respectively are to be taken. We wish to consider the limit case in which the length of the column tends to zero in such a way that the discontinuity remains inside the column. When we do so the integral on the right-hand side of (10.6.3) tends to zero and we obtain

$$(10.6.4) \qquad \lim_{a_1 \to a_0} \frac{dI}{dt} = \psi_1 v_1 - \psi_0 v_0$$

in which v_1 and v_0 are the relative velocities given by

$$(10.6.5) \qquad \begin{cases} v_1 = u_1 - \dot{\xi}, \\ v_0 = u_0 - \dot{\xi} \end{cases}$$

and ψ_1 and ψ_0 refer to the limit values of ψ to the right and to the left of the discontinuity, respectively. The important quantities v_0 and v_1 are obviously the flow velocities relative to the moving discontinuity.

Upon making use of (10.6.4) and (10.6.5) for the limit cases which arise from (10.6.1) and (10.6.2) we obtain the following conditions

$$(10.6.6) \qquad \varrho(\eta_1 + h)v_1 - \varrho(\eta_0 + h)v_0 = 0$$

and

$$(10.6.7) \quad \varrho(\eta_1+h)u_1 v_1 - \varrho(\eta_0+h)u_0 v_0 = \tfrac{1}{2}\varrho g(\eta_0+h)^2 - \tfrac{1}{2}\varrho g(\eta_1+h)^2.$$

If we introduce, as in section 10.1, the quantities $\bar{\varrho}$ and $\bar{p}$ (which are the analogues of the density and pressure in gas dynamics) by the relations (cf. (10.1.3) and (10.1.4))

$$(10.6.8) \qquad \bar{\varrho} = \varrho(\eta + h)$$

and

$$(10.6.9) \qquad \bar{p} = \frac{g\varrho}{2}(\eta + h)^2 = \frac{g}{2\varrho}\bar{\varrho}^2,$$

we obtain in place of (10.6.6) and (10.6.7) the discontinuity conditions

$$(10.6.10) \qquad \bar{\varrho}_1 v_1 = \bar{\varrho}_0 v_0,$$

and

$$(10.6.11) \qquad \bar{\varrho}_1 u_1 v_1 - \bar{\varrho}_0 u_0 v_0 = \bar{p}_0 - \bar{p}_1.$$

The last two relations are identical in form with the mechanical conditions for a shock wave in gas dynamics when the latter are expressed in terms of velocity, density and pressure changes.

Henceforth we shall often refer to a discontinuity satisfying (10.6.10) and (10.6.11) as a shock wave or simply as a shock even though such an occurrence is better known in fluid mechanics as a *bore*, or if it is stationary as a *hydraulic jump*.

Since $u_1 - u_0 = v_1 - v_0$ from (10.6.5) it is easily seen that the shock conditions (10.6.10) and (10.6.11) can be written in the form

$$(10.6.12) \qquad \begin{cases} \bar{\varrho}_1 v_1 = \bar{\varrho}_0 v_0 = m, \\ m(v_1 - v_0) = \bar{p}_0 - \bar{p}_1, \end{cases}$$

in which m represents the mass flux across the shock front.

To fix the motion on both sides of the shock five quantities are needed; i.e. the particle velocities u_0, u_1, the elevations η_0 and η_1 (or, what is the same, the "pressures" $\bar{p}$ or the "densities" $\bar{\varrho}$ as given by (10.6.8) and (10.6.9) on both sides of the shock), and the velocity ξ of the shock. Evidently the relative velocities v_0 and v_1 would then be determined. Since the five quantities satisfy the two relations (10.6.12) we see that in general only three of the five quantities could be prescribed arbitrarily. Since the equations to be satisfied are not linear it is not a priori clear whether solutions can be found for two of the quantities when any other three are arbitrarily prescribed or whether such solutions would be unique. We want to investigate this question in a number of important special cases.

Before doing so, however, it is important to consider the energy balance across a shock. The fact is, as we shall see shortly, that the law of conservation of energy does not hold across a shock, but rather the particles crossing* the shock must either lose or gain in energy. Since we do not wish to postulate the existence of energy sources at the shock front capable of increasing the energy of the water particles as they pass through it, we assume from now on that the *water particles do not gain energy upon crossing a shock front*. This will in effect furnish us with an inequality which in conjunction with the two shock relations (10.6.12) leads in all of our cases to unique solutions of the physical problems. We turn, then, to a consideration of the energy balance across a shock, which we can easily do by following

---

* It is important to observe that the water particles always do cross a shock front: the quantity m in (10.6.12), the mass flux through the shock front, is different from zero if there is an actual discontinuity since otherwise $v_1 = v_0 = 0$, $u_1 = u_0 = \dot{\xi}$, and $\bar{p}_0 = \bar{p}_1$ and hence $\bar{\varrho}_0 = \bar{\varrho}_1$ — in other words the motion is continuous. There is thus no analogue in our theory of what is called a contact discontinuity in gas dynamics in which velocity and pressure are continuous, but the density and temperature may be discontinuous.

the same procedure that was used to derive the shock relations (10.6. 10) and (10.6.11). For the rate of change dE/dt of the energy E in the water column of Fig. 10.6.2 we have, as one can readily verify:

(10.6.18)
$$\frac{dE}{dt} = \frac{d}{dt} \left\{ \int_{a_0(t)}^{a_1(t)} \left[\varrho(\eta + h) \frac{u^2}{2} + \frac{g\varrho}{2} (\eta + h)^2 \right] dx \right\}$$
$$+ \int_{-h}^{\eta_1} p_1 u_1 \, dy - \int_{-h}^{\eta_0} p_0 u_0 \, dy$$

and this in turn yields in the limit when $a_0 \to a_1$, through use of (10.6.5), (10.6.8), (10.6.9), and the hydrostatic pressure law, the relation

(10.6.14) $\dfrac{dE}{dt} = \frac{1}{2}\bar{\varrho}_1 u_1^2 v_1 - \frac{1}{2}\bar{\varrho}_0 u_0^2 v_0 + \bar{p}_1 v_1 - \bar{p}_0 v_0 + \bar{p}_1 u_1 - \bar{p}_0 u_0$

for the rate at which energy is created or destroyed at the shock front. If we multiply (10.6.11) by $\dot{\xi}$ on both sides and then subtract from (10.6.14) the result is an equation which can be written after some manipulation and use of (10.6.5) in the form

(10.6.15) $\dfrac{dE}{dt} = m \left\{ \frac{1}{2}(v_1^2 - v_0^2) + 2(\bar{p}_1/\bar{\varrho}_1 - \bar{p}_0/\bar{\varrho}_0) \right\}$

in which m is the mass flux through the shock front defined in (10.6.12). In this way we express dE/dt entirely in terms of the relative velocities v_0 and v_1 and the change in depth. By eliminating v_1 and v_0 through use of $v_1 = m/\bar{\varrho}_1$ and $v_0 = m/\bar{\varrho}_0$ and replacing $\bar{p}_1$ and $\bar{p}_0$ in terms of $\bar{\varrho}_1$ and $\bar{\varrho}_0$ we can express dE/dt in terms of $\bar{\varrho}_0$ and $\bar{\varrho}_1$; the result is readily found to be expressible in the simple form

(10.6.16) $\dfrac{dE}{dt} = \dfrac{mg}{\varrho} \dfrac{(\bar{\varrho}_0 - \bar{\varrho}_1)^3}{4\bar{\varrho}_1\bar{\varrho}_0}.$

We see therefore that *energy is not conserved unless* $\bar{\varrho}_0 = \bar{\varrho}_1$, *i.e. unless the motion is continuous.* Since $\bar{\varrho}_0 - \bar{\varrho}_1 = \varrho(\eta_0 - \eta_1)$ it follows from (10.6.16) that the rate of change of the energy of the particles crossing the shock is proportional to the cube of the difference in the depth of the water on the two sides of the shock, or as we could also put it in case $\eta_0 - \eta_1$ is considered to be a small quantity: the rate of change of energy is of third order in the "jump" of elevation of the water surface.

The statement that the law of conservation of energy does not hold in the case of a bore in water must be taken cum grano salis. What we mean is of course that the energy balance can not be maintained through the sole action of the mechanical forces postulated in the above theory. The results of our theory of the bore and the hydraulic jump are therefore to be interpreted as an idealization of the actual occurrences in which the losses in mechanical energy are accounted for through the production of heat due to turbulence at the front of the shock (cf. the photograph of the bore in the Tsien-Tang river shown in Fig. 10.6.3). In compressible gas dynamics the theory used

Fig. 10.6.3. Bore in the Tsien-Tang River

allows for the conversion of mechanical energy into heat so that the law of conservation of energy holds across a shock in that theory. The analogue of the loss in mechanical energy across a shock in water is the increase in entropy across a shock in gas dynamics; furthermore, both of these discontinuous changes are of third order in the differences of "density" on the two sides of the shock.

We have tacitly chosen as the positive direction of the x-axis, and hence of all velocities, the direction from the side 0 toward the side 1 (cf. Fig. 10.6.2). Suppose now that the mass flux m is assumed to be positive; it follows from (10.6.12) and the fact that $\bar{\varrho}_0$ and $\bar{\varrho}_1$ are positive that v_0 and v_1 are also positive and hence that the water particles cross the shock front in the direction from the side 0 toward the side 1. Our condition that the water particles can not gain in energy on crossing the shock then requires, as we see at once from (10.6.16) since m,

g, ϱ, $\bar{\varrho}_0$, and ϱ_1 are all positive, that $\bar{\varrho}_0 < \bar{\varrho}_1$. In other words, our energy condition requires that the *particles always move across the shock from a region of lower total depth to one of higher total depth.*[*] Since the mass flux m is not zero unless the flow is continuous, and hence there is no shock, it is possible to define uniquely the two sides of the shock by the following useful convention: the *front* and *back* sides of the shock are distinguished by the fact that the mass flux passes through the shock from front to back, or, as one could also put it, the water crosses the shock from the front side toward the back side. Our conclusion based on the assumed loss of energy across the shock can be interpreted in terms of this convention as follows: *the water level is always lower on the front side of the shock than on the back side.*

For the further discussion of the shock relations it is important to observe that all of them, including the relation (10.6.16) for the energy loss, *can be written in such a way as to involve only the velocities v_0 and v_1 of the water particles relative to the shock front and not the absolute velocities u_0 and u_1.* It follows that we may always assume one of the three velocities u_0, u_1, $\dot{\xi}$ to be zero if we wish, with no essential loss of generality, because the laws of mechanics are in any case invariant with respect to axes moving with constant velocity, and adding the same constant to u_0, u_1 and $\dot{\xi}$ does not affect the values of v_0 and v_1.

Let us assume then that $u_0 = 0$, i.e. that the water is at rest on one side of the shock. Also, we write the second of the shock conditions (10.6.12) in the form

$$(10.6.17) \qquad v_1 v_0 = \frac{\bar{p}_0 - \bar{p}_1}{\bar{\varrho}_0 - \bar{\varrho}_1},$$

which follows from $m v_1 = \bar{\varrho}_0 v_0 v_1$ and $m v_0 = \bar{\varrho}_1 v_1 v_0$ and (10.6.12). From $u_0 = 0$ we have $v_0 = -\dot{\xi}$ and $v_1 = u_1 - \dot{\xi}$ (cf. (10.6.5)) so that (10.6.17) takes the form

$$(10.6.18) \qquad -\dot{\xi}(u_1 - \dot{\xi}) = \frac{g}{2\varrho}(\bar{\varrho}_0 + \bar{\varrho}_1)$$

upon making use of $\bar{p} = g\bar{\varrho}^2/2\varrho$ (cf. (10.6.9)). The first shock condition now takes the form

$$(10.6.19) \qquad \bar{\varrho}_1(u_1 - \dot{\xi}) = -\bar{\varrho}_0\dot{\xi},$$

so that (10.6.18) can be written

$$(10.6.20) \qquad \dot{\xi}^2 = \frac{g\bar{\varrho}_1}{2\varrho}\left(1 + \frac{\bar{\varrho}_1}{\bar{\varrho}_0}\right)$$

[*] This conclusion was first stated by Rayleigh [R.3].

if u_1 is eliminated, or it may be written in the form

(10.6.21) $$- \dot{\xi}(u_1 - \dot{\xi}) = \frac{g\bar{\varrho}_1}{2\varrho} \left(1 - \frac{u_1 - \dot{\xi}}{\dot{\xi}}\right)$$

if $\bar{\varrho}_0$ is eliminated. Thus (10.6.19) together with either (10.6.20) or
(10.6.21) are ways of expressing the shock conditions when $u_0 = 0$.

We are now in a position to discuss some important special cases.
Having fixed the value of u_0, i.e. $u_0 = 0$, at most two of the remaining
quantities $\dot{\xi}$, $\bar{\varrho}_0$, $\bar{\varrho}_1$, and u_1 can be prescribed arbitrarily. For our later
discussion it is useful to single out the following two cases: *Case* 1. $\bar{\varrho}_1$
and $\bar{\varrho}_0$ are given, i.e. the depth of the water on both sides of the shock
and the velocity on one side are given. *Case* 2. $\bar{\varrho}_1$ and u_1 are given, i.e.
the velocity of the water on both sides of the shock and the depth of
the water on one side are given. We proceed to discuss these cases in
detail.

Case 1. From (10.6.20) we see that $\dot{\xi}^2$ is determined for any arbitrary
values (positive, of course) of $\bar{\varrho}_0$ and $\bar{\varrho}_1$, i.e. of the water depths. Hence
$\dot{\xi}$ is determined by (10.6.20) only within sign. Suppose now that
$\bar{\varrho}_1 > \bar{\varrho}_0$. In this case the side 0 is, as we have seen above, the front
side of the shock, and since $u_0 = 0$ the shock front must move in the
direction from the side 1 toward the side 0 in order that the mass flux
should pass through the shock from front to back.

Hence if it is once decided whether the side 0 is to the left or to the
right of the side 1 the sign of $\dot{\xi}$ is uniquely fixed. If, as in Fig. 10.6.4,

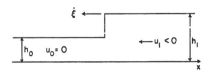

Fig. 10.6.4. Bore advancing into still water

the side 0 is chosen to the left of the side 1, and the x-direction is positive to the right, it follows that $\dot{\xi}$ is negative, as indicated. It is useful
to introduce the depths h_0 and h_1 of the water on the two sides of the
shock:

(10.6.22) $$\begin{cases} h_0 = h + \eta_0, \\ h_1 = h + \eta_1 \end{cases}$$

and to express (10.6.20) in terms of these quantities. The result for $\dot{\xi}$ in our case is

(10.6.23)
$$\dot{\xi} = -\sqrt{g\,\frac{h_1}{h_0}\left(\frac{h_1 + h_0}{2}\right)}$$

as one readily sees from $\bar{\varrho}_i = \varrho h_i$. From (10.6.23) we draw the important conclusion: Since $h_1 > h_0$, the shock speed $|\dot{\xi}|$ is greater than $\sqrt{gh_0}$ since $h_0 < (h_1 + h_0)/2 < h_1$. Also, in the case $u_0 = 0$ we have from (10.6.19)

(10.6.24)
$$u_1 = \dot{\xi}\left(1 - \frac{h_0}{h_1}\right),$$

so that *the velocity of the water behind the shock has the same sign as $\dot{\xi}$ (since $h_0/h_1 < 1$) but is less than $\dot{\xi}$ numerically.*

Finally, it is very important to consider the speed v_1 of the shock front relative to the water particles *behind* it: from (10.6.24) we have

(10.6.25)
$$v_1 = u_1 - \dot{\xi} = -\frac{h_0}{h_1}\dot{\xi},$$

and this in turn can be expressed through use of (10.6.23) in the form

(10.6.26)
$$v_1 = \sqrt{g\,\frac{h_0}{h_1}\left(\frac{h_1 + h_0}{2}\right)},$$

so that $v_1 < \sqrt{gh_1}$. In other words, *the speed of the shock relative to the water particles behind the shock is less than the wave propagation speed $\sqrt{gh_1}$* in the water behind the shock. Hence a small disturbance created *behind* a shock will eventually catch up with it. Although the conclusion was drawn for the special case $u_0 = 0$ it holds quite generally for the shock velocities *relative* to the motion of the water on both sides of a shock, in view of earlier remarks on the dependence of the shock relations on these relative velocities.

The case illustrated by Fig. 10.6.4 is that of a shock advancing into still water. The fact that $\dot{\xi}$ is in this case of necessity negative is a consequence of the assumption of an energy loss across the shock. It is worth while to restate this conclusion in the negative sense, as follows: a depression shock can not exist, i.e. a shock wave which leaves still water at reduced depth *behind* it should not be observed in nature.* The observations bear out this conclusion. Bores advancing

---

* In gas dynamics the analogous situation occurs: only compression shocks and not rarefaction shocks can exist.

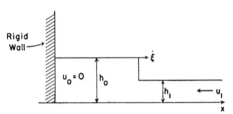

Fig. 10.6.5. Reflection from a rigid wall

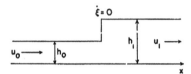

Fig. 10.6.6. Hydraulic jump

into still water are well known, but depression waves are always smooth.

Instead of assuming that $\bar{\varrho}_1 > \bar{\varrho}_0$ (or that $h_1 > h_0$) as in the case of Fig. 10.6.4 we may assume $\bar{\varrho}_1 < \bar{\varrho}_0$ (or $h_1 < h_0$), so that the side 1 is the front side. In other words the water is at rest on the back side of the shock in this case. If the front side is taken on the right, the situation is as indicated in Fig. 10.6.5. In this case $\dot{\xi}$ must be positive and u_1 negative in order that the mass flux should take place from the side 1 to the side 0. The value of u_1 is given by (10.6.24) in this case also. The case of Fig. 10.6.5 might be realized in practice as the result of reflection of a stream of water from a rigid wall so that the water in contact with the wall is brought to rest. We shall return to this case later.

In the above two cases we considered u_0 to be zero. However, we know that we may add any constant velocity to the whole system without invalidating the shock conditions. It is of interest to consider the motion which arises when the velocity $-\dot{\xi}$ is added to u_0, u_1, and $\dot{\xi}$ in the case shown in Fig. 10.6.4. The result is the motion indicated by Fig. 10.6.6 in which the shock front is stationary. This case—one of frequent occurrence in nature—is commonly referred to as the hydraulic jump. From our preceding discussion we see that the water always moves from the side of lower elevation to the side of higher elevation. The velocities u_0 and u_1 are both positive, and $u_0 > u_1$.

Also the velocity u_0 on the incoming side is greater than the wave propagation speed $\sqrt{gh_0}$ on that side while the velocity u_1 is less than $\sqrt{gh_1}$. This follows at once from the known facts concerning the relative shock velocities and the fact that u_0 and u_1 are the velocities relative to the shock front in this case. The hydraulic engineers refer to this as a transition from supercritical to subcritical speed.

Case 2. We recall that in this case $u_0 = 0$, u_1, and $\bar{\varrho}_1$ (or h_1) are assumed given and $\dot{\xi}$ and h_0 are to be determined. The value of $\dot{\xi}$ is to be determined from (10.6.21). To study this relation it is convenient to set $x = -\dot{\xi}$ and $y = u_1 - \dot{\xi}$ so that (10.6.21) can be replaced by

$$(10.6.27) \qquad \begin{cases} y = k^2 x/(x^2 - k^2), \quad k^2 = \dfrac{gh_1}{2} \\ y = u_1 + x. \end{cases}$$

In Fig. 10.6.7 we have indicated these two curves, whose intersections yield the solutions $\dot{\xi} = -x$ of (10.6.21). The first equation is represented by a curve with three branches having two asymptotes $x = \pm k$. As one sees readily, there are always three different real roots for $-\dot{\xi}$ no matter what values are chosen for the positive quantity $k^2 = gh_1/2$ and for the velocity u_1. Furthermore, one root $\dot{\xi}_+ = -x_-$ is always positive, another $\dot{\xi}_- = -x_+$ is negative, while the third $\dot{\xi} = -\bar{x}$ lies between the other two. However, the third root $\dot{\xi} = -\bar{x}$ must be rejected because it is not compatible with (10.6.19): Since $\bar{\varrho}_1$ and $\bar{\varrho}_0$ are both positive it follows that $x = -\dot{\xi}$ and $y =$

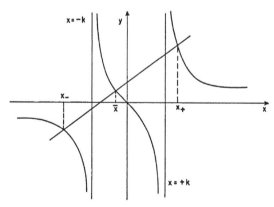

Fig. 10.6.7. Graphical solution of shock conditions

$u_1 - \dot{\xi}$ must have the same sign. But the sign of $y = \bar{y}$ corresponding to $x = \bar{x}$ is always the negative of $\bar{x}$ as one sees from Fig. 10.6.7. (If $u_1 = 0$,then $\bar{x} = \bar{y} = 0$, but there is no shock discontinuity in this case.) The other two roots, however, are such that the signs of $- \dot{\xi}$ and $u_1 - \dot{\xi}$ are the same. In the case 2, therefore, equation (10.6.21) furnishes two different values of $\dot{\xi}$ which have opposite signs and these values when inserted in (10.6.19) furnish two values of the depth h_0. The two cases are again those illustrated in Figs. 10.6.4 and 10.6.5. An appropriate choice of one of the two roots must be made in accordance with the given physical situation, as will be illustrated in one of the problems to be discussed in the next section.

Before proceeding to the detailed discussion of special problems involving shocks it is perhaps worth while to sum up briefly the main facts derived in this section concerning them: the five essential quantities defining a shock wave—$\dot{\xi}$, u_0, u_1; $\bar{\varrho}_0$, $\bar{\varrho}_1$ (or, what is the same, h_0 and h_1)—must satisfy the shock conditions (10.6.12). If it is assumed in addition that the water particles may lose energy on crossing the shock but not gain it, then it is found that the *shock wave travels always in such a direction that the water particles crossing it pass from the side of lower depth to the side of higher depth. If $h_0 < h_1$, so that the side 0 is the front side of the shock, the speeds $| v_0 |$ and $| v_1 |$ of the water relative to the shock front satisfy the inequalities*

$$(10.6.28) \qquad \begin{cases} | v_0 | > \sqrt{gh_0}, \\ | v_1 | < \sqrt{gh_1}. \end{cases}$$

In hydraulics it is customary to say that the velocity relative to the shock is *supercritical on the front side* (i.e. greater than the wave propagation speed corresponding to the water depth on that side) *and subcritical on the back side of the shock.**

10.7. Constant shocks: bore, hydraulic jump, reflection from a rigid wall

In the preceding section shock discontinuities were studied for the purpose of obtaining the relations which must hold on the two sides of the shock, and nothing was specified about the motion otherwise except that the shock under discussion should be the only disconti-

---

* In gas dynamics the analogous inequalities lead to the statement that the flow velocity relative to the shock front is supersonic with respect to the gas on the side of lower density and subsonic with respect to the gas on the other side.

nuity in a small portion of the fluid on both sides of it. In the present and following sections we wish to consider motions which are continuous except for the occurrence of a single shock. Furthermore we shall limit our investigations in this section to cases in which the motion on each of the two sides of the shock has constant velocity and depth. These motions, or flows, are evidently of a very special character, but they are easy to describe and also of frequent occurrence in nature.

It is perhaps not without interest in this connection to observe that the only *steady* and continuous wave motions (i.e., motions in which the velocity u and wave propagation speed $c = \sqrt{g(h + \eta)}$ are independent of the time) furnished by our theory for the case of constant depth h are the constant states $u = $ const., $c = $ const. This follows from the original dynamical equations (10.1.8) and (10.1.9). In fact, when u and c are assumed to be functions of x alone these equations reduce to

$$\begin{cases} u\dfrac{du}{dx} + 2c\dfrac{dc}{dx} = 0, \quad \text{and} \\[2mm] 2u\dfrac{dc}{dx} + c\dfrac{du}{dx} = 0 \end{cases}$$

for the case in which $h = $ const. (and so $H_x = 0$). These equations are immediately integrable to yield $u^2 + 2c^2 = $ const. and $uc^2 = $ const. and these two relations are simultaneously satisfied only for constant values of u and c. On the other hand, any constant values whatever could be taken for u and c. The cases we discuss in this section are motions which result by piecing together two such steady motions (each with a different constant value for the depth and velocity) through a shock which moves with constant velocity. In this case the motion as a whole would be steady if observed from a coordinate system attached to the moving shock front. In view of our above discussion it is clear that such a motion with a single shock discontinuity is the most general progressing wave which propagates unchanged in form that could be obtained from our theory.*

Let us consider now the problem referred to at the beginning of the preceding section: a vertical plate—or piston, as we have called it—at

---

* This result should not be taken to mean that the so-called "solitary wave" does not exist. (By a solitary wave is meant a *continuous* wave in the form of a single elevation which propagates unaltered in form.) It means only that our approximate theory is not accurate enough to furnish such a solitary wave. This is a point which will be discussed more fully in section 10.9.

the left end of a tank full of water at rest is suddenly pushed into the
water at constant velocity w. As we could infer from the discussion
at the beginning of the preceding section the motion must be discontinuous from the very beginning—or, as we could also put it, the
"first" point on the envelope of the characteristics would occur at
$t = 0$. Since the piston moves with constant velocity we might expect
the resulting motion to be a shock wave advancing into the still water
and leaving a constant state behind such that the water particles move
with the piston velocity w. The circumstances for such an assumed
motion are indicated in Fig. 10.7.1, which shows the x, t-plane together
with the water surface at a certain time t_0. We know that the constant
states on each side of the shock satisfy our differential equations. In
addition, we show that they can always be "connected" through a
shock discontinuity which satisfies the shock relations derived in the
preceding section. In fact, the relations (10.6.18) and (10.6.19) yield
through elimination of $\bar{\varrho}_1 = \varrho h_1$ the relation

$$(10.7.1) \qquad -\dot{\xi}(w - \dot{\xi}) = \frac{gh_0}{2}\left(1 - \frac{\dot{\xi}}{w - \dot{\xi}}\right)$$

for $\dot{\xi}$ in terms of w and the depth h_0 in the still water, when we set
$\bar{\varrho}_0 = \varrho h_0$. Equation (10.7.1) is the same as (10.6.21) except that $\bar{\varrho}_0$
replaces $\bar{\varrho}_1$, and the discussion of its roots $\dot{\xi}$ follows exactly the same
lines as for (10.6.21): for each $h_0 > 0$ and any $w \neq 0$ the cubic equation (10.7.1) has three roots for $\dot{\xi}$: one negative, another positive, and
a third which has a value between these two. In the present case the
positive root for $\dot{\xi}$ must be taken in order to satisfy our energy condition (cf. the discussion based on (10.6.27) of the preceding section)
since the side 0 is the front side of the shock. Once $\dot{\xi}_+$ has been calculated from (10.7.1) we can determine the depth of the water h_1 behind
the shock from the first shock condition

$$(10.7.2) \qquad h_1(w - \dot{\xi}_+) = -h_0\dot{\xi}_+.$$

It is therefore clear that a motion of the sort indicated in Fig. 10.7.1
can be determined in a way which is compatible with all of our conditions.*

A few further remarks about the above motion are of interest. In

---

* It should be pointed out that our discussion yields a discontinuous solution
of the differential equations, but does not prove that it is the only one which
might exist. However, it has been shown by Goldner [G.6] that our solution would
be unique under rather general assumptions regarding the type of functions
admitted as possible solutions.

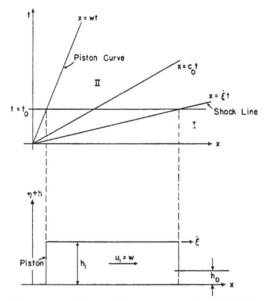

Fig. 10.7.1. A bore with constant speed and height

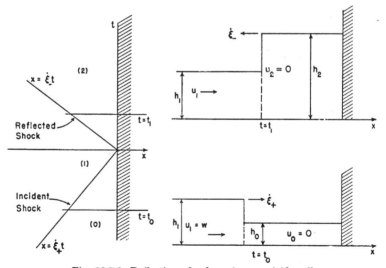

Fig. 10.7.2. Reflection of a bore from a rigid wall

Fig. 10.7.1 we have indicated the line $x = c_0 t$, $c_0 = \sqrt{gh_0}$, which would be the initial characteristic terminating the state of rest if the motion were continuous, i.e. if the disturbance proceeded into still water with the wave speed c_0 for water of the depth h_0. We know, however, from our discussion of the preceding section that the shock speed $\dot{\xi}$ is greater than c_0, which accounts for the position of the shock line $x = \dot{\xi}t$ below the line $x = c_0 t$ in Fig. 10.7.1. On the other hand we know that the velocity $w - \dot{\xi}$ of the water particles behind the shock relative to the shock is *less than the wave speed* $c_1 = \sqrt{gh_1}$ *in the water on that side*. It follows, therefore, that a new disturbance created in the water behind the shock should catch up with it since the front of such a disturbance would always move relative to the water particles with a velocity at least equal to c_1. For example, if the piston were to be decelerated at a certain moment a continuous depression wave would be created at the piston which would finally catch up with the shock front, and a complicated interaction process would then occur.

The case we have treated above corresponds to the propagation of a bore into still water. If we were to superimpose a constant velocity $- \dot{\xi}$ on the water in the motion illustrated by Fig. 10.7.1 the result would be the motion called a hydraulic jump in which the shock front is stationary. We need not consider this case further.

We treat next the problem of the reflection of a shock wave from a rigid vertical wall by following essentially the same procedure as above. The circumstances are shown in Fig. 10.7.2. We have an incoming shock moving toward the rigid wall from the left into still water of depth h_0. The shock is reflected from the wall leaving still water of depth h_2 behind it. Since the water in contact with the wall should be at rest, such an assumed motion is at least a plausible one. We proceed to show that the motion is compatible with our shock conditions and we calculate the height h_2 of the reflected wave.

We assume that h_1 and $u_1 = w$, the depth and the velocity of the water behind the shock, are known. The shock speed $\dot{\xi}_+$ is then determined by taking the largest of the three roots of the cubic equation (10.6.21), which we write down again in the form

$$(10.7.3) \qquad - \dot{\xi}(w - \dot{\xi}) = \frac{gh_1}{2}\left(1 - \frac{w - \dot{\xi}}{\dot{\xi}}\right).$$

Once $\dot{\xi}_+$ has been determined, the depth h_0 in front of the shock is fixed from the first shock condition, which is in the present case

$$(10.7.4) \qquad\qquad (w - \dot{\xi}_+)h_1 = - \dot{\xi}_+ h_0.$$

To determine the reflected shock we may once more evidently make

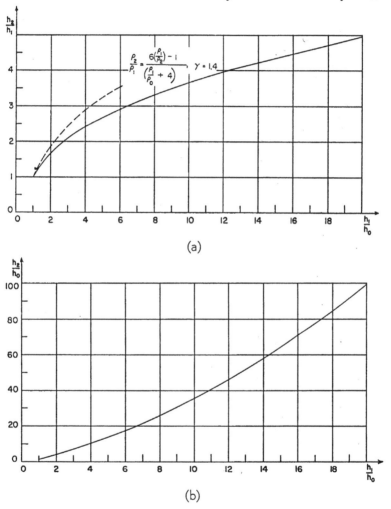

$$\frac{p_2}{p_1} = \frac{6\left(\frac{p_1}{p_2}\right) - 1}{\left(\frac{p_1}{p_0} + 4\right)}, \quad \gamma = 1.4$$

(a)

(b)

Fig. 10.7.3a, b. Reflection of a bore from a rigid wall

use of (10.7.4), since h_1 and $u_1 = w$ remain the same on one side of the shock, but we must now choose the smallest of the three roots of

(10.7.3) as the shock speed $\dot{\xi}_-$ since the side (1) is now obviously the front side of the shock. The depth h_2 of the water behind the shock after the reflection—that is, of the water in contact with the wall after reflection—is then obtained in the same way as h_0 by using (10.7.4) with $\dot{\xi}_-$ in place of $\dot{\xi}_+$ and h_2 in place of h_0:

$$(10.7.4)_1 \qquad (w - \dot{\xi}_-)h_1 = -\dot{\xi}_-h_2.$$

By taking a series of values for w we have determined the ratios h_2/h_1 and h_2/h_0 as functions of h_1/h_0. That is, the height h_2 of the reflected wave has been determined as a function of the ratio of the depth h_1 of the incoming wave to the initial depth h_0 at the wall. The results of such a calculation are shown in Figs. 10.7.3a and 10.7.3b: In Fig. 10.7.4 we give a curve showing $(h_2 - h_0)/h_0$ as a function of $(h_1 - h_0)/h_0$, that is, we give a curve showing the increase in depth after reflection as a function of the relative height $(h_1 - h_0)/h_0$ of the incoming wave.

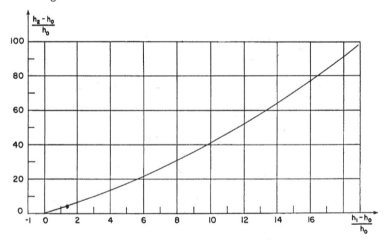

Fig. 10.7.4. Height of the reflected bore

For h_1/h_0 near to unity, i.e. for $(h_1 - h_0)/h_0$ small, it is not difficult to show that

$$(10.7.5) \qquad \frac{h_2 - h_0}{h_0} \simeq 2 \cdot \frac{h_1 - h_0}{h_0}.$$

From this relation we may write $h_2 - h_0 \simeq 2(h_1 - h_0)$ if $(h_1 - h_0)$ is small, i.e. the increase in the depth of the water after reflection is twice the height of the incoming wave when the latter is small. This is what one might expect in analogy with the reflection of acoustic waves of small amplitude. However, if h_1/h_0 is not small, the water increases in depth after reflection by a factor larger than 2. For instance, if h_1/h_0 is 2, then $h_2 - h_0 \simeq 3(h_1 - h_0)$; while if h_1/h_0 is 10, then $h_2 - h_0 \simeq 35(h_1 - h_0)$, as one sees from the graph of Fig. 10.7.4. In other words, the reflection of a shock or bore from a rigid wall results in a considerable increase in height and hence also in pressure against the wall if the incoming wave is high. In fact, for very high waves the total pressure $\bar{p}$ per unit width of the wall could be shown to vary as the cube of the depth ratio h_1/h_0.

In the upper curve of Fig. 10.7.3a we have drawn a curve for the analogous problem in gas dynamics. i.e. the reflection of a shock from the stopped end of a tube. In the case of air with an adiabatic exponent $\gamma = 1.4$ the density ratio ϱ_2/ϱ_1 as a function of ϱ_1/ϱ_0 (in an obvious notation) is plotted as a dotted curve in the figure. As we see, the density in air on reflection is higher than the corresponding quantity, the depth, in the analogous case in water. However, the curve for air ends at $\varrho_1/\varrho_0 = 6$, since it is not possible to have a shock wave in a gas with $\gamma = 1.4$ which has a higher density ratio. In water there is no such restriction. The explanation for this difference lies in the fact that the energy law is assumed to hold across a shock in gas dynamics, but not in our theory for water waves.

10.8. The breaking of a dam

At the end of section 10.5 we gave the solution to an idealized version of the problem of determining the flow which results from the sudden destruction of a dam if it is assumed that the downstream side

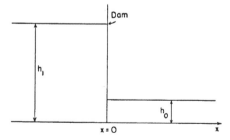

Fig. 10.8.1. Breaking of a dam

of the dam is initially free of water. In the present section we consider the more general problem which arises when it is assumed that there is water of constant depth on the downstream as well as the upstream side of the dam. Or, as the situation could also be described: a horizontal tank of constant cross section extending to infinity in both directions has a thin partition at the section $x = 0$. For $x > 0$ the water has the depth h_0 and for $x < 0$ the depth h_1, with $h_0 < h_1$, as indicated in Fig. 10.8.1. The water is assumed to be at rest on both sides of the dam initially. At the time $t = 0$ the dam is suddenly destroyed, and our problem is to determine the subsequent motion of the water for all x and t.

The special case $h_0 = 0$—the cavitation case—was treated, as we have already mentioned, at the end of section 10.5. We found there that the discontinuity for $x = 0$ and $t = 0$ was instantly wiped out and that the surface of that portion of the water in motion took the form of a parabola tangent to the x-axis (i.e. to the bottom) at the point $x = -2\sqrt{gh_1}t = -2c_1t$, in which t is the time and c_1 the wave

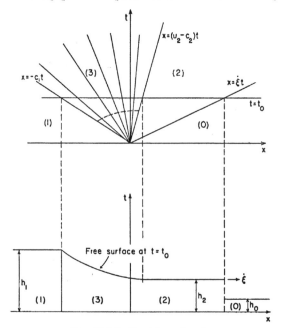

Fig. 10.8.2. Breaking of a dam

speed in water of depth h_1. If h_0 is different from zero we might therefore reasonably expect (on the basis of the discussion at the beginning
of section 10.6) that a shock wave would develop sooner or later on
the downstream side, since the water pushing down from above acts
somewhat like a piston being pushed downstream with an acceleration. In fact, since the water at $x = 0$ seems likely to acquire instantaneously a velocity different from zero it is plausible that a shock
would be created instantly on the downstream side. The simplest
assumption to make would be that the shock then moves downstream
with constant velocity $\dot\xi$ (cf. Fig. 10.8.2). If this were so, the state of
the water immediately behind the shock (i.e. on the upstream side
of it) would be constant for all time, since the velocity u_2 and depth
h_2 on the upstream side of the shock would have the constant values
determined from the shock relations for the fixed values $u_0 = 0$ and
$h = h_0$ for the velocity and depth on the downstream side and the
assumed constant value $\dot\xi$ for the shock velocity. However, it is clear
that the constant state behind the shock could not extend indefinitely
upstream since $u_2 \neq 0$ while the velocity of the water far upstream is
zero. Since we undoubtedly are dealing with a depression wave behind the shock it seems plausible to expect that the constant state
behind the shock changes eventually at some point upstream into a
centered simple wave of the type discussed in section 10.5. In Fig.
10.8.2 we indicate in an x, t-plane a motion which seems plausible as a
solution of our problem. In the following we shall show that such a
motion can be determined in a manner compatible with our theory
for every value of the ratio h_0/h_1.

As indicated in Fig. 10.8.2, we consider four different regions in the
fluid at any time $t = t_0$: the zone (0) is the zone of quiet downstream
which is terminated on the upstream side by the shock wave, or bore;
the zone (2) is a zone of constant state in which the water, however, is
not at rest; the zone (3) is a centered simple wave which connects the
constant state (2) with the constant state (1) of the undisturbed water
upstream. We proceed to show that such a motion exists and to determine it explicitly for all values of the ratio h_0/h_1 between zero and one.

For this purpose it is convenient to write the shock conditions for
the passage from the state (0) to the state (2) in the form

(10.8.1) $\qquad -\dot\xi(u_2 - \dot\xi) = \tfrac{1}{2}(c_0^2 + c_2^2);$

(10.8.2) $\qquad c_2^2(u_2 - \dot\xi) = -c_0^2\dot\xi;$

which are the same conditions as (10.6.18) and (10.6.19) with $g\bar\varrho_i/\varrho$

replaced by $c_i^2 = gh_i$, i.e. by the square of the wave propagation speed in water of depth h_i. By eliminating c_2^2 from (10.8.1) by use of (10.8.2) and then solving the resulting quadratic for u_2 one readily obtains

$$(10.8.3) \qquad u_2/c_0 = \dot{\xi}/c_0 - \frac{c_0}{4\dot{\xi}} 1 + \left(\sqrt{1 + 8(\dot{\xi}/c_0)^2} \right).$$

(The plus sign before the radical was taken in order that $u_2 - \dot{\xi}$ and $- \dot{\xi}$ should have the same sign. We observe also that only positive values of $\dot{\xi}$ and u_2 are in question throughout our entire discussion since the side of (0) is the front side of the shock and the positive x-direction is taken to the right.) It is also useful to eliminate u_2 from (10.8.3) by using (10.8.2); the result is easily put into the form

$$(10.8.4) \qquad \frac{c_2}{c_0} = \{ \tfrac{1}{2} \left(\sqrt{1 + 8(\dot{\xi}/c_0)^2} - 1 \right) \}^{\frac{1}{2}}.$$

The relations (10.8.3) and (10.8.4) yield the velocity u_2 and the wave speed c_2 behind the shock as functions of $\dot{\xi}$ and the wave speed c_0 in the undisturbed water on the downstream side of the dam. We proceed to connect the state (2) by a centered simple wave (cf. the discussion in section 10.5) with the state (1). In the present case the straight characteristics in the zone (3) (cf. Fig. 10.8.2) are those with the slope $u - c$ (rather than the slope $u + c$ as in section 10.5); hence the straight characteristics which delimit the zone (3) are the lines $x = -c_1 t$ on the left and $x = (u_2 - c_2)t$ on the right. Along each of the curved characteristics in zone (3)—one of these is indicated schematically by a dotted curve in Fig. 10.8.2—the quantity $u + 2c$ is a constant; it follows therefore that on the one hand

$$(10.8.5) \qquad\qquad\qquad u + 2c = 2c_1$$

since $u_1 = 0$, while on the other hand

$$(10.8.6) \qquad\qquad\qquad u + 2c = u_2 + 2c_2$$

throughout the zone (3). The relation

$$(10.8.7) \qquad\qquad\qquad u_2/c_0 + 2c_2/c_0 = 2c_1/c_0$$

must therefore hold. Our statement that a motion of the type shown in Fig. 10.8.2 exists for every value of the depth ratio h_1/h_0—or, what amounts to the same thing, the ratio c_1^2/c_0^2—is equivalent to the statement that the relation (10.8.7) furnishes through (10.8.3) and (10.8.4) an equation for $\dot{\xi}/c_0$ which has a real positive root for every value of c_1/c_0 larger than one. This is actually the case. In Fig. 10.8.3 we have plotted curves for u_2/c_0, $2c_2/c_0$, and $u_2/c_0 + 2c_2/c_0$ as functions of $\dot{\xi}/c_0$.

Once the curves of Fig. 10.8.3 have been obtained, our problem can be considered solved in principle: From the given value of $h_1/h_0 = c_1^2/c_0^2$ we can determine $\dot{\xi}/c_0$ from the graph (or, by solving (10.8.7)). The values of u_2/c_0 and c_2/c_0 are then also determined, either from the graph or by use of (10.8.3) and (10.8.4). The constant state in the zone (2) would therefore be known. In zone (3) the motion can now be determined exactly as in section 10.5; we would have along the straight characteristics in this zone the relations

$$\frac{dx}{dt} = \frac{x}{t} = u - c = 2c_1 - 3c = \tfrac{3}{2}u - c_1$$

from which

(10.8.8) $$c_2 = \tfrac{1}{9}\left(2c_1 - \frac{x}{t}\right)^2, \text{ and}$$

(10.8.9) $$u = \tfrac{2}{3}\left(c_1 + \frac{x}{t}\right).$$

Thus the water surface in the zone (3) is curved in the form of a parabola in all cases.* At the junctions with both zones (1) and (2) the parabola does not have a horizontal tangent, so that the slope of the water surface is discontinuous at these points.

Some interesting conclusions can be drawn from (10.8.8) and (10.8.9). By comparison with Fig. 10.8.2 we observe that the t-axis, i.e. the line $x = 0$, is a characteristic belonging to the zone (3) provided that $u_2 \geqq c_2$ since the terminal characteristic of the zone (3) on the right lies on the t-axis or to the right of it in this case. If this condition is satisfied we observe from (10.8.8) and (10.8.9)—which are then valid on the t-axis—that c and u are both independent of t at $x = 0$, which means that the depth of the water and its velocity u are both independent of t at this point, i.e. at the original location of the dam, and hence that the volume of water crossing the original dam site per unit of time (and unit of width) $dQ/dt = uh$ is independent of time although the motion as a whole is not a steady motion. In fact, $h = \tfrac{4}{9}h_1$ and $u = \tfrac{2}{3}c_1$ for all time t at this point. In addition, u and c, and thus also dQ/dt, are not only independent of t as long as $u_2 \geqq c_2$, but also independent of the undisturbed depth h_0 on the lower side of the dam if h_1 is held fixed. Of course, it is clear that h_0/h_1 must be kept under a certain value (which from section 10.5 evidently must be less than 4/9) or the condition $u_2 \geqq c_2$ could not be

* Relations (10.8.8) and (10.8.9) are exactly the same as (10.5.8) and (10.5.7) except for a change of sign which arises from a different choice of the positive x-direction.

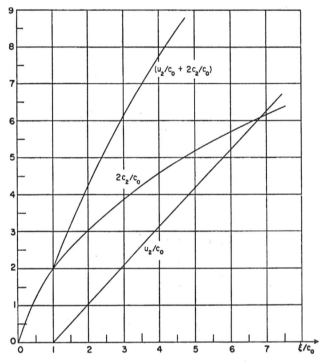

Fig. 10.8.3. Graphical solution for $\dot{\xi}$

fulfilled. In fact, the critical value of the ratio h_0/h_1 at which $u_2 = c_2$ can be determined easily by equating the right hand sides of (10.8.3) and (10.8.4) and determining the value of $\dot{\xi}/c_0$ for this case, after which $c_2/c_0 = \sqrt{h_2/h_0}$ is known from (10.8.4). Since $c_2 = \frac{2}{3} c_1$ in the critical case—either from the known fact that we still have $h_2 = \frac{4}{9} h_1$ in this case, or from (10.8.8) with $x = 0$—we thus are able to compute the critical value of $c_1^2/c_0^2 = h_1/h_0$. A numerical calculation yields for the critical value of the ratio h_1/h_0 the value 7.225, or for h_0/h_1 the value .1384. Thus if the water depth on the lower side of the dam is less than 13.8 percent of the depth above the dam the discharge rate on breaking the dam will be independent of the original depth on the lower side as well as independent of the time. However, if h_0/h_1 exceeds the critical value .1384, the depth, velocity, and discharge rate will depend on h_0; but they continue to be independent of the time since the line $x = 0$ in the x, t-plane is under the latter circumstances contained in the zone (2), which is one of constant state.

The above results, which at first perhaps seem strange, can be made understandable rather easily from the physical point of view, as follows. If the zone (3) includes the t-axis (i.e. if h_0/h_1 is below the critical value) we may apply (10.8.8) and (10.8.9) for $x = 0$ to obtain at this point $c = u = \frac{2}{3} c_1$. In other words, the flow velocity at the dam site is in this case just equal to the wave propagation speed there. For $x > 0$, i.e. downstream from the dam, we observe from (10.8.8) and (10.8.9) that u is greater than c. Since c is the speed at which the front of a disturbance propagates relative to the moving water we see that changes in conditions below the dam can have no effect on the flow above the dam since the flow velocity at all points below the dam is greater than the wave propagation speed at these points and hence disturbances can not travel upstream. However, once h_0/h_1 is taken higher than the critical value, the flow velocity at the dam will be less than the wave propagation speed at this point, as one can readily prove, and we could no longer expect the flow at that point to be independent of the initial depth assumed on the downstream side.

The discharge rate $dQ/dt = hu$ per unit width at the dam, i.e. at $x = 0$, is plotted in Fig. 10.8.4 as a function of the depth h_0. In accordance with our discussion above we observe that dQ/dt remains con-

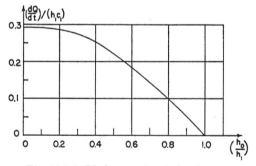

Fig. 10.8.4. Discharge rate at the dam

stant at the value $dQ/dt = .296 h_1 c_1$ until h_0/h_1 reaches the critical value .138, after which it decreases steadily to the value zero when $h_0 = h_1$, i.e. when the initial depth of the water below the dam is the same as that above the dam.

Another feature of interest in the present problem is the height of the bore, i.e. the quantity $h_2 - h_0$, as a function of the original depth ratio h_0/h_1. When $h_0 = 0$ we know that there is no bore and the water surface (as we found in section 10.5) appears as in Fig. 10.8.5. The water surface at the front of the wave on the downstream side is tangent to the bottom and moves with the speed $2c_1$. On the other hand, when h_0/h_1 approaches the other extreme value, i.e. unity, it is clear that the height $h_2 - h_0$ of the bore must again approach zero. Hence the height of the bore must attain a maximum for a certain

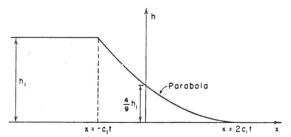

Fig. 10.8.5. Motion down the dry bed of a stream

value of h_0/h_1. In Fig. 10.8.6 we give the result of our calculations for $h_2 - h_0$ as a function of h_0/h_1. The curve rises very steeply to its maximum $h_2 - h_0 = .32h_1$ for $h_0/h_1 = .176$ and then falls to zero again when $h_0 = h_1$. It is rather remarkable that the bore can attain a height which is nearly 1/3 as great as the original depth of the water behind the dam.

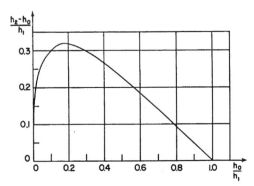

Fig. 10.8.6. Maximum height of the bore

It is instructive to describe the motion by means of the x, t-plane when h_0/h_1 is near its two limit values unity and zero. These two cases are schematically shown in Fig. 10.8.7. When $h_0 \simeq h_1$, we note that the zone (3) is very narrow and that the shock speed $\dot{\xi}$ approaches c_1, i.e. the propagation speed of small disturbances in water of depth h_1, corresponding to the fact that the height of the shock wave tends to zero as $h_0 \to h_1$ (cf. Fig. 10.8.6). The other limit situation, i.e. $h_0 \simeq 0$, is more interesting. Since we tacitly consider h_1 to remain fixed in our present discussion, and hence that h_2 is also fixed since we are in the supercritical case, it follows (for example from (10.6.23) with h_2 in place of h_1) that $\dot{\xi} \to \infty$ as $h_0 \to 0$. On the other hand as we see from

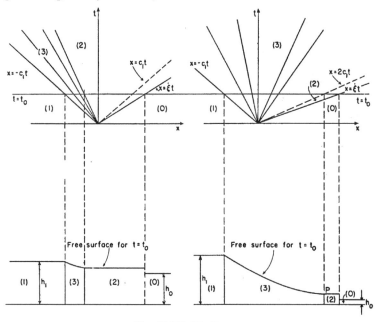

Fig. 10.8.7. Limit cases

Fig. 10.8.6, the height $h_2 - h_0$ of the shock wave tends to zero rather slowly as $h_0 \to 0$. In the limit, point P becomes the front of the wave in accordance with the motion indicated by Fig. 10.8.5. Thus as $h_0 \to 0$ the shock wave becomes very small in height but moves downstream with great speed; or, as we could also say, in the limit the water in front of the point P is pinched out and P is the front of the wave.

10.9. The solitary wave

It has long been a matter of observation that wave forms of a permanent type other than the uniform flows with an undeformed free surface occur in nature; for example, Scott Russell [R.14] reported in 1844 his observations on what has since been called the solitary wave, which is a wave having a symmetrical form with a single hump and which propagates at uniform velocity without change of form. Later on, Boussinesq [B.16] and Rayleigh [R.3] studied this problem mathematically and found approximations for the form and speed of such a solitary wave. Korteweg and de Vries [K.15] modified the method of Rayleigh in such a way as to obtain waves that are periodic in form—called cnoidal waves by them—and which tend to the solitary wave found by Rayleigh in the limiting case of long wave lengths. A systematic procedure for determining the velocity of the solitary wave has been developed by Weinstein [W.6].

At the beginning of section 10.7 we have shown that the only continuous waves furnished by the theory used so far in this chapter which progress unchanged in form are of a very special and rather uninteresting character, i.e., they are the motions with uniform velocity and horizontal free surface.* This would seem to be in crass contradiction with our intention to discuss the solitary wave in terms of the shallow water theory, and it has been regarded by some writers as a paradox.**The author's view is that this paradox—like most others— becomes not at all paradoxical when properly examined. What is involved is a matter of the range of accuracy of a given approximate theory, and also the fact that a perturbation or iteration scheme of universal applicability does not exist: one must always modify such schemes in accordance with the character of the problem. In the present case, the salient fact is that the theory used so far in the present chapter represents the result of taking only the lowest order terms in the shallow water theory as developed in section 4 of Chapter 2, and it is necessary to carry out the theory to include terms of higher order

* If motions with a discontinuity are included in the discussion, then the motion of a bore is the only other possibility up to now in this chapter with regard to waves propagating unchanged in form.

** Birkhoff [B.11, p. 23], is concerned more about the fact that the shallow water theory predicts that all disturbances eventually lead to a wave which breaks when on the other hand Struik [S.29] has proved that periodic progressing waves of finite amplitude exist in shallow water. In the next section the problem of the breaking of waves is discussed. Ursell [U.11] casts doubt on the validity of the shallow water theory in general because it supposedly does not give rise to the solitary wave.

if one wishes to obtain an approximation to the solution of the problem of the solitary wave. This has been done by Keller [K.6], who finds that the theory of Friedrichs [F.11] presented in Chapter 2, when carried out to second order,* yields both the solitary wave and cnoidal waves of the type found by Korteweg and de Vries [K.15] (thus the shallow water theory is capable of yielding periodic progressing waves of finite amplitude). As lowest order approximation to the solution of the problem, Keller finds (as he must in view of the remarks above), that the only possibility is the uniform flow with undeformed free surface, but if the speed U of the flow is taken at the critical value $U = \sqrt{gh}$ with h the undisturbed depth, then a bifurcation phenomenon occurs (that is, among the set of uniform flows of all depths and velocities, the solitary wave occurs as a bifurcation from the special flow with the critical velocity) and the second order terms in the development of Friedrichs lead to solitary and cnoidal waves with speeds in the neighborhood of this value. To clinch the matter, it has been found by Friedrichs and Hyers [F.13] that the existence of the solitary wave can be proved rigorously by a scheme which starts with the solution of Keller as the term of lowest order and proceeds by iterations with respect to a parameter in essentially the same manner as in the general shallow water theory.** In the following, we shall derive the approximation to the solution of the solitary wave problem following the method of Friedrichs and Hyers rather than the general expansion scheme which was used by Keller, and we can then state the connection between the two in more detail.

The author thus regards the nonlinear shallow water theory to be well founded and not at all paradoxical. Indeed, the linear theory of waves of small amplitude treated at such length in Part II of this book is in essentially the same position as regards rigorous justification as is the shallow water theory: we have only one or two cases so far in which the linear theory of waves of small amplitude is shown to be the lowest order term in a convergent development with respect to amplitude. We refer, in particular, to the theory of Levi-Civita [L.7] and Struik [S.29] in which the former shows the existence of periodic progressing waves in water of infinite depth and the latter the same thing (and by the same method) for waves in water of finite constant

* In order to fix all terms of second order, Keller found it necessary to make use of certain relations which result from carrying the development of some of the equations up to terms of third order.
** W. Littman, in a thesis to appear in *Communs. Pure and Appl. Math.*, has proved rigorously in the same way the existence also of cnoidal waves.

depth.* This theory will be developed in detail in Chapter 12. It might be added that those who find the nonlinear shallow water theory paradoxical in relation to the solitary wave phenomenon should by the same type of reasoning also find the linear theory paradoxical, since it too fails to yield any approximation to the solitary wave, even when carried out to terms of arbitrarily high order in the amplitude, except the uniform flow with undisturbed free surface. In fact, if one were to assume that a development exists for the solitary wave which proceeds in powers of the amplitude as in the theory discussed in the first part of Chapter 2, it is easily proved that the terms of all orders in the amplitude are identically zero. There is no paradox here, however; rather, the problem of the solitary wave is one in which the solution is not analytic in the amplitude in the neighborhood of its zero value, but rather has a singularity—possibly of the type of a branch point—there. Thus a different kind of development is needed, and, as we have seen, one such possibility is a development of the type of the shallow water theory starting with a nonlinear approximation. Another possibility has been exploited by Lavrentieff [L.4] in a difficult paper; Lavrentieff proves the existence of the solitary wave by starting from the solutions of the type found by Struik for periodic waves of finite amplitude and then making a passage to the limit by allowing the wave length to become large and, presumably, in such a way that the parameter gh/U^2 tends to unity. This procedure of Lavrentieff thus also starts with a nonlinear first approximation. The problem thus furnishes another good example of the well-known fact that it is not always easy to guess how to set up an approximation scheme for solving nonlinear boundary value problems, since the solution may behave in quite unexpected ways for particular values of the parameters. Hindsight, however, can help to make the necessity for procedures like those of Friedrichs and Hyers and of Lavrentieff in the present case more apparent: we have seen in Chapter 7.4 that a steady flow with the critical speed $U = \sqrt{gh}$ is in a certain sense highly unstable since the slightest disturbance would lead, in terms of the linear theory for waves of small amplitude, to a motion in which infinite elevations of the free surface would occur everywhere; thus the linear theory of waves of small amplitude seems quite inappropriate as the starting point for a development which begins with a

* L. Nirenberg [N.2] has recently proved the existence of steady waves of finite amplitude caused by flows over obstacles in the bed of a stream.

uniform flow at the critical speed, and one should consequently use a basically nonlinear treatment from the outset.

We turn now to the discussion of the solution of the solitary wave problem. The theory of Friedrichs and Hyers begins with a formulation of the general problem that is the same as that devised by Levi-Civita for treating the problem of existence of periodic waves of finite amplitude, and which was motivated by the desire to reformulate the problem in terms of the velocity potential φ and stream function ψ as independent variables in order to work in the fixed domain between the two stream lines $\psi = $ const. corresponding to the bottom and the free surface instead of in the partially unknown domain in the physical plane. We therefore begin with the general theory of irrotational waves in water when a free surface exists. The wave is assumed to be observed from a coordinate system which moves with the same velocity as the wave, and hence the flow can be regarded as a steady flow in this coordinate system.

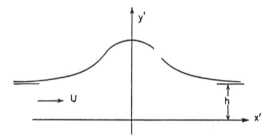

Fig. 10.9.1. The solitary wave

A complex velocity potential $\chi'(x', y') \equiv \chi'(z')$:

$$(10.9.1) \qquad \chi' = \varphi' + i\psi', \qquad z' = x' + iy'$$

is sought in an x', y'-plane (cf. Fig. 10.9.1) such that at infinity the velocity is U and the depth of the water is h. χ' is of course an analytic function of z'. The real harmonic functions φ' and ψ' represent the velocity potential and the stream function. The complex velocity $w' = d\chi'/dz'$ is given by

$$(10.9.2) \qquad w' = u' - iv',$$

in which u' and v' are the velocity components. This follows by virtue of the Cauchy-Riemann equations:

$$(10.9.3) \qquad \varphi'_{x'} = \psi'_{y'}, \qquad \varphi'_{y'} = -\psi'_{x'},$$

since $w' = \varphi'_{x'} + i\psi'_{x'}$. It is convenient to introduce new dimensionless variables:

(10.9.4) $z = z'/h$, $w = w'/U$, $\chi = \varphi + i\psi = \chi'/(hU)$,

and a parameter γ:

(10.9.5) $\gamma = gh/U^2$.

In terms of these quantities the free surface corresponds to $\psi = 1$ if the bottom is assumed given by $\psi = 0$, since the total flow over a curve extending from the bottom to the free surface is Uh. The boundary conditions are now formulated as follows:

(10.9.6) $v = - \mathcal{I}mw = 0$ at $\psi = 0$,

(10.9.7) $\tfrac{1}{2}|w|^2 + \gamma y = $ const.   at $\psi = 1$.

The second condition results from Bernoulli's law on taking the pressure to be constant at the free surface and the density to be unity, as one sees from equation (1.3.4) of Ch. 1. At ∞ we have the condition

(10.9.8) $w \to 1$ as $|x| \to \infty$.

We assume now that the physical plane (i.e. the x, y-plane) is mapped by means of $\chi(z)$ into the φ, ψ-plane in such a way that the entire flow is mapped in a one-to-one way on the strip bounded by $\psi = 0$ and $\psi = 1$.* In this case the inverse mapping function $z(\chi)$ exists, and we could regard the complex velocity w as a function of χ defined in the strip bounded by $\psi = 0$, $\psi = 1$ in the χ-plane. We then determine the analytic function $w(\chi)$ in that strip from the boundary conditions (10.9.6), (10.9.7), (10.9.8), after which $\chi(z)$ can be found by an integration and the free surface results as the curve given by $\psi = \mathcal{I}m\,\chi = 1$.

It is convenient, however, again following Levi-Civita to replace the dependent variable w by another (essentially its logarithm) through the equation

(10.9.9) $w = e^{-i(\theta + i\lambda)}$.

It follows that

(10.9.10) $|w| = e^{\lambda}$, $\theta = \arg \bar{w}$,

and thus $\lambda = \log |w|$, with $|w|$ the magnitude of the velocity vector,

---

* Our assumption that the mapping of the flow on the χ-plane is one-to-one can be shown rather easily to follow from the other assumptions and Levi-Civita carries it out. The equivalence of the various formulations of the problem is then readily seen. In Chapter 12.2 these facts are proved.

and θ is the inclination relative to the x-axis of the velocity vector. We proceed to formulate the conditions for the determination of θ and λ in the φ, ψ-plane. The condition (10.9.6) becomes, of course, $\theta = 0$ at $\psi = 0$. To transform the condition (10.9.7) we first differentiate with respect to φ along the line $\psi = 1$ to obtain

$$(10.9.11) \qquad |w| \frac{\partial |w|}{\partial \varphi} + \gamma \frac{\partial y}{\partial \varphi} = 0 \qquad \text{on } \psi = 1.$$

Since x and y are conjugate harmonic functions of φ and ψ we may write

$$(10.9.12) \qquad \begin{cases} \dfrac{\partial x}{\partial \varphi} = \dfrac{\partial y}{\partial \psi} = \dfrac{\varphi_x}{\varphi_x^2 + \varphi_y^2} = \dfrac{u}{|w|^2} \\[2ex] \dfrac{\partial y}{\partial \varphi} = -\dfrac{\partial x}{\partial \psi} = \dfrac{v}{|w|^2} \end{cases}$$

in accordance with well-known rules for calculating the derivatives of functions determined implicitly, or from

$$\frac{dz}{d\chi} = \frac{1}{\dfrac{d\chi}{dz}} = \frac{1}{u - iv} = \frac{u + iv}{u^2 + v^2}.$$

As a consequence we have from (10.9.11):

$$\frac{\partial |w|}{\partial \varphi} = -\gamma |w|^{-3} v,$$

or, since $|w| = e^{\lambda}$ and $v = -\mathscr{Im}\, e^{-i(\theta + i\lambda)} = e^{\lambda} \sin \theta$:

$$e^{\lambda} \frac{\partial \lambda}{\partial \varphi} = -\gamma e^{-2\lambda} \sin \theta,$$

and since $\partial \lambda / \partial \varphi = -\partial \theta / \partial \psi$ because λ and θ are harmonic conjugates it follows finally that

$$(10.9.13) \qquad \frac{\partial \theta}{\partial \psi} = \gamma e^{-3\lambda} \sin \theta \qquad \text{at } \psi = 1.$$

The boundary conditions $\theta = 0$ for $\psi = 0$ and (10.9.13) at $\psi = 1$ are Levi-Civita's conditions, but the condition at ∞ imposed here is replaced in Levi-Civita's and Struik's work by a periodicity condition in x, — and this makes a great difference. Levi-Civita and Struik proceed on the assumption that a disturbance of small amplitude is created relative to the uniform flow in which $w \equiv$ const.; this is interpreted to mean that $\theta + i\lambda$ is a quantity which can be developed in powers of a small parameter ε, and the convergence of the series for sufficiently small values of ε is then proved. In Chapter 12.2 we shall give a proof of the convergence of this expansion. (In lowest order, we note that the condition (10.9.13) leads for small λ and θ to the condition $\partial\theta/\partial\psi - \gamma\theta = 0$ at $\psi = 1$—in agreement with what we have seen in Part II.) In the case of the solitary wave such a procedure will not succeed, as was explained above, or rather it would not yield anything but a uniform flow. The procedure to be adopted here consists in developing, roughly speaking, with respect to the parameter γ near $\gamma = 1$; but, as in the shallow water theory in general in the version presented in section 4 of Chapter 2, we introduce a stretching of the horizontal coordinate φ which depends on γ while leaving the vertical coordinate unaltered (see equation (10.9.19)). This stretching of only one of the coordinates is the characteristic feature of the shallow water theory. (The approximating functions are then no longer harmonic in the new independent variables.) Specifically, we introduce the real parameter $\varkappa$ by means of the equation

$$(10.9.14) \qquad\qquad e^{-3\varkappa^2} = \gamma = gh/U^2.$$

This implies that $gh/U^2 < 1$, but that seems reasonable since all of the approximate theories for the solitary wave lead to such an inequality. We also introduce a new function τ, replacing λ, by the relation

$$(10.9.15) \qquad\qquad \tau = \lambda + \varkappa^2.$$

For $\theta(\varphi, \psi)$ and $\tau(\varphi, \psi)$ we then have the boundary conditions

$$(10.9.16) \qquad\qquad \theta = 0, \qquad \psi = 0$$

$$(10.9.17) \qquad\qquad \frac{\partial\theta}{\partial\psi} = e^{-3\tau}\sin\theta, \qquad \psi = 1.$$

For $\varphi \to \pm \infty$ we have the conditions imposed by the physical problem:

$$(10.9.18) \qquad\qquad \theta \to 0, \qquad \tau \to \varkappa^2,$$

the latter resulting since $\lambda \to 0$ at ∞ from $|w| = e^\lambda$ and $|w| \to 1$ at ∞. As we have already indicated, the development we use requires stretching the variable φ so that it grows large relative to ψ when $\varkappa$ is small; this is done in the present case by introducing the new independent variables

(10.9.19) $$\bar{\varphi} = \varkappa\varphi, \qquad \bar{\psi} = \psi.$$

The dependent variables θ and τ are now regarded as functions of $\bar{\varphi}$ and $\bar{\psi}$ and they are then expanded in powers of $\varkappa$:

(10.9.20)
$$\begin{cases} \tau = \varkappa^2 \tau_1(\bar{\varphi}, \bar{\psi}) + \varkappa^4 \tau_2(\bar{\varphi}, \bar{\psi}) + \cdots, \\ \theta = \varkappa^3 \theta_1(\bar{\varphi}, \bar{\psi}) + \varkappa^5 \theta_2(\bar{\varphi}, \bar{\psi}) + \cdots. \end{cases}$$

(We have omitted writing down a number of terms which in the course of the calculation would turn out to have zero coefficients.) Friedrichs and Hyers have proved that the lowest order terms in these series, as obtained formally through the use of the boundary conditions, are the lowest order terms in a convergent iteration scheme using $\varkappa$ as small parameter. Their convergence proof also involves the explicit use of the stretching process. However, the proof of this theorem is quite complicated, and consequently we content ourselves here with the determination of the lowest order terms: we remark, however, that higher order terms could also be obtained explicitly from the formal expansion.

The series in (10.9.20) are now inserted in all of the equations which serve to determine θ and τ, and relations for the coefficient functions $\tau_i(\varphi, \psi)$ and $\theta_i(\varphi, \psi)$ are obtained. The Cauchy-Riemann equations for θ and τ lead to the equations

(10.9.21) $$\theta_{\bar{\psi}} = -\varkappa\tau_{\bar{\varphi}}, \qquad \tau_{\bar{\psi}} = \varkappa\theta_{\bar{\varphi}}$$

in terms of the variables φ and ψ, and the series (10.9.20) then yield the equations

(10.9.22) $$\tau_{1\bar{\psi}} = 0, \qquad \theta_{i\bar{\psi}} = -\tau_{i\bar{\varphi}}, \qquad \tau_{2\bar{\psi}} = \theta_{1\bar{\varphi}}.$$

Thus $\tau_1 = \tau_1(\bar{\varphi})$ is independent of $\bar{\psi}$, and integration of the remaining equations gives the following results:

(10.9.23)
$$\begin{cases} \theta_1 = -\bar{\psi}\tau_1' \\ \tau_2 = -\tfrac{1}{2}\bar{\psi}^2\tau_1'' + j(\bar{\varphi}) \\ \theta_2 = \tfrac{1}{6}\bar{\psi}^3\tau_1''' - \bar{\psi}j'(\bar{\varphi}). \end{cases}$$

The primes refer to differentiation with respect to $\bar{\varphi}$. An additive arbitrary function of $\bar{\varphi}$ in the first of these equations was taken to be zero because of the boundary condition $\theta_1 = 0$ for $\psi = 0$.

Upon substitution of (10.9.20) into the boundary condition (10.9.17) we find

$$\varkappa^3\theta_{1\bar{\psi}} + \varkappa^5\theta_{2\bar{\psi}} + \cdots = \varkappa^3\theta_1 - 3\varkappa^5\tau_1\theta_1 + \varkappa^5\theta_2 + \cdots$$

and consequently we have the equations

$$(10.9.24) \qquad \begin{cases} \theta_{1\bar{\psi}} = \theta_1, \\ \theta_{2\bar{\psi}} = \theta_2 - 3\tau_1\theta_1, \end{cases} \qquad \text{at } \bar{\psi} = 1.$$

The first equation is automatically satisfied because of the first equation of (10.9.23). The second equation leads through (10.9.23) to the condition

$$(10.9.25) \qquad \tau_1''' = 9\tau_1\tau_1',$$

for τ_1, as one readily verifies. Once τ_1 has been determined, one sees that θ_1 is also immediately fixed by the first equation in (10.9.23). Boundary conditions are needed for the third order nonlinear differential equation given by (10.9.25); we assume these conditions to be

$$(10.9.26) \qquad \begin{cases} \tau_1'(0) = 0, \\ \tau_1(\infty) = 1, \\ \tau_1''(\infty) = 0. \end{cases}$$

These conditions result from our assumed physical situation: the first is taken since a symmetrical form of the wave about its crest is expected and hence $\theta_1(0) = 0$, the second arises from (10.9.18), while the third is a reasonable condition that is taken in place of what looks like the more natural condition $\tau_1'(\infty) = 0$ since the latter condition is automatically satisfied, in view of the first equation of (10.9.23) and $\theta_1(\infty) = 0$, and thus does not help in fixing τ_1 uniquely.

An integral of (10.9.25) is readily found; it is:

$$\tau_1'' = \tfrac{9}{2}\tau_1^2 + \text{const.},$$

and the boundary conditions yield

$$\tau_1'' = \tfrac{9}{2}(\tau_1^2 - 1).$$

From this one obtains, finally, the solution:

$$(10.9.27) \qquad \tau_1(\bar{\varphi}) = 1 - 3\operatorname{sech}^2(3\bar{\varphi}/2),$$

and θ_1 is then fixed by (10.9.23). From these one finds for the shape of the wave—that is, the value of y corresponding to $\psi = 1-$, and for the horizontal component u of the velocity the equations

$$(10.9.28) \qquad y = 1 + 3\varkappa^2 \operatorname{sech}^2 \frac{3\varkappa x}{2},$$

$$(10.9.29) \qquad u = 1 - 3\varkappa^2 \operatorname{sech}^2 \frac{3\varkappa x}{2}.$$

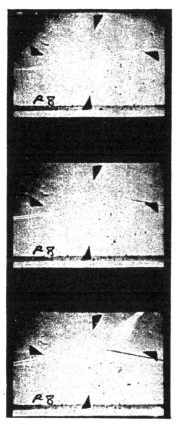

Fig. 10.9.2. A solitary wave

In calculating these quantities, higher order terms in $\varkappa$ have been neglected. The expression for the wave profile is identical with those found by Boussinesq, Rayleigh, and Keller. For the velocity u, the two former authors give $u \equiv 1$ while Keller gives the same expression as above except that the factor $3\varkappa^2$ is replaced by another which differs from it by terms of order $\varkappa^4$ or higher.

Thus a solitary wave of symmetrical form has been found with an amplitude which increases with its speed U. Careful experiments to determine the wave profile and speed of the solitary wave have been carried out by Daily and Stephan [D.1], who find the wave profile and velocity to be closely approximated by the above formulas with a maximum error in the latter of 2.5 % at the highest amplitude-depth ratio tested. Fig. 10.9.2 is a picture of a solitary wave taken by Daily and Stephan; three frames from a motion picture film are shown.

10.10. The breaking of waves in shallow water. Development of bores

In sections 10.4 and 10.6 above it has already been seen that the shallow water theory, which is mathematically analogous to the theory of compressible flows in a gas, leads to a highly interesting and significant result in cases involving the propagation of disturbances into still water that are the exact counterparts of the corresponding cases in gas dynamics involving the motions due to the action of a piston in a tube filled with gas. These cases, which are very easily described in terms of the concept of a simple wave (cf. sec. 10.3), lead, in fact to the following qualitative results (cf. sec. 10.4): there is a great difference in the mode of propagation of a depression wave and of a hump with an elevation above the undisturbed water line; in the first case the depression wave gradually smooths out, but in the second case the front of the wave becomes progressively steeper until finally its slope becomes infinite. In the latter case, the mathematical theory ceases to be valid for times larger than those at which the discontinuity first appears, but one expects in such a case that the wave will continue to steepen in front and will eventually break. This is the correct qualitative explanation, from the point of view of hydrodynamical theory, for the breaking of waves on shallow beaches. It was advanced by Jeffreys in an appendix to a book by Cornish [C.7] published in 1934. Jeffreys based his discussion on the fact that the propagation speed of a wave increases with increase in the height of a wave above the undisturbed level. Consequently, if a wave is created in such a way as to cause a rise in the water surface it follows that the higher points on the wave surface will propagate at higher speed than the lower points in front of them—in other words there is a tendency for the higher portions of the wave to overtake and to crowd the lower portions in front so that the front of the wave becomes steep and eventually curls over and breaks; the same argument indicates that a depression wave tends to flatten out and become smoother as it advances.

It is of interest to recall how waves break on a shallow beach. Figures 10.10.1, 10.10.2, and 10.10.3 are photographs* of waves on the California coast. Figure 10.10.1 is a photograph from the air, taken by the Bureau of Aeronautics of the U.S. Navy, which shows how the waves coming from deep water are modified as they move toward shore. The waves are so smooth some distance off shore that they can be seen only vaguely in the photograph, but as they move inshore the front of the waves steepens noticeably until, finally, breaking occurs. Figures 10.10.2 and 10.10.3 are pictures of the same wave, with the picture of Figure 10.10.3 taken at a slightly later time than the previous picture. The steepening and curling over of the wave are very strikingly shown.

At this point it is useful to refer back to the beginning of section 10.6 and especially to Fig. 10.6.1. This figure, which is repeated here

---

* These photographs were very kindly given to the author by Dr. Walter Munk of the Scripps Institution of Oceanography.

Fig. 10.10.1. Waves on a beach

for the sake of convenience, indicates in terms of the theory of characteristics what happens when a wave of elevation is created by

Fig. 10.10.2. Wave beginning to break

Fig. 10.10.3. Wave breaking

pushing the moveable end of a tank of water into it so that a disturbance propagates into still water of constant depth: the straight
characteristics issuing from the "piston curve" AD, along each of

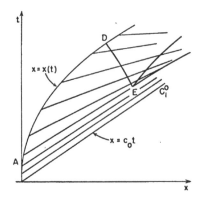

Fig. 10.10.4. Initial point of breaking

which the velocity u and the quantity $c = \sqrt{g(h + \eta)}$ are constant, eventually intersect at the point E. The point E is a cusp on the envelope of the characteristics, and represents also the point at which the slope of the wave surface first becomes infinite. The point E might thus—somewhat arbitrarily, it is true—be taken as defining the breaking point (x_b, t_b) of the wave, since one expects the wave to start curling over after this point is reached. It is possible to fix the values of x_b and t_b without difficulty once the surface elevation $\eta = \eta(0, t)$ is prescribed at $x = 0$; we carry out the calculation for the interesting case of a pulse in the form of a sine wave:

$$(10.10.1) \qquad\qquad \eta(0, t) = A \sin \omega t.$$

For $t = 0$, $x > 0$ we assume the elevation η of the water to be zero and its velocity u_0 to be constant (though not necessarily zero, since it is of interest to consider the effect of a current on the time and place of breaking).

As we know, the resulting motion is easily described in terms of the characteristics in the x, t-plane, which are straight lines emanating from the t-axis, as indicated in Figure 10.10.6. The values of u and c are constant along each such straight line. The slope dx/dt of any straight characteristic through the point $(0, \tau)$ is given by

$$(10.10.2) \qquad\qquad \frac{dx}{dt} = 3c - 2c_0 + u_0,$$

which is the same as (10.3.6). The quantity c_0 has the value $c_0 = \sqrt{gh}$, while $c = \sqrt{g(h + \eta)}$, as always. On the other hand, the slope of this characteristic is clearly also given in terms of a point (x, t) on it by $x/(t - \tau)$ so that (10.10.2) can be written in the form

$$(10.10.3) \qquad\qquad x = (t - \tau)[3c(\tau) - 2c_0 + u_0]$$

in which we have indicated explicitly that c depends only on τ since it (as well as all other quantities) is constant along any straight characteristic. Thus (10.10.3) furnishes the solution of our problem, once $c(\tau)$ is given, throughout a region of the x, t-plane which is covered by the straight lines (10.10.3) without overlapping. However, the interesting cases for us are just those in which overlapping occurs, i.e. those for which the characteristics converge and eventually cut each other, and this always happens if an elevation is created at $x = 0$. In fact, if c is an increasing function of τ, then dx/dt as given by (10.10.2) increases with τ and hence the characteristics for $x > 0$

must intersect. In this case, furthermore, the family of straight characteristics has an envelope beginning at a point (x_b, t_b), which we
have defined to be the point of breaking.

We proceed to determine the envelope of the straight lines (10.10.3).
As is well known, the envelope can be obtained as the locus resulting
from (10.10.3) and the relation

$$(10.10.4) \qquad 0 = - [3c(\tau) - 2c_0 + u_0] + 3(t - \tau)c'(\tau)$$

obtained from it by differentiation with respect to τ. For the points
(x_c, t_c) on the envelope we then obtain the parametric equations

$$(10.10.5) \qquad x_c = \frac{[3c(\tau) - 2c_0 + u_0]^2}{3c'(\tau)},$$

and

$$(10.10.6) \qquad t_c = \tau + \frac{[3c(\tau) - 2c_0 + u_0]}{3c'(\tau)}.$$

We are interested mainly in the "first" point on the envelope, that
is, the point (x_b, t_b) for which t_c has its smallest value since we identify this point as the point of breaking. To do so really requires a
proof that the water surface has infinite slope at this point. Such a
proof could be easily given, but we omit it here with the observation
that an infinite slope is to be expected since the characteristics which
intersect in the neighborhood of the first point on the envelope all
carry different values for c.

We have assumed that $\eta(0, t)$ is given by (10.10.1) and consequently
the quantity $c(\tau)$ in (10.10.5) and (10.10.6) is given by

$$(10.10.7) \qquad c(\tau) = \sqrt{g(h + A \sin \omega\tau)}.$$

If we assume $A > 0$ we see that $c'(\tau)$ is a positive decreasing function
of τ for small positive values of τ. Since $c(\tau)$ increases for small positive values of τ it follows that both x_c and t_c in (10.10.5) and (10.10.6)
are increasing functions of τ near $\tau = 0$. A minimum value of x_c and
t_c must therefore occur for $\tau = 0$, so that the breaking point is given
by

$$(10.10.8) \qquad x_b = \frac{2c_0(c_0 + u_0)^2}{3gA\omega},$$

and

$$(10.10.9) \qquad t_b = \frac{2c_0(c_0 + u_0)}{3gA\omega}$$

as one can readily verify. We note that the point (x_b, t_b) lies on the initial characteristic $x = (c_0 + u_0)t$, as it should since $\tau = 0$ for this characteristic. From the formulas we can draw a number of interesting conclusions. Since $c_0 = \sqrt{gh}$ we see that breaking occurs earlier in shallower water for a pulse of given amplitude A and frequency ω. Breaking also occurs earlier when the amplitude and frequency are larger. It follows that short waves will break sooner than long waves, since longer waves are correlated with lower frequencies. Finally we notice that early breaking of a wave is favored by small values for u_0, the initial uniform velocity of the quiet water. In fact, if u_0 is negative, i.e. if the water is flowing initially toward the point where the pulse originates, the breaking can be made to occur more quickly. Everyone has observed this phenomenon at the beach, where the breaking of an incoming wave is often observed to be hastened by water rushing down the beach from the breaking of a preceding wave.

It is of some importance to draw another conclusion from our theory for waves moving into water of constant depth: an inescapable consequence of our theory is that the maxima and minima of the surface elevation propagate into quiet water unchanged in magnitude with respect to both distance and time. This follows immediately from the fact that the values of the surface elevation are constant along the straight characteristics so that if η has a relative maximum for $x = 0$, $t = \tau$, say, then this value of η will be a relative maximum all along the characteristic which issues from $x = 0$, $t = \tau$. The waves change their form and break, but they do so without changes in amplitude.

In a report of the Hydrographic Office by Sverdrup and Munk [S.36] some results of observations of breakers on sloping beaches are given in the form of graphs showing the ratio of breaker height to deep water amplitude and the ratio of undisturbed depth at the breaking point to the deep water amplitude as functions of the "initial steepness" in deep water, the latter being defined as the ratio of amplitude to wave length in deep water. The "initial steepness" is thus essentially the quantity $A\omega$ in our above discussion, and our results indicate that it is a reasonable parameter to choose for discussion of breaking phenomena. The graphs given in the report—reproduced here in Figures 10.10.5a and 10.10.5b—show very considerable scattering of the observational data, and this is attributed in the report to errors in the observations, which are apparently difficult to make with accuracy. On the basis of our above conclusion—that the breaking of a wave in water of uniform depth occurs no matter what the amplitude of the wave may be in relation to the undisturbed depth—we could offer another explanation for the scatter of the points in Figures 10.10.5a and 10.10.5b, i.e. that the amplitude ratios are relatively independent of the initial steepness. Of course, the curves of Figures

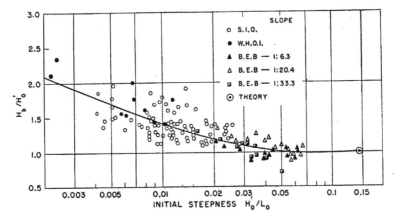

Fig. 10.10.5a. Ratio of breaker height to wave height in deep water, H_b/H_0', assuming no refraction

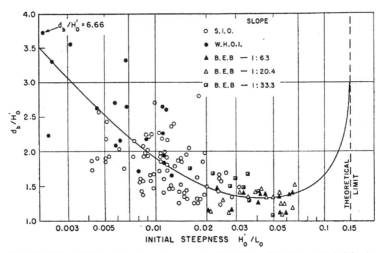

Fig. 10.10.5b. Ratio of depth of water at point of breaking to wave height in deep water, d_b/H_0', as function of steepness in deep water, H_0'/L_0, assuming no refraction

10.10.5a and 10.10.5b refer to sloping beaches and hence to cases in which the wave amplitudes increase as the wave moves toward shore; but still it would seem rather likely that the amplitude ratios would be relatively independent of the initial steepness in these cases also since the beach slopes are small. The detailed investigation of breaking of waves by Hamada [H.2], which is both theoretical and experimental in character, should be consulted for still further analysis of this and other related questions. The papers by Iversen [I.6] and Suquet [S.31] also give experimental results concerning the breaking of waves.

We continue by giving the results of numerical computations for three cases of propagation of sine pulses into still water of constant depth. The cases calculated are indicated in the following table:

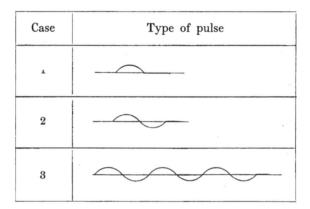

Case	Type of pulse
1	
2	
3	

Case 1 is a half-sine pulse in the form of a positive elevation, case 2 is a full sine wave which starts with a depression phase, and case 3 consists of several full sine waves.

Figure 10.10.6 shows the straight characteristics in the x, t-plane for case 1. (In all of these cases, the quantities x and y are now certain dimensionless quantities, the definitions of which are given in [S.19].) We observe that the envelope begins on the initial characteristic in this case, in accord with earlier developments in this section. The envelope has two distinct branches which meet in a cusp at the breaking point (x_b, t_b). Figure 10.10.7 gives the shape of the wave for two different times. As we see, the front of the wave steepens until it finally becomes vertical for $x = x_b$, $t = t_b$, while the back of the wave flattens out. The solution given by the characteristics in Figure

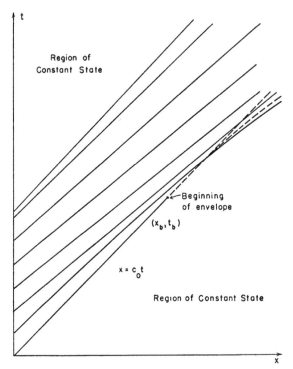

Fig. 10.10.6. Characteristic diagram in the x, t-plane

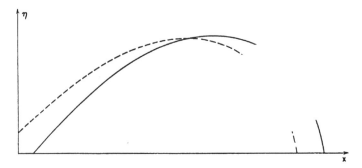

Fig. 10.10.7. Wave height versus distance for a half-sine wave of amplitude $h_0/5$ in water of constant depth at two instants, where h_n is the height of the still water level

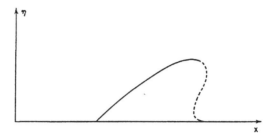

Fig. 10.10.8. Wave profile after breaking

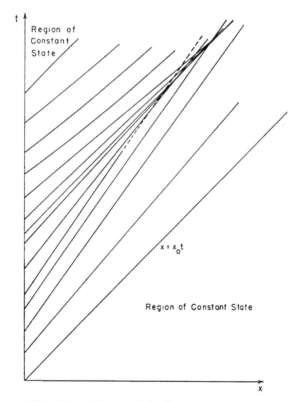

Fig. 10.10.9. Characteristic diagram in the x, t-plane

10.10.6 is not valid for $x > x_b$, $t > t_b$, and we expect breaking to ensue. However, we observe that the region between the two branches of the envelope is quite narrow, so that the influence of the developing

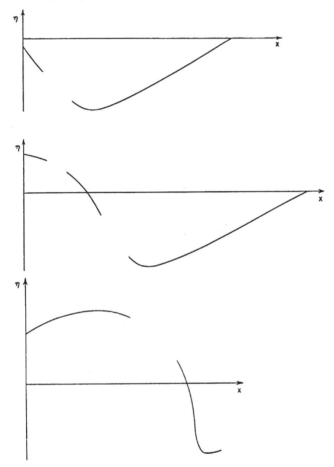

Fig. 10.10.10. Wave height versus distance for a full negative sine wave with amplitude $h_0/5$ in water of constant depth at $t = 3.0$, $t = 5.0$, and $t = 6.28$

breaker may not seriously affect the motion of the water behind it. Thus we might feel justified in considering the solution by characteristics given by Figure 10.10.6 as being approximately valid for values of

t slightly greater than t_b. (This also seems to the writer to be intuitively rather plausible from the mechanical point of view.) Figure 10.10.8 was drawn on this basis for a time considerably greater than t_b. The full portion of the curve was obtained from the characteristics outside the region between the branches of the envelope, while the dotted portion—which is of doubtful validity—was obtained by using the characteristics between the branches of the envelope in an obvious manner. In this way one is able to approximate the early stages of the curling over of a wave.

Figures 10.10.9, 10.10.10, and 10.10.11 refer to case 2, in which a depression phase precedes a positive elevation. In this case the envelope of the characteristics does not begin, of course, on the initial characteristic but rather in the interior of the simple wave region, as indicated in Figure 10.10.9. Figure 10.10.10 shows three stages in the progress of the pulse into still water. The steepening of the wave front is very marked by the time the breaking point is reached—much more marked than in the preceding case for which no depression phase occurs in front. Figure 10.10.11 shows the shape of the wave a short time after passing the braking point. This curve was obtained, as in the preceding case, by using the characteristics between the branches of the envelope. Although this can yield only a rough approximation, still

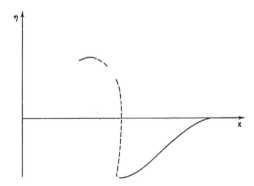

Fig. 10.10.11. η versus x at $t = 7$ for non-sloping bottom where the pulse is an entire negative sine-wave. The dotted part of the curve represents η in the region between the branches of the envelope

one is rather convinced that the wave really would break very soon after the point we have somewhat arbitrarily defined as the breaking point.

Figure 10.10.12 shows the water surface in case 3 for a time well beyond the breaking point.

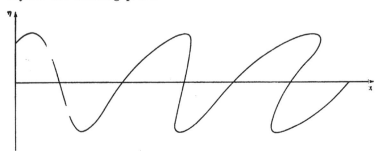

Fig. 10.10.12. Water profile after breaking

In gas dynamics where u and c represent the velocity and sound speed throughout an entire cross section of a tube containing the gas, it clearly is not possible to give a physical interpretation to the region between the two branches of the envelope in the cases analogous to that shown in Figure 10.10.6, since the velocity and propagation speed must of necessity be single-valued functions of x. However, in our case of water waves u and c refer to values *on the water surface* so that there is no reason a priori to reject solutions for u and c which are not single-valued in x. Thus we might be tempted to think that the dotted part of the curve in Figure 10.10.8 is valid within the general framework of our theory, but this is, unfortunately, not the case: our fundamental differential equations are not valid in the "overhanging" part of the wave, simply because that part is not resting on a rigid bottom. It may be that one could pursue the solutions beyond the point where the breaking begins by using the appropriate differential equations in the overhanging part of the wave and then piecing together solutions of the two sets of differential equations so that continuity is preserved, but this would be a problem of considerable difficulty. In this connection, however, it is of interest to report the results of a calculation by Biesel [B.10] for the change of form of progressing waves over a beach of small slope. Not the least interesting aspect of Biesel's results is the fact that they are based essentially on the theory of waves of small amplitude, i.e. on the type of theory which forms the basis for the discussions in Part II of this book. However, in Part II only the so-called Eulerian representation was used, in which the dependent quantities such as velocity, pressure, etc., are all obtained at fixed points in space. As a result, when linearizations are introduced the free surface elevation η, for example, is a function of x and t and must of necessity be single-valued. Biesel, however, observes that one can also use the Lagrangian representation* just about as conveniently as the Eulerian when a development with respect to amplitude is contemplated. In this approach, all quantities are fixed in terms of the initial positions of the water particles (and the time, of course). In particular, the displacements (ξ, η) of the water particles on the free surface would be given as functions of a parameter, i.e. $\xi = \xi(a, t)$, $\eta = \eta(a, t)$, and there would be no necessity a priori to require that η should be a single-valued function of x. Biesel has carried out this program with the results shown in Figs. 10.10.13 to 10.10.16 inclusive. A sinusoidal progressing wave in deep water is assumed. The first two figures refer to the theory when carried out only to first order terms in the displacements relative to the rest position of equilibrium. The second figure is a detail of the motion in a neighborhood of the location shown by the dotted circle in the first figure. Fig. 10.10.15 and Fig. 10.10.16 treat the same problem, but the solution is carried to second order terms. In both cases the development of a breaker is strikingly shown. A comparison of the results of the first order and second order theories is of interest; the main conclusions are: if second order corrections are made the breaking is seen to occur earlier (i.e. in deeper water), the height of the wave at breaking is much greater, and the tendency of the wave to plunge downward after curling over at the top is considerably lessened.

Actually, our shallow water theory cannot be expected to yield a good approximation near the breaking point since the curvature of the water surface is likely to be large there. However, since the motion should be given with good accuracy at points outside the immediate vicinity of the breaking point it might be possible to refine the treatment of the breaking problem along the following lines: consider the motion of a fixed portion of the water between a pair of planes located some distance in front and in back of the breaking point. The motion of the water particles outside the bounding planes can be considered as given by our shallow water theory. We might then seek to determine the motion of the water between these two planes by making use of a refinement of the shallow water theory or by reverting to the original exact formulation of the problem in terms of a potential function with

* In Chapter 12.1 this representation is explained and used to solve other problems involving unsteady motions.

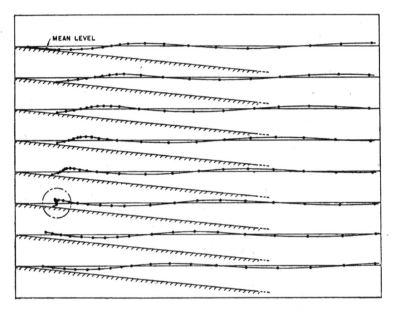

Fig. 10.10.13. Progression and breaking of a wave on a beach of 1 in 10 slope.
First-order theory

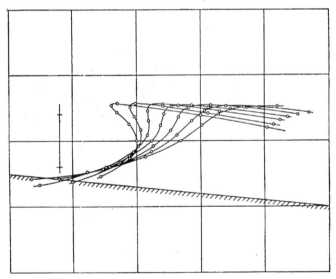

Fig. 10.10.14. Details of breaking of wave shown in Fig. 10.10.13. First-order
theory

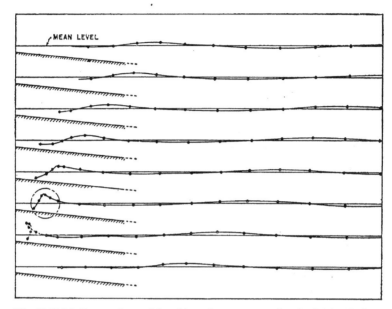

Fig. 10.10.15. Progression and breaking of a wave on a beach of 1 in 10 slope. Second-order theory

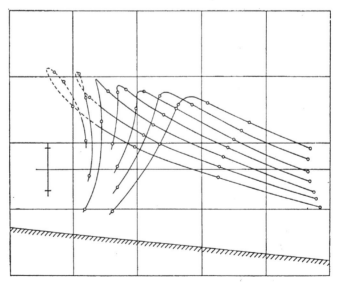

Fig. 10.10.16. Details of breaking of wave shown in Fig. 10.10.15. Second-order theory

the nonlinear free surface condition and determine it by using finite difference methods in a bounded region.

It is of interest now to return to the problem with which we opened the discussion of the present section, i.e. to the problem of a tank with a moveable end which is pushed into the water. As we have seen, the wave which arises will eventually break. Suppose now we assume that the end of the tank continues to move into the water with a uniform velocity. The end result after the initial curling over and breaking will be the creation of a steady progressing wave front which is steep and turbulent, behind which the water level is constant and the

Fig. 10.10.17. The bore in the Tsien Tang River

water has everywhere the constant velocity imparted to it by the end of the tank. Such a steady progressing wave with a steep front is called a bore. It is the exact analogue of a steady progressing shock wave in a gas. In Figure 10.10.17 we show a photograph, taken from the book by Thorade [T.4], of the bore which occurs in the Tsien-Tang River as a result of the rising tide, which pushes the water into a narrowing estuary at the mouth of the river. The height of this bore apparently is as much as 20 to 30 feet. According to the theory presented above, this bore should have been preceded by an unsteady phase during which the smooth tidal wave entering the estuary first curled over and broke. Methods for the treatment of problems involving the gradual development of a bore in an otherwise smooth flow have been worked out by A. Lax [L.5] (see also Whitham [W.12]).

We have, so far, used our basic theory—the nonlinear shallow water theory—to interpret the solutions of only one type of problem, i.e. the problem of the change of form of a pulse moving into still water of constant depth. The theory, however, can be used to study the propagation of a wave over a beach with decreasing depth just as well (cf. the author's paper [S.19]), but the calculations are made much more difficult because of the fact that no family of straight characteristics exists unless the depth is constant. This problem, in fact, brings to the fore the difficulties of a computational nature which occur in important problems involving the propagation of flood waves and other surges in rivers and open channels in general. Such problems will be discussed in the next chapter.

On an actual beach on which waves are breaking, the motion of the water, of course, does not consist in the propagation of a single pulse into still water, but rather in the occurrence of an approximately periodic train of waves. However, experiments indicate that only a slight reflection of the wave motion from the shore occurs. The incoming wave energy seems to be destroyed in turbulence due to breaking or to be converted into the energy of flow of the undertow. In other words, each wave propagates, to a considerable degree, in a manner unaffected by the waves which preceded it. Consequently the above treatment of breaking, in which propagation of a wave into still water was assumed, should be at least qualitatively reasonable. Another objection to our theory has already been mentioned, i.e. that large curvatures of the water surface near the breaking point seem sure to make the results inaccurate. Nevertheless, the theory should be valid, except near this point, in many cases of waves on sloping beaches, since the wave lengths are usually at least 10 to 20 or more times the depth of the water in the breaker zone, hence the theory presented above should certainly yield correct qualitative results and perhaps also reasonably accurate quantitative results.

Waves do not by any means always break in the manner described up to this point. In Fig. 10.10.18a, b we show photographs (given to the author by Dr. Walter Munk) of waves breaking in a fashion considerably at variance with the results of the theory presented here. We observe that the waves break, in this instance, by curling over slightly at the crest, but that the wave remains, as a whole, symmetrical in shape, while the theory presented here yields a marked steepening of the wave front and a very unsymmetrical shape for the wave at breaking.

Observation of cases like that shown in Figure 10.10.18 doubtlessly led to the formulation of the theory of breaking due to Sverdrup and

(a)

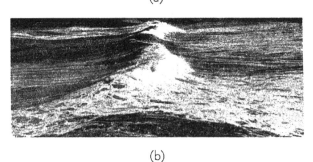

(b)

Fig. 10.10.18a, b. Waves breaking at crests

Munk [S.33]; their theory is based on results taken from the study of the solitary wave, which has been discussed in the preceding section.* The solitary wave is, by definition, a wave of finite amplitude con-

* An interesting mathematical treatment of breaking phenomena from this point of view was given some time ago by Keulegan and Patterson [K.13].

sisting of a single elevation of such a shape that it can propagate unchanged in form. At first sight, this would seem to be a rather curious
wave form to take as a basis for a discussion of the phenomena of
breaking, since it is precisely the change in form resulting in breaking
that is in question. On the other hand, the waves often look as in
Figure 10.10.18 and do retain, on the whole, a symmetrical shape,*
with some breaking at the crest. Actually, the situation regarding the
two different theories of breaking from the mathematical point of
view is the following, as we can infer from the discussion of the preceding section: Both theories are shallow water theories. In fact, as
Keller [K.6], and Friedrichs and Hyers [F.13], have shown, *the theory
of the solitary wave can be obtained from the approximation of next higher
order above that used in the present section, if the assumption is made that
the motion is a steady motion.* In other words, the theory used by
Sverdrup and Munk is a shallow water theory of higher order than
the theory used in this section, which furnishes in principle the constant state as the only continuous wave which can propagate unaltered in form. On the other hand, the theory presented here makes

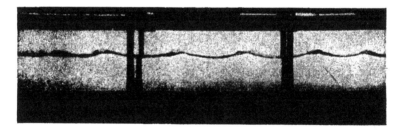

Fig. 10.10.19. Symmetrical waves breaking at crests

it possible to deal directly with the unsteady motions, while Sverdrup
and Munk are forced to approximate these motions by a series of
different steady motions. One could perhaps sum up the whole matter
by saying that waves break in different ways depending upon the
individual circumstances (in particular, the depth of the water compared with the wave length is very important), and the theory which
should be used to describe the phenomena should be chosen accord-
ingly. In fact, Figures 10.10.17 and 10.10.18 depicting a bore and

* Sverdrup and Munk, like the author, assume that, when considering breaking
phenomena, each wave in a train can be treated with reasonable accuracy as
though it were uninfluenced by the presence of the others.

waves breaking only at the crests of otherwise symmetrical waves perhaps represent extremes in a whole series of cases which include the breaker shown in Figures 10.10.2 and 10.10.8 as an intermediate case. Some pertinent observations on this point have been made by Mason [M.4]. A theory has been developed by Ursell [U.11] which differs from the theories discussed here and which may well be appropriate in cases not amenable to treatment by the shallow water theory. The paper by Hamada [H.2] referred to above should also be mentioned again in this connection. In particular, Fig. 10.10.19, taken from that paper, shows waves created in a tank which break by curling at the crest but still preserving a symmetrical form. It is interesting to observe that the wave length in this case is almost the same as the depth of the water. It is also interesting to add that in this case a current of air was blown over the water in the direction of travel of the waves. Fig. 10.10.20 shows a similar case, but with somewhat greater wave length. The two waves were both generated by a wave

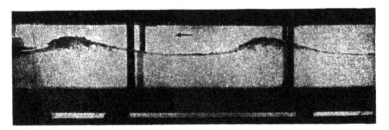

Fig. 10.10.20. Breaking induced by wind action

making apparatus at the right; the only difference is that a current of air was blown from right to left in the case shown by the lower photograph. The breaking thus seems due entirely to wind action in

this case. Finally, Fig. 10.10.21 shows two stages in the breaking of a
wave in shallow water, when marked dissymmetry and the formation
of what looks like a jet at the summit of the wave are seen to occur.

It is of interest, historically and otherwise, to refer once more to the
case of symmetrical waves breaking at their crests. The wave crests
in such cases are quite sharp, as can be seen in the photograph shown
in Fig. 10.10.18. Stokes [S.28] long ago gave an argument, based on
quite reasonable assumptions, that steady progressing waves with an
angular crest of angle 120° could occur; in fact, this follows almost at
once from the Bernoulli law at the free surface when the free surface is
assumed to be a stream line with an angular point. There is another
fact pertinent to the present discussion, i.e. that the exact theory for
steady periodic progressing waves of finite amplitude, as developed
in Chapter 12.2, shows that with increasing amplitude the waves
flatten more and more in the troughs, but sharpen at the crests.

Fig. 10.10.21. Breaking of a long wave in shallow water

In fact, the terms of lowest order in the development of the free surface
amplitude η as given by that theory can easily be found; the result is

$$\eta(x) = - a \cos x + a^2 \cos 2x$$

for a wave of wave length 2π. Fig. 10.10.22 shows the result of superimposing the second-order term $a^2 \cos 2x$ on the wave $- a \cos x$ which would be given by the linear theory; as one sees, the effect is as indicated. It would be a most interesting achievement to show rigorously that the wave form with a sharp crest of angle 120° is attained with increase in amplitude. An interesting approximate treatment of the problem has been given by Davies [D.5]. However, the problem thus posed is not likely to be easy to solve; certainly the method of Levi-Civita as developed in Ch. 12.2 does not yield such a wave form since it is shown there that the free surface is analytic. Presumably,

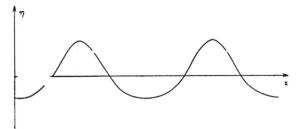

Fig. 10.10.22. Sharpening of waves at the crest

any further increase in amplitude would lead to breaking at the crests —hence no solutions of the exact problem would exist for amplitudes greater than a certain value.

10.11. Gravity waves in the atmosphere. Simplified version of the problem of the motion of cold and warm fronts

In practically all of this book we assume that the medium in which waves propagate is water. It is, however, a notable fact that some motions of the atmosphere, such as tidal oscillations due to the same cause as the tides in the oceans, i.e. gravitational effects of the sun and moon, as well as certain large scale disturbances in the atmosphere such as wave disturbances in the prevailing westerlies of the middle latitudes, and motions associated with disturbances on certain discontinuity surfaces called fronts, are all phenomena in which the air can be treated as a gravitating incompressible fluid. In addition, one of the best-founded laws in dynamic meteorology is the hydrostatic pressure law, which states that the pressure at any point in the atmosphere is very accurately given by the static weight of the column of air above it. When we add that the types of motions enumerated above are all such that a typical wave length is large compared with an average thickness (on the basis of an average density, that is) of the layer of air over the earth, it becomes clear that these problems fall into the general class of problems treated in the present chapter. Of course, this means that thermodynamic effects are ignored, and with them the ingredients which go to make up the local weather, but it seems that these effects can be regarded with a fair approximation as small perturbations on the large scale motions in question.

There are many interesting problems, including very interesting unsolved problems, in the theory of tidal oscillations in the atmosphere. These problems have been treated at length in the book by Wilkes [W.2]; we shall not attempt to discuss them here. The problems involved in studying wave propagation in the prevailing westerlies will also not be discussed here, though this interesting theory, for which papers by Charney [C.15] and Thompson [T.10] should be consulted, is being used as a basis for forecasting the pressure in the atmosphere by numerical means. In other words, the dynamical theory is being used for the first time in meteorology, in conjunction with modern high speed digital computing equipment, to predict at least one of the elements which enter into the making of weather forecasts.

In this section we discuss only one class of meteorological problems, i.e. motions associated with frontal discontinuities, or, rather, it would be better to say that we discuss certain problems in fluid dynamics which are in some sense at least rough approximations to the actual situations and from which one might hope to learn something about the dynamics of frontal motions. The problems to be treated here—unlike the problems of the type treated by Charney and Thompson referred to above—are such as to fit well with the preceding material in this chapter; it was therefore thought worthwhile to include them in this book in spite of their somewhat speculative character from the point of view of meteorology. Actually, the idea of using methods of the kind described in this chapter for treating certain special types of motions in the atmosphere has been explored by a number of meteorologists (cf. Abdullah [A.7], Freeman [F.10], Tepper [T.11]).

One of the most characteristic features of the motion of the atmosphere in middle latitudes and also one which is of basic importance in determining the weather there is the motion of wave-like disturbances which propagate on a discontinuity surface between a thin wedge-shaped layer of cold air on the ground and an overlying layer of warmer air. In addition to a temperature discontinuity there is also in general a discontinuity in the tangential component of the wind velocity in the two layers. The study of such phenomena was initiated long ago by Bjerknes and Solberg [B.20] and has been continued since by many others. In considering wave motions on discontinuity surfaces it was natural to begin by considering motions which depart so little from some constant steady motion (in which the discontinuity surface remains fixed in space) that linearizations can be performed, thus bringing the problems into the realm of the classical linear mathematical physics. Such studies have led to valuable insights, particularly with respect to the question of stability of wave motions in relation to the wave length of the perturbations. (The problems being linear, the motions in general can be built up as a combination, roughly speaking, of simple sine and cosine waves and it is the wave length of such components that is meant here, cf. Haurwitz [H.5, p. 234].) One conjecture is that the cyclones of the middle latitudes are initiated because of the occurrence of such unstable waves on a discontinuity surface.

A glance at a weather map, or, still better, an examination of weather maps over a period of a few days, shows clearly that the wave motions on the discontinuity surfaces (which manifest themselves as the so-called fronts on the ground) develop amplitudes so rapidly and of such a magnitude that a description of the wave motions over a period of, say, a day or two, by a linearization seems not feasible with any accuracy. The object of the present discussion is to make a first step in the direction of a nonlinear theory, based on the exact hydrodynamical equations, for the description of these motions, that can be attacked by numerical or other methods. No claim is made that the problem is solved here in any general sense. What is done is to start with the general hydrodynamical equations and make a series of assumptions regarding the flow; in this way a sequence of three nonlinear problems (we call them Problems I, II, III), each one furnishing a consistent and complete mathematical problem, is formulated. One can see then the effect of each additional assumption in simplifying the mathematical problem. The first two problems result from a series of assumptions which would probably be generally accepted by meteorologists as reasonable, but unfortunately even Problem II is still pretty much unmanageable from the point of view of numerical analysis. Further, and more drastic, assumptions lead to a still simpler Problem III which is formulated in terms of three first order partial differential equations in three dependent and three independent variables (as contrasted with eight differential equations in four independent variables in Problem I). The three differential equations of Problem III are probably capable of yielding reasonably accurate approximations to the frontal motions under consideration, but they are still rather difficult to deal with, even numerically, principally because they involve three independent variables*: such equations are well known to be beyond the scope of even the most modern digital computing machines as a rule. Consequently, still further simplifying assumptions are made in order to obtain a theory capable of yielding some concrete results through calculation.

At this point, two different approaches to the problem are proposed. One of them, by Whitham [W.12], deals rather directly with the three differential equations of Problem III. Two of these equations are essentially the same as those of the nonlinear shallow water theory treated in the preceding sections of this chapter. These two equations —which refer to motions in vertical planes—can therefore be integrated. Afterwards the transverse component of the velocity is found by integrating a linear first order partial differential equation. In this way a quite reasonable qualitative description of the dynamics of frontal motions can be achieved, at least in special cases, which is in good agreement with many of the observed phenomena. However, this theory has a disadvantage in that it does not permit a complete numerical integration because of a peculiar difficulty at cold fronts. (The difficulty stems from the fact that a cold front corresponds in this theory to what amounts to the propagation of a bore down the dry bed of a stream—a mathematical impossibility. If one had a means of taking care of turbulence and friction at the ground, it would perhaps be possible to overcome this difficulty.) Nevertheless, the qualitative agreement with the observed phenomena is an indication that the three differential equations furnishing the basic approximate theory from which we start—i.e. those of our Problem III—have in them the possibility of furnishing reasonable results.

The author's method (cf. [S.24]) of treating the three basic differential equations is quite different from that of Whitham, but it unfortunately involves a further assumption which has the effect of limiting the applicability of the theory. The guiding principle was that

---

* The work of Freeman [F.9, 10] is based on a theory which could be considered as a special case of Problem III in which the Coriolis terms due to the rotation of the earth are neglected and the motion is assumed at the outset to depend on only one space variable and the time. The idea of deriving the theory resulting in Problem III occurred to the author while reading Freeman's paper and, indeed, Freeman indicates the desirability of generalizing his theory.

differential equations in only two independent variables should be found, but that the number of dependent variables need not be so ruthlessly limited. Finally, it is highly desirable to obtain differential equations of hyperbolic type in order that the theory embodied in the method of characteristics becomes available in formulating and solving concrete problems. These objectives can be attained by making quite a few further simplifying assumptions with respect to the mechanics of the situation. The result is what might be called Problem IV. The theory formulated in Problem IV is embodied in a system of four nonlinear first order partial differential equations of hyperbolic type in four dependent and two independent variables. A numerical integration of these equations is possible, but the labor of integrating the equations is so great that only meagre results are so far available.

Once Whitham's theory and Problem IV have been formulated, one is led once more to consider dealing with Problem III numerically in spite of the fact that there are three independent variables in this case; in Problem IV, and also in the theory by Whitham, for that matter, the basic idea is that variations in the y-direction are less rapid than those in the x-direction, and thus a finite difference scheme in two space variables and the time might be possible under such special circumstances.

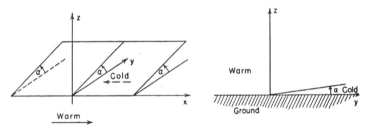

Fig. 10.11.1. A stationary front

We proceed to the derivation of the basic approximate theory. To begin with, a certain steady motion (called a stationary front) is taken as an initial state, and this consists of a uniform flow of two superimposed layers of cold and warm air, as indicated in Figure 10.11.1. The z-axis is taken positive upward* and the x, y-plane is a tangent

---

* Here we deviate from our standard practice of taking the y-axis as the vertical axis, in order to conform to the usual practice in dynamic meteorology. This should cause no confusion, since this section can be read to a large extent independently of the rest of the book.

plane to the earth. The rotation of the earth is to be taken into account but, for the sake of simplicity, not its sphericity—a common practice in dynamic meteorology. The coordinate system is assumed to be rotating about the z-axis with a constant angular velocity $\Omega = \omega \sin \varphi$, with ω the angular velocity of the earth and φ the latitude of the origin of our coordinate system. (The motivation for this is that the main effects one cares about are found if the Coriolis terms are included, and that neglect of the curvature of the earth has no serious qualitative effect.) As indicated in Figure 10.11.1, the cold air lies in a wedge under the warm air and the discontinuity surface between the two layers is inclined at angle α to the horizontal. The term "front" is always applied to the intersection of the discontinuity surface with the ground, and in the present case we have therefore as initial state a stationary front running along the x-axis. The wind velocity in the two layers is parallel to the x-axis (otherwise the discontinuity surface could not be stationary), but it will in general be different in magnitude and perhaps even opposite in direction in the two layers. The situation shown in Figure 10.11.1 is not uncommon. For instance, the x-axis might be in the eastward direction, the y-axis in the northward direction and the warm air would be moving in the direction of the prevailing westerlies. The origin of the cold air at the ground is, of course, the cold polar regions. We shall see later that such configurations are dynamically correct and that the angle α is uniquely determined (and quite small, of the order of $\frac{1}{2}°$) once the state of the warm air and cold air is given. (The discontinuity surface is not horizontal because of the Coriolis force arising from the rotation of the earth.)

We proceed next to describe what is observed to happen in many cases once such a stationary front starts moving. In Figure 10.11.2 a sequence of diagrammatic sketches is given which indicate in a general way what can happen. All of the sketches show the intersection of the moving discontinuity surface (cf. Figure 10.11.1) with the ground (the x, y-plane with the y-axis taken northward, the x-axis taken eastward). The shaded area indicates the region on the ground covered by cold air, while the unshaded region is covered at the ground by warm air. Of course, the cold air always lies in a thin wedge under a thick layer of warm air. In Figure 10.11.2a the development of a bulge in the stationary front toward the north is indicated.* Such a bulge then

* What agency serves to initiate and to maintain such motions appears to be a mystery. Naturally such an important matter has been the subject of a great

(footnote continued)

frequently deepens and at the same time propagates eastward with a
velocity of the order of 500 miles per day. It now becomes possible to
define the terms *cold front* and *warm front*. As indicated in Figure 10.
11.2*b*, the cold front is that part of the whole front at which cold air is
taking the place of warm air at the ground, and the warm front is the

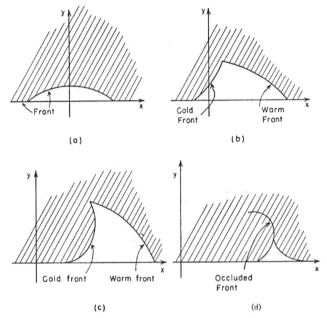

Fig. 10.11.2. Stages in the motion of a frontal disturbance

portion of the whole front where cold air is retreating with warm air
taking its place at the ground. Since such cold and warm fronts are
accompanied by winds, and by precipitation in various forms—in
fact, by all of the ingredients that go to make up what one calls

deal of discussion and speculation, but there seems to be no consistent view about
it among meteorologists. In applying the theory derived here no attempt is made
to settle this question a priori: we would simply take our dynamical model,
assume an initial condition which in effect states that a bulge of the kind just
described is initiated, and then study the subsequent motion by integrating the
differential equations subject to appropriate initial and boundary conditions.
However, if the approximate theory is really valid, such studies might perhaps
be used, or could be modified, in such a way as to throw some light on this important and vexing question.

weather—it follows that the weather at a given locality in the middle latitudes is largely conditioned by the passage of such frontal disturbances. Cold fronts and warm fronts behave differently in many ways. For example, the cold front in general moves faster than the warm front and steepens relative to it, so that an originally symmetrical disturbance or wave gradually becomes distorted in the manner indicated in Figure 10.11.2c. This process sometimes—though by no means always—continues until the greater portion of the cold front has overrun the warm front; an occluded front, as indicated in Figure 10.11.2d, is then said to occur. The prime object of what follows is to derive a theory—or perhaps better, to invent a simplified dynamical model—capable of dealing with fluid motions of this type that is not on the one hand so crude as to fail to yield at least roughly the observed motions, and on the other hand is not impossibly difficult to use for the purpose of mathematical discussion and numerical calculation.

Since it is desired that this section should be as much as possible self contained, we do not lean on the basic theory developed earlier in this book. Thus we begin with the classical hydrodynamical equations. The equations of motion in the Eulerian form are taken:

$$(10.11.1) \quad \begin{cases} \varrho \dfrac{du}{dt} = -\dfrac{\partial p}{\partial x} + \varrho F_{(x)} \\[2mm] \varrho \dfrac{dv}{dt} = -\dfrac{\partial p}{\partial y} + \varrho F_{(y)} \\[2mm] \varrho \dfrac{dw}{dt} = -\dfrac{\partial p}{\partial z} + \varrho F_{(z)} - \varrho g \end{cases}$$

with d/dt (the particle derivative) defined by the operator $\partial/\partial t + u\,\partial/\partial x + v\,\partial/\partial y + w\,\partial/\partial z$. In these equations u, v, w are the velocity components relative to our rotating coordinate system, p is the pressure, ϱ the density, $\varrho F_{(x)}$ etc. the components of the Coriolis force due to the rotation of the coordinate system, and ϱg is the force of gravity (assumed to be constant). These equations hold in both the warm air and the cold air, but it is preferable to distinguish the dependent quantities in the two different layers; this is done here throughout by writing u', v', w' for the velocity components in the warm air and similarly for the other dependent quantities.

We now introduce an assumption which is commonly made in dynamic meteorology in discussing large-scale motions of the atmosphere, i.e. that the air is incompressible. In spite of the fact that such

an assumption rules out thermodynamic processes, it does seem rather
reasonable since the pressure gradients which operate to create the
flows of interest to us are quite small and, what is perhaps the decisive
point, the propagation speed of the disturbances to be studied is very
small compared with the speed of sound in air (i.e. with disturbances
governed by compressibility effects). It would be possible to consider
the atmosphere, though incompressible, to be of variable density.
However, for the purpose of obtaining as simple a dynamical model as
possible it seems reasonable to begin with an atmosphere having a
constant density in each of the two layers. As a consequence of these
assumptions we have the following equation of continuity:

$$(10.11.2) \qquad\qquad u_x + v_y + w_z = 0.$$

The equations (10.11.1) and (10.11.2) together with the conditions
of continuity of the pressure and of the normal velocity components
on the discontinuity surface, the condition $w = 0$ at the ground,
appropriate initial conditions, etc. doubtlessly yield a mathematical
problem—call it Problem I—the solution of which would furnish a
reasonably good approximation to the observed phenomena. Unfortunately, such a problem is still so difficult as to be far beyond the
scope of known methods of analysis—including analysis by numerical
computation. Thus still further simplifications are in order.

One of the best-founded empirical laws in dynamic meteorology is
the hydrostatic pressure law, which states that the pressure at any
point in the atmosphere is very closely equal to the static weight of
the column of air vertically above it. This is equivalent to saying that
the vertical acceleration terms and the Coriolis force in the third
equation of (10.11.1) can be ignored with the result

$$(10.11.3) \qquad\qquad \frac{\partial p}{\partial z} = -\varrho g.$$

This is also the basis of the long-wave or shallow water theory of
surface gravity waves, as was already mentioned above. Since the
vertical component of the acceleration of the particles is thus ignored,
it follows on purely kinematical grounds that the horizontal components of the velocity will remain independent of the vertical coordinate
z for all time if that was the case at the initial instant $t = 0$. Since we
do in fact assume an initial motion with such a property, it follows
that we have

(10.11.4) $u = u(x, y, t), \qquad v = v(x, y, t), \qquad w = 0.*$

The first two of the equations of motion (10.11.1) and the equation of continuity (10.11.2) therefore reduce to

(10.11.5)
$$\begin{cases} u_t + uu_x + vu_y = -\dfrac{1}{\varrho} p_x + F_{(x)} \\[2mm] v_t + uv_x + vv_y = -\dfrac{1}{\varrho} p_y + F_{(y)} \\[2mm] u_x + v_y = 0, \end{cases}$$

where we use subscripts to denote partial derivatives and subscripts enclosed in parentheses to indicate components of a vector. The Coriolis acceleration terms are now given by

(10.11.6)
$$\begin{cases} F_{(x)} = 2\omega \sin \varphi \cdot v = \lambda v \\ F_{(y)} = -2\omega \sin \varphi \cdot u = -\lambda u \end{cases}$$

when use is again made of the fact that $w = 0$. (The latitude angle φ, as was indicated earlier, is assumed to be constant.) We observe once more that all of these relations hold in both the warm and cold layers, and we distinguish between the two when necessary by a prime on the symbols for quantities in the warm air. It is perhaps also worth mentioning that the equations (10.11.5) with $F_{(x)}$ and $F_{(y)}$ defined by (10.11.6) are valid for all orientations of the x, y-axes; thus it is not necessary to assume (as we did earlier, for example) that the original stationary front runs in the east-west direction.

We have not so far made full use of the hydrostatic pressure law (10.11.3). To this end it is useful to introduce the vertical height $h = h(x, y, t)$ of the discontinuity surface between the warm and cold layers and the height $h' = h'(x, y, t)$ of the warm layer itself (see Figure 10.11.3). Assuming that the pressure p' is zero at the top of the warm layer we find by integrating (10.11.3):

(10.11.7) $p'(x, y, z, t) = \varrho' g(h' - z)$

for the pressure at any point in the warm air. In the cold air we have, in similar fashion:

(10.11.8) $p(x, y, z, t) = \varrho' g(h' - h) + \varrho g(h - z)$

---

* It would be wrong, however, to infer that we assume the vertical *displacements* to be zero. This is a peculiarity of the shallow water theory in general which results, when a formal perturbation series is used, because of the manner in which the independent variables are made to depend on the depth (cf. Ch. 2 and early parts of the present chapter).

when the condition of continuity of pressure, $p' = p$ for $z = h$, is used. (The formula (10.11.8) is the starting point of the paper by Freeman [F.10] which was mentioned earlier.) Insertion of (10.11.8)

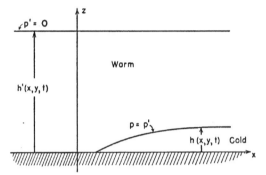

Fig. 10.11.3. Vertical height of the two layers

in (10.11.5) and of (10.11.7) in (10.11.5)' yields the following six equations for the six quantities u, v, h, u', v', h':

(10.11.9)
(cold air)
$$
\begin{cases}
u_t + uu_x + vu_y = -g\left[\frac{\varrho'}{\varrho}h'_x + \left(1 - \frac{\varrho'}{\varrho}\right)h_x\right] + \lambda v \\[2ex]
v_t + uv_x + vv_y = -g\left[\frac{\varrho'}{\varrho}h'_y + \left(1 - \frac{\varrho'}{\varrho}\right)h_y\right] - \lambda u \\[2ex]
u_x + v_y = 0
\end{cases}
$$

(10.11.10)
(warm air)
$$
\begin{cases}
u'_t + u'u'_x + v'u'_y = -gh'_x + \lambda v' \\[1ex]
v'_t + u'v'_x + v'v'_y = -gh'_y - \lambda u' \\[1ex]
u'_x + v'_y = 0.
\end{cases}
$$

These equations together with the kinematic conditions appropriate at the surfaces $z = h$ and $z = h'$, and initial conditions at $t = 0$, would again constitute a reasonable mathematical problem—call it Problem II—which could be used to study the dynamics of frontal motions. The Problem II is much simpler than the Problem I formulated above in that the number of dependent quantities is reduced from eight to six and, probably still more important, the number of independent variables is reduced from four to three. These simplifications, it should be noted, come about solely as a consequence of assuming the hydrostatic pressure law, and since meteorologists have much evidence supporting the validity of such an assumption, the Problem II should then furnish a reasonable basis for discussing the problem of frontal motions. Unfortunately, Problem II is just about as inaccessible as Problem I from the point of view of mathematical and numerical analysis. Consequently, we make still further hypotheses leading to a simpler theory.

As a preliminary to the formulation of Problem III we write down the kinematic free surface conditions at $z = h$ and $z = h'$ (the dynamical free surface conditions, $p = 0$ at $z = h'$ and $p = p'$ at $z = h$, have already been used.) These conditions state simply that the particle derivatives of the functions $z - h(x, y, t)$ and $z - h'(x, y, t)$ vanish, since any particle on the surface $z - h = 0$ or the surface $z - h' = 0$ remains on it. We have therefore the conditions

$$(10.11.11) \qquad \begin{cases} uh_x + vh_y + h_t = 0 \\ u'h_x + v'h_y + h_t = 0 \\ u'h'_x + v'h'_y + h'_t = 0, \end{cases}$$

in view of the fact that w vanishes everywhere. It is convenient to replace the third equations (the continuity equations) in the sets (10.11.9) and (10.11.10) by

$$(10.11.12) \qquad (uh)_x + (vh)_y + h_t = 0, \qquad \text{and}$$

$$(10.11.13) \qquad [u'(h' - h)]_x + [v'(h' - h)]_y + (h' - h)_t = 0,$$

which are readily seen to hold because of (10.11.11). In fact, the last two equations simply state the continuity conditions for a vertical column of air extending (in the cold air) from the ground up to $z = h$, and (in the warm air) from $z = h$ to $z = h'$.

We now make a really trenchant assumption, i.e. that *the motion of the warm air layer is not affected by the motion of the cold air layer.* This assumption has a rather reasonable physical basis, as might be argued in the following way: Imagine the stationary front to have developed a bulge in the y-direction, say, as in Figure 10.11.4a. The warm air can adjust itself to the new condition simply through a slight change in its vertical component, without any need for a change in u' and v', the horizontal components. This is indicated in Figure 10.11.4b, which is a vertical section of the air taken along the line AB in Figure 10.11.4a; in this figure the cold layer is shown with a quite small height—which is what one always assumes. Since we ignore changes in the vertical velocity components in any case, it thus seems reasonable to make our assumption of unaltered flow conditions in the warm air. However, in the cold air one sees readily—as indicated in Figure 10.11.4c—that quite large changes in the components u, v of the velocity in the cold air may be needed when a frontal disturbance is created. Thus we assume from now on that u', v', h' have for all time the known values they had in the initial steady state in

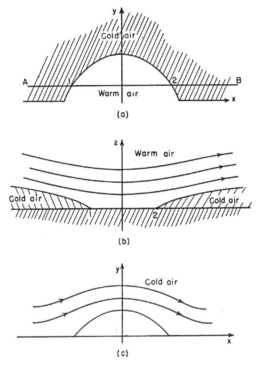

Fig. 10.11.4. Flows in warm and cold air layers

which $v' = 0$, $u' = $ const. The differential equations for our Problem III can now be written as follows:

$$(10.11.14) \quad \begin{cases} u_t + uu_x + vu_y = -g \left[\dfrac{\varrho'}{\varrho} h'_x + \left(1 - \dfrac{\varrho'}{\varrho}\right) h_x \right] + \lambda v \\[2mm] v_t + uv_x + vv_y = -g \left[\dfrac{\varrho'}{\varrho} h'_y + \left(1 - \dfrac{\varrho'}{\varrho}\right) h_y \right] - \lambda u \\[2mm] h_t + (uh)_x + (vh)_y = 0, \end{cases}$$

in which h'_x and h'_y are known functions given in terms of the initial state in the warm air. The initial state, in which $v' = v = 0$, $u = $ const., $u' = $ const., must satisfy the equations (10.11.9) and (10.11.10); this leads at once to the conditions

$$(10.11.15) \quad \begin{cases} h'_y = -\dfrac{\lambda}{g} u', & h'_x = 0 \\[2ex] h_y = \dfrac{-\dfrac{\lambda}{g} u - \dfrac{\varrho'}{\varrho} h'_y}{\left(1 - \dfrac{\varrho'}{\varrho}\right)} = \dfrac{\dfrac{\lambda}{g}\left(\dfrac{\varrho'}{\varrho} u' - u\right)}{\left(1 - \dfrac{\varrho'}{\varrho}\right)}, & h_x = 0 \end{cases}$$

for the slopes of the free surfaces initially. The slope h_y of the discontinuity surface between the two layers is nearly proportional to the velocity difference $u' - u$ since ϱ'/ϱ differs only slightly from unity, and it is made quite small under the conditions normally encountered because of the factor λ, which is a fraction of the angular velocity of the earth. The relation for the slope h_y of the stationary discontinuity surface is an expression of the law of Margules in meteorology. The differential equations for Problem III can, finally, be expressed in the form:

$$(10.11.16) \atop \text{Problem III} \quad \begin{cases} u_t + uu_x + vu_y + g\left(1 - \dfrac{\varrho'}{\varrho}\right)h_x = \lambda v \\[2ex] v_t + uv_x + vv_y + g\left(1 - \dfrac{\varrho'}{\varrho}\right)h_y = \lambda\left(\dfrac{\varrho'}{\varrho} u' - u\right) \\[2ex] h_t + (uh)_x + (vh)_y = 0, \end{cases}$$

by using the formulas for h'_x and h'_y given in (10.11.15). We note that the influence of the warm air expresses itself through its density ϱ' and its velocity u'. The three equations (10.11.16) undoubtedly have uniquely determined solutions once the values of u, v, and h are given at the initial instant $t = 0$, together with appropriate boundary conditions if the domain in x, y is not the whole space, and such solutions might reasonably be expected to furnish an approximate description of the dynamics of frontal motions.* Unfortunately, these equations are still quite complicated. They could be integrated numerically

---

* These equations are in fact quite similar to the equations for two-dimensional unsteady motion of a compressible fluid with h playing the role of the density of the fluid.

only with great difficulty even with the aid of the most modern high-
speed digital computers—mostly because there are three independent
variables.

Consequently, one casts about for still other possibilities, either of
specialization or simplification, which might yield a manageable
theory. One possibility of specialization has already been mentioned:
if one assumes no Coriolis force and also assumes that the motion is
independent of the y-coordinate, one obtains the pair of equations

$$(10.11.17) \qquad \begin{cases} u_t + uu_x + g\left(1 - \dfrac{\varrho'}{\varrho}\right)h_x = 0 \\[2mm] h_t + (uh)_x = 0 \end{cases}$$

which are identical with the equations of the one-dimensional shallow
water gravity wave theory. These equations contain in them the
possibility of the development of discontinuous motions—called
bores in sec. 10.7—and this fact lies at the basis of the discussions by
Freeman [F.10] and Abdullah [A.7]. In such one-dimensional treatments, it is clear that it is in principle not possible to deal with the
bulges on fronts and their deformation in time and space, since such
problems depend essentially on both space variables x and y. Another
possibility would be a linearization of the differential equations
(10.11.16) based on assuming small perturbations of the frontal surface and of the velocities from the initial uniform state. This procedure
might be of some interest, since such a formulation would take care of
the boundary condition at the ground, while the existing linear treatments of this problem do not. However, our interest here is in a nonlinear treatment which permits of large displacements of the fronts.
One such possibility, devised by Whitham [W.12], involves essentially
the integration of the first and third equations for u and h as functions
of x and t, regarding y as a parameter, and derivatives with respect to
y as negligible compared with derivatives with respect to x, and assuming initial values for v; this is feasible by the method of characteristics.
Afterwards, v would be determined by integrating the second equation considering u and h as known, and this can in principle be done
because the equation is a linear first order equation under these conditions. As stated earlier, this procedure furnishes qualitative results
which agree with observations. In addition, the discussion can be
carried through explicitly in certain cases, by making use of solutions
of the type called simple waves, along exactly the same lines as in
sec. 10.8 above. We turn, therefore, to this first of two proposed

approximate treatments of Problem III, as embodied in equations
(10.11.16).

The basic fact from which Whitham starts is that the slope $\alpha = h_y$
of the discontinuity surface is small initially, as we have already seen
in connection with the second equation of (10.11.15), and the fact that
λ is a fraction of the earth's angular velocity, and is expected to remain
in general small throughout the motions considered. Since the Coriolis
forces are of order α also (since they are proportional to λ) it seems
clear that derivatives of all quantities with respect to y will be small of
a different order from those with respect to x; it is assumed therefore
that u_y, h_y and v_y are all small of order α, but that u_x and h_x are finite.
Furthermore we can expect that the main motion will continue to be
a motion in the x-direction, so that the y-component v of the velocity
will be small of order α while the x-component u remains of course
finite. Under these circumstances, the equations (10.11.16) can be
replaced by simpler equations through neglect of all but the lowest
order terms in α in each equation; the result is the set of equations

$$(10.11.18) \qquad \begin{cases} h_t + uh_x + hu_x = 0 \\[4pt] u_t + uu_x + kh_x = 0 \\[4pt] v_t + uv_x = -kh_y + \lambda\left(\dfrac{\varrho'}{\varrho}u' - u\right) \end{cases}$$

with the constant k defined by

$$(10.11.19) \qquad\qquad k = g\left(1 - \frac{\varrho'}{\varrho}\right).$$

A considerable simplification has been achieved by this process, since
the variable y enters into the first two equations of (10.11.18) only as
a parameter and these two equations are identical with the equations
of the shallow water theory developed in the preceding sections of this
chapter if k is identified with g and h with η. This means that the
theory developed for these equations now becomes available to discuss our meteorological problems. Of course, the solutions for h and
u will depend on the variable y through the agency of initial and
boundary conditions. Once $u(x, y, t)$ and $h(x, y, t)$ have been obtained,
they can be inserted in the third equation of (10.11.18), which then
is a first order linear partial differential equation which, in principle
at least, can be integrated to obtain v when arbitrary initial conditions
$v = v(x, y, 0)$ are prescribed. The procedure contemplated can thus be
summed up as follows: the motion is to be studied first in each vertical·

plane $y =$ constant by the same methods as in the shallow water theory for two-dimensional motions (which means gas dynamics methods for one-dimensional unsteady motions), to be followed by a determination of the "cross-component" v of the velocity through integration of a first order linear equation which also contains the variable y, but only as a parameter.

This is in principle a feasible program, but it presents problems too complicated to be solved in general without using numerical computations. On the other hand we know from the earlier parts of this chapter that interesting special solutions of the first two equations of (10.11.18) exist in the form of what were called simple waves, and these solutions lend themselves to an easy discussion of a variety of motions in an explicit way through the use of the characteristic form of the equations. In order to preserve the continuity of the discussion it is necessary to repeat here some of the facts about the characteristic theory and the theory of simple waves; for details, secs. 10.2 and 10.3 should be consulted.

By introducing the new function $c^2 = kh$, replacing h, we obtain instead of the first two equations in (10.11.18) the following equations:

$$(10.11.20) \qquad \begin{cases} 2c_t + 2uc_x + cu_x = 0 \\ u_t + uu_x + 2cc_x = 0. \end{cases}$$

Thus the quantity $c = \sqrt{kh}$, which has the dimensions of a velocity, is the propagation speed of small disturbances, or wavelets—in analogy with the facts derived in sec. 10.2. These equations can in turn be written in the form

$$\left[\frac{\partial}{\partial t} + (u \pm c) \frac{\partial}{\partial x} \right] (u \pm 2c) = 0,$$

which can be interpreted to mean that the quantities $u \pm 2c$ are constant along curves $C_\pm$ in the x, t-plane such that $dx/dt = u \pm c$:

$$(10.11.21) \qquad \begin{cases} u + 2c = \text{const. along } C_+ : \dfrac{dx}{dt} = u + c \\[2mm] u - 2c = \text{const. along } C_- : \dfrac{dx}{dt} = u - c. \end{cases}$$

These relations hold in general for any solutions of (10.11.20). Under special circumstances it may happen that $u - 2c$, for example, has the same constant value on all C_- characteristics in a certain region; in that case since $u + 2c$ is constant along each C_+ characteristic it follows that u and c would separately by constant along each of the C_+ characteristics, which means that these curves would all be straight lines. Such a region of the flow (the term region here being applied with respect to some portion of an x, t-plane) is called a simple wave. It is then a very important general fact that any flow region adjacent to a region in which the flow is uniform, i.e. in which both c and u are everywhere constant (in both space and time, that is), is a simple wave, provided that u and c are continuous in the region in question.

It is reasonable to suppose that simple waves would occur in cases of interest to us in our study of the dynamics of frontal motions, simply because we do actually begin with a flow in which u and h (hence also c) are constant in space and time, and it seems reasonable to suppose that disturbances are initiated, not everywhere in the flow region, but only in certain areas. In other words, flows adjacent to uniform flows would occur in the nature of things. Just how in detail initial or boundary conditions, or both, should be prescribed in order to conform with what actually occurs in nature is, as has already been pointed out, something of a mystery; in fact one of the principal objects of the ideas presented here could be to make a comparison of calculated motions under prescribed initial and boundary conditions with observed motions in the hope of learning something by inference concerning the causes for the initiation and development of frontal disturbances as seen in nature.

One fairly obvious and rather reasonable assumption to begin with might be that u, v, and h are prescribed at the time $t = 0$ to have values over a certain bounded region of the upper half ($y > 0$) of the x, y-plane (cf. Fig. 10.11.1) in such a fashion that they differ from the constant values in the original uniform flow with a stationary front. According to the approximate theory based on equations (10.11. 18), this means, in particular, that in each vertical plane $y = y_0$—an x, t-plane—initial conditions for $u(x, y_0, t)$ and $h(x, y_0, t)$ would be prescribed over the entire x-axis, but in such a way that u and h are constant with values $u = u_0 > 0$, $h = h_0 \geqq 0$ (hence $c = c_0 = \sqrt{kh_0}$)* everywhere except over a certain segment $x_1 \leqq x \leqq x_2$, as indicated in Fig. 10.11.5. The positive characteristics C_+ are drawn in full lines, the characteristics C_- with dashed lines in this diagram,

* It should, however, always be kept in mind in the discussion to follow that c_0, particularly, will usually have different values in different vertical planes $y = $ const.

which is to be interpreted as follows. Simple waves exist everywhere in the x, t-plane except in the triangular region bounded by the C_+ characteristic through A and the C_- characteristic through B and terminating at point C; in this region the flow could be determined

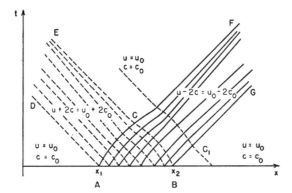

Fig. 10.11.5. Simple waves arising from initial conditions

numerically, for example by the method indicated in sec. 10.2 above (in connection with Fig. 10.2.1). The disturbance created over the segment AB propagates both "upstream" and "downstream" after a certain time in the form of two simple waves, which cover the regions bounded by the straight (and parallel) characteristics issuing from A, B, and C. In other words the disturbance eventually results in two distinct simple waves, one propagating upstream, the other downstream, and separated by a uniform flow identical with the initial state. In our diagram it is tacitly assumed that $c > |u|$, i.e. that the flow is subcritical in the terminology of water waves (subsonic in gas dynamics)—otherwise no propagation upstream could occur. We have supposed u to be positive, i.e. that the x-component of the flow velocity in the cold air layer has the same direction as the velocity in the warm air, which in general flows from the west, but it can be (and not infrequently is) in the westward rather than the eastward direction. Since the observed fronts seem to move almost invariably to the eastward, it follows, for example, that it would be the wave moving upstream which would be important in the case of a wind to the westward in the cold layer, and a model of the type considered here—in which the disturbance is prescribed by means of an initial condition and the flow is subcritical—implies that the initial disturbances are always of such a special character that the downstream wave has a negligible amplitude. For a wind to the eastward, the reverse would be the case. All of this is, naturally, of an extremely hypothetical character, but nevertheless one sees that certain important elements pertinent to a discussion of possible motions are put in evidence.

The last remarks indicate that a model based on such an initial disturbance may not be the most appropriate in the majority of cases. In fact, such a formulation of the problem is open to an objection which is probably rather serious. The objection is that such a motion has its origin in an initial impulse, and this provides no mechanism by which energy could be constantly fed into the system to "drive" the wave. Of course, it would be possible to introduce external body forces in various ways to achieve such a purpose, but it is not easy to see how to do that in a rational way from the point of view of mechanics. Another way to introduce energy into the system would be to feed it in through a boundary—in other words formulate appropriate boundary conditions as well as initial conditions. For the case of fronts moving eastward across the United States, a boundary condition might be reasonably applied at some point to the east of the high mountain system bordering the west coast of the continent, since these mountain ranges form a rather effective north-south barrier between the motions at the ground on its two sides. In fact, a cold front is not infrequently seen running nearly parallel to the mountains and to the east of them—as though cold air had been deflected southward at this barrier. Hence a boundary condition applied at some point on the west seems not entirely without reason. In any case, we seek models from which knowledge about the dynamics of fronts might be obtained, and a model making use of boundary conditions should be studied. We suppose, therefore, that a boundary condition is applied at $x = 0$, and that the initial condition for $t = 0$, $x > 0$ is that the flow is undisturbed, i.e. $u = u_0 = $ const., $c = c_0 = $ const.. (Again we remark that we are considering the motion in a definite vertical plane $y = y_0$.) In this case we would have only a wave propagating eastward—in effect, we replace the influence of the air to the westward by an assumed boundary condition. The general situation is indicated in Fig. 10.11.6. There is again a simple wave in the region of the x, t-plane above the straight line $x = (u_0 + c_0)t$ which marks the boundary between the undisturbed flow and the wave arising from disturbances created at $x = 0$. This is exactly the situation which is treated at length in sec. 10.8; in particular, an explicit solution of the problem is easily obtained (cf. the discussion in sec. 10.4) for arbitrarily prescribed disturbances in the values of either of the two quantities u or c. Through various choices of boundary conditions it is possible to supply energy to the system in a variety of ways.

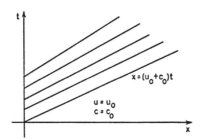

Fig. 10.11.6. Wave arising from conditions applied at a boundary

We proceed next to discuss qualitatively a few consequences which result if it is assumed that frontal disturbances can be described in terms of simple waves in all vertical planes $y = y_0 =$ const. at least over some ranges in the values of the y-coordinate. (We shall see later that simple waves are not possible for all values of y.) In this discussion we do not specify *how* the simple wave was originated—we simply assume it to exist. Since we consider only waves moving eastward (i.e. in the positive x-direction) it follows that the straight character-istics are C_+ characteristics, and hence that $u - 2c$ is constant (in each plane $y =$ const.) throughout the wave; we have therefore

$$(10.11.22) \qquad u - 2c = A(y),$$

with $A(y)$ fixed by the values u_0 and $c_0(y)$ in the undisturbed flow:

$$(10.11.22)_1 \qquad A(y) = u_0 - 2c_0(y).$$

In addition, as explained before, we know that $u + 2c$ is a function of y alone on each positive characteristic $dx/dt = u + c$; hence u and c are individually functions of y on each of these characteristics. There-fore, the characteristic equation may be integrated to yield

$$(10.11.23) \qquad x = \xi + (u + c)t,$$

where ξ is the value of x at $t = 0$. (The time $t = 0$ should be thought of as corresponding to an arbitrary instant at which simple waves exist in certain planes $y =$ const.) Now, the values of u and c on the characteristic given by (10.11.23) are exactly the same as the values (for the same value of y) at the point $t = 0$, $x = \xi$; therefore, if we suppose, for example, that c is a given function $C(x, y)$ at $t = 0$, the value of c in (10.11.23) is $C(\xi, y)$ and the value of u is, from (10.11.22), $A(y) + 2C(\xi, y)$. Thus the simple wave solution can be described by the equations

$$(10.11.24) \qquad \begin{cases} c = C(\xi, y), \\ u = A(y) + 2C(\xi, y), \\ x = \xi + \{A(y) + 3C(\xi, y)\}t. \end{cases}$$

(Although the arbitrary function occurring in a simple wave could be specified in other ways, it is convenient for our purposes to give the value of h, and hence c, at $t = 0$.)

We could write down the solution for the "cross component", or north-south component, v of the velocity in this case; by standard methods (cf. the report of Whitham [W.12]) it can be obtained by integrating the linear first order partial differential equation which occurs third in the basic equations (10.11.18). To specify the solution of this equation uniquely an initial condition is needed; this might reasonably be furnished by the values $v = v(x, y)$ at the time $t = 0$. The result is a rather complicated expression from which not much can be said in a general way. One of the weaknesses of the present attack on our problem through the use of simple waves now becomes apparent: it is necessary to know values of v some time subsequent to the initiation of a disturbance in order to predict them for the future.

It is possible, however, to draw some interesting conclusions from the simple wave motions without considering the north-south component of the velocity. For example, suppose we consider a motion after a bulge to the northward in an initially stationary front had developed as indicated schematically in Fig. 10.11.2. In a plane $y =$ const. somewhat to the north of the bulge we could expect the top of the cold air layer (the discontinuity surface, that is) as given by $h(x, t)$ to appear, for $t = 0$ say, as indicated in Fig. 10.11.7. The main features of the graph are that there is a depression in the discontinuity surface, but that $h > 0$ so that this surface does not touch the ground. (The latter possibility will be discussed later.) Assuming that the motion is described as a simple wave, we see from (10.11.24) that the value of $c = \sqrt{kh}$ at the point $x = x_1$ is equal to the value of c which was at the point $x = \xi_1$ at $t = 0$, where $\xi_1 = x_1 - \{A(y) + 3C(\xi_1, y)\}t_1$. That is, the value $c = C(\xi_1, y)$ has been displaced to the right by an

amount $\{A(y) + 3C(\xi_1, y)\}t_1$. Since this quantity is greater for greater
values of C, the graph of h becomes distorted in the manner shown in
Fig. 10.11.7: the "negative region" (where $h_x < 0$) steepens whilst the
"positive region" (where $h_x > 0$) flattens out. The positive region
continues to smooth out, but, if the steepening of the negative region

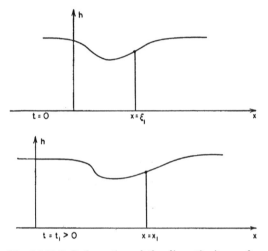

Fig. 10.11.7. Deformation of the discontinuity surface

were to continue indefinitely, there would ultimately be more than
one value of C at the same point, and the wave, as in our discussion
of water waves (cf. sec. 10.6 and 10.7), starts to break. Clearly the
latter event occurs when the tangent at a point of the curve in
Fig. 10.11.7 first becomes vertical. At this time, the continuous solution breaks down (since c and u would cease to be single-valued
functions) and a discontinuous jump in height and velocity must be
permitted. In terms of the description of the wave by means of the
characteristics, what happens is that the straight line characteristics
converge and eventually form a region with a fold. Such a discontinuous "bore" propagates faster than the wavelets ahead of it (the
paths of the wavelets in the x, t-plane are the characteristics) in a
manner analogous to the propagation of shock waves in gas dynamics
and bores in water.

In the above paragraph we supposed that h and c were different
from zero, and hence the discussion does not apply to the fronts, which

are by definition the intersection of the discontinuity surface with the ground. When $c = 0$ there are difficulties, especially at cold fronts, but nevertheless a few pertinent observations can be made, assuming the motion to be a simple compression wave with $u - 2c$ constant. When $c = 0$, it follows that $u = u_0 - 2c_0$, and since $c_0 = \sqrt{kh_0}$ and $h_0 = \alpha y_0$ with α the initial inclination of the top of the cold air layer, it follows that $u = u_0 - 2\sqrt{\alpha k y}$ in this case. But u then measures the speed of the front itself in the x-direction, since a particle once on the front stays there; consequently for the speed u_f of the front we have

$$(10.11.25) \qquad u_f = u_0 - 2\sqrt{\alpha k y}.$$

Thus the speed of the front decreases with y, and on this basis it follows that a northward bulge would become distorted in the fashion indicated by Fig. 10.11.8, and this coincides qualitatively with observations of actual fronts.

Actually, things are not quite as simple as this. If $c = 0$, it follows from the first equation of (10.11.20) that $c_t + uc_x = 0$ on such a locus, and this in turn means that $c = 0$ on the particle path defined by $dx/dt = u$. At the same time the C_+ and C_- characteristics have the same direction, since they are given by $dx/dt = u \pm c$. On the other hand, we have, again from (10.11.20), $u_t + uu_x = -kh_x$ and we see that the relation $u =$ const. along a characteristic for which $c = 0$ cannot be satisfied unless $h_x = 0$. In connection with Fig. 10.11.7 we have seen that the rising portion toward the east of a depression in the discontinuity surface tends to flatten out, while the falling part from the west tends to steepen and break because the higher portions tend to move faster and crowd the lower portions. Thus when h, and hence c, tends to zero the tendency will be for breaking to occur at the cold front, but not at the warm front. The slope of the discontinuity surface at the cold front will then be infinite. However, a bore in the sense described above cannot occur since there must always be a mass flux through a bore: the motion of the cold front is analogous to what would happen if a dam were broken and water rushed down the dry bed of a stream. Without considering in some special way what happens in the turbulent motion caused by such continuous breaking at the ground, it is not possible to continue our discussion of the motion of a cold front along the present lines, although such a problem is susceptible to an approximate treatment. Nevertheless, this discussion has led in a rational way to a qualitative explanation for the well-known fact that a warm front does indeed progress in a relatively smooth fashion as compared with the turbulence which is commonly observed at cold fronts. Thus near a cold front the height of the cold air layer may be considerably greater than in the vicinity of the warm front, where $h \simeq 0$; consequently the speed of propagation of the cold front could be expected to be greater than

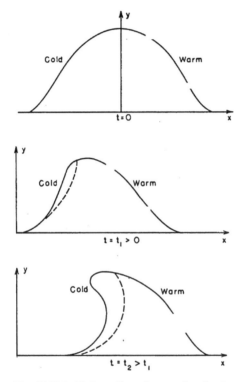

Fig. 10.11.8. Deformation of a moving front

near the warm front (as indicated by the dotted modifications of the shape of the cold front in Fig. 10.11.8), with the consequence that the gap between the two tends to close, and this hints at a possible explanation for the occlusion process. One might also look at the matter in this way: Suppose $c \neq 0$, but is small in the trough of the wave shown in Fig. 10.11.7. If breaking once begins, it is well known that the resulting bore moves with a speed that is greater than the propagation speed of wavelets in the medium in front (to the right) of it, although slower than the propagation speed in the medium behind it. Again one sees that the tendency for the wave on the cold front side to catch up with the wave on the warm front side is to be expected on the basis of the theory presented here.

Finally we observe that the velocity of the wave near the undisturbed stationary front is u_0, but well to the north it is given roughly by $u_f = u_0 - 2\sqrt{\alpha k y}$, which is less than u_0. There is thus a tendency to produce what is called in meteorology a cyclonic rotation around the center of the wave disturbance.

To sum up, it seems fair to say that the approximate theory embodied in equations (10.11.18), even when applied to a very special type of motions (i.e. simple waves in each plane $y = $ const.), yields a variety of results which are at least qualitatively in accord with observations of actual fronts in the atmosphere. Among the phenomena given correctly in a qualitative way are: the change in shape of a wave as it progresses eastward, the occurrence of a smooth wave at a warm front but a turbulent wave at a cold front, and a tendency to produce the type of motion called a cyclone.

It therefore seems reasonable to suppose that the differential equations of our Problem III, which were the starting point of the discussion just concluded, contain in them the possibility of dealing with motions which have the general characteristics of frontal motions in the atmosphere, and that numerical solutions of the equations of Problem III might well furnish valuable insights. This is a difficult task, as has already been mentioned. However, an approximate theory different from that of Whitham is possible, which has the advantage that no especial difficulty arises at cold fronts, and which would permit a numerical treatment. This approximate theory might be considered as a new Problem IV.

The formulation of Problem IV was motivated by the following considerations. If one looks at a sequence of weather maps and thinks of the wave motion in our thin wedge of cold air, the resemblance to the motion of waves in water which deform into breakers (especially in the case of frontal disturbances which develop into occluded fronts) is very strong. The great difference is that the wave motion in water takes place in the vertical plane while the wave motion in our thin layer of cold air takes place essentially in the horizontal plane. When the hydrostatic pressure assumption is made in the case of water waves the result is a theory in exact analogy to gas dynamics, and thus wave motions with an appropriate "sound speed" become possible even though the fluid is incompressible—the free surface permits the introduction of the depth of the water as a dependent quantity, this quantity plays the role of the density in gas dynamics, and thus a dynamical model in the form of a compressible fluid is obtained. It would seem therefore reasonable to try to invent a similar theory for frontal motions in the form of a long-wave theory suitable for waves which move essentially in the horizontal, rather than the vertical, plane, and in which the waves propagate essentially parallel to the edge of the original stationary front, i.e. the x-axis. In this way one might hope to be rid of the dependence on the variable y at right angles to the stationary front, thus reducing the independent variables to two, x and t; and if one still could obtain a hyperbolic system of differential equations then numerical treatments by finite differences would be feasible. This program can, in fact, be carried out in such a way as to yield a system of four first order nonlinear differential equations in two independent and four dependent variables which are of the hyperbolic type.

Once having decided to obtain a long-wave theory for the horizontal plane, the procedure to be followed can be inferred to a large extent from what one does in developing the same type of theory for gravity waves in water, as we have seen in Chapter 2 and at the beginning of the present chapter. To begin with it seems clear that the displacement $\eta(x, t)$ of the front itself in the y-direction should be introduced as one of the dependent quantities—all the more since this quantity is anyway the most obvious one on the weather maps. To have such a "shallow water" theory in the horizontal plane requires—unfortunately—a rigid "bottom" somewhere (which is, of course, vertical in this case), and this we simply postulate, i.e. we assume that the y-component v of the velocity vanishes for all time on a vertical plane $y = \delta = $ const. parallel to the stationary front along the x-axis (see Figure 10.11.9). The velocity $v(x, y, t)$ is then assumed to vary linearly* in y, and its value at the front, $y = \eta(x, t)$, is called $\bar{v}(x, t)$. The intersection of the discontinuity surface $z = h(x, y, t)$ with the plane $y = \delta$ is a curve given by $z = \bar{h}(x, t)$, and we assume that the discontinuity surface is a ruled surface having straight line generators running from the front, $y = \eta(x, t)$, to the curve $z = \bar{h}(x, t)$, and

---

* The analogous statement holds also in the long-wave theory in water (to lowest order in the development parameter, that is).

parallel to the y, z-plane. Finally, we assume (as in the shallow water theory) that u, the x-component of the velocity, depends on x and t

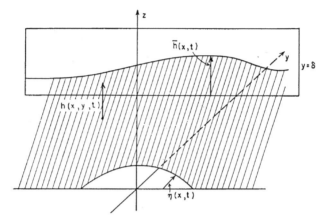

Fig. 10.11.9. Notations for Problem IV

only: $u = u(x, t)$. The effect of these assumptions is to yield the relations

$$(10.11.26) \qquad h(x, y, t) = \frac{y - \eta(x, t)}{\delta - \eta(x, t)} \cdot \bar{h}(x, t),$$

$$(10.11.27) \qquad v(x, y, t) = \frac{\delta - y}{\delta - \eta(x, t)} \cdot \bar{v}(x, t),$$

as one readily sees. In addition, we assume that a particle that is once on the front $y - \eta(x, t) = 0$ always remains on it, so that the relation:

$$(10.11.28) \qquad \bar{v}(x, t) = \eta_t + u\eta_x$$

must hold. The four quantities $u(x, t)$, $\eta(x, t)$, $\bar{h}(x, t)$, and $\bar{v}(x, t)$ are our new dependent variables. Differential equations for them will be obtained by integrating the basic equations (10.11.16) of Problem III with respect to y from $y = \eta$ to $y = \delta$—which can be done since the dependence of u, v, and h on y is now explicitly given—and these three equations together with (10.11.28) will yield the four equations we want.

Before writing these equations down it should be said that the most trenchant assumption made here is the assumption concerning the existence of the rigid boundary $y = \delta$. One might think that as long as the velocity component v dies out with sufficient rapidity in the y-direction such an assumption would yield a good approximation, but the facts in the case of water waves indicate this to be not sufficient for the accuracy of the approximation: with water waves in very deep water the vertical component of the velocity (corresponding to our v here) dies out very rapidly in the depth, but it is nevertheless essential for a good approximation to assume that the ratio of the depth down to a rigid bottom to the wave length is small. However, such a rigid vertical barrier to the winds does exist in some cases of interest to us in the form of mountain ranges, which are often much higher than the top of the cold surface layer (i.e. higher than the curve $z = \bar{h}(x, t)$ in Figure 10.11.9). In any case, severe though this restriction is, it still seems to the author to be worth while to study the motions which are compatible with it since something about the dynamics of frontal motions with large deformations may be learned in the process. In particular, one might learn something about the kind of perturbations that are necessary to initiate motions of the type observed, and under what circumstances the motions can be maintained.

In carrying out the derivation of the differential equations of our theory according to the plan outlined above, we calculate first a number of integrals. The first of these arise from (10.11.26) and (10.11.27):

$$\int_{\eta}^{\delta} h \, dy = \frac{\bar{h}}{\delta - \eta} \int_{\eta}^{\delta} (y - \eta) \, dy = \frac{1}{2} \bar{h}(\delta - \eta),$$

$$\int_{\eta}^{\delta} v \, dy = \frac{\bar{v}}{\delta - \eta} \int_{\eta}^{\delta} (\delta - y) \, dy = \frac{1}{2} \bar{v}(\delta - \eta).$$

From these we derive by differentiations with respect to x and t another set of relations:

$$\int_{\eta}^{\delta} h_x \, dy = \frac{1}{2} \bar{h}_x(\delta - \eta) - \frac{1}{2} \bar{h}\eta_x,$$

$$\int_{\eta}^{\delta} v_x \, dy = \frac{1}{2} \bar{v}_x(\delta - \eta) + \frac{1}{2} \bar{v}\eta_x,$$

$$\int_{\eta}^{\delta} v_t \, dy = \frac{1}{2} \bar{v}_t(\delta - \eta) + \frac{1}{2} \bar{v}\eta_t,$$

$$\int_{\eta}^{\delta} h_t \, dy = \frac{1}{2} h_t(\delta - \eta) - \frac{1}{2} \bar{h}\eta_t.$$

(In deriving these relations, it is necessary to observe that the lower limit η is a function of x and t.) One additional relation is needed, as follows:

$$\int_{\eta}^{\delta} (hu)_x dy = \frac{\partial}{\partial x}\left\{ u \int_{\eta}^{\delta} h\, dy \right\} = \frac{1}{2}(\bar{h}_x u + \bar{h}u_x)(\delta - \eta) - \frac{1}{2}\bar{h}u\eta_x.$$

We now integrate both sides of the equations (10.11.16) with respect to y from η to δ, make use of the above integrals, note that $u = u(x, t)$ is independent of y, and divide by $\delta - \eta$. The result is the equations

$$(10.11.29) \quad \begin{cases} u_t + u\dot{u}_x + \dfrac{1}{2} k\bar{h}_x - \dfrac{1}{2}\dfrac{k\bar{h}}{\delta - \eta}\eta_x = \dfrac{1}{2}\lambda\bar{v}, \\[2ex] \bar{v}_t + u\bar{v}_x + \dfrac{u\bar{v}\eta_x}{\delta - \eta} + \dfrac{\bar{v}}{\delta - \eta}\eta_t = \dfrac{\bar{v}^2}{\delta - \eta} - \dfrac{2k\bar{h}}{\delta - \eta} - 2\lambda\left(u - \dfrac{\varrho'}{\varrho}u'\right). \\[2ex] u\bar{h}_x + \bar{h}u_x - \dfrac{\bar{h}u}{\delta - \eta}\eta_x + \bar{h}_t - \dfrac{\bar{h}}{\delta - \eta}\eta_t = 0, \end{cases}$$

with k a constant replacing the quantity $g(1 - \varrho'/\varrho)$. These equations, together with (10.11.28), form a system of four partial differential equations for the four functions u, η, $\bar{v}$, and $\bar{h}$. By analogy with gas dynamics and the nonlinear shallow water theory, it is convenient to introduce a new dependent quantity c (which will turn out to be the propagation speed of wavelets) through the relation

$$(10.11.80) \qquad c^2 = \frac{1}{2} k\bar{h} = \frac{1}{2} g\left(1 - \frac{\varrho'}{\varrho}\right)\bar{h}.$$

The quantity c is real if ϱ' is less than ϱ, and this holds since the warm air is lighter than the cold air. In terms of this new quantity the equations (10.11.28) and (10.11.29) take the form

$$(10.11.81) \quad \begin{cases} u_t + uu_x + 2cc_x - \dfrac{c^2}{\delta - \eta}\eta_x = \dfrac{1}{2}\lambda\bar{v}, \\[2ex] \bar{v}_t + u\bar{v}_x = -\dfrac{4c^2}{\delta - \eta} - 2\lambda\left(u - \dfrac{\varrho'}{\varrho}u'\right), \\[2ex] 2c_t + cu_x + 2uc_x = \dfrac{c\bar{v}}{\delta - \eta}, \\[2ex] \eta_t + u\eta_x = \bar{v}. \end{cases}$$

It is now easy to write the equations (10.11.81) in the characteristic form simply by replacing the first and third equations by their sum and by their difference. The result is:

$$(10.11.32)\begin{cases} u_t+(u+c)u_x+2\{c_t+(u+c)c_x\}-\dfrac{c}{\delta-\eta}\{\eta_t+(u+c)\eta_x\}=\tfrac{1}{2}\lambda\bar v, \\[2mm] u_t+(u-c)u_x-2\{c_t+(u-c)c_x\}+\dfrac{c}{\delta-\eta}\{\eta_t+(u-c)\eta_x\}=\tfrac{1}{2}\lambda\bar v, \\[2mm] \bar v_t+u\bar v_x=-\dfrac{4c^2}{\delta-\eta}-2\lambda\left(u-\dfrac{\varrho'}{\varrho}u'\right), \\[2mm] \eta_t+u\eta_x=\bar v. \end{cases}$$

As one sees, the equations are in characteristic form: the characteristic curves satisfy the differential equations

$$(10.11.33)\qquad \frac{dx}{dt}=u+c,\quad \frac{dx}{dt}=u-c,\quad \frac{dx}{dt}=u,$$

and each of the equations (10.11.32) contains only derivatives in the direction of one of these curves. The characteristic curves defined by $dx/dt=u$ are taken twice. Thus one sees that the quantity c is indeed entitled to be called a propagation speed, and small disturbances can be expected to propagate with this speed in both directions relative to the stream of velocity u. (In the theory by Whitham, in which the motion in each vertical plane $y=$ const. is treated separately, the propagation or sound speed of small disturbances is given by $\sqrt{kh}$. The sound speed in the theory given here thus represents a kind of average with respect to y of the sound speeds of Whitham's theory.) Since the propagation speed depends on the height of the discontinuity surface, it is clear that the possibility of motions leading to breaking is inherent in this theory.

Once the dynamical equations have been formulated in characteristic form it becomes possible to see rather easily what sort of subsidiary initial and boundary conditions are reasonable. In fact, there are many possibilities in this respect. One such possibility is the following. At time $t=0$ it is assumed that $u=$ const., $\eta=0$, $\bar h=$ const. (as in a stationary front), but that $\eta_t=f(x)$ over a segment of the x-axis. In other words, it is assumed that a transverse impulse is given to the stationary front over a portion of its length. The subsequent motion is uniquely determined and can be calculated numerically. Another possibility is to prescribe a stationary front at $t=0$

for $x > 0$, say, and then to give the values of all dependent quantities*
at $x = 0$ as arbitrary functions of the time; i.e. to prescribe a boundary
condition which allows energy to be introduced gradually into the
system. One might visualize this case as one in which, for example,
cold air is being added or withdrawn at a particular point ($x = 0$ in
the present case). This again yields a problem with a uniquely determined solution, and various possibilities are being explored numerically.

It was stated above that the most objectionable feature of the
present theory is the assumption of a fixed vertical barrier back of the
front. There is, however, a different way of looking at the problem as
a whole which may mitigate this restriction somewhat. One might try
to consider the motion of the entire cap of cold air that lies over the
polar region, using polar coordinates (θ, φ) (with θ the latitude angle,
say). One might then consider motions once more that depend
essentially only on φ and t by getting rid of the dependence on θ
through use of the same type of assumptions (linear behavior in θ,
say) as above. Here the North Pole itself would take the place of the
vertical barrier ($v = 0$!). The result is again a system of nonlinear
equations—this time with variable coefficients. Of course, it would
be necessary to begin with a stationary flow in which the motion takes
place along the parallels of latitude.

All in all, the ideas presented here would seem to yield theories
flexible enough to permit a good deal of freedom with regard to initial
and other conditions, so that one might hope to gain some insight into
the complicated dynamics of frontal motions by carrying out numerical solutions in well-chosen special cases.

10.12. Supercritical steady flows in two dimensions. Flow around bends. Aerodynamic applications

The title of this section is a slight misnomer, since the flows in
question are really three-dimensional in nature; however, since we
consider them here only in terms of the shallow water theory, the
depth dimension is left out. Thus the velocity is characterized by the
two components (u, w) in the horizontal plane (the x, z-plane); and
these quantities, together with the depth h of the water at any point
constitute the quantities to be determined in any given problem. By

---

* In the numerical cases so far considered we have had $|c| < |u|$ so that
even on the t-axis all four dependent quantities can be prescribed.

specializing the general equations (2.4.18), (2.4.19), (2.4.20), of the shallow water theory as derived in Chapter 2 for the case of a steady flow, the differential equations relevant for this section result. They can also be derived readily from first principles, as follows: Assuming that the hydrostatic pressure law holds and that the fluid starts from rest (or any other motion in which the vertical component of the velocity of the water is zero) it follows that the vertical component of the velocity remains zero and that u and w are independent of the vertical coordinate. The law of continuity can thus be readily derived for a vertical column; for a steady flow it is

$$(10.12.1) \qquad (hu)_x + (hw)_z = 0.$$

We assume that the flows we study are irrotational, and hence that

$$(10.12.2) \qquad u_z - w_x = 0.$$

The Bernoulli law then holds and can be written in the form

$$(10.12.3) \qquad (u^2 + w^2) + 2gh = \text{const.}$$

In these equations h is the depth of the water at any point. By using (10.12.3) to express h in terms of u and w, and introducing the quantity c by the relation

$$(10.12.4) \qquad c^2 = gh$$

we obtain the equation

$$(10.12.5) \qquad (c^2 - u^2)u_x - uw(w_x + u_z) + (c^2 - w^2)w_z = 0,$$

and this equation together with (10.12.2), with c defined in terms of u and w through (10.12.4) and (10.12.3), constitute a pair of first order partial differential equations for the determination of $u(x, z)$ and $w(x, z)$.

The theory of these latter equations can be developed, as in the cases treated previously in this chapter, by using the method of characteristics, provided that the quantity c remains always less than the flow speed everywhere, i.e. provided that

$$(10.12.6) \qquad c^2 < u^2 + w^2.$$

The flow is then said to be supercritical. (In hydraulics the contrast subcritical—supercritical is commonly expressed as tranquil-shooting.) Only then do real characteristics exist. We shall not develop this theory here, but rather indicate some of the problems which have been treated by using the theory. Complete expositions of the characteristic theory can be found in the paper by Preiswerk [P.16], and in Chapter IV of the book by Courant and Friedrichs [C.9]. The theory is, of course, again perfectly analogous to the theory of steady two-dimensional supersonic flows in gas dynamics.

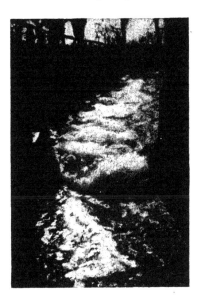

Fig. 10.12.1. Hydraulic jump

We have already encountered an interesting example of a flow which is in part supercritical, in part subcritical, i.e. the case of a hydraulic jump in which the character of the flow changes on passage through the discontinuity. Figure 10.12.1 is a photograph, taken from the paper of Preiswerk, of such a hydraulic jump. Figure 10.12.2, also taken from the paper of Preiswerk, shows a more complicated case in which hydraulic jumps occur at oblique angles to the direction of the flow. The picture shows a flow through a sluice in a dam, with conditions (i.e. depth differences above and below the dam) such that supercritical flow develops in the sluice, and changes in level take place so abruptly that they might well be treated as discontinuities (as was done in earlier sections in the treatment of bores). The two disconti-nuities at the sides of the sluice (marked 1 and 2 in the figure) are turned toward each other and eventually intersect to form a still higher one (marked $1 + 2$). Such oblique discontinuities can be treated mathematically; the details can be found in the works cited above.

Another interesting problem of the category considered here is the

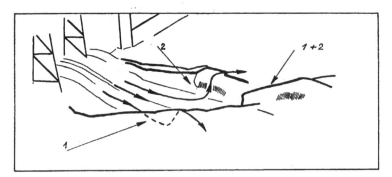

Fig. 10.12.2. Hydraulic jumps at oblique angles to the direction of the flow

problem of supercritical flow around a bend in a stream. This type of problem is relevant not only for flows in water, but also for certain flows in the atmosphere (for which see Freeman [F.9]). It is possible in these cases to have flows of the type which are mathematically of the kind called simple waves in earlier sections. This means that one of the families of characteristics is a set of straight lines along each of which u and w (hence also h) are constant. Even the notion of a centered simple wave can be realized in these cases. Suppose that the flow comes with constant supercritical velocity along a straight wall (cf. Fig. 10.12.3) until a smooth bend begins at point A. The straight characteristics are denoted by C_+ in the figure; they form a set of

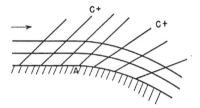

Fig. 10.12.3. Supercritical flow around a smooth bend

parallel lines in the region of constant flow, which then terminates along the C_+ characteristic through the point A, where the bend begins; beyond that characteristic a variable regime begins. The straight characteristics themselves are called Mach lines; they have physical significance and would be visible to the eye: the Mach lines are lines along which infinitesimal disturbances of a supercritical flow are propagated; in an actual flow they would be made visible because of the existence of small irregularities on the surface of the wall of the bend. If the bend contracts into a sharp corner, the straight characteristics, or Mach lines, which lie in the region in which the flow is variable, all emanate from the corner, as indicated in Fig. 10.12.4; the flow as a whole consists of two different uniform flows

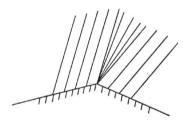

Fig. 10.12.4. Supercritical flow around a sharp corner

connected through a centered simple wave. If the bend in the stream is concave toward the flow, rather than convex as in the preceding two cases, the circumstances are quite different, since the Mach lines would now converge, rather than diverge, in certain portions of the flow, as indicated in Fig. 10.12.5. Overlapping of the characteristics would mean mathematically that the depth and velocity would be multi-valued at some points in the flow; this being physically impossible it is to be expected that something new happens and, in fact, the development of a hydraulic jump is to be expected. If the bend

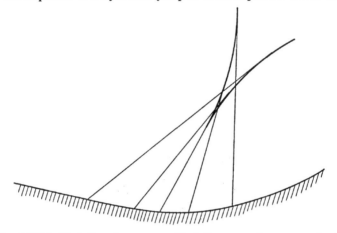

Fig. 10.12.5. Mach lines for a supercritical flow around a concave bend

is a sharp angle, as in Fig. 10.12.6, the configuration consisting of two uniform flows parallel to the walls of the bend and connected by an oblique hydraulic jump is mathematically possible, and it occurs in practice.

Having considered flows delimited on one side only by a wall, it is natural to consider next flows between two walls as in a sluice or channel of variable breadth. (Such flows are analogous to two-dimen-

Fig. 10.12.6. Oblique hydraulic jump

sional steady flows through nozzles in gas dynamics.) The possibilities here are very numerous, and most of them lead to cases not describable solely in terms of simple waves. They are of considerable importance in practice. For example, v. Kármán [K.2] was led to the study of particular flows of this type because of their occurrence in bends in the concrete spillways designed to carry the flows of the Los Angeles

river basin through the city of Los Angeles; the seasonal rainfall is so
heavy and the terrain so steep that supercritical flows are the rule
rather than the exception during the rainy season. Experiments for
sluices of special form were carried out by Preiswerk; Fig. 10.12.7,

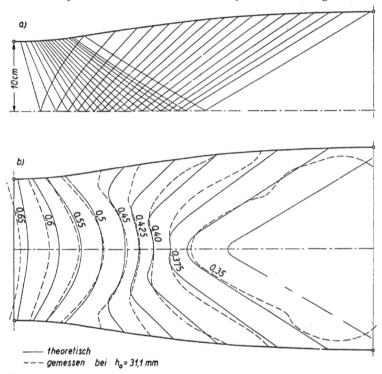

Fig. 10.12.7. Laval nozzle a) Mach lines b) contour lines of the water surface

for example, shows the result of an experiment in a particular case.
The upper figure shows the Mach lines, the lower figure shows the
contour lines of the water surface as given by the theory as well as by
experiment; as one sees, the agreement is quite good.

Finally, we discuss briefly some applications of interest because of
their connection with aerodynamics. Because of the analogy of the
shallow water theory with compressible gas dynamics, it is of course
possible to interpret experiments with flows in shallow water in terms
of the analogous flows in gases. Since it is much cheaper and simpler

to obtain supercritical flows experimentally in water than it is to obtain supersonic flows in gases, it follows that "water table" experiments (as they are sometimes called) may have considerable importance for those whose principle interest is in aerodynamics. There is a considerable literature devoted to this subject; we mention, for example, papers by Crossley [C.12], Einstein and Baird [E.5], Harleman

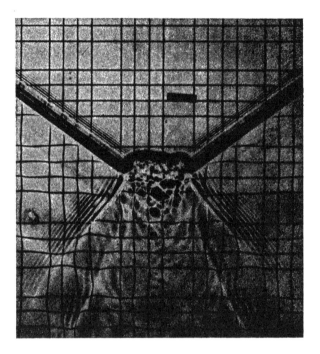

Fig. 10.12.8. Photogram of hydraulic-jump intersection

[H.3], Laitone [L.1], Bruman [B.19]. Figure 10.12.8 is a photograph, taken from the paper by Crossley, showing the interaction of two hydraulic jumps; this is a case essentially the same as that shown by Fig. 10.12.2. The ripples with short wave lengths constitute an effect due to surface tension, and the discontinuities are smoothed out so that a hydraulic jump does not really occur; the changes in depth are quite abrupt, however. Another important case that has been studied by means of the hydraulic analogy is, as a matter of course, the flow pattern which results when a rigid body (simulating a projectile or an airfoil) is immersed in a stream. Figure 10.12.9 shows a photograph of

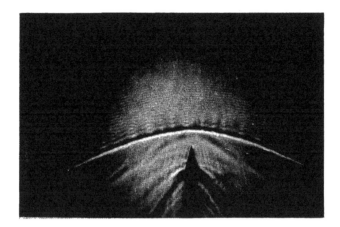

Fig. 10.12.9. Shock wave in front of a projectile

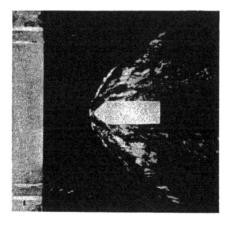

Fig. 10.12.10. Flow pattern of a projectile

such a flow (taken from the paper by Laitone). The shock wave in front of the projectile is well shown. Figure 10.12.10 is another photograph made by Preiswerk; here, Mach lines are clearly visible.

10.13. Linear shallow water theory. Tides. Seiches. Oscillations in harbors. Floating breakwaters

Up to now in this chapter we have considered problems of wave motion in water sufficiently shallow to permit of an approximation in terms of what we call the shallow water theory. This theory is non-linear in character, and consequently presents difficulties which are often quite formidable. By making the assumption that the wave amplitudes in the motions under study are small in addition to the assumption that the water is shallow, it is possible to obtain a theory which is linear—and thus attackable by many known methods— and which is also applicable with good approximation in a variety of interesting physical situations. We begin by deriving the linear shallow water theory under conditions sufficiently general to permit us to discuss the cases indicated in the heading of this section. (A brief mention of the linear shallow water theory was made in Chapter 2 and in Chapter 10.1.)

The linear shallow water theory could of course be derived by appropriate linearizations of the nonlinear shallow water theory. It is, however, more convenient—and perhaps also interesting from the standpoint of method—to proceed by linearizing first the basic general theory as developed in Chapter 1, and afterwards making the approximations arising from the assumption that the water is shallow. In other words, we shall begin with the exact linear theory of Chapter 2.1, and proceed to derive the linear shallow water theory from it. One of the advantages of this procedure is that the error terms involved in the shallow water approximation can be exhibited explicitly.

We suppose the water to fill a region lying above a fixed surface

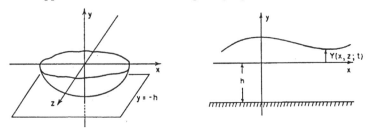

Fig. 10.13.1. Linear shallow water theory

(the bottom) $y = - h(x, z)$, and beneath a surface $y = Y(x, z; t)$, the motion of which is for the time being supposed known (cf. Fig.

10.13.1). The y-axis is taken vertically upward, and the x, z-plane is horizontal. The upper surface of the water given by $y = Y(x, z; t)$ will consist partly of the free surface (to be determined, for example, by the condition that the pressure vanishes there) and partly of the surfaces of immersed bodies; it is, however, not necessary to specify more about this surface for the present than that it should represent always a *small displacement from a rest position of equilibrium of the combined system* consisting of water and immersed bodies.

We recapitulate the equations of the exact linear theory as derived in Chapter 2.1. The velocity components are determined as the derivatives of the velocity potential $\Phi(x, y, z; t)$ which satisfies the Laplace equation

$$(10.13.1) \qquad \Phi_{xx} + \Phi_{yy} + \Phi_{zz} = 0$$

in the space filled by the water. It is legitimate to assume that all boundary conditions at the upper surface of the water are to be satisfied at the equilibrium position; this position is supposed given by

$$(10.13.2) \qquad y = \bar{\eta}(x, z).$$

(The bar over the quantity η points to the fact that $\bar{\eta}$ could also be interpreted as the *average* position of the water in the important special case in which the motion is a simple harmonic motion in the time.) The x, z-plane is taken in the undisturbed position of the free surface, and this in turn means that $\bar{\eta}$ in (10.13.2) has the value zero there. Under any immersed bodies the value of $\bar{\eta}$ will be fixed by the static equilibrium position of the given bodies. Thus $\bar{\eta}$ is in all cases a given function of x and z; for a floating rigid body, for example, it would be determined by hydrostatics.

The condition to be satisfied at the upper surface is the kinematic condition:

$$(10.13.3) \qquad \Phi_x Y_x + \Phi_z Y_z - \Phi_y + Y_t = 0,$$

which states that a particle once on the surface remains on it. At the bottom surface, the condition to be satisfied is

$$(10.13.3)_1 \qquad \Phi_x h_x + \Phi_z h_z + \Phi_y = 0 \qquad \text{at } y = -h(x, z).$$

Bernoulli's law for determining the pressure at any point in the water is

$$(10.13.4) \qquad \frac{p}{\varrho} + \Phi_t + gy - W = 0.$$

Here we have assumed that there may be other external forces beside gravity, and these forces are assumed to be determined by a work function $W(x, y, z; t)$ whose space derivatives furnish the force components; in this case it is known that the motion can be irrotational and that Bernoulli's law holds in the above form (cf. the derivations in Chapter 1). We now write the equation of the moving upper surface in the form

$$(10.13.5) \qquad y = Y(x, z; t) = \bar{\eta}(x, z) + \eta(x, z; t)$$

and assume in accordance with our statement above that $\eta(x, z; t)$ represents a small vertical displacement from the equilibrium position given by $y = \bar{\eta}$. Upon insertion in (10.13.3) and (10.13.4) we find after ignoring quadratic terms in η and Φ and their derivatives:

$$(10.13.6) \quad \Phi_x \bar{\eta}_x + \Phi_z \bar{\eta}_z - \Phi_y + \eta_t = 0 \quad\Big\}$$
$$(10.13.7) \quad \frac{p}{\varrho} + \Phi_t + g(\bar{\eta} + \eta) - W = 0 \quad\Big\} \quad \text{at } y = \bar{\eta}(x, z)$$

as boundary conditions to be satisfied at the equilibrium position of the upper surface of the water. At points corresponding to a free surface where $p = 0$ we would have, for example, $\bar{\eta} = 0$ and hence

$$(10.13.8) \qquad\qquad - \Phi_y + \eta_t = 0 \quad\Big\}$$
$$(10.13.9) \qquad\qquad \Phi_t + g\eta - W = 0 \quad\Big\} \quad \text{at } y = 0.$$

A special case might be that in which the motion of a portion of the upper surface is prescribed, i.e. $\eta(x, z; t)$ as well as $\bar{\eta}$ would be presumed known; in such a case the condition (10.13.6) alone would suffice as a boundary condition for the harmonic function Φ. In some of the problems to be treated here, however, we do not wish to assume that the motion of some immersed body, for example, is known in advance; rather, it is to be determined by the interaction with the water which exerts a pressure $p(x, z; t)$ on it in accordance with (10.13.7). Thus the exact formulation of our problems would require the determination of a harmonic function $\Phi(x, y, z; t)$ in the space between $y = - h(x, z)$ and $y = \bar{\eta}$ which satisfies the conditions (10.13.6) and (10.13.7) at the upper surface (in particular the conditions (10.13.8) and (10.13.9) on the free surface) and $(10.13.3)_1$ at the bottom. Additional conditions where immersed bodies occur (to be obtained from the appropriate dynamical conditions for such bodies) would be necessary to determine the pressure p, which provides the "coupling" between the water on the one hand and the immersed bodies on the other. Finally, appropriate initial conditions for the

water and the immersed bodies at the initial instant would be needed
if one were to study non-steady motions, or—as will be the case here
—conditions at ∞ of the radiation type would be needed if simple
harmonic motions (that is, steady vibrations instead of transients)
are studied. It need hardly be said that the difficulties of carrying
out the solutions of such problems are very great indeed (cf. Chapter
9, for example)—so much so that we turn to an approximate theory
which is based on the assumption that the depth of the water is
sufficiently small and that the immersed bodies are rather flat.*

In the derivation of the shallow water theory we start from the
Laplace equation (10.13.1) for Φ and integrate it with respect to y
from the bottom to the equilibrium position** of the top surface
$y = \bar{\eta}(x, z)$ to obtain, after integration by parts:

$$(10.13.10) \quad \int_{-h}^{\bar{\eta}} \Phi_{yy}\, dy = \overline{\Phi}_y - \underline{\Phi}_y = - \int_{-h}^{\bar{\eta}} (\Phi_{xx} + \Phi_{zz})\, dy$$

$$= - \frac{\partial}{\partial x} \int_{-h}^{\bar{\eta}} \Phi_x\, dy - \frac{\partial}{\partial z} \int_{-h}^{\bar{\eta}} \Phi_z\, dy + \bar{\eta}_x \overline{\Phi}_x + \bar{\eta}_z \overline{\Phi}_z$$

$$+ h_x \underline{\Phi}_x + h_z \underline{\Phi}_z.$$

Here, and in what follows, a bar over the quantity Φ means that it is
to be evaluated at the equilibrium position of the upper surface of the
water, i.e. for $y = \bar{\eta}(x, z)$, and a bar under the quantity means that it
is to be evaluated at the bottom $y = - h(x, z)$. From the kinematic
surface condition (10.13.6) and the condition (10.13.3)$_1$ at the bottom,
we have therefore (due regard being paid to the fact that a bar should
now be put over Φ in (10.13.6) and under Φ in (10.13.3)$_1$):

$$(10.13.11) \qquad \eta_t = - \frac{\partial}{\partial x} \int_{-h}^{\bar{\eta}} \Phi_x\, dy - \frac{\partial}{\partial z} \int_{-h}^{\bar{\eta}} \Phi_z\, dy.$$

This condition—really a continuity condition—expresses the fact
that the water is incompressible. Consider next the result of integrating by parts the right hand side of (10.13.11); in particular:

$$(10.13.12) \qquad \int_{-h}^{\bar{\eta}} \Phi_x\, dy = \bar{\eta}\overline{\Phi}_x + h\underline{\Phi}_x - \int_{-h}^{\bar{\eta}} y\Phi_{xy}\, dy.$$

Since we have

---

* In the course of the derivation the terms neglected are given explicitly so
that a precise statement about them can be made.

** One sees readily that carrying out the integration to $y = \bar{\eta}$ rather than
to $y = \bar{\eta} + \eta$ yields the same results within terms of second order in small
quantities.

(10.13.13) $\int_{-h}^{\bar{\eta}} h\Phi_{xy}\, dy = h\overline{\Phi}_x - h\underline{\Phi}_x$

we may eliminate $\underline{\Phi}_x$ from (10.13.12) to obtain:

(10.13.14) $\int_{-h}^{\bar{\eta}} \Phi_x\, dy = (\bar{\eta} + h)\overline{\Phi}_x - \int_{-h}^{\bar{\eta}} (h + y)\Phi_{xy}\, dy.$

Indeed, we have quite generally for any function $F(x, y, z; t)$ the formula:

(10.13.14)₁ $\int_{-h}^{\bar{\eta}} F\, dy = (\bar{\eta} + h)\overline{F} - \int_{-h}^{\bar{\eta}} (h + y)F_y\, dy.$

Making use of the analogous expression for the integral of Φ_z we obtain from (10.13.11) the relation

(10.13.15) $\eta_t = - [(\bar{\eta} + h)\overline{\Phi}_x]_x - [(\bar{\eta} + h)\overline{\Phi}_z]_z + I_x + J_z$

in which

(10.13.16) $I = \int_{-h}^{\bar{\eta}} (h + y)\Phi_{xy}\, dy, \quad J = \int_{-h}^{\eta} (h + y)\Phi_{zy}\, dy.$

In addition, we have from (10.13.10) in combination with (10.13.14) the condition:

(10.13.17) $\overline{\Phi}_y = - (\bar{\eta} + h)[\overline{\Phi}_{xx} + \overline{\Phi}_{zz}] - h_x\overline{\Phi}_x - h_z\overline{\Phi}_z + I_x + J_z,$

as one can readily verify.

Up to this point we have made no approximations other than those arising from linearizing. The essential step in obtaining our approximate theory is now taken in neglecting the terms I_x and J_z. This in turn is justified if it is assumed that certain second and third derivatives of Φ are bounded when h is small and that $\bar{\eta}$ and its first derivatives are small* of the same order as h: one sees that the terms I_x and J_z in the right hand sides of (10.13.15) and (10.13.17) are then of order h^2 while the remaining terms are of order h. Under the free surface in the case of a simple harmonic oscillation one can show that this approximation requires the depth to be small in comparison with the wave length.

Upon differentiating the relation (10.13.7) for the pressure at the upper surface of the water with respect to t (again noting that a bar should be placed over the term Φ_t in (10.13.7)) and using (10.13.15) we find the equation

---

* This means that the theory developed here applies to immersed bodies which are flat.

(10.13.18) $\quad \dfrac{1}{g}\overline{\varPhi}_{tt} + \dfrac{\bar{p}_t}{g\varrho} - \dfrac{1}{g}\overline{W}_t = [(\bar{\eta} + h)\overline{\varPhi}_x]_x + [(\bar{\eta} + h)\overline{\varPhi}_z]_z$

after dropping the terms I_x and J_z. This is the basic differential equation for the function $\overline{\varPhi}(x, z; t)$ which holds everywhere on the upper surface of the water. In particular, we have at the free surface where $p = 0$ and $\bar{\eta} = 0$ the equation

(10.13.19) $\quad (h\overline{\varPhi}_x)_x + (h\overline{\varPhi}_z)_z - \dfrac{1}{g}\overline{\varPhi}_{tt} = \dfrac{1}{g}\overline{W}_t.$

We recall that $W(x, y, z; t)$ represents the work function for any external forces in addition to gravity (tide generating forces, for example), so that $\overline{W}$, its value on the free surface, would be given by $W(x, 0, z; t)$. If, in addition, it is assumed that h is a constant, i.e. that the depth of the water is uniform, and that gravity is the only external force, we would have the equation

(10.13.19)$_1$ $\quad\quad \varPhi_{xx} + \varPhi_{zz} - \dfrac{1}{gh}\varPhi_{tt} = 0,$

that is, the linear wave equation in the two space variables x, z and the time t. As a consequence, all disturbances propagate in such a case with the constant speed $c = \sqrt{gh}$, as is well known for this equation.

If there is an immersed object in the water, the equation (10.13.19) holds everywhere in the x, z-plane exterior to the curve C which defines the water line on the immersed body in its equilibrium position. The curve C is supposed given by the equations

(10.13.20) $\quad\quad x = x(s), \quad\quad z = z(s)$

in terms of a parameter s. We must have boundary, or perhaps it is better to say, transition conditions at the curve C which connect the solutions of (10.13.19) in the exterior of C in an appropriate manner with the motion of the water under the immersed body. Reasonable conditions for this purpose can be obtained from the laws of conservation of mass and energy. In deriving these conditions we assume $W = 0$, since we wish to deal only with gravity as the external force when considering problems involving immersed bodies. Consider an element of length ds of the curve C representing the water line of the immersed body (cf. Fig. 10.13.2). The expression

$$\left\{ \varrho \int_{-h}^{\bar{\eta}} \Phi_n \, dy \right\} ds$$

represents the mass flux through a vertical strip having the normal **n**

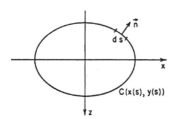

Fig. 10.13.2. Boundary at water line of an immersed body

and extending from the bottom to the top of the water. From (10.13. 14)$_1$ applied for $F = \Phi_n$ we have

$$(10.13.21) \quad \varrho \int_{-h}^{\bar{\eta}} \Phi_n \, dy = \varrho(\bar{\eta} + h)\bar{\Phi}_n - \varrho \int_{-h}^{\bar{\eta}} (h + y)\Phi_{ny} \, dy$$
$$\simeq \varrho(\bar{\eta} + h)\bar{\Phi}_n$$

where the second term is ignored because it is of order h^2. Thus it would be reasonable to require that $(\bar{\eta} + h)\bar{\Phi}_n$ should be continuous on C since this is the same as requiring that the mass of the water is conserved within terms of the order retained otherwise in our theory. For the flux of energy across a vertical strip with the normal **n** we have

$$(10.13.22) \quad \left\{ \int_{-h}^{\bar{\eta}} p\Phi_n \, dy \right\} ds = \left\{ - \varrho \int_{-h}^{\bar{\eta}} \Phi_t \Phi_n \, dy - g\varrho \int_{-h}^{\bar{\eta}} y\Phi_n \, dy \right\} ds$$

upon making use of the Bernoulli law (10.13.4) for the pressure p (when $W = 0$). Once more we may ignore the second term in the brackets since it is of order h^2. Upon applying (10.13.14)$_1$ with $F = \Phi_t \Phi_n$ and again ignoring a term of order h^2 we find

$$(10.13.23) \quad \left\{ \int_{-h}^{\bar{\eta}} p\Phi_n \, dy \right\} ds = \{ - \varrho(\bar{\eta} + h)\bar{\Phi}_n \bar{\Phi}_t \} \, ds.$$

Since we have already required that $(\bar{\eta} + h)\bar{\Phi}_n$ should be continuous, we see that the additional requirement, $\bar{\Phi}_t$ continuous, ensures the continuity of the energy flux.

As reasonable transition conditions on the curve C delimiting the immersed body at its water line we have therefore

$$(10.13.24) \quad (\bar{\eta} + h)\bar{\Phi}_n, \quad \bar{\Phi}_t \text{ continuous on } C.$$

Of course, if $\bar{\eta}$ is continuous (e.g. if the sides of the body in contact with the water do not extend vertically below the undisturbed free surface) it follows that $\bar{\Phi}_n$ would then be continuous.

In order to make further progress it would be necessary to specify the properties of the immersed body. However, we have succeeded in obtaining the equation (10.13.18) which is generally valid and of basic importance for our theory together with the transition conditions (10.13.24) valid at the edge of immersed bodies; in particular, we have the definitive equation for the free surface itself in the form of the linear wave equation (10.13.19). The idea behind this method of approximation is to get rid of the depth variable by an integration over the depth so that the problems then are considered only in the x, z-plane. As a result, the problems are no longer problems in potential theory in three space variables, but rather problems involving the wave equation with only two space variables, and hence they are more open to attack by known methods. How this comes about will be seen in special cases in the following.

As a first example of the application of the above theory we consider briefly the problem of the tides in the oceans, with a view to indicating where this theory fits into the theory of gravity waves in general, but not with the purpose of giving a detailed exposition. (For details, the long Chapter VIII in Lamb [L.3] should be consulted.) To begin with, it might seem incredible at first sight that the shallow water theory could possibly be accurate for the oceans, since depths of five miles or more occur. However, it is the depth in relation to the wave length of the motions under consideration which is relevant. The tides are forced oscillations caused by the tide-generating attractions of the moon and the sun, and hence have the same periods as the motions of the sun and moon relative to the earth. These periods are measured in hours, and consequently the tidal motions in the water result in waves having wave lengths of hundreds of miles;* the depth-wave length ratio is thus quite small and the shallow water theory should be amply accurate to describe the tides. This means, in effect, that the differential equation (10.13.19), or rather, its analogue for the case of water lying on a rotating spheroid (with Coriolis terms put in if a coordinate system rotating with the earth is

* For example, in water of depth 10,000 feet (perhaps a fairly reasonable average value for the depth of the oceans) a steady progressing wave having a length of 10,000 feet has a period of only 44.2 sec. (cf. Lamb [L.3], p. 369). Since the wave length varies as the square of the period, the correctness of our statement is obvious.

used), would serve as a basis for calculating tidal motions. Of course, the function $\overline{W}_t$ would be defined in terms of the gravitational forces due to the attraction of the sun and moon. The variable depth of the water in the oceans would come into play, as well as boundary conditions at the shore lines of the continents. Presumably, $\overline{W}_t$ would be analyzed into its harmonic components (which could be obtained from astronomical data), the response to each such harmonic would be calculated, and the results superimposed. Such a problem constitutes a linear vibration problem of classical type—it is essentially the same as the problem of transverse forced oscillations of a tightly stretched non-uniform membrane with an irregular boundary. If it were not for one essential difficulty, to be mentioned in a moment, such a problem would in all likelihood be solvable numerically by using modern high speed computational equipment. The difficulty mentioned was pointed out to the author in a conversation with H. Jeffreys, and it is that there are difficulties in prescribing an appropriate boundary condition in coastal regions where there is dissipation of energy in the tidal motions (in the bay of Fundy, for example, to take what is probably an extreme case). At other coastal regions the correct boundary condition would of course often be simply that the component of the velocity normal to the coast line vanishes. Of course, there would also be a difficulty in using a differential equation like (10.13.19) near any shores where $h = 0$, since the differential equation becomes singular at such points. Nevertheless, a computation of the tides on a dynamical basis would seem to be a worthwhile problem—perhaps it could be managed in such a way as to help, in conjunction with observations of the actual tides, in providing information concerning the dissipation of energy in such motions.

These remarks might be taken to imply that the dynamical theory is not at present used to compute the tides. This is not entirely correct, since the tide tables for predicting the tides in various parts of the world are based on fundamental consequences of the assumption that the tides are indeed governed by a differential equation of the same general type as (10.13.19). The point is that the oceans are regarded as a linear vibrating system under forced oscillations due to excitation from the periodic forces of attraction of the sun and moon. It is assumed that all free vibrations of the oceans were long ago damped out, and hence, as remarked above, that the tidal motions now existing in the oceans are a superposition of simple harmonic oscillations having periods which are very accurately known from astronomical observations. To obtain tide tables for any given point a superposition of oscillations of these frequencies is taken with undetermined amplitudes and phases which are then fixed by comparing them with a harmonic analysis of actual tidal observations made at the point in question. The tide predictions are then made by using the result of such a calculation to prepare tables for future times. The dynamical theory is thus used only in a qualitative way. An interesting additional point might be mentioned, i.e. that tides of observable amplitudes are sometimes measured which have as frequency the sums or differences (or also other linear combinations with integers) of certain of the astronomical frequencies, which means from the point of view of vibration theory that observable nonlinear effects must be present.

Another type of phenomenon in nature which can be treated by the theory derived here concerns periodic motions of rather long period, called seiches, which occur in lakes in various parts of the world. The first observations of this kind seem to have been made by Forel [F.7] in the lake at Geneva in Switzerland, in which oscillations having a period of the order of an hour and amplitudes of up to six feet have been observed. In larger lakes still larger periods of oscillation are observed—about fifteen hours in Lake Erie, for example. A rather destructive oscillation, generally supposed to be of the type of a seiche, occurred in Lake Michigan in June 1954; a wave with an amplitude of the order of ten feet occurred and swept away a number of people who were fishing from piers and breakwaters. What the mechanism is that gives rise to seiches in lakes has been the object of considerable discussion, but it seems rather clear that the motions represent free vibrations of the water in a lake which are excited by external forces of an impulsive character, the most likely type arising from sudden differences in atmospheric pressure over various portions of the water surface. Bouasse [B.15, p. 158] reports, however, that the Lisbon earthquake of 1755 caused oscillations in Loch Lomond with a period of about 5 minutes and amplitudes of several feet. In any case, the periods observed seem to correspond to those calcu- lated on the basis of the linear shallow water theory, which should be quite accurate for the study of seiches because of their long periods and small amplitudes. It follows, therefore, that the differential equation (10.13.18) is applicable; we suppose that $\overline{W}_t = 0$ (since tidal forces play no role in this case), and also set $\bar{\eta} = 0$ since there are no immersed bodies to be considered. The differential equation for $\Phi(x, z; t)$ is thus

$$(10.13.25) \qquad (h\overline{\Phi}_x)_x + (h\overline{\Phi}_z)_z - \frac{1}{g}\overline{\Phi}_{tt} = \frac{\bar{p}_t}{g\varrho}.$$

The free natural vibrations of the lake are investigated by setting $\bar{p}_t = 0$ and $\overline{\Phi}(x, z; t) = \varphi(x, z)e^{i\sigma t}$ in (10.13.25) with the result

$$(10.13.26) \qquad (h\varphi_x)_x + (h\varphi_z)_z + \frac{\sigma^2}{g}\varphi = 0.$$

As boundary condition along the shore of the lake we would have

$$(10.13.27) \qquad\qquad \varphi_n = 0.$$

The problem thus posed is one of the classical eigenvalue problems of mathematical physics. Solutions φ other than the trivial solution $\varphi \equiv 0$ of (10.13.26) under the homogeneous boundary condition (10.13.27) are wanted; such solutions exist only for special values of the circular frequency σ, and these values yield the natural frequencies corresponding to the natural modes $\varphi(x, z)$ which are correlated with them. In general, an infinite set of such natural frequencies occurs. For particular shapes and depths h—rectangular or circular lakes of constant depth, for example—it is possible to solve such problems more or less explicitly. In practice however, lakes have such irregular outlines and depths that the determination of the natural frequencies and modes requires numerical computation. A reasonable and generally applicable method of carrying out such computations is furnished here, as in other instances in this and the subsequent chapter, by the method of finite differences.* In this method, the derivatives in the differential equation and boundary conditions are replaced by difference quotients defined by means of the values of the function at the discrete points of a net in the domain of the independent variables. The resulting finite equations are then solved to yield approximate values for the unknown function at the net points. The difference approximation will be more accurate for a closer spacing of the net points. We proceed to illustrate the method for the case of a lake of constant depth in the form of a square of length l on each side, with a view to comparing the result with the exact solution which is easy to write down in this case. The differential equation (10.13.26) can be written in the form

---

* A different method was used by Chrystal [C.2] to calculate the periods of the free oscillations of Loch Earn; he found good agreement with the observations for the first six modes of oscillation.

$(10.13.26)_1 \qquad \varphi_{xx} + \varphi_{zz} + m^2\varphi = 0, \qquad m^2 = \sigma^2/gh$

in this case. A division of the square in a mesh with mesh width $\delta = l/7$ is taken, as indicated in Fig. 10.13.3. In numbering the net points it has been assumed that only modes of oscillation that are symmetrical with respect to the center lines parallel to the sides and to to the diagonals are sought, which, however, is not the case for the

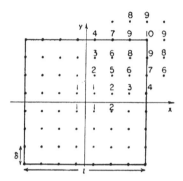

Fig. 10.13.3. Finite differences for a seiche

mode having the lowest frequency. The boundary condition $\varphi_n = 0$ is satisfied approximately by supposing that the solution is reflected over the boundaries in such a way as to yield values which are equal at the mirror images in the boundaries, as is also indicated in Fig. 10.13.3. The formulas used for approximating the derivatives are defined as follows (cf. Fig. 10.13.3):

$$\frac{\partial \varphi}{\partial x} \simeq \frac{\varphi_{m+1,\,n} - \varphi_{m-1,\,n}}{2\delta} \qquad \qquad \frac{\partial \varphi}{\partial z} \simeq \frac{\varphi_{m,\,n+1} - \varphi_{m,\,n-1}}{2\delta}$$

$$\frac{\partial^2 \varphi}{\partial x^2} \simeq \frac{-2\varphi_{m,\,n} + \varphi_{m+1,\,n} + \varphi_{m-1,\,n}}{\delta^2}, \qquad \frac{\partial^2 \varphi}{\partial z^2} \simeq \frac{-2\varphi_{m,\,n} + \varphi_{m,\,n+1} + \varphi_{m,\,n-1}}{\delta^2}.$$

Consequently, the differential equation $(10.13.26)_1$ is replaced at each net point (m, n) by the difference equation

$(10.13.28) \qquad -4\varphi_{m,n} + \varphi_{m,n+1} + \varphi_{m+1,n} + \varphi_{m,n-1} + \varphi_{m-1,n} + \delta^2 m^2 \varphi_{m,n} = 0.$

Such an equation is written for each of the net points in Fig. 10.13.3. The results for points 1,6,9, for example, are:

$$(10.13.29) \quad \begin{cases} 1: \ -2\varphi_1 + 2\varphi_2 + (\delta m)^2\varphi_1 = 0 \\ 6: \ -4\varphi_6 + \varphi_3 + \varphi_7 + \varphi_8 + \varphi_5 + (\delta m)^2\varphi_6 = 0 \\ 9: \ -4\varphi_9 + 2\varphi_8 + \varphi_7 + \varphi_{10} + (\delta m)^2\varphi_9 = 0. \end{cases}$$

These homogeneous linear equations of course have always the solution $\varphi_i = 0$, $i = 1, 2, \ldots, 10$ unless their determinant vanishes, and this condition is a tenth degree equation in the quantity $(\delta m)^2$, the smallest root of which furnishes an approximation to the lowest frequency. The exact solution of the differential equation $(10.13.26)_1$ which satisfies the boundary condition is, in the present case, $\varphi = A \cos (k\pi x/l) \cos (j\pi z/l)$, with k and j any integers, provided that $m^2 = \pi^2(k^2 + j^2)/l^2$. A numerical comparison of the lowest value of m for the mode having double symmetry—i.e. the value for $k = 1$, $j = 1$—with the value computed from the determinantal equation shows the approximate value of m to be too low by 6.5 %.

However, this mode corresponds to one of the higher eigenvalues, so that the accuracy of the finite difference method is rather good. The error for the lowest mode is very much smaller, but because of the lack of symmetries the amount of calculation needed to determine the corresponding frequency would be much greater for the present case. If one were to treat a long narrow lake, the calculation would be simpler. It could also be advantageous to employ the Rayleigh-Ritz method. In principle, similar calculations could be made in more complicated cases (for many examples of problems solved along these lines see the book by Southwell [S.14]).

Wave motions in harbors are often of a type suitable for discussion in terms of the linear shallow water theory: they are indeed often of the type called seiches above. In these cases oscillations of the water in the harbor are also commonly excited by the motion at the harbor mouth, which in its turn is due, of course, to wave motions generated in the open sea. An experimental and theoretical investigation of such waves in a model has been carried out by McNown [M.7]. The model was in the form of a circle 3.2 meters in diameter with vertical walls. The depth of the water in this idealized harbor was 16 cm. An opening of angle $\pi/8$ radians in the harbor wall permitted a connection with a large tank in which waves (simulating the open sea) were produced. Figures 10.13.4 and 10.13.5 are photographs of the model (taken from the paper by McNown), which also show two specific cases of symmetrical oscillations. The free vibrations again are governed by equation (10.13.26). (It might be noted that McNown makes use of the exact linear theory rather than the shallow water theory. The only difference is that the relation between σ^2 and m is $\sigma^2 = gm \tanh mh$, instead of $\sigma^2 = ghm^2$, as given above: the differential equation for the velocity potential $\Phi(x, y, z; t)$ in the exact linear theory treated in Part I is

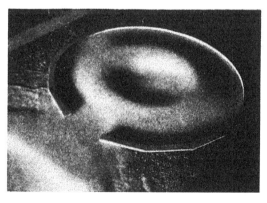

Fig. 10.13.4. and 10.13.5. Waves in a harbor model

written in the form $\Phi = A \cosh m(y + h)e^{i\sigma t}\varphi(x, z)$, and $\varphi(x, z)$ then satisfies $\nabla^2\varphi + m^2\varphi = 0$.) Solutions of the differential equation are sought in the form

$$(10.13.30) \qquad \varphi(r, \theta) = J_n(mr) \cos n\theta$$

in polar coordinates (r, θ), under the assumption that the port is closed, i.e. that its boundary is the whole circle $r = R$. As is well known, $\varphi(r, \theta)$ is a solution of $(10.13.26)_1$ only if $J_n(mr)$ is a Bessel function of order n, and since it is reasonable to look only for solutions that are bounded we choose the Bessel functions of the first kind which are regular at the origin. The boundary condition requires that

$$(10.13.31) \qquad \frac{\partial \varphi}{\partial r} = 0 \qquad \text{at } r = R$$

and this in turn leads to the condition $dJ_n/dr = 0$ for $r = R$. For each n this transcendental equation has infinitely many roots $m_k^{(n)}$, each corresponding to a mode of oscillation with various nodal diameters and circles, and with a definite frequency which is fixed by $\sigma^2 = g m_k^{(n)}$ (or, more accurately, by $\sigma^2 = g m_k^{(n)} \tanh h m_k^{(n)}$). Figure 10.13.6, obtained by McNown, shows a comparison of observed and calculated amplitudes for two modes of oscillation; the upper curve is drawn for a motion having no diametral nodes and two nodal circles, while the lower is for a motion having two nodal diameters and one

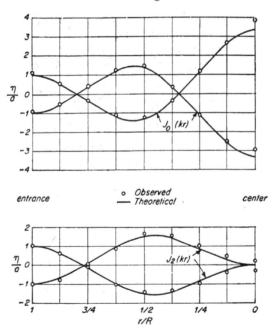

Fig. 10.13.6. Comparison of results of experiment and theory for resonant movements in a circular port

nodal circle. The motions were excited by making waves in the tank, and providing an opening for communication with the harbor, as noted above. The figures were drawn assuming that the amplitudes would agree at the entrance to the harbor—the experimental check

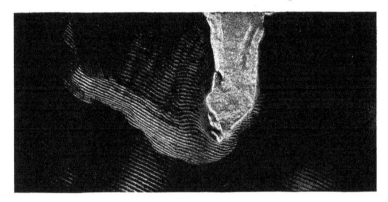

Fig. 10.13.7. Model of a harbor without breakwater

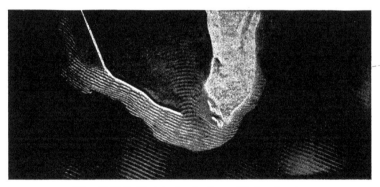

Fig. 10.13.8. Model of a harbor with breakwater

thus applies only to the shapes of the curves. As one sees, the experimental and theoretical values are remarkably close. The amplitudes used were large enough so that nonlinear effects were observed: the troughs are flatter than the crests by measurable amounts. Of course, having an opening in the harbor wall violates the boundary condition assumed, but this effect apparently is slight: changing the angle of the opening at the harbor mouth had practically no effect on the waves produced, and, in addition, it was found that very little wave energy radiates outward through the harbor entrance.

Problems of harbor design, involving construction of breakwaters, location of docks, etc. are commonly studied by constructing models. Figs. 10.13.7 and 10.13.8 show two photographs of a model of a harbor,* the first before a breakwater was constructed, the second afterward. As one sees, the breakwater has a quite noticeable effect. Fig. 10.13.9. shows the same model, with the waves approaching the harbor mouth at a different angle, however; as one sees the break-

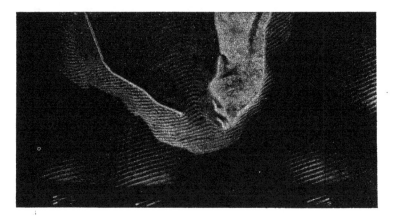

Fig. 10.13.9. Model of a harbor with breakwater

water seems to be on the whole less effective when the wave fronts are less oblique to the breakwater. The diamond-shaped pattern, due to reflection, of the waves on the sea side of the breakwater is inter-

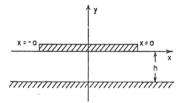

Fig. 10.13.10. Floating plane slab

* These photographs were given to the author by the Hydrodynamics Laboratory at California Institute of Technology.

esting. Model studies are rather expensive, and consequently it might well be reasonable to explore the possibilities of numerical solution of the problems, perhaps by using appropriate modifications of the method of finite differences outlined above for a simple case.

We turn next to a discussion of the effect of floating bodies on waves in shallow water, on the basis of the theory presented in this section. Only two-dimensional motions will be considered (so that all quantities are independent of the variable z). The first case to be studied is that of the motion of a floating rigid body in the form of a thin plane slab (cf. Fig. 10.13.10) in water of uniform depth. Such problems have been treated by F. John [J.5]. The ends of the slab are at $x = \pm a$. In accordance with the theory presented above we must determine the surface value $\overline{\Phi}(x, t)$ and the displacement $\eta(x, t)$ of the board from the differential equations (cf. (10.13.19), (10.13.15) with $\bar{\eta} = 0$, and dropping I_x and J_z):

(10.13.32) $$\Phi_{xx} = \frac{1}{gh} \Phi_{tt}, \qquad |x| > a$$

(10.13.33) $$\eta_t = -h\Phi_{xx}, \qquad |x| < a.$$

We have dropped the bar over the quantity Φ. We have also assumed that $\bar{\eta}(x)$ for $|x| < a$, the rest position of equilibrium of the board, is zero; this is an approximation that is justified because we assume that the board is so light that it does not sink appreciably below the water surface when in equilibrium. (This assumption is by no means necessary—it would not be difficult to deal with the problem if this simplifying assumption were not made.)

Since $\bar{\eta}$ is zero, it follows (cf. (10.13.24)) that the transition conditions at the ends of the board are

(10.13.34) $$\Phi_x, \Phi_t \text{ continuous at } x = \pm a.$$

We are interested in the problem of the effectiveness of the floating board as a barrier to a train of waves coming from the right ($x = +\infty$). The equation (10.13.32) has as its general solution

$$\Phi(x, t) = F(x - ct) + G(x + ct), \qquad c = \sqrt{gh}$$

in terms of two arbitrary functions F and G (as one can readily verify) which clearly represent a superposition of two progressing waves moving to the right and to the left, respectively, with the speed $\sqrt{gh}$. It is natural, in our present problem, to expect that for $x > a$ there would exist in general both an incoming and an outgoing wave because of reflection from the barrier, while for $x < -a$ we would prescribe only a wave going outward (i.e. to the left). We shall see that these qualitative requirements lead to a unique solution of our problem.

We consider only simple harmonic waves; it is thus natural to write

(10.13.35) $\Phi(x, t) = \varphi(x)e^{i\sigma t},$ $|x| > a,$

(10.13.36) $\eta(x, t) = v(x)e^{i\sigma t},$ $-a < x < a,$

with the stipulation that the real part is to be taken at the end. (It is necessary also to permit $\varphi(x)$ and $v(x)$ to be complex-valued functions of the real variable x.) The conditions (10.13.32) and (10.13.33) now become

(10.13.37) $$\frac{d^2\varphi}{dx^2} + \frac{\sigma^2}{gh}\varphi = 0,$$ $|x| > a$

(10.13.38) $$\frac{d^2\varphi}{dx^2} + \frac{i\sigma}{h}v = 0,$$ $|x| < a.$

The equation (10.13.37) has as general solution

(10.13.39) $$\varphi(x) = Ae^{-ikx} + Be^{ikx}$$

with k given by

(10.13.40) $$k = \sigma/\sqrt{gh}.$$

For $\Phi(x, t)$ we have therefore

(10.13.41) $$\Phi(x, t) = Ae^{-i(kx - \sigma t)} + Be^{i(kx + \sigma t)},$$

the first term representing a progressing wave moving to the right, the second a wave moving to the left. In our problems we prescribe the incoming wave from the right, and hence for $\varphi(x)$ we write

(10.13.42) $$\varphi(x) = Be^{ikx} + Re^{-ikx},$$ $x > a,$

in which B is prescribed, while R—the amplitude of the reflected wave (more precisely, $|R|$ is its amplitude)—is to be determined. At the left we write

(10.13.43) $$\varphi(x) = Te^{ikx},$$

with T—the amplitude of the transmitted wave—to be determined.

To complete the formulation of the problem it is necessary to consider the dynamics of our floating rigid body. We shall treat two cases: a) the board is held rigidly fixed in a horizontal position, b) the board floats freely in the water.

a) *Rigidly Fixed Board.*

If the board is rigidly fixed we have $\eta(x, t) \equiv 0$, and hence (cf. (10.13.36)) $v(x) \equiv 0$. It follows from (10.13.38) that φ_{xx} vanishes identically under the board and hence that $\varphi(x)$ is a linear function:

(10.13.44) $\varphi(x) = \gamma x + \delta,$ $-a < x < +a.$

Since $\Phi_x(x, t)$ furnishes the horizontal velocity component of the water, it follows from (10.13.44) and (10.13.35) that the velocity under the board is given by $\gamma e^{i\sigma t}$, i.e. it is constant everywhere under the board at each instant—a not unexpected result.

We now write down the transition conditions at $x = \pm a$ from (10.13.34), making use of (10.13.35) and of (10.13.42) at $x = +a$ and (10.13.43) at $x = -a$; the result is:

(10.13.45)
$$\begin{cases} Be^{i\kappa a} + Re^{-i\kappa a} = \gamma a + \delta \\ Be^{ika} - Re^{-ika} = \gamma/ik \\ Te^{-ika} = -\gamma a + \delta \\ Te^{-ika} = \gamma/ik. \end{cases}$$

Once the real number B—which fixes the amplitude of the incoming wave—has been prescribed, these four equations serve to fix the constants R, T, γ, and δ and hence the functions $\Phi(x, t)$ and $\eta(x, t)$. The pressure under the board can then be determined (cf. the expression (10.13.4) for Bernoulli's law) from

(10.13.46) $p(x, t) = -\varrho\Phi_t = -\varrho i\sigma\varphi(x)e^{i\sigma t}.$

(Observe that the quantity y in (10.13.4) is zero in the present case.)

In terms of the dimensionless parameter

(10.13.47) $\theta = 2a/\lambda,$

the ratio of the length of the board to the wave length λ on the free surface, given by (cf. (10.13.41))

(10.13.48) $\lambda = 2\pi/k,$

the solution of (10.13.45) is

(10.13.49)
$$\begin{cases} R = \dfrac{\theta\pi i\, Be^{2\theta\pi i}}{\theta\pi i + 1} \\[2mm] T = \dfrac{Be^{2\theta\pi i}}{\theta\pi i + 1} \\[2mm] \gamma = \dfrac{\theta\pi i}{a} \cdot \dfrac{Be^{\theta\pi i}}{\theta\pi i + 1} \\[2mm] \delta = Be^{\theta\pi i}. \end{cases}$$

The reflection and transmission coefficients are obtained at once:

$$\begin{cases} C_r = \left| \dfrac{R}{B} \right| = \dfrac{\theta\pi}{\sqrt{1 + \theta^2\pi^2}}, \\[2ex] C_t = \left| \dfrac{T}{B} \right| = \dfrac{1}{\sqrt{1 + \theta^2\pi^2}}. \end{cases}$$

(10.13.50)

They depend only upon the ratio $\theta = 2a/\lambda$, as one would expect. They also satisfy the relation $C_r^2 + C_t^2 = 1$, as they should: this is an expression of the fact that the incoming and outgoing energies are the same. The following table gives a few specific values for these coefficients:

θ	C_t	C_r
0.5	0.54	0.85
1.0	0.30	0.95
2.0	0.157	0.986

Thus a fixed board whose length is half the incoming wave length has the effect of reducing the amplitude behind it by about 50 percent and of reflecting about 72 percent of the incoming energy. One should, however, remember that the theory is only for long waves in shallow water, and, in addition, it seems likely that the length of the board will also play a role in determining the accuracy of the approximation. This question has been investigated by Wells [W.10] by deriving the shallow water theory in such a way as to include all terms of third order in the depth h and studying the magnitude of the neglected terms in special cases; in particular, the present case of a floating rigid body is investigated. Wells finds that if h/λ is small and if a/λ (the ratio of the half-length of the board to the wave length) is not smaller than 1, the neglected higher order terms are indeed negligible, but if a/λ is less than 1, the higher order terms need not be small. In other words, floating obstacles ought to have lengths of the order of the wave length of the incoming waves if the shallow water theory to lowest order in h is expected to furnish a good approximation.

It is of interest to study the pressure variation under the board. This is given in the present case (cf. (10.13.46)) by

(10.13.51) $\qquad p(x, t) = - i\sigma\varrho(\gamma x + \delta)e^{i\sigma t}$,

the real part only to be taken. Thus the pressure varies linearly in x, but it is a different linear function at different times since γ and δ are complex constants. The result of taking the real part of the right hand side of (10.13.51) can be readily put in the form

$$(10.13.52) \qquad p(x, t) = p_1(x) \cos \sigma t - p_2(x) \sin \sigma t$$

with

$$(10.13.53) \qquad \begin{cases} p_1(x) = \sigma \varrho B(b_1(x) \sin r + b_2(x) \cos r) \\ p_2(x) = \sigma \varrho B(b_2(x) \sin r - b_1(x) \cos r) \end{cases} \qquad r = \theta \pi$$

and

$$(10.13.54) \qquad \begin{cases} b_1(x) = \dfrac{r^2}{1 + r^2} \dfrac{x}{a} + 1 \\ b_2(x) = \dfrac{r}{1 + r^2} \dfrac{x}{a} . \end{cases}$$

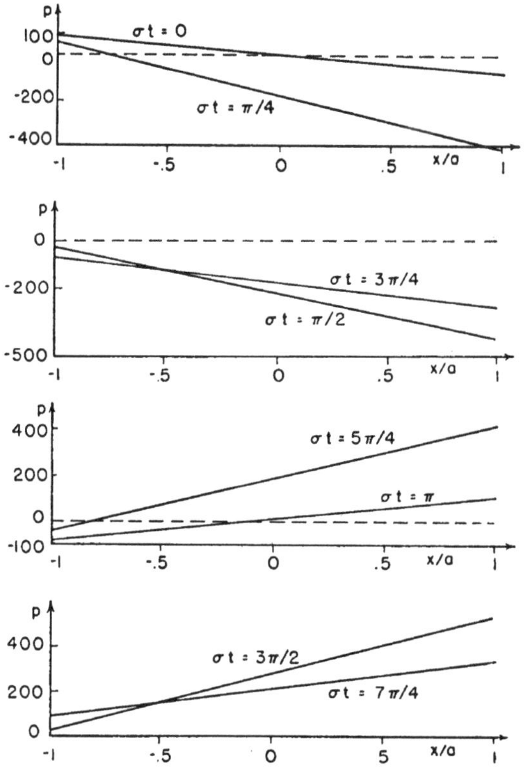

Fig. 10.13.11. Pressure variations for a stationary board. $\theta = 1$, p in *pounds*/$(ft)^2$

We have assumed in making these calculations, as stated above, that B, which represents the amplitude of the incoming wave, is a real number.

In Fig. 10.13.11 the results of computations for the pressure distribution for time intervals of 1/4 cycle over the full period are given for a special numerical case in which the parameter θ has the value $\theta = 1$, i.e. the length of the board is the same as the wave length. One observes that the pressure variation is greater at the right end than at the left, which is not surprising since the board has a damping effect on the waves. One observes also that the pressure is sometimes less than atmospheric (i.e. it is negative at times, while $p = 0$ is the assumed pressure at the free surface).

b) Freely Floating Board.

In Fig. 10.13.12 the notation for the present case is indicated: $\bar{u}(t)$, $\bar{v}(t)$ represent the coordinates of the center of gravity of the board in the displaced position, and $\omega(t)$ the angular displacement. As before, we consider only simple harmonic oscillations and thus take $\bar{u}$, $\bar{v}$, and ω in the form

$$(10.13.55) \qquad \bar{u} = \bar{x}e^{i\sigma t}, \qquad \bar{v} = \bar{y}e^{i\sigma t}, \qquad \omega = \bar{w}e^{i\sigma t},$$

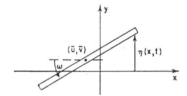

Fig. 10.13.12. A freely floating board

in which $\bar{x}$, $\bar{y}$, and $\bar{w}$ are constants representing the complex amplitudes of these components of the oscillation. For $\eta(x, t)$ we have, therefore

$$(10.13.56) \qquad \begin{aligned} \eta(x, t) &= [\bar{y} + (x - \bar{x})\bar{w}]e^{i\sigma t} \\ &= (\bar{y} + x\bar{w})e^{i\sigma t} \end{aligned}$$

when terms of first order in $\bar{x}$, $\bar{y}$, and $\bar{w}$ only are considered. (The horizontal component of the oscillation is thus seen to yield only a second order effect.) The relation (10.13.56) now yields (cf. (10.13.38)):

$$(10.13.57) \qquad \varphi_{xx} = -\frac{i\sigma}{h}(\bar{y} + x\bar{w})$$

in which φ is the complex amplitude of the velocity potential $\Phi(x, t) = \varphi(x)e^{i\sigma t}$. Hence φ is the following cubic polynomial:

$$(10.13.58) \qquad \varphi(x) = -\frac{i\sigma}{h}\left(\frac{\bar{w}x^3}{6} + \frac{\bar{y}x^2}{2}\right) + \gamma x + \delta.$$

Since the pressure is given by $p = -\varrho\Phi_t - \varrho g\eta$ we have in the present case

$$(10.13.59) \qquad p(x) = [-i\sigma\varrho\varphi(x) - \varrho g(\bar{y} + x\bar{w})]e^{i\sigma t}.$$

The transition conditions (10.13.34) at $x = \pm a$ now lead, in the same way as above from (10.13.42) and (10.13.43), to the equations

$$(10.13.60) \begin{cases} Be^{ika} + Re^{-ika} = -\dfrac{i\sigma}{h}\left(\dfrac{\bar{w}a^3}{6} + \dfrac{\bar{y}a^2}{2}\right) + a\gamma + \delta \\[2ex] Be^{ika} - Re^{-ika} = -\dfrac{\sigma}{kh}\left(\dfrac{\bar{w}a^2}{2} + \bar{y}a\right) - \dfrac{i}{k}\gamma \\[2ex] Te^{-ika} = -\dfrac{i\sigma}{h}\left(-\dfrac{\bar{w}a^3}{6} + \dfrac{\bar{y}a^2}{2}\right) - a\gamma + \delta \\[2ex] Te^{-ika} = -\dfrac{\sigma}{kh}\left(\dfrac{\bar{w}a^2}{2} - \bar{y}a\right) - \dfrac{i}{k}\gamma. \end{cases}$$

These four equations are not sufficient to determine the six constants $R, T, \bar{w}, \bar{y}, \gamma$, and δ. We must make use of the dynamical equations of motion of the floating rigid body for this purpose. We have the equations of motion

$$(10.13.61) \qquad\qquad F = M\ddot{v}, \qquad \text{and} \qquad L = I\alpha$$

at our disposal. In the first equation F and M are the total vertical force on the board and its mass, per unit width, and $\ddot{v}$ is the vertical acceleration of its center of gravity, I the moment of inertia, L the torque, and α the angular acceleration. These dynamical conditions then yield the following relations:

$$(10.13.62) \qquad\qquad \int_{-a}^{a} p\,dx = M\ddot{v}, \qquad \int_{-a}^{a} px\,dx = I\ddot{\omega},$$

and these in turn lead to the equations

$$(10.13.63) \begin{cases} \displaystyle\int_{-a}^{a}\left\{-i\sigma\varrho\left[-\frac{i\sigma}{h}\left(\frac{\bar{w}x^3}{6} + \frac{\bar{y}x^2}{2}\right) + \gamma x + \delta\right] - \varrho g(\bar{y} + x\bar{w})\right\}dx = -M\sigma^2\bar{y} \\[3ex] \displaystyle\int_{-a}^{a}\left\{-i\sigma\varrho\left[-\frac{i\sigma}{h}\left(\frac{\bar{w}x^4}{6} + \frac{\bar{y}x^3}{2}\right) + \gamma x^2 + \delta x\right] - \varrho g(x\bar{y} + x^2\bar{w})\right\}dx = -I\sigma^2\bar{w} \end{cases}$$

In the first equation we have ignored the weight of the board, since it is balanced by the hydrostatic pressure. The equations (10.13.60) and (10.13.63) now determine all of the unknown complex amplitudes.

We omit the details of the calculations, which can be found in the paper by Fleishman [F.5]. In Fig. 10.13.13 the results of calculations for the pressure distribution in a numerical case are given. The parameters were chosen as follows:

$$\theta = 1, \qquad h = 1 \ ft, \qquad B = 1 \ ft^2/sec, \qquad a = 4 \ ft,$$

$$M = 18.72 \ pounds/ft, \qquad \sigma = 4.46 \ rad/sec, \qquad \lambda = 8 \ ft.$$

It might be added that the value chosen for M is such that the structure sinks down 0.0375 feet when in equilibrium.

A few observations should be made. First of all, we note that in both cases the pressure variation at the right end ($x = + a$), where the incoming wave is incident, is greater than at the left end. This is to be expected, since the barrier exercises a damping effect on the wave going under it. The pressure distribution in the case of the floating board is quadratic in x, in contrast with the case of the fixed board in which the distribution of pressure was linear in x. Next, we note that the pressure variation near the right end of the stationary board is greater than at the same end of the floating one; this too might be expected since the fixed board receives the full impact of the incident wave, while the floating one yields somewhat. Finally, we see that at the left end the opposite effect occurs: there the pressure variation under the stationary barrier is less than that under the floating barrier. This is not surprising either, since the fixed board should damp the wave more successfully than the movable board.

Finally, we take up the case of a floating elastic beam (cf. [F.5]). The beam is assumed to extend from $x = - l$ to $x = 0$ and, as in the above cases, to be in simple harmonic motion due to an incoming wave from $x = + \infty$. The basic relations for $\Phi(x, t)$ on the free surface, and for $\eta(x, t)$ under the beam are the same as before:

$$(10.13.64) \qquad \Phi_{xx} = \frac{1}{gh} \Phi_{tt}, \qquad x > 0, \qquad x < - l,$$

$$(10.13.65) \qquad \eta_t = - h\Phi_{xx}, \qquad - l < x < 0.$$

We assume once more that the beam sinks very little below the water surface when in equilibrium (i.e. very little in relation to the depth of

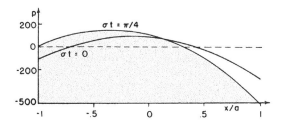

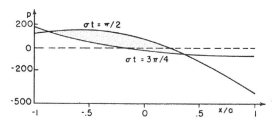

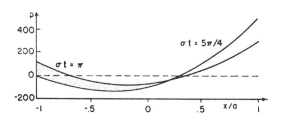

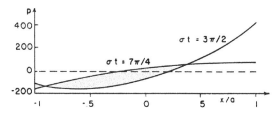

Fig. 10.13.13. Pressure variations for floating board. $\theta = 1$, p in *pounds/(ft)²*

the water), so that the coefficient of Φ_{xx} in (10.13.65) can be taken as h rather than $(h + \eta)$ (cf. (10.13.15)), and also the transition conditions at the ends of the beam are

(10.13.66) Φ_x and Φ_t continuous at $x = 0$, $x = -l$.

After writing

(10.13.67) $\Phi(x, t) = \varphi(x)e^{i\sigma t}$, $\eta(x, t) = v(x)e^{i\sigma t}$

we find, as before:

(10.13.68) $\varphi_{xx} + \dfrac{\sigma^2}{gh} \varphi = 0$, $x > 0$, $x < -l$

(10.13.69) $\varphi_{xx} + \dfrac{i\sigma}{h} v = 0$, $-l < x < 0$.

The conditions at ∞ have the effect that (cf. (10.13.41) et seq.):

(10.13.70) $\varphi(x) = Be^{ikx} + Re^{-ikx}$, $x > 0$,

(10.13.71) $\varphi(x) = Te^{ikx}$, $x < -l$,

with $k = \sigma/\sqrt{gh}$. All of this is the same as for the previous cases. We turn now to the conditions which result from the assumption that the floating body is a beam.

The differential equation governing small transverse oscillations of a beam is

(10.13.72) $EI\eta_{xxxx} + m\eta_{tt} = p$,

in which E is the modulus of elasticity, I the moment of inertia of a cross section of unit breadth (or, perhaps better, EI is the bending stiffness factor), m the mass per unit area, and p is the pressure. We ignore the weight of the beam and at the same time disregard the contribution of the hydrostatic pressure term in p corresponding to the equilibrium position of the beam—i.e. the pressure here is that due entirely to the dynamics of the situation. Thus

(10.13.73) $\begin{aligned} p &= -\varrho\Phi_t - \varrho g\eta \\ &= (-i\sigma\varrho\varphi - \varrho gv)e^{i\sigma t}. \end{aligned}$

Insertion of this relation in (10.13.72) and use of (10.13.69) leads at once to the differential equation for $\varphi(x)$:

(10.13.74) $\dfrac{d^6\varphi}{dx^6} + \dfrac{\varrho g - m\sigma^2}{EI}\dfrac{d^2\varphi}{dx^2} + \dfrac{\sigma^2\varrho}{EIh}\varphi = 0$, $-l < x < 0$,

that is valid under the beam. The case of greatest importance for us—
that of a floating beam used as a breakwater—leads obviously to the
boundary conditions for the beam which correspond to free ends, i.e.
to the conditions that the shear and bending moments should vanish
at the ends of the beam. These conditions in turn mean that η_{xx} and
η_{xxx} should vanish at the ends of the beam, and from (10.13.67) and
(10.13.69) we thus have for φ the boundary conditions

$$(10.13.75) \qquad \frac{d^4\varphi}{dx^4} = \frac{d^5\varphi}{dx^5} = 0 \qquad \text{at } x = 0, \quad x = -l.$$

The transition conditions (10.13.66) require that φ and φ_x be continuous at $x = 0$, $x = -l$, and this, in view of (10.13.70) and (10.13.
71), requires that

$$(10.13.76) \qquad \begin{cases} \varphi(0) = B + R, & \varphi_x(0) = ikB - ikR \\ \varphi(-l) = Te^{-ikl}, & \varphi_x(-l) = ikTe^{-ikl}. \end{cases}$$

We remark once more that the constant B is assumed to be real, but
that R and T will in general be complex constants, and that the real
parts of Φ and η as given by (10.13.67) are to be taken at the end.

In order to solve our problem we must solve the differential equation (10.13.74) subject to the conditions (10.13.75) and (10.13.76).
A count of the relations available to determine the solution should be
made: The general solution of (10.13.74) contains six arbitrary constants, and we wish to determine the constants R and T (the amplitudes of the reflected and transmitted waves) occurring in (10.13.
76) once the constant B (the amplitude of the incoming wave) has been
fixed. In all there are thus eight constants to be found, and we have in
(10.13.75) and (10.13.76) eight relations to determine them. Once
these constants have been determined, the reflection and transmission
coefficients are known, and the deflection of the beam can be found
from (10.13.69). The maximum bending stresses in the beam can then
be calculated from the usual formula: $s = Mc/I$, with $M = EI\eta_{xx}$
and c the distance from the neutral axis to the extreme outer fibres of
the beam.

In principle, therefore, the solution of the problem is straightforward. However, the carrying out of the details in the case of the beam
of finite length is very tedious, involving as it does a system of eight
linear equations for eight unknowns with complex coefficients. In
addition, one must determine the roots of a sixth degree algebraic
equation in order to find the general solution of (10.13.74). These

roots are in general complex numbers and they involve the essential parameters of the mechanical system. Thus it is clear that a discussion of the behavior of the system in general terms with respect to arbitrary values of the parameters of the system is not feasible, and one must turn rather to concrete cases in which most of the parameters have been given specific numerical values. The results of some calculations of this kind, for a case proposed as a practical possibility, will be given a little later.

The case of a semi-infinite beam—i.e. a beam extending from $x = 0$ to $x = -\infty$—is simpler to deal with in that the conditions in the second line of (10.13.76) fall away, and the conditions (10.13.75) at $x = -\infty$ can be replaced by the requirement that φ be bounded at $x = -\infty$. The number of constants to be fixed then reduces to four instead of eight, but the determination of the deflection of the beam still remains a formidable problem; we shall consider this case as well as the case of a beam of finite length.

We begin the program indicated with a discussion of the general solution of the differential equation (10.13.74). Since it is a linear differential equation with constant coefficients we proceed in the standard fashion by setting $\varphi = e^{\varkappa}$, inserting in (10.13.74), to find for $\varkappa$ the equation

(10.13.77) $$\varkappa^6 + a\varkappa^2 + b = 0$$

with

(10.13.78) $$a = \frac{\varrho g - m\sigma^2}{EI}, \quad b = \frac{\sigma^2 \varrho}{EIh}.$$

This is a cubic equation in $\varkappa^2 = \beta$, which happens to be in the standard form to which the Cardan formula for the roots of a cubic applies directly. For the roots β_i of this equation one has therefore

(10.13.79) $$\begin{cases} \beta_1 = u + v \\ \beta_2 = \varepsilon u + \varepsilon^2 v \\ \beta_3 = \varepsilon^2 u + \varepsilon v \end{cases}$$

with u and v defined by

(10.13.80) $$u = \left(-\frac{b}{2} + \left(\frac{b^2}{4} + \frac{a^3}{27}\right)^{\frac{1}{2}}\right)^{\frac{1}{3}}, \quad v = \left(-\frac{b}{2} - \left(\frac{b^2}{4} + \frac{a^3}{27}\right)^{\frac{1}{2}}\right)^{\frac{1}{3}}$$

and ε the following cube root of unity:

(10.13.81) $$\varepsilon = \frac{-1 + i\sqrt{3}}{2}.$$

The constant a is positive, since σ, the frequency of the incoming wave, is so small in the cases of interest in practice that ϱg is much larger than $m\sigma^2$. The constant b is obviously positive. Consequently the root β_1 is real and negative since $|u| < |v|$ and v is negative. Thus the roots $\varkappa_1 = +\beta_1^{1/2}$, $\varkappa_2 = -\beta_1^{1/2}$ are pure imaginary. The quantities β_2 and β_3 are complex conjugates, and their square roots yield two pairs of complex conjugates

$$\varkappa_3 = +\beta_2^{1/2}, \quad \varkappa_4 = -\beta_2^{1/2}, \quad \varkappa_5 = +\beta_3^{1/2}, \quad \varkappa_6 = -\beta_3^{1/2}.$$

For β_2 and β_3 we have

$$(10.13.82) \qquad \beta_2 = -\frac{1}{2}(u+v) + i\frac{\sqrt{3}}{2}(u-v),$$

$$(10.13.83) \qquad \beta_3 = -\frac{1}{2}(u+v) - i\frac{\sqrt{3}}{2}(u-v).$$

Thus β_2 and β_3 both have positive real parts. We suppose the roots $\varkappa_3$, $\varkappa_4$, $\varkappa_5$, $\varkappa_6$ to be numbered to that $\varkappa_3$ and $\varkappa_5$ are taken to have positive real parts, while $\varkappa_4$ and $\varkappa_6$ have negative real parts. The general solution of (10.13.74) thus is

$$(10.13.84) \quad \varphi(x) = a_1 e^{\varkappa_1 x} + a_2 e^{\varkappa_2 x} + a_3 e^{\varkappa_3 x} + a_4 e^{\varkappa_4 x} + a_5 e^{\varkappa_5 x} + a_6 e^{\varkappa_6 x}.$$

In the case of a beam covering the whole surface of the water, i.e. extending from $-\infty$ to $+\infty$, the condition that φ be bounded at $x = \pm\infty$ would require that $a_3 = a_4 = a_5 = a_6 = 0$ since the exponentials in the corresponding terms have non-vanishing real parts. The remaining terms yield progressing waves traveling in opposite directions; their wave lengths are given by $\lambda = 2\pi/|\varkappa_1| = 2\pi/|\varkappa_2|$ and thus by

$$(10.13.85) \qquad \lambda = 2\pi/\sqrt{|u+v|},$$

with $u+v$ defined by (10.13.80). The wave length and frequency are thus connected by a rather complicated relation, and, unlike the case of waves in shallow water with no immersed bodies or constraints on the free surface, the wave length is not independent of the frequency and the wave phenomena are subject to dispersion.

In the case of a beam extending from the origin to $-\infty$ while the water surface is free for $x > 0$, the boundedness conditions for φ at $-\infty$ requires that we take $a_4 = a_6 = 0$ since $\varkappa_4$ and $\varkappa_6$ have negative real parts and consequently $e^{\varkappa_4 x}$ and $e^{\varkappa_6 x}$ would yield exponentially unbounded contributions to φ at $x = -\infty$. We know that $\varkappa_1$ and $\varkappa_2$

are pure imaginary with opposite signs, with $\varkappa_2$, say, negative imaginary. Since no progressing wave is assumed to come from the left, we must then take $a_2 = 0$. Thus the term $a_1 e^{\varkappa_1 x}$ yields the transmitted wave and the terms involving a_3 and a_5 yield disturbances which die out exponentially at ∞. The conditions (10.13.70) and (10.13.71) at $x = 0$ now yield the following four linear equations:

$$(10.13.86) \quad \begin{cases} \varkappa_1^4 a_1 + \varkappa_3^4 a_3 + \varkappa_5^4 a_5 = 0 \\ \varkappa_1^5 a_1 + \varkappa_3^5 a_3 + \varkappa_5^5 a_5 = 0 \\ a_1 + a_3 + a_5 = B + R \\ \varkappa_1 a_1 + \varkappa_3 a_3 + \varkappa_5 a_5 = ik(B - R) \end{cases}$$

for the constants a_1, a_3, a_5, R. For the amplitude R of the reflected wave one finds

$$(10.13.87) \quad \frac{R}{B} = \frac{\begin{vmatrix} -1 & -1 & -1 & -1 \\ \varkappa_1 & \varkappa_3 & \varkappa_5 & ik \\ \varkappa_1^4 & \varkappa_3^4 & \varkappa_5^4 & 0 \\ \varkappa_1^5 & \varkappa_3^5 & \varkappa_5^5 & 0 \end{vmatrix}}{\begin{vmatrix} -1 & -1 & -1 & +1 \\ \varkappa_1 & \varkappa_3 & \varkappa_5 & ik \\ \varkappa_1^4 & \varkappa_3^4 & \varkappa_5^4 & 0 \\ \varkappa_1^5 & \varkappa_3^5 & \varkappa_5^5 & 0 \end{vmatrix}}.$$

Even in this relatively simple case of the semi-infinite beam the reflection coefficient is so complicated a function of the parameters (even though it is algebraic in them) that it seems not worthwhile to write it down explicitly. The results of numerical calculations based on (10.13.87) will be given shortly.

In the case of the beam of finite length extending from $x = -l$ to $x = 0$ the eight conditions given by (10.13.75) and (10.13.76) must be satisfied by the solution (10.13.84) of the differential equation (10.13.74), and these conditions serve to determine the six constants of integration and the amplitudes R and T of the reflected and transmitted waves. The problem thus posed is quite straightforward but extremely tedious as it involves solving eight linear equations for eight complex constants. For details reference is again made to the work of Wells [F.5].

This case of a floating beam was suggested to the author by J. H.

Carr of the Hydraulics Structures Laboratory at the California Institute of Technology as one having practical possibilities; at his suggestion calculations in specific numerical cases were carried out in order to determine the effectiveness of such a breakwater. The reason for considering such a structure for a breakwater as a means of creating relatively calm water between it and the shore is the following: a structure which floats on the surface without sinking far into the water need not be subjected to large horizontal forces and hence would not necessarily require a massive anchorage. However, in order to be effective as a reflector of waves such a floating structure would probably have to be built with a fairly large dimension in the direction of travel of the incoming waves. As a consequence of the length of the structure, it would be bent like a beam under the action of the waves and hence could not in general be treated with accuracy as a rigid body in determining its effectiveness as a barrier. This brings with it the possibility that the structure might be bent so much that the stresses set up would be a limiting feature in the design. The specifications (as suggested by Carr) for a beam having a width of one foot (parallel to the wave crest, that is) were:

Weight: 85 *pounds/ft²*
Moment of inertia (of area of cross-section): 0.2 *ft⁴*
Modulus of elasticity: 437 × 10⁷ *pounds/ft²*.

The depth of the water is taken as 40 feet. Simple harmonic progressing waves having periods of 8 and of 15 secs. were to be considered, and these correspond to wave lengths of 287 and 539 feet, and to circular frequencies σ of 785×10^{-3} and 418×10^{-3} cycles per second, respectively. The problem is to determine the reflecting power of the beam under these circumstances when the length of the beam is varied. In other words, we assume a wave train to come from the right hand side of the beam and that it is partly transmitted under the beam to the left hand side and partly reflected back to the right hand side. The ratio R/B of the amplitude R of the reflected wave and the amplitude B of the incoming wave is a measure of the effectiveness of the beam as a breakwater.

Before discussing the case of beams of finite length it is interesting and worthwhile to consider semi-infinite beams first. Since the calculations are easier than for beams of finite length it was found possible to consider a larger range of values of the parameters than was given

above. The results are summarized in the following tables (taken
from [F.5]):

TABLE A

λ (ft)	$\sigma\left(\dfrac{1}{sec}\right)$	W (pounds)	I (ft⁴)	$E\left(\dfrac{pounds}{ft^2}\right)$	h (ft)	R/B
539	0.418	85	0.20	487×10^7	40	0.14
287	0.785	85	0.20	487×10^7	40	0.19
225	1.0	85	0.20	487×10^7	40	0.23
150	1.5	85	0.20	487×10^7	40	0.32
113	2.0	85	0.20	487×10^7	40	0.43

In Table A the beam design data are as given above. At the two specified circular frequencies of 0.418 and 0.785 one sees that the floating
beam is quite ineffective as a breakwater since the reflected wave has
an amplitude of less than 1/5 of the amplitude of the incoming wave,
even for the higher frequency (and hence shorter wave length), which
means that less than 4 % of the incoming energy is reflected back. At
higher frequencies, and hence smaller wave lengths, the breakwater
is more effective, as one would expect. However the approximate
theory used to calculate the reflection coefficient R/B can be expected
to be accurate only if the ratio λ/h of wave length to depth is sufficiently large, and even for the case $\lambda = 287$ ft. ($\sigma = .785$) the reflection coefficient of value 0.19 may be in error by perhaps 10 % or
more since λ/h is only about 7, and the errors for the shorter wave
lengths would be greater. Calculations for still other values of the
parameters are shown in Table B. The only change as compared with

TABLE B

λ	σ	W	I	E	h	R/B
539	0.418	384	0.20	487×10^7	40	0.51
287	0.785	384	0.20	487×10^7	40	0.75

the first two rows of Table A is that the weight per foot of the beam
has been increased by a factor of more than 4 from a value of 85
pounds/ft² to a value of 384 *pounds/ft²*. The result is a decided increase
in the effectiveness of the breakwater, especially at the shorter wave
length, since more than half (i.e. $(.75)^2$) of the incoming energy would
be reflected back. However, this beneficial effect is coupled with a
decided disadvantage, since quadrupling the weight of the beam

would cause it to sink deeper in the water in like proportion and hence
might make heavy anchorages necessary. Table C is the same as the

TABLE C

λ	σ	W	I	E	h	R/B
539	.418	85	2.0	437×10^7	40	.26
287	.785	85	2.0	437×10^7	40	.32
			∞			1

first two rows of Table A except that the bending stiffness has been
increased by a factor of 10 by increasing the moment of inertia of the
beam cross-section from 0.2 ft^4 to 2.0 ft^4. Such an increase in stiffness
results in a noticeable increase in the effectiveness of the breakwater,
but by far not as great an increase as is achieved by multiplying the
weight by a factor of four. If the stiffness were to be made infinite
(i.e. if the beam were made rigid) the reflection coefficient could be
made unity, and no wave motion would be transmitted. This is
evidently true for a semi-infinite beam, but would not be true for a
rigid body of finite length.

TABLE D

λ	σ	W	I	E	h	R/B
539	.418	85	0	0	40	.001
287	.785	85	0	0	40	.007

In Table D the difference as compared with Table A is that the
beam stiffness is taken to be zero. This means that the surface of the
water is assumed to be covered by a distribution of inert particles
weighing 85 pounds per foot. (Such cases have been studied by Goldstein and Keller [G.1].) As we observe, there is practically no reflection and this is perhaps not surprising since the mass distribution per
unit length has such a value that the beam sinks down into the water
only slightly.

One might perhaps summarize the above results as follows: A very
long beam can be effective as a floating breakwater if it is stiff enough.
However, a reasonable value for the stiffness (the value 0.2 given
above) leads to an ineffective breakwater unless the weight of the
beam per square foot is a fairly large multiple (say 8 or 10) of the
weight of water.

In practice it seems unlikely that beams long enough to be considered semi-infinite would be practicable as breakwaters. (The term "long enough" might be interpreted to mean a sufficiently large multiple of the wave length, but since the wave lengths are of the order of 200 feet or more the correctness of this statement seems obvious.) It therefore is necessary to investigate the effectiveness of beams of finite length. Such an investigation requires extremely tedious calculations —so much so that only a certain number of numerical cases have been treated. These are summarized in the following tables.

$\sigma = .785, \quad \lambda = 287$			$\sigma = .418, \quad \lambda = 539$	
$l\,(ft)$	R/B		$l\,(ft)$	R/B
17.5	0		145.9	.17
49.2	.93		196.9	.53
72.9	0		291.8	.13
98.5	.75		443.0	.90
145.9	.10		583.6	.74
196.9	0		656.2	.62
291.8	.33		874.9	.07
450.4	.32		875.4	.08
583.6	.12		948.3	.54
656.3	.13		∞	.14
875.4	.32			
∞	.19			

In these tables the parameters have values the same as in the first two rows of Table A, except that now lengths other than infinite length are considered. The most noticeable feature of the results given in the tables is their irregularity and the fact that at certain lengths —even certain rather short lengths—the beam proposed by Carr seems to be quite effective. For example, when the wave length is 287 ft. a beam less than 50 ft. long reflects more than 80 % of the incoming energy. A beam of length 443 ft. is also equally effective at the longer wave length of 539 ft.*

* It might not be amiss to consider the physical reason why it is possible that a beam of finite length could be more effective as a breakwater than a beam of infinite length. Such a phenomenon comes about, of course, through multiple reflections that take place at the ends of the beam. Apparently the phases sometimes arrange themselves in the course of these complicated interactions in such a way as to yield a small amplitude for the transmitted wave. That such a process might well be sensitive to small changes in the parameters, as is noted in the discussion, cannot be wondered at.

However, the maximum effectiveness of any such breakwater occurs for a specific wave length within a certain range of wave lengths; thus the reflection of a given percentage of the incoming wave energy would involve changing the length (or some other parameter) of the structure in accordance with changes in the wave length of the incoming waves. Also, the reflection coefficient seems to be rather sensitive to changes in the parameters, particularly for the shorter structures (a relatively slight change in length from an optimum value, or a slight change in frequency, leads to a sharp decrease in the reflection coefficient). It is also probable — as was indicated earlier on the basis of calculations by Wells [W.10] — that the shallow water approximation used here as a basis for the theory is not sufficiently accurate for a floating beam whose length is too much less than the wave length. Nevertheless, it does seem possible to design floating breakwaters of reasonable length which would be effective at a given wave length. Perhaps it is not too far-fetched to imagine that sections could be added to or taken away from the breakwater in accordance with changing conditions.

Another consequence of the theory — which is also obvious on general grounds — is that there is always the chance of creating a large standing wave between the shore and the breakwater because of reflection from the shore, unless the waves break at the shore; this effect is perhaps not important if the main interest is in breakwaters off beaches of not too large slope, since breaking at the shore line then always occurs. (The theory developed here could be extended to cases in which the shore reflects all of the incoming energy, it might be noted.) In principle, the calculation of the deflection curve of the structure, and hence also of the bending stresses in it, as given by the theory is straightforward, but it is very tedious; consequently only the reflection coefficients have been calculated.

CHAPTER 11

Mathematical Hydraulics

In this chapter the problems to be treated are, from the mathematical point of view, much like the problems of the preceding chapter, but the emphasis is on problems of rather concrete practical significance. Aside from this, the essential difference is that external forces other than gravity, such as friction, for example, play a major role in the phenomena. Problems of various types concerning flows and wave motions in open channels form the contents of the chapter. The basic differential equations suitable for dealing with such flows under rather general circumstances are first derived. This is followed by a study of steady motions in uniform channels, and of progressing waves of uniform shape, including roll waves in inclined channels. Flood waves in rivers are next taken up, including a discussion of numerical methods appropriate in such cases; the results of such calculations for a flood wave in a simplified model of the Ohio River and for a model of its junction with the Mississippi are given. This discussion follows rather closely the two reports made to the Corps of Engineers of the U.S. Army by Stoker [S.23] and by Isaacson, Stoker, and Troesch [I.4]. These methods of dealing with flood waves have been applied, with good results, to a 400-mile stretch of the Ohio as it actually is for the case of the big flood of 1945, and also to a flood through the junction of the Ohio and the Mississippi; these results will be discussed toward the end of this chapter.

There is an extensive literature devoted to the subject of flow in open channels. We mention here only a few items more or less directly connected with the material of this chapter: the famous Essai of Boussinesq [B.17], the books of Bakhmeteff [B.3] and Rouse [R.10, 11] (in particular, the article by Gilcrest in [R.11]), the Enzyklopädie article of Forchheimer [F.6] and the booklet by Thomas [T.2].

11.1. Differential equations of flow in open channels

It has already been stated that the basic mathematical theory to be used in this chapter does not differ essentially from the theory derived in the preceding chapter. However, there are additional complications due to the existence of significant forces beside gravity, and we wish to permit the occurrence of variable cross-sections in the channels. Consequently the theory is derived here again, and a somewhat different notation from that used in previous chapters is employed both for the sake of convenience and also to conform somewhat with notations used in the engineering literature.

The theory is one-dimensional, i.e. the actual flow in the channel is assumed to be well approximated by a flow with uniform velocity over each cross-section, and the free surface is taken to be a level line in each cross-section. The channel is assumed also to be straight enough so that its course can be thought of as developed into a straight line without causing serious errors in the flow. The flow velocity is denoted by v, the depth of the stream (commonly called the stage in the engineering literature) by y, and these quantities are functions of the

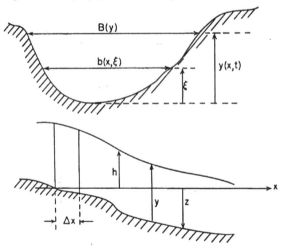

Fig. 11.1.1. River cross-section and profile

distance x down the stream and of the time t (cf. Fig. 11.1.1). The vertical coordinates of the bottom and of the free surface of the stream, as measured from the horizontal axis x, are denoted by $z(x)$ and

$h(x, t)$, with z positive downward, h positive upward; thus $y = h + z$. The slope of the bed is therefore counted positive in the positive x-direction, i.e. downward. The breadth of the free surface at any section of the stream is denoted by B.

The differential equations governing the flow are expressions of the laws of conservation of mass and momentum. In deriving them the following assumptions, in addition to those mentioned above, are made *: 1) the pressure in the water obeys the hydrostatic pressure law, 2) the slope of the bed of the river is small, 3) the effects of friction and turbulence can be accounted for through the introduction of a resistance force depending on the square of the velocity v and also, in a certain way to be specified, on the depth y.

We first derive the equation of continuity from the fact that the mass $\varrho A \Delta x$ included in a layer of water of density ϱ, thickness Δx, and cross-section area A, changes in its flow along the stream only through a possible inflow along the banks of the stream, say at the rate ϱq per unit length along the river. The total flow out of the element of volume $A \Delta x$ is given by the net contributions $\varrho (Av)_x \Delta x$ from the flow through the vertical faces plus the contribution $\varrho B h_t \Delta x$ due to the rise of the free surface, with B the width of the channel; since $B h_t$ represents the area change A_t it follows that the sum $[(Av)_x + A_t] \Delta x$ equals the volume influx $q \Delta x$ over the sides of the channel, with q the influx per unit length of channel. The subscripts x and t refer, of course, to partial derivatives with respect to these variables. The equation of continuity therefore has the form

(11.1.1) $$(Av)_x + A_t = q.$$

It should be observed that $A = A(y(x, t), x)$ is in the nature of things a given function of y and x, although $y(x, t)$ is an unknown function to be determined; in addition, $q = q(x, t)$ depends in general on both x and t in a way that is supposed given. In the important special case of a rectangular channel of constant breadth B, so that $A = By$, the equation of continuity takes the form

(11.1.2) $$v_x y + v y_x + y_t = q/B.$$

The equation of motion is next derived for the same slice of mass $m = \varrho A \Delta x$ by equating the rate of change of momentum $d(mv)/dt$

---

* These assumptions are not the minimum number necessary: for example, assumption 1) has as a consequence the independence of the velocity on the vertical coordinate if that were true at any one instant (cf. the remarks on this point in Ch. 2 and Ch. 10).

to the net force on the element. We write the equation of motion for the horizontal direction:

$$(11.1.3) \quad \varrho \frac{d}{dt} (Av\Delta x) = H\Delta x - F_f\Delta x \cos \varphi + \varrho gA\Delta x \sin \varphi.$$

In this equation H represents the unbalanced horizontal pressure force at the surface of the element. The angle φ is the slope angle of the bed of the channel, reckoned positive downward. The quantity F_f represents the friction force along the sides and bottom of the channel, and the term $\varrho gA\Delta x \sin \varphi$ represents the effect of gravity in accelerating the slice down-hill as manifested through the normal reaction of the stream bed. Since φ was assumed small we may replace $\sin \varphi$ by the slope $S = dz/dx$ and $\cos \varphi$ by 1. In the frictional resistance term we set

$$(11.1.4) \qquad\qquad F_f = \frac{\varrho gAv|v|}{\gamma R^{4/3}}.$$

This is an empirical formula called Manning's formula. The resistance is thus proportional to the square of the velocity and is opposite to its direction; in addition, the friction is inversely proportional to the 4/3-power of the hydraulic radius R, defined as the ratio of the cross-section area A to the wetted perimeter (thus $R = By/(B + 2y)$ for a rectangular channel and $R = y$ for a very wide rectangular channel), and inversely proportional to γ, a roughness coefficient.

We calculate next the momentum change $\varrho d(Av\Delta x)/dt$. In doing so, we observe that the symbol d/dt must be interpreted as the particle derivative (cf. Chapter 1.1 and equation (1.1.8)) $\partial/\partial t + v\partial/\partial x$ since Newton's law must be applied in following a given mass particle along its path $x = x(t)$. However, the law of continuity (11.1.1) derived above is clearly equivalent to writing $d(A\Delta x)/dt = q\Delta x$, with d/dt again interpreted as the particle derivative. Since

$$\frac{d}{dt} (Av\Delta x) = v \frac{d}{dt} (A\Delta x) + A\Delta x \frac{dv}{dt}$$

it follows that

$$\frac{d}{dt} (Av\Delta x) = A\Delta x(vv_x + v_t) + qv\Delta x.$$

Finally, the net contribution $H\Delta x$ of the pressure forces over the surface of the slice is calculated as follows: The total pressure over a vertical face of the slab is given by $\int_0^y \varrho g[y(x, t) - \xi]b(x, \xi)\, d\xi$ from the hydrostatic pressure law (cf. Fig. 11.1.1); while the component in the x-direction of the total pressure over the part of the slice in contact with the banks of the river is given by $\left\{ \int_0^y \varrho g[y - \xi] b_x(x, \xi) \, d\xi \right\} \Delta x$, we have for $H \Delta x$ the following equation:

$$(11.1.5) \qquad H \Delta x = - \frac{\partial}{\partial x} \left\{ \int_0^y \varrho g[y(x, t) - \xi] b(x, \xi) d\xi \right\} \Delta x$$

$$+ \left\{ \int_0^y \varrho g[y - \xi] b_x(x, \xi) d\xi \right\} \Delta x$$

$$= - \int_0^y \varrho g y_x b(x, \xi) d\xi = - \varrho g A y_x.$$

In this calculation the integrals involving b_x cancel out, and we have used the fact that y_x is independent of ξ.

Adding all of the various contributions we have

$$(11.1.6) \qquad v_t + v v_x + \frac{q}{A} v = Sg - S_f g - g y_x$$

upon defining what is called the friction slope S_f by the formula

$$(11.1.7) \qquad S_f = \frac{1}{\varrho g A} F_f,$$

with F_f defined by (11.1.4). It should perhaps be mentioned that the term qv/A on the left hand side of (11.1.6) arises because of the tacit assumption that flows enter the main stream from tributaries or by flow over the banks at zero velocity in the direction of the main stream; if such flows were assumed to enter with the velocity of the main stream, the term would not be present—it is, in any case, a term which is quite small. If we introduce $A = A(y(x, t), x)$ in (11.1.1) the result is

$$(11.1.8) \qquad A_y y_x v + A_x v + A v_x + A_y y_t = q.$$

The two differential equations (11.1.6) and (11.1.8), which serve to determine the two unknown functions, the depth $y(x, t)$ and the velocity $v(x, t)$, are the basic equations for the study of flood waves in rivers and flows in open channels generally. For any given river or channel it is thus necessary to have data available for determining the cross-section area A and the quantities γ and R in the resistance term F_f as functions of x and y, and of the slope S of its bed as a function of x in order to have the coefficients in the differential equations (11.1.6) and (11.1.8) defined. Three of these quantities are purely geometrical in character and could in principle be determined

from an accurate contour map of the river valley, but the determination of the roughness coefficient γ of course requires measurements of
actual flows for its determination.

11.2. Steady flows. A junction problem

We define a steady flow in the usual fashion to be one for which the
velocity v and depth y are independent of the time, that is, $v_t = y_t = 0$.
In this section channels of constant rectangular cross-section and
constant slope will be considered for the most part. It follows from the
equation of continuity (cf. (11.1.2)):

$$y_t + vy_x + yv_x = 0,$$

that for steady flow

(11.2.1) $(vy)_x = 0$ whence $vy = D$ (D a constant),

when no flow into the channel from its sides occurs (i.e. $q = 0$ in
(11.1.2)). Similarly, the equation of motion (cf. (11.1.6))

$$v_t + vv_x + gy_x + g(S_f - S) = 0$$

yields

(11.2.2) $vv_x + gy_x + g(S_f - S) = 0.$

It follows from equation (11.2.1) that

$$v = \frac{D}{y} \quad \text{and} \quad v_x = -\frac{D}{y^2}\, y_x,$$

so that equation (11.2.2) becomes

(11.2.3) $$\left(g - \frac{D^2}{y^3}\right)y_x + g\left(\frac{D^2}{\gamma y^2 \left\{y/\left(1 + \dfrac{2y}{B}\right)\right\}^{4/3}} - S\right) = 0.$$

Here the hydraulic radius is given by $R = y/(1 + 2y/B)$ because the
channel is assumed to be rectangular in cross-section.

For a channel with given physical parameters such as cross-section,
resistance coefficient, etc. the steady flows would provide what are
called backwater curves. In general, one could in principle always
find steady solutions $y = y(x)$ and $v = v(x)$ for a non-uniform channel. The explicit determination of the stage y and discharge rate BD
as functions of x would be possible by numerical integration of the
pair of first order ordinary differential equations arising from (11.1.6)
and (11.1.8) when time derivatives are assumed to vanish.

We note that equation (11.2.3) has the simple solution $y = $ constant for y satisfying

(11.2.4)
$$\frac{D^2}{\gamma S} = y^2 \left\{ y / \left(1 + \frac{2y}{B} \right) \right\}^{4/3}.$$

This means that we can find a flow of uniform depth and velocity having a constant discharge rate BD (B is, as in the preceding section, the width of the channel). Conversely, by fixing the depth y we can find the discharge from (11.2.4) appropriate to the corresponding uniform flow. Physically this means that the flow velocity is chosen so that the resistance due to turbulence and friction and the effect of gravity down the slope of the stream just balance each other. We remark that if (11.2.4) is satisfied at any point where the coefficient $g - D^2/y^3$ of y_x in (11.2.3) does not vanish, then $y = $ constant is the only solution of (11.2.3) because of the fact that the solution is then uniquely determined by giving the value of y at any point x. We note that $g - D^2/y^3 = 0$ corresponds to $v = \sqrt{gy}$, i.e. to a flow at critical speed (a term to be discussed in the next section), since $D = vy$. Furthermore, the differential equation (11.2.3) can be integrated to yield x as a function of y:

(11.2.5)
$$x = \int_{y_0}^{y} \frac{1}{g} \left(\frac{-D^2}{\gamma y^2 \left\{ y / \left(1 + \frac{2y}{B} \right) \right\}^{4/3}} + S \right)^{-1} \left(g - \frac{D^2}{y^3} \right) dy$$

when $x = 0$ for $y = y_0$.

We proceed to make use of (11.2.5) in order to study a problem involving a steady flow at the junction of two rivers each having a rectangular channel. Later on, the same problem will be treated but for an unsteady motion resulting from a flood wave traveling down one of the branches, and such that the steady flow to be treated here is expected to result as a limit state after a long time. The numerical data are chosen here for the problem in such a way as to correspond roughly with the actual data for the junction of the Ohio River with the Mississippi River. Thus the Ohio is assumed to have a rectangular channel 1000 feet in width and a constant slope of .5 feet/mile. In Manning's formula for the resistance the constant γ is assumed given by $\gamma = (1.49/n)^2$ in terms of Manning's roughness coefficient n, and n is given the value 0.03. The upstream branch of the Mississippi was taken the same in all respects as the Ohio, but the downstream branch is assumed to have twice the breadth, i.e. 2000 feet, and its slope to have a slightly smaller value, i.e. 0.49 feet/mile instead of 0.5 feet/mile. With these values of the parameters, a flow having the same uniform depth of 20 feet in all three branches is possible—the choice of the

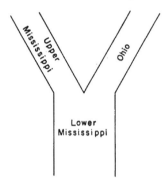

Fig. 11.2.1. Junction of Ohio and Mississippi Rivers

value 0.49 feet/mile for the slope of the downstream branch of the Mississippi River was in fact made in order to ensure this. Later on we intend to calculate the progress of a flood which originates at a moment when the flow is such a uniform flow of depth 20 feet. The flood wave will be supposed to initiate at a point 50 miles up the Ohio from the junction and to be such that the Ohio rises rapidly at that point from the initial depth of 20 feet to a depth of 40 feet in 4 hours. A wave then moves down the Ohio to the junction and creates waves which travel both upstream and downstream in the Mississippi as well as a reflected wave which travels back up the Ohio. After a long time we would expect a steady state to develop in which the depth at the point 50 miles up the Ohio is 40 feet, while the depth far upstream in the Mississippi would be the original value, i.e. 20 feet (since we would not expect a retardation of the flow far upstream because of an inflow at the junction). Downstream in the Mississippi we expect a change in the flow extending to infinity. It is the steady flow with these latter characteristics that we wish to calculate in the present section. (See Fig. 11.2.1)

We remark first of all that the stream velocities in all of the three branches will always be subcritical—in fact, they are of the order of a few miles per hour while the critical velocities $\sqrt{gy}$ are of the order of 15 to 25 miles per hour. It follows that the quantity $g - D^2/y^3$ in the integrand of the basic formula (11.2.5) for the river profiles (i.e. the curve of the free surface) is always positive. The integrand $I(y)$ in that formula has the general form indicated by Fig. 11.2.2 in the case of flows at subcritical velocities. The vertical asymptote corresponds to the value of y for which a steady flow of constant depth exists

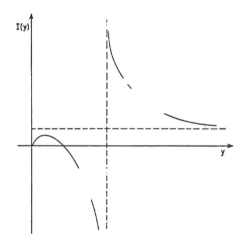

Fig. 11.2.2. The integrand in the wave profile formula

(cf. (11.2.4)), since the square bracket (the denominator in the integrand) vanishes for this value. It follows that x can become positive infinite for finite values of y only if y takes on somewhere this value; but in that case we have seen that the whole flow is then one with constant depth everywhere. Consequently the downstream side of the Mississippi carries a flow of constant speed and depth, though the values of these quantities are not known in advance. However, in the upstream branch of the Mississippi the flow need not be constant, and of course we do not expect it to be constant in the Ohio: in these branches x must be taken to be decreasing on going upstream and consequently the negative portion of $I(y)$ indicated in Fig. 11.2.2 comes into use since we may, and do, set $x = 0$ at the junction.

For the sake of convenience we use subscripts 1, 2, and 3 to refer to all quantities in the Ohio, the upstream branch of the Mississippi, and the downstream branch of the Mississippi respectively. The conditions to be satisfied at the junction are chosen to be

(11.2.6) $\qquad y_1 = y_2 = y_3 = y_j$
(11.2.7) $\qquad D_1 + D_2 = 2D_3$ $\qquad$ for $x = 0$.

The first condition simply requires the water level to have the same value y_j (which is, however, not known in advance) in all three branches, while the second states, upon taking account of the first condition, that the combined discharge of the two tributaries makes up the total discharge in the main stream. The quantity D_2, the discharge in the upper Mississippi, is known since the flow far upstream in this branch is supposed known—i.e. it is a uniform flow of depth 20 feet.

By using (11.2.7) in (11.2.4) as applied to the lower branch of the Mississippi (in which the flow is known to be constant) we have

$$(11.2.8) \qquad \frac{(D_1+D_2)^2}{4\gamma S_3} = y_j^2 \left(\frac{y_j}{1+\dfrac{2y_j}{B_3}} \right)^{4/3}.$$

Next, we write equation (11.2.5) for the 50-mile stretch of the Ohio which ends at the point where the depth in that branch was prescribed to be 40 feet (and which was the point of initiation of a flood wave); the result is

$$(11.2.9) \qquad 50 = \int_{y_j}^{\bar{y}} I(y, D_1, B_1) \, dy$$

in which it is indicated that D and B (as well as all other parameters) are to be evaluated for the Ohio; the quantity $\bar{y}$ has the value 40/5280 in miles. Equations (11.2.8) and (11.2.9) are two equations containing y_j and D_1 as unknowns, since D_2 is known. They were solved by an iterative process, i.e. by taking for D_1 an estimated value, determining a value for y_j from (11.2.9), reinserting this value in (11.2.8) to determine a new value for D_1, etc. The results obtained by such a calculation are as follows:

$$y_1 = y_2 = y_3 = y_j = 31.2 \text{ feet}$$

$$v_1 = 4.83 \text{ miles/hour}, \quad v_2 = 1.53 \text{ miles/hour}, \quad v_3 = 3.18 \text{ miles/hour}.$$

The profiles of the river surface can now be computed from (11.2.5); the results are given in Fig. 11.2.3.

The solution of the mathematical problem has the features we would expect in the physical problem. The flow velocity and stage are increased at the junction, even quite noticeably, by the influx from the Ohio. Upstream in the Mississippi the stage decreases rather rapidly on going away from the junction, and very little backwater effect is noticeable at distances greater than 50 miles from the junction. This illustrates a fact of general importance, i.e. that backwater

effects in long rivers arising from even fairly large discharges of tributaries into the main stream do not persist very far upstream,
but such an influx has an influence on the flow far downstream.
For unsteady motions this general observation also holds, and is in
fact one of the basic assumptions used by hydraulics engineers in

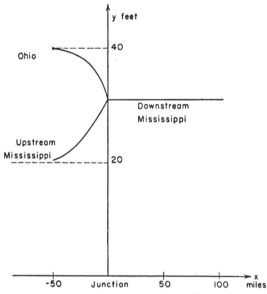

Fig. 11.2.3. Steady flow profile in a model of the Ohio and Mississippi Rivers

computing the passage of flood waves down rivers (a process called
flood routing by them). Later on, in sec. 6 of this chapter, we shall
deal with the unsteady motion described above in our model of the
Ohio-Mississippi system, and we will see that the unsteady motion
approaches the steady motion found here as the time increases.

11.3. Progressing waves of fixed shape. Roll waves

In addition to the uniform steady flows treated above there also
exist a variety of possible flows in uniform channels in the form of
progressing waves moving downstream at constant speed without
change in shape. Such waves are expressed mathematically by depths
$y(x, t)$ and velocities $v(x, t)$ in the form

(11.3.1) $y(x, t) = y(x - Ut), \quad v(x, t) = v(x - Ut), \quad U = \text{const.}$

The constant U is of course the propagation speed of the wave as viewed from a fixed coordinate system; if viewed from a coordinate system moving downstream with constant velocity U the wave profile would appear fixed, and the flow would appear to be a steady flow relative to the moving system. It is convenient to introduce the variable ζ by setting

$$(11.3.2) \qquad \zeta = x - Ut$$

so that y and v are functions of ζ only. In this case the equations of continuity and motion given by (11.1.6) and (11.1.8) become, for a rectangular channel of fixed breadth and slope:

$$(11.3.3) \qquad \begin{cases} (v - U)y_\zeta + yv_\zeta = 0, \\ (v - U)v_\zeta + gy_\zeta + g(S_f - S) = 0, \end{cases}$$

with S the slope of the channel and S_f defined, as before, by

$$(11.3.4) \qquad S_f = \frac{v|v|}{\gamma \left\{ y / \left(1 + \dfrac{2y}{B} \right) \right\}^{4/3}}.$$

The first equation of (11.3.3) can be integrated to yield

$$(11.3.5) \qquad (v - U)y = D = \text{const.}$$

as one readily verifies, and the second equation then takes the form

$$(11.3.6) \qquad \left(g - \frac{D^2}{y^3} \right) y_\zeta + g(S_f - S) = 0.$$

The first order differential equation (11.3.6) has a great variety of solutions, which have been studied extensively, for example by Thomas [T.1], but most of them are not very interesting from the physical point of view. However, one type of solution of (11.3.6) is particularly interesting from the point of view of the applications, and we therefore proceed to discuss it briefly. The solution in question furnishes the so-called monoclinal rising flood wave in a uniform channel (see the article by Gilcrest in the book of Rouse [R.11, p. 644]). This, as its name suggests, is a progressing wave the profile of which tends to different constant values (and the flow velocity v also to different constant values) downstream and upstream, with the lower depth downstream, connected by a steadily falling portion, as indicated schematically in Fig. 11.3.1. In this wave the propagation speed U is larger than the flow velocity v. It is always a possible type of solution of (11.3.6) if the speed of propagation of the wave relative to the flow is

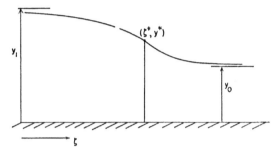

Fig. 11.3.1. Monoclinal rising flood wave

subcritical, i.e. if $(U-v)^2$ is less than gy, in which case the coefficient of the first derivative term in (11.3.6) is seen to be positive. This can be verified along the following lines. The differential equation can be solved explicitly for ζ as a function of y:

$$(11.3.7) \qquad \zeta = \int_{y^*}^{y} I(y)dy,$$

with the integrand $I(y)$ defined in the obvious manner; here y^* is the value of y corresponding to $\zeta = 0$. The function $I(y)$ has the general

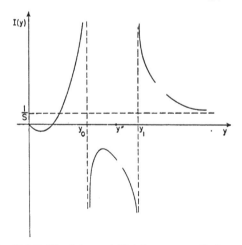

Fig. 11.3.2. The integrand $I(y)$ for a monoclinal wave

form shown in Fig. 11.3.2 if the propagation speed U and the constant D in (11.3.5) are chosen properly. The main point is that the curve has two vertical asymptotes at $y = y_0$ and $y = y_1$ between which $I(y)$ is negative. By choosing y^* between y_0 and y_1 we can hope that $\zeta \to + \infty$ as $y \to y_0$, while $\zeta \to - \infty$ as $y \to y_1$: all that is necessary is that $I(y)$ becomes infinite at y_0 and y_1 of sufficiently high order. This is, in fact, the case; we can select values of D and U in such a way that $I(y)$ becomes infinite at y_0 and y_1 through setting the quantity $S_f - S$ in (11.3.6) equal to zero, i.e. by choosing D and U such that

$$(11.3.8) \quad \begin{cases} (Uy_0+D)^2 = S\gamma y_0{}^2 \left(\dfrac{y_0}{1+\dfrac{2y_0}{B}} \right)^{4/3}, \\[4ex] (Uy_1+D)^2 = S\gamma y_1{}^2 \left(\dfrac{y_1}{1+\dfrac{2y_1}{B}} \right)^{4/3}. \end{cases}$$

For given positive values of y_0 and y_1 these are a pair of linear equations (after taking a square root) which determine U and D uniquely. An elementary discussion of the possible solutions of these equations shows that U must be positive and D negative, and this means, as we see from (11.3.5), that U is larger than v, i.e. the propagation speed of the wave is greater than the flow speed.

By taking the numerical data for the model of the Ohio given in the preceding section and assuming the depth far upstream to be 40 feet, far downstream 20 feet, it was found that the corresponding monoclinal flood wave in the Ohio would propagate with a speed of 5 miles/hour. The shape of the wave will be given later in sec. 6 of this chapter, where it will be compared with an unsteady wave obtained by gradually raising the level in the Ohio at one point from 20 feet to 40 feet, then holding the level fixed there at the latter value, and calculating the downstream motion which results. We shall see that the motion tends to the monoclinal flood wave obtained in the manner just now described. Thus the unsteady wave tends to move eventually at a speed of about 5 miles/hour, while on the other hand, as we know from Chapter 10 (and will discuss again later on in this chapter), the propagation speed of small disturbances relative to the stream is $\sqrt{gy}$ and hence is considerably larger in the present case, i.e. of the order of 15 to 25 miles/hour. This important and interesting point will be discussed in sec. 6 below.

Fig. 11.3.3. Roll waves, looking down stream (The Grünnbach, Switzerland)

We turn next to another type of progressing waves in a uniform channel which can be described with the aid of the differential equation (11.3.6), i.e. the type of wave called a roll wave. A famous example of such waves is shown in Fig. 11.3.3, which is a photograph taken from a book of Cornish [C.7], and printed here by the courtesy of the Cambridge University Press. As one sees, these waves consist of a series of bores (cf. Chapter 10.7) separated by stretches of smooth flow. The sketch of Fig. 11.3.4 indicates this more specifically. Such

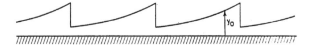

Fig. 11.3.4. Roll waves

waves frequently occur in sufficiently steep channels as, for example, spill-ways in dams or in open channels such as that of Fig. 11.3.3. Roll-waves sometimes manifest themselves in quite unwanted places, as for example in the Los Angeles River in California. The run-off from the steep drainage area of this river is carried through the city of Los Angeles by a concrete spill-way; in the brief rainy season a large amount of water is carried off at high velocity. It sometimes happens that roll waves occur with amplitudes high enough to cause spilling over the banks, though a uniform flow carrying the same total amount of water would be confined to the banks. The phenomenon of roll waves thus has some interest from a practical as well as from a theoretical point of view; we proceed to give a brief treatment of it in the remainder of this section following the paper of Dressler [D.12]. In doing so, we follow Dressler in taking what is called the Chézy formula for the resistance rather than Manning's formula, as has been done up to now. The Chézy formula gives the quantity S_f the following definition:

$$(11.3.9) \qquad S_f = \frac{r^2 \cdot v|v|}{gR}$$

in which r^2 is a "roughness coefficient" and R is, as before, the hydraulic radius. For a very broad rectangular channel, the only case we consider, $R = y$. Under these circumstances the differential equation (11.3.6) takes the form

$$(11.3.10) \qquad \frac{dy}{d\zeta} = \frac{gS - \dfrac{r^2(Uy+D)|Uy+D|}{y^3}}{g - \dfrac{D^2}{y^3}}$$

as can be readily seen.

It is natural to inquire first of all whether (11.3.10) admits of solutions which are continuous periodic functions of ζ since this is the general type of motion we seek. There are, however, no such periodic and continuous solutions (cf. the previously cited paper of Thomas [T.1]) of the equations; in fact, since the right hand side of (11.3.10) can be expressed as the quotient of cubic polynomials in y the types of functions which arise on integrating it are linear combinations of the powers, the logarithm, and the arc tangent function and one hardly expects to find periodic functions on inverting solutions $\zeta(y)$ of this type. This fact, together with observations of roll waves of the kind shown in Fig. 11.3.3, leads one to wonder whether there might not be *discontinuous* periodic solutions of (11.3.10) with discontinuities in the form of bores, which should be fitted in so that the discontinuity or shock conditions described in sec. 6 of the preceding chapter * are satisfied. This Dressler shows to be the case; he also gives a complete quantitative analysis of the various possibilities.

The starting point of the investigation is the observation, due to Thomas [T.1], that only quite special types of solutions of (11.3.10) come in question once the roll wave problem has been formulated in terms of a periodic distribution of bores. In fact, we know from Chapter 10 that the flow relative to a bore must be subcritical behind a bore but supercritical in front of it; consequently there must be an intermediate point of depth y_0, say, (cf. Fig. 11.3.4) where the smooth flow has the critical speed, i.e. where

$$(11.3.11) \qquad (v_0 - U)^2 = gy_0,$$

since U, the speed of the bore, coincides with the propagation speed of the wave. At such a point the denominator on the right hand side of (11.3.10) vanishes, since $D=(v - U)y$, and hence $dy/d\zeta$ would be infinite there—contrary to the observations—unless the numerator of the right hand side also vanishes at that point. The right hand side can now be written as a quotient of cubic polynomials, and we know

---

* The shock conditions were derived in Chapter 10 under the assumption that no resistances were present. As one would expect, the resistance terms play no role in shock conditions, as Dressler [D.12] verifies in his paper.

that numerator and denominator have y_0 as a common root; it follows that a factor $y - y_0$ can be cancelled and the differential equation then can be put in the form

(11.3.12)
$$\frac{dy}{d\zeta} = S \frac{y^2 + \left(y_0 - \dfrac{U^2 r^2}{gS}\right) y + \dfrac{r^2 y_0^2}{S}}{y^2 + y_0 y + y_0^2}$$

after a little algebraic manipulation. Since the denominator on the right hand side is positive and since we seek solutions for which $dy/d\zeta$ is everywhere (cf. Fig. 11.3.4) positive, it follows in particular that the quadratic in the numerator must be positive for $y = y_0$. This leads to the following necessary condition for the formation of roll waves

(11.3.13) $4r^2 < S,$

obtained by using (11.3.11) and other relations. A practically identical inequality was derived by Thomas on the basis of the same type of reasoning. The inequality states that the channel roughness, which is larger or smaller with r^2, must not be too great in relation to the steepness of the channel, and this corroborates observations by Rouse [R.10] that roll waves can be prevented by making a channel sufficiently rough. Dressler also shows in his paper that it is important for the formation of roll waves that the friction force for the same roughness coefficient and velocity should increase when the depth decreases; he finds, in fact, that roll waves would not occur if the hydraulic radius R in the Chézy formula (11.3.9) were to be assumed independent of the depth y.

Dressler goes on in his paper to show that smooth solutions of (11.3.12) can be pieced together through bores in such a way that the conditions referring to continuity of mass and momentum across the discontinuity are satisfied as well as the inequality requiring a loss rather than a gain in energy. For the details of the calculations and a quantitative analysis in terms of the parameters, the paper of Dressler should be consulted, but a few of the results might be mentioned here. Once the values of the slope S and the roughness coefficient r^2 are prescribed by the physical situation, and the wave propagation speed U is arbitrarily given, there exists a one-parameter family of possible roll-waves. As parameter the wave length λ, i.e. the distance between two successive bores, can be chosen; if this parameter is also fixed, the roll wave solution is uniquely determined. A specific solution would also be fixed if the time period of the oscillation were to be fixed together with one other parameter—the average discharge rate, say. Perhaps it is in this fashion that the roll waves are definitely fixed in some cases—for example, the roll waves down the spill-way of a dam are perhaps fixed by the period of surface waves in the dam at the crest of the spill-way. Schonfeld [S.4a] discusses the problem from the point of view of stability and arrives at the conclusion that only one of the solutions obtained by Dressler would be stable, and hence it would be the one likely to be observed.

11.4. Unsteady flows in open channels. The method of characteristics

In treating unsteady flows it becomes necessary to integrate the nonlinear partial differential equations (11.1.1) and (11.1.6) for prescribed initial and boundary conditions. It has already been mentioned that such problems fall into the same category as the problems treated in the preceding chapter, since they are of hyperbolic type in two independent variables and thus amenable to solution by the method of characteristics. It is true that the equations (11.1.1) and (11.1.6) are more complicated than those of Chapter 10 because of the occurrence of the variable coefficient A and of the resistance term, so that solutions of the type called simple waves (cf. Ch. 10.3) do not exist for these equations. Nevertheless the theory of characteristics is still available and leads to a variety of valuable and pertinent observations regarding the integration theory of equations (11.1.1) and (11.1.6) which are very important. The essential facts have already been stated in Chapter 10.2, but we repeat them briefly here for the sake of preserving the continuity of the discussion. Our emphasis in this chapter is on numerical solutions, which can be obtained by operating with the characteristic form of the differential equations, but since we shall not actually use the characteristic form for such purposes we shall base the discussion immediately following on a special case, although the results and observations are applicable in the most general case. The special case in question is that of a river of constant rectangular section and uniform slope, with no flow into the river from the banks (i.e. $q = 0$ in (11.1.2) and (11.1.6)). In this case the differential equations can be written as follows:

$$(11.4.1) \qquad v_x y + v y_x + y_t = 0,$$

$$(11.4.2) \qquad v_t + v v_x + g y_x + E = 0.$$

We have introduced the symbol E for the external forces per unit mass:

(11.4.3) $E = - gS + gS_f$, $S = $ const.

The term E differs from the others in that it contains no derivatives of y or v.

The theory of characteristics for these equations can be approached very directly * in the present special case by introducing a new quantity c to replace y, as follows:

(11.4.4) $c = \sqrt{gy}$.

This quantity has great physical significance, since it represents—as we have seen in Chapter 10—the propagation speed of small disturbances in the river. From (11.4.4) we obtain at once the relations

(11.4.5) $2cc_x = gy_x$, $2cc_t = gy_t$,

and the differential equations (11.4.1) and (11.4.2) take the form

(11.4.6) $\begin{cases} 2cc_x + v_t + vv_x + E = 0, \\ cv_x + 2vc_x + 2c_t = 0. \end{cases}$

These equations are next added, then subtracted, to obtain the following equivalent pair of equations:

(11.4.7) $\begin{cases} 2\left\{ (c+v)\dfrac{\partial}{\partial x} + \dfrac{\partial}{\partial t} \right\} c + \left\{ (c+v)\dfrac{\partial}{\partial x} + \dfrac{\partial}{\partial t} \right\} v + E = 0, \\ -2\left\{ (-c+v)\dfrac{\partial}{\partial x} + \dfrac{\partial}{\partial t} \right\} c + \left\{ (-c+v)\dfrac{\partial}{\partial x} + \dfrac{\partial}{\partial t} \right\} v + E = 0. \end{cases}$

We observe that the derivatives in these equations now have the form of directional derivatives—indeed, to achieve that was the purpose of the transformation—so that c and v in the first equation, for example, are both subject to the operator $(c+v)\partial/\partial x + \partial/\partial t$, which means that these functions are differentiated along curves in the x, t-plane which satisfy the differential equation $dx/dt = c + v$. In similar fashion, the functions c and v in the second equation are both subject to differentiation along curves satisfying the differential equation $dx/dt = - c + v$.

It is entirely feasible to develop the integration theory of equations (11.4.7) quite generally on the basis of these observations (as is done, for example, in Courant-Friedrichs [C.9, Ch. 2]), but it is simpler, and leads to the same general results, to describe it for the special case in

---

* For a treatment which shows quite generally how to arrive at the formulation of the characteristic equations, see Courant-Friedrichs [C.9, Ch. 2].

which the resistance force F_f is neglected so that the quantity E in (11.4.7) is a constant (see (11.4.3)). In this case the equations (11.4.7) can be written in the form

(11.4.7)$_1$
$$\left\{ \frac{\partial}{\partial t} + (v+c)\frac{\partial}{\partial x} \right\}(v+2c+Et)=0,$$

$$\left\{ \frac{\partial}{\partial t} + (v-c)\frac{\partial}{\partial x} \right\}(v-2c+Et)=0,$$

as one can readily verify. But the interpretation of the operations defined in (11.4.7)$_1$ has just been mentioned: the relations state that the functions $(v \pm 2c + Et)$ are constant for a point moving through the fluid with the velocity $(v \pm c)$, or, as we may also put it, for a point whose motion in the x, t-plane is characterized by the ordinary differential equations $dx/dt = v \pm c$. That is, we have the following situation in the x, t-plane: There are two sets of curves, C_1 and C_2, called characteristics, which are the solution curves of the ordinary differential equations

(11.4.8)
$$\begin{cases} C_1: \dfrac{dx}{dt}=v+c, \text{ and} \\[2mm] C_2: \dfrac{dx}{dt}=v-c, \end{cases}$$

and we have the relations

(11.4.9) $\begin{cases} v+2c+Et=k_1=\text{const. along a curve } C_1 \text{ and} \\ v-2c+Et=k_2=\text{const. along a curve } C_2. \end{cases}$

Of course the constants k_1 and k_2 will be different on different curves in general. It should also be observed that the two families of characteristics determined by (11.4.8) are really distinct because of the fact that $c = \sqrt{gy} \neq 0$ since we suppose that $y > 0$, i.e. that the water surface never touches the bottom.

By reversing the above procedure it can be seen rather easily that the system of relations (11.4.8) and (11.4.9) is completely equivalent to the system of equations (11.4.6) for the case of constant bottom slope and zero resistance, so that a solution of either system yields a solution of the other. In fact, if we set $f(x, t) = v + 2c + Et$ and observe that $f(x, t) = k_1 = \text{const.}$ along any curve $x = x(t)$ for which $dx/dt = v + c$ it follows that along such curves

(11.4.10)
$$f_t + f_x \frac{dx}{dt} = f_t + (v+c)f_x = 0.$$

In the same way the function $g(x, t) = v - 2c + Et$ satisfies relation

(11.4.11) $g_t + (v - c)g_x = 0$

along the curves for which $dx/dt = v - c$. Thus wherever the curve families C_1 and C_2 cover the x, t-plane in such a way as to furnish a curvilinear coordinate system the relations (11.4.10) and (11.4.11) hold. If now equations (11.4.10) and (11.4.11) are added and the definitions of $f(x, t)$ and $g(x, t)$ are recalled it is readily seen that the first of equations (11.4.6) results. By subtracting (11.4.11) from (11.4.10) the second of equations (11.4.6) is obtained. In other words, any functions v and c which satisfy the relations (11.4.8) and (11.4.9) will also satisfy (11.4.6) and the two systems of equations are therefore now seen to be completely equivalent.

As we would expect on physical grounds, a solution of the original dynamical equations (11.4.6) could be shown to be uniquely determined when appropriate initial conditions (for $t = 0$, say) and boundary conditions are prescribed; it follows that any solutions of (11.4.8) and (11.4.9) are also uniquely determined when such conditions are prescribed since we know that the two systems of equations are equivalent.

At first sight one might be inclined to regard the relations (11.4.8) and (11.4.9) as more complicated than the original pair of partial differential equations, particularly since the right hand sides of (11.4.8) are not known and hence the characteristic curves are also not known. Nevertheless, the formulation in terms of the characteristics is quite useful in studying properties of the solutions and also in studying questions referring to the appropriateness of various boundary and initial conditions. In Chapter 10.2 a detailed discussion along these lines is given; we shall not repeat it here, but will summarize the conclusions. The description of the properties of the solution is given in the x, t-plane, as indicated in Fig. 11.4.1. In the first place, *the values of v and c at any point $P(x, t)$ within the region of existence of the solution are determined solely by the initial values prescribed on the segment of the x-axis which is subtended by the two characteristics issuing from P.* In addition, the two characteristics issuing from P are themselves also determined solely by the initial values on the segment subtended by them. Such a segment of the x-axis is often called the domain of dependence of the point P. Correspondingly we may define the range of influence of a point Q on the x-axis as the region of the x, t-plane in which the values of v and c are influenced by the initial values assigned to point Q, i.e., it is the region between the two characteristics issuing from Q. In Fig. 11.4.1 we indicate these two regions.

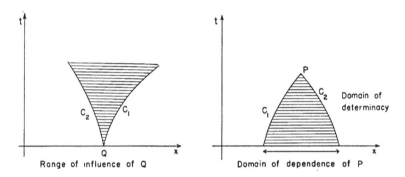

Fig. 11.4.1. The role of the characteristics

We are now in a position to understand the role of the characteristics as curves along which discontinuities in the first and higher derivatives of the initial data are propagated, since it is reasonable to expect (and could be proved) that those points P whose domains of dependence do not contain such discontinuities are points at which the solutions v and c also have continuous derivatives. On the other hand, it could be shown that a discontinuity in the initial data at a certain point does not in general die out along the characteristic issuing from that point. Such a discontinuity (or disturbance in the water) therefore spreads in both directions over the surface of the water with the speed $v + c$ in one direction and $v - c$ in the other in view of the interpretation given to the characteristics through the relations $(11.4.7)_1$. Since v is the velocity of the water particles we see that c represents quite generally the speed at which a discontinuity in a derivative of the initial data propagates relative to the moving water. We are therefore justified in referring to the quantity $c = \sqrt{gy}$ as the wave speed or propagation speed.

We considered above a problem in which only initial conditions, and no boundary conditions, were prescribed. In the problems we consider later, however, such boundary conditions will occur in the form of conditions prescribed at a certain fixed point of the river in terms of the time: for example, the height, or stage, of the river might be given at a certain station as a function of the time. In other words, conditions would be prescribed not only along the x-axis of our x, t-plane, but also along the t-axis (in general only for $t > 0$) for a certain fixed value of x. The method of finite differences used in Chapter 10.2 to discuss the initial value problem, with the general result given above, can be modified in an obvious way to deal with cases in which boundary conditions are also imposed. In doing so, it would also become clear just what kind of boundary conditions could and should be imposed. For example, in the great majority of rivers—in fact, for all in which the flow is subcritical, i.e. such that v is everywhere less than the wave speed $\sqrt{gy}$—it is possible to prescribe only *one condition* along the t-axis, which might be either the velocity v or the depth y, in contrast with the necessity to impose *two* conditions along the x-axis. This fact would become obvious on setting up the finite difference scheme, and examples of it will be seen later on.

Finally, it should be stated that the role of the characteristics, and also the method of finite differences applied to them could be used with reference to the general case of the characteristic equations as embodied in the equations (11.4.7) and (11.4.8) in essentially the same way as was sketched out above for the system comprised of (11.4.8) and (11.4.9) which referred to a special case. In particular, the role of the characteristics as curves along which small disturbances propagate, and their role in determining the domain of dependence, range of influence, etc. remain the same.

11.5. Numerical methods for calculating solutions of the differential equations for flow in open channels

It has already been stated that while the formulation of our problems by the method of characteristics is most valuable for studying many questions concerned with general properties of the solutions of the differential equations, it is in most cases not the best formulation to use for the purpose of calculating the solutions numerically. That is not to say that the device of replacing derivatives by difference quotients should be given up, but rather that this device should be used in a different manner. The basic idea is to operate with finite differences by using a *fixed rectangular net* in the x, t-plane, in contrast with the method outlined in Chapter 10.2, in which the net of points in the x, t-plane at which the solution is to be approximated is determined only gradually in the course of the computation. In the latter procedure it is thus necessary to calculate not only the values of the unknown functions v and c, but also the values of the coordinates x, t of the net points themselves, whereas a procedure making use of a fixed net would require the calculation of v and c only, and it would also have the advantage of furnishing these values at a convenient set of points.

However, the question of the convergence of the approximate solution to the exact solution when the mesh width of a rectangular net is made to approach zero is more delicate than it is when the method of characteristics is used. For example, it is not correct, in general, to choose a net in which the ratio of the mesh width Δt along the t-axis and the mesh width Δx along the x-axis is kept constant independent of the solution: such a procedure would not in general yield approximations converging to the solution of the differential equation problem. The reason for this can be understood with reference to one of the basic facts about the solution of the differential equations which was brought out in the discussion of the preceding section. The basic fact in question is the existence of what was called there the domain of dependence of the solution. For example, suppose the solution were to be approximated at the points of the net of Fig. 11.5.1a by advancing from one row parallel to the x-axis to the next row a distance Δt from it. In addition, suppose this were to be done by determining the approximate values of v and c at any point such as P (cf. Fig. 11.5.1b)

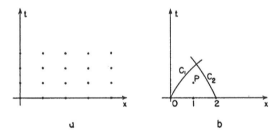

11.5.1. Approximation by using a rectangular net

by using the values of these quantities at the nearest three points 0, 1, 2 in the next line below, replacing derivatives in the two differential equations by difference quotients, and then solving the resulting algebraic equations for the two unknowns v_P and c_P. It seems reasonable to suppose that such a scheme would be appropriate only if P were in the triangular region bounded by the characteristics drawn from points 0 and 2 to form the region within which the solution is determined solely by the data given on the segment $0-2$: otherwise it seems clear that the initial values at additional points on the x-axis ought to be utilized since our basic theory tells us that the initial data at some of them would indeed influence the solution at point P. On the other hand, the characteristic curves themselves depend upon the values of the unknown functions v and c—their slopes, in fact, are given (cf. (11.4.8)) by $dx/dt = v \pm c$ and thus the interval Δt must be chosen small enough in relation to a fixed choice of the interval Δx so that the points such as P will fall within the appropriate domains of determinacy relative to the points used in calculating the solution at P. In other words, the theory of characteristics, even if it is not used directly, comes into play in deciding the relative values of Δt and Δx which will insure convergence (for rigorous treatments of these questions see the papers by Courant, Isaacson, and Rees [C.11], and by Keller and Lax [K.4]).

We shall introduce two different schemes employing the method of finite differences in a fixed rectangular net of the x, t-plane. The first of these makes use of the differential equations in the form given by (11.4.7), and we no longer suppose that the function E is restricted in any way. (It might be noted that the slopes of the characteristics as given by (11.4.8) are determined by the quantities $v \pm c$, no matter how the function E is defined, and in fact also for the most general case of a river having a variable cross section A, etc., and hence we are in a position to determine appropriate lengths for the t-intervals, in accord with the above discussion, in the most general case. This is also a good reason for working with the quantity c in place of y.) At the same time, the calculation is based on assuming that the ap-

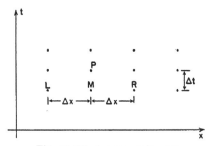

Fig. 11.5.2. A rectangular net

proximate values of c and v have been calculated at the net points L, M, R (cf. Fig. 11.5.2) and that the differential equations are to be used to advance the approximate solution to the point P. The differential equations to be solved are thus

(11.5.1) $\qquad 2\{(c + v)c_x + c_t\} + \{(c + v)v_x + v_t\} + E = 0,$

(11.5.2) $\qquad -2\{(-c + v)c_x + c_t\} + \{(-c + v)v_x + v_t\} + E = 0,$

and the characteristic directions are determined by $dx/dt = v \pm c$. The characteristic with slope $v + c$ we call the *forward* characteristic, and that with slope $v - c$ the *backward* characteristic. We shall replace the derivatives in the equations by difference quotients which approximate the values of the derivatives at the point M. In order to advance the values of v and c from the points L, M, R to the point P by using (11.5.1) and (11.5.2) it is natural to replace the time derivatives v_t and c_t by the following difference quotients

(11.5.3) $\qquad v_t = \dfrac{v_P - v_M}{\Delta t}, \qquad c_t = \dfrac{c_P - c_M}{\Delta t}$

in *both* equations. However, in order to insure the convergence of the approximations to the exact solution when $\Delta x \to 0$ and $\Delta t \to 0$ (see Courant, Isaacson, and Rees [C.11] for a proof of this fact) it is necessary to replace the derivatives v_x and c_x by difference quotients which are defined differently for (11.5.1) than for (11.5.2), as follows:

(11.5.4) $\qquad v_x = \dfrac{v_M - v_L}{\Delta x}, \qquad c_x = \dfrac{c_M - c_L}{\Delta x} \qquad$ in (11.5.1),

(11.5.5) $\qquad v_x = \dfrac{v_R - v_M}{\Delta x}, \qquad c_x = \dfrac{c_R - c_M}{\Delta x} \qquad$ in (11.5.2).

The reason for this procedure is, at bottom, that (11.5.1) is an equation associated with the forward characteristic, while (11.5.2) is associated with the backward characteristic. The coefficients of the derivatives and the function E are, of course, to be evaluated at the point M. The difference equations replacing (11.5.1) and (11.5.2) are thus given by

(11.5.6) $\quad 2\left\{ (c_M + v_M) \dfrac{c_M - c_L}{\Delta x} + \dfrac{c_P - c_M}{\Delta t} \right\}$

$$+ \left\{ (c_M + v_M) \dfrac{v_M - v_L}{\Delta x} + \dfrac{v_P - v_M}{\Delta t} \right\} + E(v_M, c_M) = 0,$$

(11.5.7) $\quad -2\left\{ (-c_M + v_M) \dfrac{c_R - c_M}{\Delta x} + \dfrac{c_P - c_M}{\Delta t} \right\}$

$$+ \left\{ (-c_M + v_M) \dfrac{v_R - v_M}{\Delta x} + \dfrac{v_P - v_M}{\Delta t} \right\} + E(v_M, c_M) = 0.$$

We observe that the two unknowns, v_P and c_P, occur linearly in these equations; hence they are easily found by solving the equations. The result is

$$(11.5.8) \qquad v_P = v_M + \frac{\Delta t}{\Delta x} \left[(c_M + v_M)(\tfrac{1}{2}v_L - \tfrac{1}{2}v_M + c_L - c_M) \right.$$

$$\left. - (c_M - v_M)(\tfrac{1}{2}v_M - \tfrac{1}{2}v_R - c_M + c_R) - \Delta x E_M \right],$$

$$(11.5.9) \qquad c_P = c_M + \tfrac{1}{2}\frac{\Delta t}{\Delta x} \left[(c_M + v_M)(\tfrac{1}{2}v_L - \tfrac{1}{2}v_M + c_L - c_M) \right.$$

$$\left. + (c_M - v_M)(\tfrac{1}{2}v_M - \tfrac{1}{2}v_R - c_M + c_R) \right].$$

In accordance with the remarks made above, we must also require that the ratio of Δt to Δx be taken small enough so that P lies within the triangle formed by drawing lines from L and R in the directions of the forward and backward characteristics respectively, i.e. lines with the slopes $v_L + c_L$ at L and $v_R - c_R$ at R: a condition that is well-determined since the values of v and c are presumably known at L and R.

One can now see in general terms how the initial value problem starting at $t = 0$ can be solved approximately: One starts with a net along the x-axis with spacing Δx. Since c and v are known at all of these points, the values of c and v can be advanced through use of (11.5.8) and (11.5.9) to a parallel row of points on a line distant Δt along the t-axis from the x-axis. However, the mesh width Δt must be chosen small enough so that the convergence condition is satisfied at all net points where new values of v and c are computed.

We can now see also how to take care of boundary conditions, i.e. of conditions imposed at a fixed point (say at the origin, $x = 0$) as given functions of the time. For example, the depth y (corresponding to the stage of the river) or the velocity v (which together with the cross-section area A fixes the rate of discharge) might be given in terms of the time. Initial conditions downstream from this point (i.e. for $x > 0$) might also be prescribed. Suppose, for example, that the stage of the river is prescribed at $x = 0$, i.e. that $y(0, t)$ is known, and that the calculation had already progressed so far as to yield values of v and c at net points along a certain line parallel to the x-axis and containing the points L, M, R, as indicated in Fig. 11.5.3. It is clear that the determination of the values of v and c at point P can be obtained from their values at L, M, R by using (11.5.8) and (11.5.9), as in the above discussion of the initial value problem, and similarly for points P_1, P_2, etc. However, the value of v at Q must be determined in a different manner; for this purpose we use the equation (11.5.7) with the subscript Q replacing P, L replacing M, and M replacing R. Since v_M, c_M, v_R, c_R are supposed known, and c_Q is also

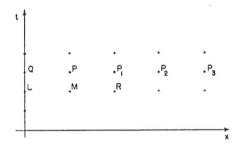

Fig. 11.5.3. Satisfying boundary conditions

known since the values of y are prescribed on the t-axis, it follows that equation (11.5.7) contains v_Q as the only unknown; in fact it is given by the equation

$$(11.5.10) \quad v_Q = v_L + \varDelta t \left\{ \frac{1}{\varDelta x} (c_L - v_L)(2c_L - 2c_M - v_L + v_M) - E_L \right\} + 2(c_Q - c_L).$$

The reason for using (11.5.7) instead of (11.5.6) is, of course, that the points M and Q are associated with the backward characteristic, and hence (11.5.2) should be used to approximate the x-derivatives at point L (where the differential equations are replaced by difference equations). It is quite clear that the same general procedure could be used to calculate c_Q if the values of v had been assumed given along the t-axis. If, on the other hand, we had a boundary condition on the right of our domain instead of on the left, as above, we could make use of (11.5.6) for the forward characteristic as a basis for obtaining the formula for advancing the solution along the t-axis.

The above discussion would seem to imply that under all circumstances only one boundary condition could be imposed — that is, that either v or c could be prescribed at a fixed point on the river, but not both — since prescribing one of these quantities leads to a unique determination of the other. This is, indeed, true in any ordinary river, but not necessarily in all cases. In fact, we made a tacit assumption in the above discussion, i.e. that of the two characteristics issuing from any point of the t-axis only the forward characteristic goes into the region $x > 0$ to the right of the t-axis, and this in turn implies that $v + c$ and $v - c$, which fix the slopes of the characteristics, are opposite in sign. The physical interpretation of this is that the value of v (which is positive here) must be less than $c = \sqrt{gy}$, i.e. that the flow must be what is called tranquil, or subcritical.* Otherwise, as we see from Fig. 11.5.4, we should expect to determine the values of v and c at points close to and to the right of the t-axis, say at K, by

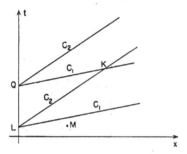

Fig. 11.5.4. A case of super-critical flow

utilizing values for both v and c along the segment LQ, its domain of dependence. The scheme outlined above would therefore have to be modified in a proper way under such circumstances. One sees, however, how useful the theory based on the characteristics can be even though no direct use of it is made in the numerical calculations (aside from decisions regarding the maximum permissible size of the t-interval).

The procedure sketched out above, while it is recommended for use

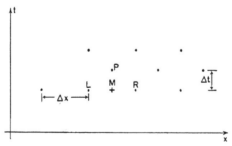

Fig. 11.5.5. A staggered net

* In gas dynamics the flow in an analogous case would be called subsonic.

when a boundary condition is to be satisfied, is not always the best one to use for advancing the solution to such points as P, P_1, P_2, ... in Fig. 11.5.3. For such "interior points" a staggered rectangular net, as indicated in Fig. 11.5.5, and a difference equation scheme based on the original differential equations (11.4.6) may be preferable (cf. Keller and Lax [K.4] for a discussion of this scheme). The equations (11.4.6) were

(11.5.11)
$$\begin{cases} 2cc_x + v_t + vv_x + E = 0, \\ cv_x + 2vc_x + 2c_t = 0. \end{cases}$$

The values v_M and c_M at the mid-point M (which is, however, not a net point) of the segment LR are defined by the averages:

(11.5.12)
$$v_M = \frac{v_L + v_R}{2}, \qquad c_M = \frac{c_L + c_R}{2},$$

after which the derivatives at M are approximated in a quite natural way by the difference quotients

(11.5.13)
$$\begin{cases} v_x = \dfrac{v_R - v_L}{\varDelta x}, & c_x = \dfrac{c_R - c_L}{\varDelta x}, \\ v_t = \dfrac{v_P - v_M}{\varDelta t}, & c_t = \dfrac{c_P - c_M}{\varDelta t}. \end{cases}$$

Upon substitution of these quantities into (11.5.11), evaluation of the coefficients c, v, and E at point M, and subsequent solution of the two equations for v_P and c_P, the result is

(11.5.14)
$$\begin{cases} v_P = v_M + \dfrac{\varDelta t}{\varDelta x} \left[2(c_L - c_R)c_M + (v_L - v_R)v_M - \varDelta x E_M \right], \\ c_P = c_M + \tfrac{1}{2} \dfrac{\varDelta t}{\varDelta x} \left[2(c_L - c_R)v_M + (v_L - v_R)c_M \right]. \end{cases}$$

As we see on comparison with (11.5.8) and (11.5.9'), these equations are simpler than the earlier ones. The criterion for convergence remains the same as before, i.e. that P should lie within a triangle formed by the segment LR and the two characteristics issuing from its ends.

11.6. Flood prediction in rivers. Floods in models of the Ohio River and its junction with the Mississippi River

The theory developed in the preceding sections can be used to make predictions of floods in rivers on the basis of the observed, or estimated, flow into the river from its tributaries and from the local run-off, together with the state of the river at some initial instant. Hydraulics engineers have developed a procedure, called flood-routing, to accomplish the same purpose. The flood-routing procedure can be deduced as an approximation in some sense to the solution of the basic differential equations for flow in open channels (cf. the article by B. R. Gilcrest in the book by Rouse [R.11]), but it makes no direct use of the differential equations. However, the flood-routing procedure in question seems not to give entirely satisfactory results in cases other than that of determining the progress of a flood down a long river — for example, the problem of what happens at a junction, such as that of the Ohio and Mississippi Rivers, or the problem of calculating the transient effects resulting from regulation at a dam, such as the Kentucky dam at the mouth of the Tennessee River, seem to be difficult to treat by methods that are modifications of the more or less standard flood-routing procedures. Even for a long river like the Ohio, the usual procedure fails occasionally to yield the observed river stages at some places. On the other hand, the basic differential equations for flow in open channels are in principle applicable in all cases and can be used to solve the problems once the appropriate data describing the physical characteristics of the river and the appropriate initial and boundary conditions are known.

The idea of using the differential equations directly as a means of treating problems of flow in open channels is not at all new. In fact, it goes back to Massau [M.5] as long ago as 1889. Since then the idea has been taken up by many others (mostly in ignorance of the work of Massau)—for example, by Preiswerk [P.16], von Kármán [K.2], Thomas [T.2], and Stoker [S.19]. Thomas, in particular, attacked the flood-routing problem in his noteworthy and pioneering paper and outlined a numerical procedure for its solution based on the idea of using the method of finite differences. However, his method is very laborious to apply and would also not necessarily furnish a good approximation to the desired solution even if a large number of divisions of the river into sections were to be taken. In general, the amount of numerical work to be done in a direct integration of the differential equations looked too formidable for practical purposes until rather recently.

During and since the late war new developments have taken place which make the idea of tackling flood prediction and other similar problems by numerical solution of the relevant differential equations quite tempting. There have been, in fact, developments in two different directions, both motivated by the desire to solve difficult problems in compressible gas dynamics: 1) development of appropriate numerical procedures—for the most part methods using finite differences —for solving the differential equations, and 2) development of computing machines of widely varying characteristics suitable for carrying out the numerical calculations. As we have seen, the differential equations for flood control problems are of the same type as those for compressible gas dynamics, and consequently the experience and calculating equipment developed for solving problems in gas dynamics can be used, or suitably modified, for solving flood control problems.

In carrying out such a study of an actual river it is necessary to make use of a considerable bulk of observational data—cross-sections and slopes of the channels, measurements of river depths and discharges as functions of time and distance down the river, drainage areas, observed flows from tributaries, etc.—in order to obtain the information necessary to fix the coefficients of the differential equations and to fix the initial and boundary conditions. This is a task with many complexities. For the purposes of this book it is more reasonable to carry out numerical solutions for problems which are simplified versions of actual problems. The present section has as its purpose the presentation of the solutions in a few such special cases, together with an analysis of their bearing on the concrete problems for actual rivers. In any case, the general methods for an actual river would be the same—there would simply be greater numerical complications.

The simplified models chosen correspond in a rough general way (a) to two types of flow for the Ohio River and (b) to the Ohio and Mississippi Rivers at their junction. Rivers of constant slope, with rectangular cross-sections having a uniform breadth, and with constant roughness coefficients are assumed. In this way differential equations with constant coefficients result. The values of these quantities are, however, taken to correspond in order of magnitude with those for the actual rivers. In the model of the Ohio, for example, the slope of the channel was assumed to be 0.5 ft/mile, the quantity n

(the roughness coefficient in Manning's formula) was given the value 0.03, and the breadth of the river was taken as 1000 feet. It is assumed that a steady uniform flow with a depth of 20 ft existed at the initial instant $t = 0$, and that for $t > 0$ the depth of the water was increased at a uniform rate at the point $x = 0$ from 20 ft to 40 ft within 4 hours and was then held fixed at the latter value. (These depths are the same as for the problem of a steady progressing wave treated in sec. 11.2 above.) The problem is to determine the flow downstream, i.e. the depth y and the flow velocity v as functions of x (for $x > 0$) and t.

The methods used to obtain the solution of this problem of a flood in a model of the Ohio River, together with a discussion of the results, will be given in detail later on in this section. Before doing so, a few general remarks and observations about them should be made at this point. In the first place, it was found possible to carry out the solution numerically by hand computation over a considerable range of distances and times (values at 900 net points in the x, t-plane were determined by finite differences), and this in itself shows that the problems are well within the capacity of modern calculating equipment. It might be added that the special case chosen for a flood in the Ohio was one in which the rate of rise at the starting point upstream was extremely high (5 feet per hour, in comparison with the rate of rise during the flood of 1945—one of the biggest ever recorded in the Ohio—which was never larger than 0.7 feet per hour at Wheeling, West Virginia), so that a rather severe test of the finite difference method was made in view of the rapid changes of the basic quantities in space and time. The decisive point in estimating the magnitude of the computational work in using finite differences is the number of net points needed; for a river such as the Ohio it is indicated that an interval Δx of the order of 10 miles along the river and an interval Δt of the order of 0.3 hours in time in a rectangular net in the x, t-plane will yield results that are sufficiently accurate. (Of course, a problem for the Ohio in its actual state involves empirical coefficients in the differential equations and other empirical data, which must be coded for calculating machines, but this would have no great effect on these estimates for Δx and might under extreme flood conditions reduce Δt by a factor of $1/2$.)

As we know from sec. 11.3 above, there is a case in which an exact solution of the differential equations is known, i.e. the case of a steady progressing wave with two different depths at great distances upstream and downstream. The exact solution obtained in sec. 11.3 for the case of a wave of depth 20 ft far downstream and 40 ft far upstream was taken as furnishing the *initial conditions* at $t = 0$ for a wave motion in the river. With the initial conditions prescribed in this way the finite difference method was used to determine the motion at later times; of course the calculation,if accurate, should furnish a wave profile and velocity distribution which is the same at time t as at the initial instant $t = 0$ except that all quantities are displaced downstream a distance Ut, with U the speed of the steady progressing wave. In this way an opportunity arises to compare the approximate solution with an exact solution. In the present case the phase velocity U is approximately 5 mph. Interval sizes of $\Delta x = 5$ miles in a "staggered" finite difference scheme (cf. equations (11.5.14)) with $\Delta t = .08$ hr were taken and a numerical solution was worked out. We report the results here. After 12 hours, the calculated values for the stage y agreed to within .5 per cent with the exact values. The discharge and the velocity deviated by less than .8 per cent from the exact values.

One of the valuable insights gained from working out the solution of the flood problem in a model of the Ohio was an insight into the relation between the methods used by engineers—for example, by the engineers of the Ohio River Division of the Corps of Engineers in Cincinnati—for predicting flood stages, and the methods explained here, which make use of the basic differential equations. At first sight the two methods seem to have very little in common, though both, in the last analysis, must be based on the laws of conservation of mass and momentum; indeed, in one important respect they even seem to be somewhat contradictory. The methods used in engineering practice (which make no direct use of our differential equations) tacitly assume that a flood wave in a long river such as the Ohio propagates only in the downstream direction, while the basic theory of the differential equations we use tells us that a disturbance at any point in a river flowing at subcritical speed (the normal case in general and always the case for such a river as the Ohio) will propagate as a wave traveling upstream as well as downstream. Not only that, the speed of propagation of small disturbances relative to the flowing stream, as defined by the differential equations, is $\sqrt{gy}$ for small disturbances and this is a good deal larger (by a factor of about 4 in our model of the Ohio) than the propagation speed used by the engineers for their flood wave traveling downstream. There is, however, no real discrepancy. The method used by the engineers can be interpreted as a method which yields solutions of the differential equations, with certain terms neglected, that are good approximations (though not under all circumstances, it seems) to the actual solutions in some cases, among them that of flood waves in a river such as the Ohio. The neglect of terms in the differential equations in this approximate theory is so drastic as to make the theory of characteristics, from which the properties of the solutions of the differential equations were derived here, no longer available. The numerical solution presented here of the differential equations for a flood wave in a model of the Ohio yields, as we have said, a wave the front of which travels downstream at the speed $\sqrt{gy}$; but the amplitude of this forerunner is quite small,* while the portion of the wave with an amplitude in the range of practical interest is found by this method to travel with essentially the same speed as would be determined by the engineers' approximate method. What seems to happen is the following: small forerunners of a disturbance travel with the speed $\sqrt{gy}$ relative to the flowing stream, but the resistance forces act in such a way as to decrease the speed of the main portion of the disturbance far below the values given by $\sqrt{gy}$, i.e. to a value corresponding closely to the speed of a steady progressing wave that travels unchanged in form. (One could also interpret the engineering method as one based on the assumption that the waves encountered in practice differ but little from steady progressing waves). As we shall see a little later, our unsteady flow tends to the configuration of a steady progressing wave of depth 40 ft upstream and 20 ft downstream.

This analysis of the relation between the methods discussed here and those commonly used in engineering practice indicated why it may be that the latter methods, while they furnish good results in many important cases, fail to mirror the observed occurrences in other cases. For example, the problem of what happens at a junction of two major streams, and various problems arising in connection with the operation of such a dam as the Kentucky Dam in the Tennessee River seem to be cases in which the engineering methods do not furnish accurate results. These would seem to be cases in which the motions of interest depart too much from those of steady progressing waves, and cases in which the propagation of waves *upstream* is as vital as the propagation downstream. Thus at a major junction it is clear that

* In an appendix to this chapter an exact statement on this point is made.

considerable effects on the upstream side of a main stream are to be expected when a large flow from a tributary occurs. In the same way, a dam in a stream (or any obstruction, or change in cross-section, etc.) causes reflection of waves upstream, and neglect of such reflections might well cause serious errors on some occasions.

The above general description of what happens when a flood wave starts down a long stream—in particular, that it has a lengthy front portion which travels fast, but has a small amplitude, while the main part of the disturbance moves much more slowly—has an important bearing on the question of the proper approach to the numerical solution by the method of finite differences. It is, as we shall see shortly, necessary to calculate—or else estimate in some way—the motion up to the front of the disturbance in order to be in a position to calculate it at the places and times where the disturbances are large enough to be of practical interest. This means that a large number of net points in the finite difference mesh in the x, t-plane lie in regions where the solution is not of much practical interest. Since the fixing of the solution in these regions costs as much effort as for the regions of greater interest, the differential equation method is at a certain disadvantage by comparison with the conventional method in such a case. However, it is possible in simple cases to determine analytically the character of the front of the wave and thus estimate accurately the places and times at which the wave amplitude is so small as to be negligible; these regions can then be regarded as belonging to the regions of the x, t-plane where the flow is undisturbed, with a corresponding reduction in the number of net points at which the solutions must be calculated. A method which can be used for this purpose has been derived by G. Whitham and A. Troesch, and a description of it is given in an appendix to this chapter. If a modern high speed digital computer were to be used to carry out the numerical work, however, it would not matter very much whether the extra net points in the front portion of the wave were to be included or not: many such machines have ample capacity to carry out the necessary calculations.

We proceed to give a description of the calculations made for our model of the Ohio, including a discussion of various difficulties which occurred for the flood wave problem near the front of the disturbance, and particularly at the beginning of the wave motion (i.e. near $x = 0$, $t = 0$), and an enumeration of the features of the calculation which must play a similar role in the more complicated cases presented by rivers in their actual state. This will be followed by a description of the method used and the calculations made for a problem simulating a flood coming down the Ohio and its effect on passing into the Mississippi. This problem and its solution give rise to further general observations which will be made later on.

The differential equations to be solved are

(11.6.1)
$$\begin{cases} 2cc_x + v_t + vv_x + E = 0, \\ 2c_t + 2vc_x + cv_x = 0, \end{cases}$$

with $v(x, t)$ the velocity, and $c = \sqrt{gy}$ the propagation speed of small disturbances. The assumption of a uniform cross-section and the assumption that no flow over the banks occurs (i.e. $q = 0$ in the basic differential equations (11.1.1) and (11.1.6)) have already been used. The quantity E is given by

$$E = -gS + gS_f,$$

with S the slope of the river bed and S_f, the friction slope, given by Manning's formula

$$S_f = \frac{v\,|\,v\,|}{\gamma\left\{y/\left(1 + \dfrac{2y}{B}\right)\right\}^{4/3}}.$$

Here we assume the channel to be rectangular with breadth B.

The numerical data for the problem of a flood in a model of the Ohio River are as follows. For the slope S a value of 0.5 ft/mi was chosen, and B is given the value 1000 ft. For γ a value of 2500 was taken (in foot-sec units), corresponding to a value of Manning's constant n (in the formula $\gamma = (1.49/n)^2$) of 0.03. The special problem considered was then the following: At time $t = 0$, a steady flow of depth 20 ft is assumed. At the "headwaters" of the river, corresponding to $x = 0$, we impose a linear increase of depth with time which brings the level to 40 ft in 4 hours. For subsequent times the level of 40 ft at $x = 0$ is maintained. The initial velocity of the water corresponding to a uniform flow of depth $y_0 = 20$ ft is calculated from $S_f = S$ to be

$$v_0 = 2.38 \text{ mph};$$

the propagation speed of small disturbances corresponding to the depth of 20 ft is

$$c_0 = \sqrt{gy_0} = 17.3 \text{ mph.}$$

The problem then is to determine the solution of (11.6.1) for $v(x, t)$, $c(x, t)$ for all later times $t \geq 0$ along the river $x \geq 0$. Figures 11.6.1 and 11.6.2 present the result of the computation in the form of stage and discharge curves plotted as functions of distance along the river at various times.

In order to indicate how the solution was calculated it is convenient to refer to diagrams in the (x, t) plane given by Figs. 11.6.3 and 11.6.4. According to the basic theory, we know that for $x \geq (v_0 + c_0)t = 19.7t$, called region O in Fig. 11.6.3, the solution is given

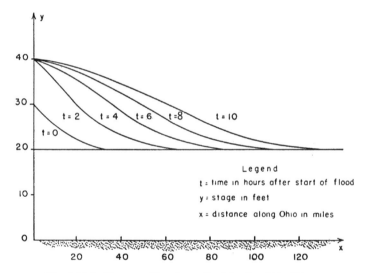

Fig. 11.6.1. Stage profiles for a flood in the Ohio River

by the unchanged initial data, $v(x, t) = v_0$, $c(x, t) = c_0$ (since the forerunner of the disturbance travels at the speed $v_0 + c_0 = 19.7$ mph).

Experiments were made with various interval sizes and finite difference schemes in order to try to determine the most efficient way to calculate the progress of the flood. We proceed to describe the various schemes tried and the regions in which they were used on the basis of Figs. 11.6.3 and 11.6.4.

Region I, $0 \leq x \leq 19.7t$, $0 \leq t \leq .4$. Quite small intervals of

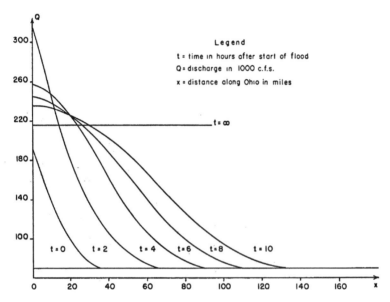

Fig. 11.6.2. Discharge records for a flood in the Ohio River

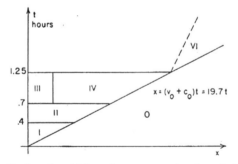

Fig. 11.6.3. Regions in which various computational methods were tried

$\varDelta x = 1$ mile and $\varDelta t = .048$ hours were required owing to the sudden increase of depth at $x = 0$, $t = 0$. The finite difference formulas given above in equations (11.5.8), (11.5.9) were used.

In Region II, $0 \leqq x \leqq 19.7t$, $.4 \leqq t \leqq .7$, with $\varDelta x = 1$ mile,

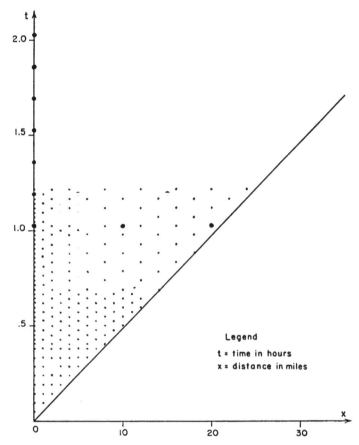

Fig. 11.6.4. Net points used in the finite difference schemes

$\Delta t = .024$ hr, the "staggered" scheme was used. The formulas for this scheme have been given above in equations (11.5.14). In order to calculate $v(0, t)$, the velocity at the upstream boundary of the river, the formula associated with the backward characteristic, namely equation (11.5.10), has to be used twice in succession: for the triangles FBM and MRP (cf. Figs. 11.5.3 and 11.6.5). The values c_B and v_B are simply determined by linear interpolation from the values at the points F and G.

Region III, $0 \leq x \leq 5$, $.7 \leq t \leq 1.25$, with $\Delta x = 1$ mile, $\Delta t = .024$ hr. The same procedure was used as in Region II.

Region IV, $5 \leq x \leq 19.7t$, $.7 \leq t \leq 1.25$, with $\Delta x = 2$ miles, $\Delta t = .048$ hr. The values at the boundary between Regions III and

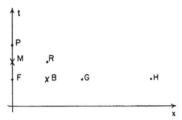

Fig. 11.6.5. Net point arrangement used at boundary in "staggered" scheme

IV were obtained by linear interpolation from the neighboring values. Other quantities were computed by the "staggered" scheme as in Regions II and III.

Region V, $0 \leq x \leq Ut$, $1.25 \leq t \leq 10$, $\Delta x = 5$ miles, $\Delta t = .17$ hr. U represents a variable speed which marks the downstream end of what might be called the observable disturbance ($U \approx 10$ mph). That is, by using an expansion scheme (see the appendix to this chapter) we obtain the solution in

Region VI, defined by $Ut \leq x \leq 19.7t$, back of the forerunner of the disturbance, in which the flow is essentially undisturbed for all practical purposes. The expansion valid near the front of the wave and referred to above was used to calculate the various quantities in Region VI, and a staggered scheme was used to compute the values in Region V.

A number of conclusions reached on the basis of the experience gained from these calculations of a flood in a model of the Ohio River can be summarized as follows:

(a) The rate of rise of the flood—5 feet per hour—is extreme, and such a case exaggerates the way in which errors in the finite difference methods are propagated. For example, slight inaccuracies at the head, $x = 0$, were found to develop upon increasing the size of the Δx interval. In spite of the exceptionally high rate of rise of the flood, the fluctuations created by using finite difference methods were damped out rather strongly (in about $8-10$ time steps). It is possible to control these inaccuracies simply by using small interval sizes. The process by which the small errors of the finite difference scheme are caused to die out may be described as follows: A value of v which is too large produces a correspondingly larger friction force which slows down the motion and produces at a later time a smaller velocity. The lower velocity in a similar way then operates through the resistance to create a larger velocity and the process repeats in an oscillatory fashion with a steady decrease in the amplitude of variation.

(b) The accuracy of our computation (as a function of the interval size) was checked by repeating the calculation for two different interval sizes over the same region in space and time.

(c) A linearized theory of wave propagation, obtained by assuming a small perturbation about the uniform flow with 20 ft depth, is easily obtained, and the problem was solved using such a theory. However, it does not give an accurate description of the solution of our problem. It was found that the stage was predicted too low by the linear theory by as much as 2 feet after only 2 hours—a very large error.

(d) It would be convenient to be in possession of a safe estimate for the maximum value of the particle velocity, in order to select an appropriate safe value for the time interval Δt, since we must have $\Delta t \leq \Delta x/(v + c)$ in order to make sure that the finite difference scheme converges. The calculations in our special case indicate that this may not be easy to obtain in a theoretical way, since the maximum velocity at $x = 0$, for example, greatly exceeds its asymptotic value, as indicated in Fig. 11.6.6. In a

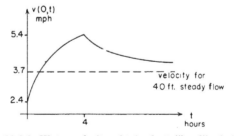

Fig. 11.6.6. Water velocity obtained at "head" of river

computation for an actual river, however, no real difficulty is likely to result, since c is in general much larger than v and is determined by the depth alone.

(e) As was already indicated above, the curves of constant stage turn out to have slopes which are closer to 5 mph (the speed with which a steady progressing flow, 40 ft upstream and 20 ft downstream, moves) than they are to the 19.7 mph speed of propagation of small disturbances. This is shown by Fig. 11.6.7.

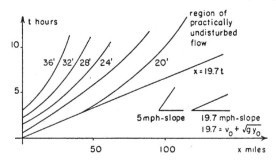

Fig. 11.6.7. Curves of constant stage—comparison with first characteristic and steady progressing flow velocity

The region of practically undisturbed flow (determined by an expansion about the "first" characteristic $x = 19.7t$, for which see the appendix to this chapter) is shown above. In an actual river, we would of course expect the local runoff discharges and the non-uniform flow conditions to eliminate largely the region of practically undisturbed flow. For this reason it is not feasible to use analytic expansion schemes as a means of avoiding computational labor.

We turn next to our model of the junction of the Ohio and Mississippi Rivers and the problem of what happens when a flood wave comes down the Ohio and passes through the junction.* The physical data chosen are the same as were used above in sec. 11.2 in discussing the problem of a steady flow at a junction.

We suppose the upstream side of the Mississippi to be identical with the Ohio River—i.e. that it has a rectangular cross-section 1000 ft wide, a slope of .5 ft/mile, and that Manning's constant n has the value .03. The downstream Mississippi is also taken to be rectangular, but twice as wide, i.e. 2000 ft in width, Manning's constant is again assumed to have the value .03, but the slope of this branch is given the value .49 ft/mile. This modification of the slope was made

---

* The analogous problem in gas dynamics would be concerned with the propagation of a wave at the junction of two pipes containing a compressible gas.

in order to make possible an initial solution corresponding to a uniform flow of 20 ft depth in all three branches. (Such a change is necessary in order to overcome the decrease in wetted perimeter which occurs on going downstream through the junction.) Figure 11.6.8 shows a schematic plan of the junction. The concrete problem to be solved is formulated as follows. A flood is initiated in the Ohio

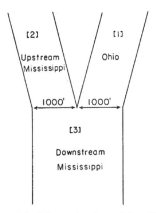

Fig. 11.6.8. Schematic plan of junction

at a point 50 miles above the junction by prescribing a rise in depth of the stream at that point from 20 ft to 40 ft in 4 hours—in other words, the same initial and boundary conditions were assumed as for the case of the flood in the Ohio treated in detail above. After about 2.5 hours the forerunner, or front, of the wave in the Ohio caused by the disturbance 50 miles upstream reaches the junction; up to this instant nothing will have happened to disturb the Mississippi, and the numerical calculations made above for the Ohio remain valid during the first 2.5 hours. Once the disturbance created in the Ohio reaches the junction, it will cause disturbances which travel both upstream and downstream in the Mississippi, and of course also a reflected wave will start backward up the Ohio. The finite difference calculations therefore were begun in all three branches from the moment that the junction was reached by the forerunner of the Ohio flood, and the solution was calculated for a period of 10 hours.

We proceed to describe the method of determining the numerical solution. Let $v_{(1)}$, $c_{(1)}$, $v_{(2)}$, $c_{(2)}$, $v_{(3)}$, $c_{(3)}$ represent the velocity v and the propagation speed c for the Ohio, upstream Mississippi, and downstream Mississippi, respectively. A "staggered" scheme was used with intervals $\Delta x = 5$ miles and $\Delta t = .17$ hr as indicated in Fig. 11.6.9. The junction point is denoted by $x = 0$, the region of the Ohio and

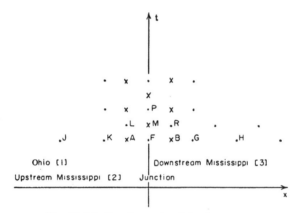

Fig. 11.6.9. Junction net point scheme

the upstream Mississippi are represented by $x \leq 0$, while the downstream Mississippi is described for $x \geq 0$. The time $t = 2.5$ hrs, as explained above, corresponds to the instant that the forerunner of the flood reaches the junction.

The values of the quantities v and c at the junction were determined as follows: Assume that the values of v and c have been obtained at all net points for times preceding that of the boundary net point P, which represents a point at the junction. We use at this point the relations

$$c_{(1)} = c_{(2)} = c_{(3)},$$

since $c = \sqrt{gy}$ and the water level is the same in the three branches at the junction. In addition, we have

$$y_{(1)}v_{(1)} + y_{(2)}v_{(2)} = 2y_{(3)}v_{(3)},$$

since what flows into the junction from the upstream side of the Mississippi and from the Ohio must flow out of the junction into the downstream branch of the Mississippi. If the values of v and c were known at the point M in Fig. 11.6.9 in the respective branches of the rivers, we could find the values at P from equation (11.5.6) for the Ohio and the upstream side of the Mississippi, and equation (11.5.7) for the downstream side of the Mississippi. We rewrite the equations for convenience, as follows:

$$c_{P(1)} = c_{P(2)} = c_{P(3)}, \qquad \text{(with } c = \sqrt{gy}\text{)},$$

$$v_{P(1)} + v_{P(2)} = 2v_{P(3)}, \qquad \text{(since } y_{(1)} = y_{(2)} = y_{(3)}\text{)},$$

$$(11.6.2)_2 \qquad 2\left\{ \frac{c_{P(j)} - c_{M(j)}}{\Delta t} + (c_{M(j)} + v_{M(j)}) \frac{(c_{M(j)} - c_{L(j)})}{\Delta x} \right\}$$

and

$$+ \left\{ \frac{v_{P(j)} - v_{M(j)}}{\Delta t} + (c_{M(j)} + v_{M(j)}) \left(\frac{v_{M(j)} - v_{L(j)}}{\Delta x} \right) \right\} + E_{M(j)} = 0, \qquad j = 1, 2,$$

$$(11.6.2)_3 \qquad -2\left\{ \frac{c_{P(3)} - c_{M(3)}}{\Delta t} + (v_{M(3)} - c_{M(3)}) \frac{(c_{R(3)} - c_{M(3)})}{\Delta x} \right\}$$

$$+ \left\{ \frac{v_{P(3)} - v_{M(3)}}{\Delta t} + (v_{M(3)} - c_{M(3)}) \frac{(v_{R(3)} - v_{M(3)})}{\Delta x} \right\} + E_{M(3)} = 0.$$

The above system of six linear equations determines uniquely the values $v_{(1)}, c_{(1)}, v_{(2)}, c_{(2)}, v_{(3)}, c_{(3)}$ at P in terms of their values at the preceding points L, M and R. The equations can be solved explicitly. The values of the relevant quantities at M are determined in the same way from the preceding values at A, F and B. The values at A and B are determined by interpolation between the neighboring points (K, F) and (F, G) respectively (see Fig. 11.6.9). Of course, it is necessary to treat the motions in each of the branches away from the junction by the same methods as were described for the problem of the Ohio treated above, and this is feasible once the values of v and c have been obtained at the junction.

The results of the calculations are shown in Fig. 11.6.10, which furnishes the river profiles, i.e. the depths as functions of the location in each of the three branches, for times $t = 0$, 2.5, 4, and 10 hours after the beginning of the flood 50 miles up the Ohio. The curves for $t = \infty$ are those for the steady flow which was calculated above in sec. 11.2 (cf. Fig. 11.2.3). The calculations indicate that the unsteady flow does tend to the steady flow as the time increases. Another noticeable effect is the backwater effect in the upper branch of the Mississippi. For example, the stage is increased by about 2 feet at a point in the Mississippi 20 miles above the junction and 7.5 hours after the flood wave from the Ohio first reaches the junction.

It might be mentioned that the forerunners of the flood in all three branches were computed by using the expansion scheme which is explained in the appendix to this chapter.

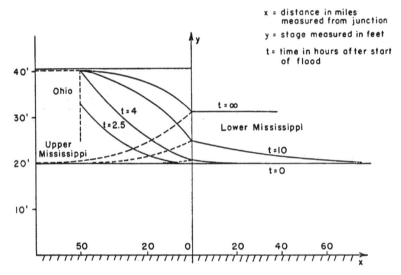

Fig. 11.6.10 River profiles for the junction

11.7. Numerical prediction of an actual flood in the Ohio, and at its junction with the Mississippi. Comparison of the predicted with the observed floods

The methods for numerical analysis of flood wave problems in rivers developed above and applied to simplified models of the Ohio and its junction with the Mississippi have been used to predict the progress of a flood in the Ohio as it actually is, and likewise to predict the progress of a flood coming from the Ohio and passing through the junction with the Mississippi. The data for the flood in the Ohio were taken for the case of the big flood of 1945, and predictions were made numerically for periods up to sixteen days for the 400-mile long stretch of the Ohio extending from Wheeling, West Virginia, to Cincinnati, Ohio. For the flood through the junction. the data for the 1947 flood were used, and predictions were made in all three branches for distances of roughly 40 miles from the junction along each branch. In each case the state of the river, or river system, was taken from the observed flood at a certain time $t = 0$; for subsequent times the inflows from tributaries and the local run-off in the main river valley were taken from the actual records, and then the differential equations were integrated numerically with the use of the UNIVAC digital computer in order to obtain the river stages and discharges at future times. The flood predictions made in this way were then compared with the actual records of the flood.

A comparison of observed with calculated flood stages will be given later on; however, it can be said in general that there is no doubt that this method of dealing with flood waves in rivers is entirely feasible since it gives accurate results without the necessity for unduly large amounts of expensive computing time on a machine such as the UNIVAC. For example, a prediction for six days in the 400-mile stretch of the Ohio requires less than three hours of machine time. This amount of calculating time—which is anyway not unreasonably large—could almost certainly be materially reduced by modifying appropriately the basic methods; so far, no attention has been given to this aspect of the problem, since it was thought most important first of all to find out whether the basic idea of predicting floods by integrating the complete differential equations is sound. The fact that such problems can be solved successfully in this way is, of course, a matter of considerable practical importance from various points of view. For example, this method of dealing with flood problems in rivers is far less expensive than it is to build models of a long river or a river system, and it appears to be accurate. Actually, the two methods—empirically by a model, or by calculation from the theory —are in the present case basically similar, since the models are really huge and expensive calculating machines of the type called analogue computers, and the processes used in both methods are at bottom the same, even in details. An amplification of these remarks will be made later on.

It would require an inordinate amount of space in this book to deal in detail with the methods used to convert the empirical data for a river into a form suitable for computations of the type under discussion here, and with the details of coding for the calculating machine; for this, reference is made to a report [I.4]. Instead, only a brief outline of the procedures used will be given here.

In the first place, it is necessary to have records of past floods with stages up to the maximum of any to be predicted. It would be ideal to have records of flood stages and discharges (or, what comes to the same thing, of average velocities over a cross-section) at points closely spaced along the river—at ten mile intervals, say. Unfortunately, measurements of this kind are available only at much wider intervals *—of the order of 50 to 80 miles or more—even in the Ohio River, for which the data are more extensive than for most rivers in the United States. From such records, it is possible to obtain the coefficient of the all-important resistance term in the differential equation expressing the law of conservation of momentum. This coefficient depends on both the location of the point along the river and the stage. The other essential quantity, the cross-section area, also as a function of location along the river and of stage, could in principle be determined from contour maps of the river valley; this is, in fact, the method used in building models, and it could have been used in setting up the problem for numerical calculation in the manner under discussion here. If that had been done, the results obtained would probably have been more accurate; however, such a procedure is extremely laborious and time consuming, and since the other equally important empirical element, i.e. the resistance coefficient, is known only as an average over each of the reaches (this applies equally to the models of a river), it seems reasonable to make use of an average cross-section area over each reach also. Such an average cross-section area was obtained by analyzing data from past floods in such a way as to determine the water storage volumes in each reach, and from them an average cross-section area as a function of the river stages was calculated. In this way the coefficients of the differential equations are obtained as numerically tabulated functions of x and y. (It might perhaps be reasonable to remark at this point that the carrying out of this program is a fairly heavy task, which requires close cooperation with the engineers who are familiar with the data and who understand also what is needed in order to operate with the differential equations).

In Fig. 11.7.1 a diagrammatic sketch of the Ohio River between Wheeling and Cincinnati is shown, together with the reaches and observation stations at their ends. What we now have are resistance coefficients and cross-section areas that represent averages over any given reach. However, the reaches are too long to serve as intervals for the method of finite differences—which is basic for the numerical integration of the differential equations. Rather, an interval between

---

* Each such interval is called a *reach* by those who work practically with river regulation problems.

net points (in the staggered scheme described in the preceding section) of 10 miles was taken in order to obtain a sufficiently accurate approximation to the exact solution of the problem. A time interval of

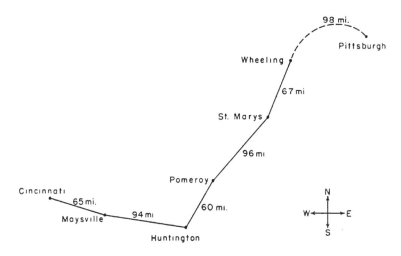

Fig. 11.7.1. Reaches in the Ohio

9 minutes was used. Actually, calculations were first made using a 5-mile interval along the river, but it was found on doubling the interval to 10 miles that no appreciable loss in accuracy resulted.

To begin with, flood predictions for the 1945 flood were made, starting at a time when the river was low and the flow was practically a steady flow. Calculations were first made for a 36 hour period during which the flood was rising; as stated earlier, these were made using the measured inflows from tributaries, and the estimated run-off in the main valley. Upon comparison with the actual records, it was found that the predicted flood stages were systematically higher than the observed flood stages, and that the discrepancy increased steadily with increase in the time. It seemed reasonable to suppose that the error was probably due to an error in the resistance coefficient. Consequently a series of calculations was made on the UNIVAC in which this coefficient was varied in different ways; from these results, corrected coefficients were estimated for each one of the reaches. Actually this was done rather roughly, with no attempt to make corrections that would require a modification in the shape of these curves in their dependence on the stage. The new coefficients, thus corrected on the basis of 36-hour predictions (and thus for flood stages far under the maximum), were then used to make predictions for various 6-day periods, as well as some 16-day periods, with quite good results, on the whole.

  It might be said at this point that making such a correction of the resistance ocefficient on the basis of a comparison with an actual flood corresponds exactly to what is done in making model studies. There, it is always necessary to make a number of verification runs after the model is built in order to compare the observed floods in the model with actual floods. In doing so, the first run is normally made without making any effort to have the resistance correct—in fact, the roughness of the concrete of the model furnishes the only resistance at the start. Of course it is then observed that the flood stages are too low because the water runs off too fast. Brass knobs are then screwed into the bed of the model, and wire screen is placed at some parts of the model, until it is found that the flood stages given by the model agree with the observations. This is, in effect, what was done in making numerical calculations. In other words, the resistance cannot be scaled properly in a model, but must be taken care of in an empirical way. The model is thus not a true model, but, as was stated earlier, it is rather a calculating machine of the class called analogue computers. It is, however, a very expensive calculating machine which can, in addition, solve only one very restricted problem. A model of two fair sized rivers, for example, consisting of two branches perhaps 200 miles in length upwards from their junction, together with a short portion below the junctions. could cost more than a UNIVAC.

  It has already been stated that average cross-section areas for the individual reaches were used in making the numerical computations, while in the model the cross-sections are obtained from the contour maps. In operating numerically it is possible to change the local cross-section areas without any difficulty, and this might be necessary at certain places along the river.

  Some idea of the results of the calculations for the 1945 flood in the Ohio is given by Fig. 11.7.2. The graph shows the river stage at Pomeroy as a function of the time. At the other stations the results were on the whole more accurate. The graph marked "computation with original data", and which covers a 36 hour period. was computed on

the basis of the resistance coefficients as estimated from the basic
flow data for the river. As one sees, these coefficients resulted in much
too high stages, and corrections to them were made along the river

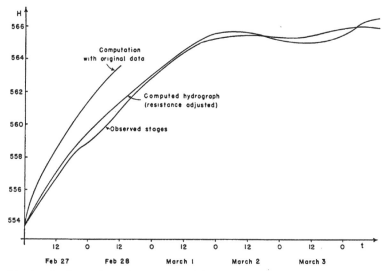

Fig. 11.7.2. Comparison of calculated with observed stages at Pomeroy for the
1945 flood in the Ohio River

on the basis of the results of this computation. Afterwards, flood
predictions were made for periods up to 16 days without further

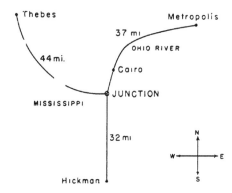

Fig. 11.7.3. The junction of the Ohio and the Mississippi

correction of these coefficients. The graph indicates results for a 6 day period during which the flood was rising. Evidently, the calculated and observed stages agree very well.

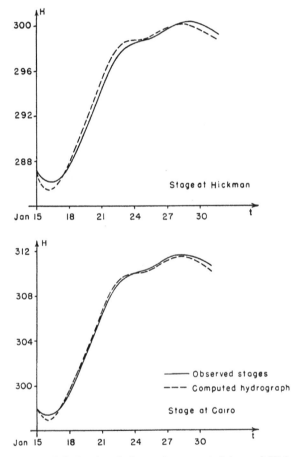

Fig. 11.7.4. Calculated and observed stages at Cairo and Hickman

In Fig. 11.7.3 a diagrammatic sketch of the junction of the Ohio and the Mississippi is shown indicating the portions of these rivers which entered into the calculation of a flood coming down the Ohio and passing through the junction. The flood in question was that of

1947. It was assumed that the stages at Metropolis in the Ohio (about 40 miles above Cairo) and at Thebes in the upper Mississippi (also about 40 miles above Cairo) were given as a function of the time. At Hickman in the lower Mississippi (about 40 miles below Cairo) the stage-discharge relation at this point, as known from observations, was used as a boundary condition. The results of a calculation for a 16 day period are shown in Fig. 11.7.4, which gives the stages at Cairo, and at the terminating point in the lower Mississippi, i.e. at Hickman. As one sees, the accuracy of the prediction is very high, the error never exceeding 0.6 foot. It might be mentioned that a prediction for 6 days requires about one hour of calculating time on the UNIVAC, so that the calculating time for the 16 day period was under 3 hours, which seems reasonable. This problem of routing a flood through a junction is, as has been mentioned before, one which has not been dealt with successfully by the engineering methods used for flood routing in long rivers.*

Appendix to Chapter 11

Expansion in the neighborhood of the first characteristic

It has been mentioned already that whereas the forerunner of a disturbance initiated at a certain point in a river at a moment when the flow is uniform travels downstream with the speed $v + \sqrt{gy}$, the main part of the flood wave travels more slowly (cf. Deymié [D.9]), depending strongly on the resistance of the river bed. An investigation of the motion near the head of the wave, i.e. near the first characteristic (cf. the first part of sec. 11.6) with the equation $x = (v_0 + c_0)t$, shows immediately why the main part of the disturbance will in general fall behind the forerunners of the wave.

The motion is investigated in this Appendix by means of an expansion in terms of a parameter that has been devised by G. Whitham and A. Troesch and carried out to terms of the two first orders for the model of the Ohio River, and to the lowest order in the much more

---

* *Added in proof:* In the meantime, calculations have been completed (see [I.4a]) for the case of floods through the Kentucky Reservoir at the mouth of the Tennessee River. The calculated and observed stages differed only by inches for a flood period of three weeks over the 186 miles of the resevoir.

complicated case of the junction problem. The results obtained make it possible to improve the accuracy of the solution near the first characteristic which separates the region of undisturbed flow from that of the flood wave. It turns out that the finite difference scheme yields river depths which are too large, as indicated by Fig. 11.A.1.

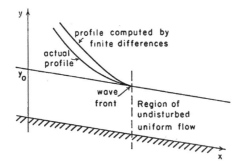

Fig. 11.A.1. Error introduced by finite difference scheme in neighborhood of first characteristic of a rapidly rising flood wave

In order to expand the solution in the neighborhood of the wave front, we introduce new coordinates ξ and τ as follows:

$$\xi = x \text{ and } \tau = (v_0 + c_0)t - x$$

such that the ξ-axis (i.e. $\tau = 0$) coincides with the first characteristic. Near the front of the wave τ will be small, and the expansion will be carried out by developing v and c in powers of τ. The basic system of equations is restated for convenience:

$$2cc_x + v_t + vv_x - gS + gS_f = 0,$$
$$cv_x + 2vc_x + 2c_t = 0.$$

Upon substitution of the new variables ξ and τ we find

$$(11.A.1) \quad \begin{cases} 2c(c_\xi - c_\tau) + v(v_\xi - v_\tau) + (v_0 + c_0)v_\tau - gS + gS_f = 0, \\ 2v(c_\xi - c_\tau) + c(v_\xi - v_\tau) + 2(\dot{v}_0 + c_0)c_\tau = 0, \end{cases}$$

where the friction slope S_f for a rectangular channel of width B is given by

$$S_f = \frac{v\,|\,v\,|}{\gamma \left\{ y / \left(1 + \dfrac{2y}{B}\right) \right\}^{4/3}} = \frac{g^{4/3}}{\gamma} v\,|\,v\,| \left\{ \frac{1}{c^2} + \frac{2}{gB} \right\}^{4/3}.$$

We expand v and c as power series in τ with coefficients that are functions of ξ as follows:

$$v = v_0 + v_1(\xi)\tau + v_2(\xi)\tau^2 + \ldots,$$
$$c = c_0 + c_1(\xi)\tau + c_2(\xi)\tau^2 + \ldots.$$

This expansion is to be used for $\tau > 0$ only, since for $\tau < 0$ we are in the undisturbed region and all the functions $v_1(\xi)$, $v_2(\xi)$, $\ldots$, $c_1(\xi)$, $c_2(\xi)$, $\ldots$ vanish identically. If we insert the series for v and c into equations (11.A.1) and collect terms of the same order in τ, we get ordinary differential equations for $v_1(\xi)$, $c_1(\xi)$, $\ldots$. The equations resulting from the terms of zero order in τ yield $v_1 = 2c_1$. The first order terms become, after thus eliminating v_1,

11.A.2)
$$\begin{cases} 2(v_0 + c_0)\dfrac{dc_1}{d\xi} - 6c_1^2 + 2c_0 v_2 - 4c_0 c_2 + 4c_1 gS\left(\dfrac{1}{v_0} - \dfrac{2}{3c_0}\cdot\dfrac{1}{1 + \dfrac{2c_0^2}{gB}}\right) = 0 \\[4mm] 2(v_0 + c_0)\dfrac{dc_1}{d\xi} - 6c_1^2 - 2c_0 v_2 + 4c_0 c_2 \hspace{3.2cm} = 0. \end{cases}$$

By adding these two equations and removing the common factor 4, we find the differential equation for $c_1(\xi)$ is:

$$(v_0 + c_0)^2\cdot\frac{dc_1}{d\xi} - 3c_1^2 + c_1 gS\left(\frac{1}{v_0} - \frac{2}{3c_0}\cdot\frac{1}{1 + \dfrac{2c_0^2}{gB}}\right) - 0.$$

Although the solution of this differential equation for $c_1(\xi)$ is easily obtained, the result expressed in general terms is complicated, and it is preferable to give it only for the case of the model of the Ohio River

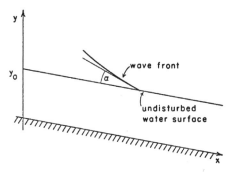

Fig. 11.A.2. Behavior near the front of a wave

using the parameter values introduced above. In this case we find: $c_1 = (1.05 + 8.06 \ e^{0.146\xi})^{-1}$, with c_1 and ξ in miles and hours. This result has the following physical meaning: The angle α of the profile measured between the wave front and the undisturbed water surface dies out exponentially: $\alpha \sim 1/(1 + a e^{bx})$, with a and b constants depending on the river and the boundary condition at $x = 0$. Theoretically, α could also increase exponentially downstream so that a bore would eventually develop, but only if the increase in level at $x = 0$ is extremely fast; in our example no bore will develop unless the water rises at the extremely rapid rate of at least 1 ft per minute.

Unfortunately, the evaluation of $c_2(\xi)$, which yields the curvature of

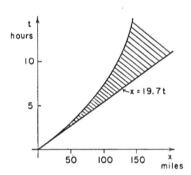

Fig. 11.A.3. Region of practically undisturbed flow

the profile at the wave front, is already very cumbersome. The curvature is found to decrease for large x like xe^{-bx}, b being a positive constant. With the two highest order terms in the expansion known, it is possible to estimate the region adjacent to the first characteristic where the flow is practically undisturbed. It is remarkable how far behind the forerunner the first measurable disturbance travels (see Fig. 11.A.3).

In a similar way, an expansion as a power series in τ has been carried out for the problem of the junction of the Ohio and Mississippi, as described in earlier sections. Here even the lowest order term was obtained only after a complicated computation, since it was necessary to work simultaneously in three different x, t-planes, with boundary conditions at the junction. The differential equations for c_1 are, in all three branches, of the same type as for the Ohio, and their solution for the junction problem with the parameters of section 11.6

are $c_1 = .00084 \ e^{.145\xi}$ for the upstream branch of the Mississippi, and $c_1 = .00084 \ e^{-.229\xi}$ for the downstream branch of the Mississippi, c_1 and ξ both being given in miles and hours. This means that the angle α also dies out exponentially in the Mississippi, a little faster downstream than upstream, as might have been expected, since the oncoming water in the upstream branch has the affect of making the wave front steeper.

In the problem of the idealized Ohio River and of the idealized problem of its junction with the Mississippi River the expansions were carried out numerically in full detail and were used to avoid computation by finite differences in a region of practically undisturbed flow.*

* This would become more and more important if the flow were to be computed beyond 10 hours.

PART IV

CHAPTER 12

Problems in which Free Surface Conditions are Satisfied Exactly. The Breaking of a Dam. Levi-Civita's Theory

This concluding chapter constitutes Part IV of the book. In Part I the basic general theory and the two principal approximate theories were derived. Part II deals with problems treated by means of the linearized theory arising from the assumption that the motion is a small deviation from a state of rest or from a uniform flow. Part III is concerned with the approximate nonlinear theory which arises when the depth of the water is small, but the amplitude of the waves need not be small. Finally, in this chapter we deal with a few problems in which no assumptions other than those involved in the basic general theory are made. In particular, the nonlinear free surface conditions are satisfied exactly.

The first type of problem considered in this chapter belongs in the category of problems concerned with motions in their early stages after initial impulses have been applied. A typical example is the motion of the water in a dam when the dam is suddenly broken. This problem will be treated along lines worked out by Pohle [P.11], [P.12]. Similar problems involving the collapse of a column of liquid in the form of a circular half-cylinder or of a hemisphere resting on a rigid bottom have been treated by Penney and Thornhill [P.2] by a method different from that used by Pohle.

The second section of the chapter deals with the theory of steady progressing waves of finite amplitude. The existence of exact solutions of this type is proved, following in the main the theory worked out by Levi-Civita [L.7].

12.1. Motion of water due to breaking of a dam, and related problems

With the exception of the present section we employ throughout this book the so-called Euler representation in which the velocity and pressure fields are determined as functions of the space variables and the time. In this section it is convenient to make use of what is commonly called the Lagrange representation, in which the displacements of the individual fluid particles are determined with respect to the time and to parameters which serve to identify the particles. Usually the parameters used to specify individual particles are the initial positions of the particles, and we shall conform here to that practice. Only a two-dimensional problem will be treated in detail here; consequently we choose the quantities a, b, and t as independent variables, with a and b representing Cartesian coordinates of the initial positions of the particles at the time $t = 0$. The displacements of the particles are denoted by $X(a, b; t)$ and $Y(a, b; t)$, and the pressure by $p(a, b; t)$. The equations of motion are

$$X_{tt} = -\frac{1}{\varrho} p_X$$

$$Y_{tt} = -\frac{1}{\varrho} p_Y - g$$

in accord with Newton's second law. We assume gravity to be the only external force. These equations are somewhat peculiar because of the fact that derivatives of the pressure p with respect to the dependent variables X and Y occur. To eliminate them we multiply by X_a and Y_a, respectively, and add, then also by X_b, Y_b, and add; the result is

$$(12.1.1) \quad \begin{cases} X_{tt}X_a + (Y_{tt} + g)Y_a + \dfrac{1}{\varrho} p_a = 0, \\[2mm] X_{tt}X_b + (Y_{tt} + g)Y_b + \dfrac{1}{\varrho} p_b = 0, \end{cases}$$

and these are the equations of motion in the Lagrangian form. These equations are not often used because the nonlinearities occur in an awkward way; however, they have the great advantage that a solution is to be found in a fixed domain of the a, b-plane even though a free surface exists. For an incompressible fluid—the only case considered here—the continuity condition is expressed by requiring that the Jacobian of X and Y with respect to a and b should remain unchanged during the flow (since an area element composed always of the same particles has this property); but since $X = a$ and $Y = b$ initially, it follows that

$$(12.1.2) \quad X_a Y_b - X_b Y_a = 1$$

is the condition of continuity. If the pressure p is eliminated from (12.1.1) by differentiation the result is

(12.1.3) $(X_a X_{bt} + Y_a Y_{bt})_t = (X_b X_{at} + Y_b Y_{at})_t.$

Integration with respect to t leads to

(12.1.4) $(X_a X_{bt} + Y_a Y_{bt}) - (X_b X_{at} + Y_b Y_{at}) = f(a, b)$

with f an arbitrary function. It can easily be shown by a calculation using the Eulerian representation that the left hand side of this equation represents the vorticity; consequently the equation is a verification of the law of conservation of vorticity. If the fluid starts from rest, or from any other state with vanishing vorticity, the function $f(a, b)$ would be zero.

The method used by Pohle [P.11], [P.12] to solve the equations (12.1.1) and (12.1.2)—which furnish the necessary three equations for the three functions X, Y, and p—consists in assuming that solutions exist in the form of power series developments in the time, with co-efficients which depend on a and b:

(12.1.5) $\begin{cases} X(a, b; t) = a + X^{(1)}(a, b) \cdot t + X^{(2)}(a, b) \cdot t^2 + \ldots, \\ Y(a, b; t) = b + Y^{(1)}(a, b) \cdot t + Y^{(2)}(a, b) \cdot t^2 + \ldots, \\ p(a, b; t) = p^{(0)}(a, b) + p^{(1)}(a, b) \cdot t + p^{(2)}(a, b) \cdot t^2 + \ldots. \end{cases}$

In these expansions we observe that the terms of order zero in X and Y are a and b—in accordance with the basic assumption that these quantities fix the initial positions of the particles. It should also be noted that $X^{(1)}$ and $Y^{(1)}$ are the components of the initial velocity, and $X^{(2)}$ and $Y^{(2)}$ similarly for the acceleration; in general, we would therefore expect that $X^{(1)}$ and $Y^{(1)}$ would be prescribed in advance as part of the initial conditions. Of course, boundary conditions imposed on X, Y, and p would lead to boundary conditions for the coefficient functions in the series developments. The convergence of the series for the cases discussed below has not been studied, but it seems likely that the series would converge at least for sufficiently small values of the time. The convergence of developments of this kind in some simpler problems in hydrodynamics has been proved by Lichtenstein [L.12].

The series (12.1.5) are inserted first in equation (12.1.2) and the coefficient of each power of t is equated to zero with the following result for the first two terms:

$$(12.1.6) \qquad \begin{cases} X_a^{(1)} + Y_b^{(1)} = 0, \\ X_a^{(2)} + Y_b^{(2)} = - \left(X_a^{(1)} Y_b^{(1)} - X_b^{(1)} Y_a^{(1)} \right). \end{cases}$$

We observe that $X^{(1)}$ and $Y^{(1)}$ are subject to the above relation and hence cannot both be prescribed arbitrarily; however, if the fluid starts from rest so that $X^{(1)} = Y^{(1)} = 0$, the condition is automatically satisfied. The equation for $X^{(2)}$ and $Y^{(2)}$ is linear in these quantities, but nonlinear in $X^{(1)}$ and $Y^{(1)}$. This would be the situation in general: $X^{(n)}$ and $Y^{(n)}$ would satisfy an equation of the form

$$X_a^{(n)} + Y_b^{(n)} = F(X^{(1)}, Y^{(1)}, X^{(2)}, Y^{(2)}, \ldots, X^{(n-1)}, Y^{(n-1)}),$$

with F a nonlinear function in $X^{(i)}, Y^{(i)}, i = 1, 2, \ldots, n - 1$. In the following we shall consider only motions starting from rest. Consequently, we have $X^{(1)} = Y^{(1)} = 0$, and equation (12.1.4) holds with $f \equiv 0$; a substitution of the series in powers of t in equation (12.1.4) yields (for the lowest order term):

$$(12.1.7) \qquad\qquad X_b^{(2)} - Y_a^{(2)} = 0.$$

The higher order coefficients satisfy an equation of the form $X_b^{(n)} - Y_a^{(n)} = G(X^{(2)}, Y^{(2)}, \ldots, X^{(n-1)}, Y^{(n-1)})$, with G a nonlinear function of $X^{(i)}, Y^{(i)}, i = 2, 3, \ldots, n - 1$. Thus we observe that $X^{(2)}$ and $Y^{(2)}$ satisfy the Cauchy-Riemann equations and are therefore conjugate harmonic functions of a and b. The higher order coefficients would satisfy Poisson's equation with a right hand side a known function fixed by the coefficient functions of lower order. Thus the coefficients in the series for X and Y can be determined step-wise by solving a sequence of Poisson equations. Once the functions $X^{(i)}$ and $Y^{(i)}$ have been determined, the coefficients in the series for the pressure p can also be determined successively by solving a sequence of Poisson equations. To this end we of course make use of equations (12.1.1); the result for $p^{(o)}(a, b)$ is

$$(12.1.8) \qquad p_{aa}^{(o)} + p_{bb}^{(o)} = - 2\varrho(X_a^{(2)} + Y_b^{(2)}) = 0,$$

from (12.1.6) and $X^{(1)} = Y^{(1)} = 0$. Thus $p^{(o)}(a, b)$ is a harmonic function. For $p^{(n)}(a, b)$ one would find a Poisson equation with a right hand side determined by $X^{(i)}$ and $Y^{(i)}$ for $i = 2, 3, \ldots, n + 2$.

It would be possible to consider boundary conditions in a general way, but such a procedure would not be very useful because of its complexity. Instead, we proceed to formulate boundary conditions for the special problem of breaking of a dam, which is in any case typical for the type of problems for which the present procedure is recommended. We assume therefore that the region occupied initially by the water (or rather, a vertical plane section of that region) is the half-strip $0 \leq a < \infty$, $0 \leq b \leq h$, as indicated in Fig. 12.1.1. The

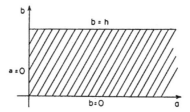

Fig. 12.1.1. The breaking of a dam

dam is of course located at $a = 0$. Since we assume that the water is initially at rest when filling the half-strip we have the conditions

(12.1.9) $X(a, b; 0) = a,$ $Y(a, b; 0) = b,$

and

(12.1.10) $X_t(a, b; 0) = 0,$ $Y_t(a, b; 0) = 0.$

When the dam is broken, the pressure along it will be changed suddenly from hydrostatic pressure to zero; it will of course be prescribed to be zero on the free surface. This leads to the following boundary conditions for the pressure:

(12.1.11) $\begin{cases} p(a, h; t) = 0, & 0 \leq a < \infty, & t > 0, \\ p(0, b; t) = 0, & 0 \leq b \leq h, & t > 0. \end{cases}$

Finally the boundary condition at the bottom $b = 0$ results from the assumption that the water particles originally at the bottom remain in contact with it; as a result we have the boundary condition

(12.1.12) $Y(a, 0; t) = 0,$ $0 \leq a < \infty,$ $t > 0.$

The conditions (12.1.9) are automatically satisfied because of the form (cf. (12.1.5)) chosen for the series expansion. The conditions (12.1.10) are satisfied by taking $X^{(1)}(a, b) = Y^{(1)}(a, b) = 0$.

In order to determine the functions $X^{(2)}(a, b)$ and $Y^{(2)}(a, b)$, it is necessary to obtain boundary conditions in addition to the differential equations given for them by (12.1.6) and (12.1.7). Such boundary conditions can be obtained by using (12.1.11) and (12.1.12) in conjunction with (12.1.1) and the power series developments. Thus from

(12.1.12) we find $Y^{(2)}(a, 0) = 0$ for $0 \leq a < \infty$ (indeed, $Y^{(n)}(a, 0)$ would be zero for all n). Insertion of the series (12.1.5) and use of the boundary conditions for $b = h$ yields

(12.1.13) $X^{(2)}(a, h) = 0,$

upon using the first of the equations in (12.1.1). The second equation of (12.1.1) leads to the condition

(12.1.14) $Y^{(2)}(0, b) = -\dfrac{g}{2}.$

We know that $Z(z) = Y^{(2)} + iX^{(2)}$ is an analytic function of the complex variable $z = a + ib$ in the half-strip, and we now have prescribed values for either its real or its imaginary part on each of the three sides of the strip; it follows that the function Z can be determined by standard methods—for example by mapping conformally on a halfplane. In fact, the solution can be given in closed form, as follows: Since $X^{(2)}(a, h) = 0$, we see that $X_a^{(2)}(a, h) = 0$, and hence that $Y_a^{(2)}(a, h) = 0$ since $X^{(2)}$ and $Y^{(2)}$ are harmonic conjugates. Therefore the harmonic function $Y^{(2)}(a, b)$ can be continued over the line $b = h$ by reflection into a strip of width $2h$, as indicated in Fig. 12.1.2; the boundary values for $Y^{(2)}$ are also shown. Thus a complete-

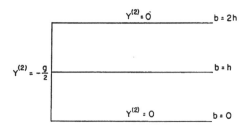

Fig. 12.1.2. Boundary value problem for $Y^{(2)}(a, b)$

ly formulated boundary value problem for $Y^{(2)}(a, b)$ in a half-strip has been derived. To solve this problem we map the half-strip on the upper half of a w-plane by means of the function $w = \cosh(\pi z/2h)$ —either by inspection or by using the Schwarz-Christoffel mapping formula—and observe that the vertices $z = 0$ and $z = 2ih$ of the half-strip map into the points $w = \pm 1$ of the w-plane, as indicated in Fig. 12.1.3. The appropriate boundary values for $Y^{(2)}(w)$ on the real axis of the w-plane are indicated. The solution for $Y^{(2)}(w)$ under

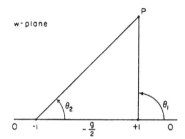

Fig. 12.1.3. Mapping on the w-plane

these conditions is well known; it is the function $Y^{(2)}(P) = - (g/2\pi)(\theta_2 - \theta_1)$, with θ_1 and θ_2 the angles marked in Fig. 12.1.3. The analytic function of which this is the real part is well known; it is

$$Y^{(2)} + iX^{(2)} = - \frac{ig}{2\pi} \log \frac{w - 1}{w + 1},$$

as can in any case be easily verified. Transferring back to the z-plane we have

$$Z(z) = Y^{(2)} + iX^{(2)} = - \frac{ig}{2\pi} \log \left\{ \frac{\cosh \dfrac{\pi z}{2h} - 1}{\cosh \dfrac{\pi z}{2h} + 1} \right\},$$

and upon separation into real and imaginary parts we have finally:

$$(12.1.15) \quad \begin{cases} X^{(2)}(a, b) = - \dfrac{g}{2\pi} \log \left\{ \dfrac{\cos^2 \dfrac{\pi b}{4h} + \sinh^2 \dfrac{\pi a}{4h}}{\sin^2 \dfrac{\pi b}{4h} + \sinh^2 \dfrac{\pi a}{4h}} \right\}, \\[4ex] Y^{(2)}(a, b) = - \dfrac{g}{\pi} \arctan \left\{ \dfrac{\sin \dfrac{\pi b}{2h}}{\sinh \dfrac{\pi a}{2h}} \right\}. \end{cases}$$

One checks easily that the boundary conditions $X^{(2)}(a, h) = 0$, $Y^{(2)}(a, 0) = 0$ are satisfied, and that $Y^{(2)}(0, b) = - g/2$. The initial pressure distribution $p^{(0)}(a, b)$ can be calculated, now that $X^{(2)}(a, b)$ is known, by using the first equation of (12.1.1), which yields

(12.1 16) $$p_a^{(o)} = -2\varrho X^{(2)}.$$

In the present case there are advantages in working first with the pressure $p(a, b; t)$ and determining the coefficient of the series for it directly by solving appropriate boundary value problems; afterwards the coefficients of the series for X and Y are easily found. The main reason for basing the calculation on the pressure in the first instance is that the boundary conditions at $b = h$ and $a = 0$ are very simple, i.e. $p = 0$ and hence $p^{(i)} = 0$ for all indices i. The boundary conditions at the bottom $b = 0$ involve the displacements Y. For instance, one finds readily in the same general way as above that $p_b^{(o)} = -\varrho g$, $p_b^{(1)} = 0$, and $p_b^{(2)} = -\varrho g Y_b^{(2)}$ as boundary conditions at $b = 0$. Since $p^{(0)}$ is harmonic, it is found at once without reference to displacements—an interesting fact in itself. Once $p^{(o)}$ is found, $X^{(2)}$ and $Y^{(2)}$ can be calculated without integrations (cf. (12.1.16), for example). Since $p_b^{(1)} = 0$ for $b = 0$, and $p^{(1)}$ is also harmonic, it follows that $p^{(1)}(a, b) = 0$. Since $Y^{(2)}$ is now known, it follows that a complete set of boundary conditions for $p^{(2)}(a, b)$ is known, and $p^{(2)}(a, b)$ is then determined by solving the differential equation

(12.1.17) $\nabla^2 p^{(2)} = p_{aa}^{(2)} + p_{bb}^{(2)}$

$$= \varrho \left[\frac{\partial(X^{(2)}, Y^{(2)})}{\partial(a, b)} - \nabla^2 \{(X^{(2)})^2 + (Y^{(2)})^2 + gY^{(2)}\} \right]$$

$$= \frac{-8\varrho g^2}{h^2 \left(\cosh \dfrac{\pi a}{h} - \cos \dfrac{\pi b}{h} \right)}$$

whose right hand side is obtained after a certain amount of manipulation. This process can be continued. One would find next that $X^{(3)} = Y^{(3)} = 0$, and that $X^{(4)}$ and $Y^{(4)}$ can be found once $p^{(2)}$ is known. However, the boundary condition at the bottom, and the right hand sides in the Poisson equations for the functions $p^{(i)}(a, b)$ become more and more complicated.

The initial pressure $p^{(o)}(a, b)$ can be discussed more easily on the basis of a Fourier series representation than from the solution in closed form obtainable from (12.1.16); this representation is

(12.1.18) $p^{(o)}(a, b)$

$$= \varrho g(h - b) - \frac{8\varrho g h}{\pi^2} \sum_{n=0}^{\infty} \frac{1}{(2n + 1)^2} e^{-\frac{(2n+1)\pi a}{2h}} \cos \frac{(2n + 1)\pi b}{2h}.$$

We note that the first term represents the hydrostatic pressure, and that the deviation from hydrostatic pressure dies out exponentially as $a \to \infty$ and also as $h \to 0$, i.e. on going far away from the dam and also on considering the water behind the dam to be shallow (or, better, considering a/h to be large). This is at least some slight evidence of the validity of the shallow water theory used in Chapter 10 to discuss this same problem of the breaking of a dam — at least at points not too close to the site of the dam.

The shape of the free surface of the water can be obtained for small times from the equations

$$(12.1.19) \qquad X = a + X^{(2)}t^2, \qquad Y = b + Y^{(2)}t^2$$

evaluated for $a = 0$ (for the particles at the face of the dam) and for $b = h$ on the upper free surface. The results of such a calculation for the specific case of a dam 200 feet high are shown in Fig. 12.1.4.

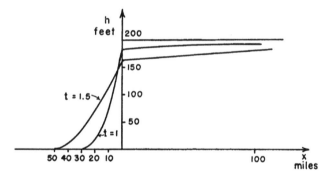

Fig. 12.1.4. Free water surface after the breaking of a dam

One of the peculiarities of the solution is a singularity at the origin $a = 0$, $b = 0$ which is brought about by the discontinuity in the pressure there. In fact, $X^{(2)}$ has a logarithmic singularity for $a = 0$, $b = 0$, as one sees from (12.1.15) and X is negative infinite for all $t \neq 0$. This, of course, indicates that the approximation is not good at this point; in fact, there would be turbulence and continuous breaking at the front of the wave anyway so that any solution ignoring these factors would be unrealistic for that part of the flow.

In the thesis by Pohle [P.11], the solution of the problem of the collapse of a liquid half-cylinder and of a hemisphere on a rigid plane are treated by essentially the same method as has been explained for the problem of the breaking of a dam. These problems have also been treated by Penney and Thornhill [P.2], who also use power series in the time but work with the Eulerian rather than the Lagrangian representation, which leads to what seem to the author to be more complicated calculations than are needed when the Lagrangian representation is used.

12.2. The existence of periodic waves of finite amplitude

In this section a proof, in detail, of the existence of two-dimensional periodic progressing waves of finite amplitude in water of infinite depth will be given. This problem was first solved by Nekrassov [N.1, 1a] and later independently by Levi-Civita [L.7]; Struik [S.29] extended the proof of Levi-Civita to the same problem for water of finite constant depth. A generalization of the same theory to liquids of variable density has been given by Dubreuil-Jacotin [D.15, 15a]. Lichtenstein [L.11] has given a different method of solution based on E. Schmidt's theory of nonlinear integral equations. Davies [D.5] has considered the problem from still a different point of view. Gerber [G.5] has recently derived theorems on steady flows in water of variable depth by making use of the Schauder-Leray theory.

We shall start from the formulation of the problem given by Levi-Civita (and already derived in 10.9 above), but, instead of proving directly, as he does, the convergence of a power series in the amplitude to the solution of the problem, an iteration procedure devised by W. Littman and L. Nirenberg will be used to establish the existence of the solution. The two procedures are not, however, essentially different.

It is convenient to break up this rather long section into sub-sections as a means of focusing attention on separate phases of the existence proof.

12.2a. Formulation of the problem

As in sec. 10.9, the problem of treating a progressing wave which moves unchanged in form and with constant velocity is reduced to a problem of steady flow by observing the motion from a coordinate system which moves with the wave. A complex velocity potential (see sec. 10.9 for details) $\chi(z)$ is therefore to be found in the x, y-plane (cf. Fig. 12.2.1):

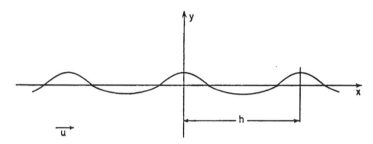

Fig. 12.2.1. Periodic waves of finite amplitude

(12.2.1) $$\chi = \varphi + i\psi = \chi(z), \qquad z = x + iy.$$

The velocity at $y = -\infty$ should be U. The real harmonic functions $\varphi(x, y)$ and $\psi(x, y)$ represent the velocity potential and the stream function. The complex velocity w is given by

(12.2.2) $$w = \frac{d\chi}{dz} = u - iv$$

with u, v the velocity components. This follows at once from the Cauchy-Riemann equations:

(12.2.3) $$\varphi_x = \psi_y = u, \qquad \varphi_y = -\psi_x = v,$$

since $w = \varphi_x + i\psi_x$.

We proceed to formulate the boundary conditions at the free surface. The kinematic free surface condition can be expressed easily because the free surface is a stream line, and we may choose $\psi(x, y) = 0$ along it. The dynamic condition expressed in Bernoulli's law is given by

(12.2.4) $$\tfrac{1}{2} \mid w \mid^2 + gy = \text{const.} \qquad \text{at } \psi = 0,$$

as one can readily verify. The problem of satisfying this nonlinear condition is of course the source of the difficulties in deriving an existence proof. At ∞ the boundary condition is

(12.2.5) $$w \to U \quad \text{uniformly as } y \to -\infty,$$

and w is in addition supposed to be nowhere zero and to be uniformly bounded. We seek waves which are periodic in the x-coordinate and thus we require χ to satisfy the condition

(12.2.6) $$\chi(z + h) - \chi(z) = 0,$$

with h a real constant.

Following Levi-Civita, we assume that the region of flow in the z-plane is mapped into the φ, ψ-plane by means of $\chi(z)$. The free surface in the physical plane corresponds to the real axis $\psi = 0$ of the χ-plane, and we assume that the entire region of the flow in the z-plane is mapped in a one-to-one way on the lower half of the χ-plane. (We shall prove shortly that a function $\chi(z)$ satisfying the conditions given above would have this property.) In this case the inverse mapping $z(\chi)$ exists, and we may regard the complex velocity $w(z)$ as an analytic function of χ defined in the lower half of the χ-plane. In this way we are enabled to work with a domain in the φ, ψ-plane that is fixed in advance instead of with an unknown domain of the x, y-plane. Levi-Civita goes a step further by introducing a new dependent variable ω, replacing w, by the relation

$$(12.2.7) \qquad\qquad w = Ue^{-i\omega}, \qquad \omega = \theta + i\tau;$$

so that ω is an analytic function of $\varphi + i\psi$. Consequently we have (cf. (12.2.2))

$$(12.2.8) \qquad\qquad |w| = Ue^{\tau}, \qquad \theta = \arg \bar{w}.$$

Thus $\tau = \log(|w|/U)$, while θ is the inclination of the velocity vector. In the same way as in sec. 10.9 (cf. the equations following (10.9.11)) the boundary condition (12.2.4) can be put in the form

$$(12.2.9) \qquad\qquad \theta_{\psi} = \lambda' e^{-3\tau} \sin \theta, \qquad \text{for } \psi = 0,$$

with λ' defined by

$$(12.2.10) \qquad\qquad\qquad \lambda' = g/U^3.$$

Our problem now is to determine an analytic function $\omega(\chi) = \theta(\varphi, \psi) + i\tau(\varphi, \psi)$ in the lower halfplane $\psi < 0$ and a constant λ' in (12.2.9) such that a) ω is analytic for $\psi < 0$, continuous for $\psi \leq 0$, b) θ_{ψ} is continuous for $\psi \leq 0$ and the nonlinear boundary condition (12.2.9) is satisfied, c) ω has the period Uh in φ, d) $\omega(\chi) \to 0$ as $\psi \to -\infty$, e) $|\omega(\chi)| \leq \frac{1}{2}$. The last two conditions are motivated by the conditions imposed on w at ∞ and the condition $w \neq 0$: the condition d) from (12.2.5) and (12.2.7), while the condition e) is imposed in order to ensure that w is uniformly bounded away from both zero and infinity. As we shall see, the condition c) leads to the periodicity condition (12.2.6) on χ.

We proceed to show briefly (again following Levi-Civita) that a solution of the problem we have formulated for ω would lead through (12.2.7) to a function $w(\chi)$ and then to a function $\chi(z)$ through the differential equation $d\chi(z)/dz = w(\chi)$ which satisfies all of the conditions formulated above. The essential items requiring verification are the periodicity condition and the one-to-one character of the mapping $z(\chi)$ defined by

$$\int_0^\chi \frac{d\chi}{w(\chi)} = z(\chi) = x(\varphi, \psi) + iy(\varphi, \psi)$$

over the halfplane $\psi < 0$.

We proceed to investigate the second property. From (12.2.7) we have

$$1/w(\chi) = \frac{1}{U} e^{-\tau} (\cos \theta + i \sin \theta),$$

and hence that

$$\mathscr{R}e \left(\frac{1}{w}\right) = \frac{1}{U} e^{-\tau} \cos \theta.$$

Since $|\omega| \leq \frac{1}{2}$, it follows that w is bounded away from zero, so that the integral $\int_0^\chi \frac{d\chi}{w}$ converges. Since both $|\tau| \leq \frac{1}{2}$ and $|\theta| \leq \frac{1}{2}$, it follows that $\mathscr{R}e(1/w)$ is positive (we assume U to be positive) and bounded away from 0 and ∞. We have $x_\psi + iy_\psi = i/w$, $x_\varphi + iy_\varphi = 1/w$, so that $y_\psi = \mathscr{R}e(1/w)$ and $x_\varphi = \mathscr{R}e(1/w)$; since $\mathscr{R}e(1/w) > 0$ it follows therefore that y is a strictly monotonic increasing function of ψ, and x similarly in φ. Consequently the mapping $z(\chi)$ is one-to-one, and in addition $y \to -\infty$ when $\psi \to -\infty$, while $x \to \pm\infty$ when $\varphi \to \pm\infty$ since $\mathscr{R}e(1/w)$ is positive and bounded away from zero; thus the flow is mapped onto the entire halfplane $\psi < 0$.

We consider the periodicity condition next. We have $\left.\dfrac{dz}{d\chi}\right|_{\chi+Uh} = \left.\dfrac{dz}{d\chi}\right|_\chi$ since ω has the period Uh by assumption. This implies that $z(\chi + Uh) - z(\chi) = $ const. This constant is easily seen to have the value h by letting $\psi \to -\infty$ in the formula

$$\int_\chi^{\chi+Uh} \frac{d\chi}{w(\chi)} = z(\chi + Uh) - z(\chi),$$

since $w \to U$ uniformly when $\psi \to -\infty$. Consequently we have

$$x(\chi + Uh) - x(\chi) = h,$$
$$y(\chi + Uh) - y(\chi) = 0.$$

We know from (12.2.8) and $| \theta | \leq \frac{1}{2}$ that the stream lines $\psi = $ const.
have no vertical tangents, hence they can be represented in the form
$y = y(x)$, and the last two equations show that they are periodic in
x of period h. The problem of determining $\omega(\chi)$ subject to the conditions a)—e) is therefore equivalent to the problem formulated for $\chi(z)$.

12.2b. Outline of the procedure to be followed in proving the existence of the function $\omega(\chi)$

The proof of the existence of the analytic function $\omega(\chi)$ which
solves our problem will be carried out as follows. First of all, we observe that the problem has always the solution $\omega(\chi) \equiv 0$, corresponding to the uniform flow $w \equiv U$ with undisturbed free surface. We
shall begin by assuming that a solution $\omega(\chi) \not\equiv 0$ exists, and will then
proceed, through the use of the properties assumed for ω, to derive a
functional equation for the values $\omega(\varphi, 0)$ of ω on the boundary $\psi = 0$,
$- \infty < \varphi < \infty$. It will then be shown that the functional equation
has a solution $\omega(\varphi, 0) \not\equiv 0$ in the form of a complex-valued continuous
function $\hat{\omega}(\varphi)$, and this function will be used to determine an analytic
function $\omega(\varphi, \psi)$ in $- \infty < \psi < 0$, $- \infty < \varphi < \infty$, with $\hat{\omega}(\varphi)$ as
boundary values, which is then shown to satisfy all of the conditions
a)—e).

It will occasion no surprise to remark at this point that the solution
we obtain will give a motion in a neighborhood of the uniform flow
with horizontal free surface, i.e. with an amplitude in a neighborhood
of the zero amplitude. Also, it should be remarked that the problem
in perturbation theory which thus arises involves a bifurcation phenomenon, since the desired solution of the nonlinear problem, once the
wave length is fixed, requires that the perturbations take place in the
neighborhood of a definite value of the velocity U. In other words, the
desired solution bifurcates from a definite one of the infinitely many
possible flows with uniform velocity which are exact solutions of the
nonlinear problem.

The decisive relation in the process just outlined is the nonlinear
boundary condition (12.2.9). It is convenient to introduce at this point
some notations which refer to it, to recast it in a different and more
convenient form, and also to derive a number of consequences which
flow out of it. At the same time, some factors which motivate all that
follows will be put in evidence.

Since we wish to concentrate attention on the boundary values $\omega(\varphi, 0)$ of ω, it is useful to introduce the notation

(12.2.11) $\omega(\varphi, 0) \equiv \hat{\omega}(\varphi) = \hat{\theta}(\varphi) + i\hat{\tau}(\varphi)$,

and then to introduce the operator $f[\hat{\omega}]$ defined by

(12.2.12) $f[\hat{\omega}] = \lambda(e^{-3\hat{\tau}} \sin \hat{\theta} - \hat{\theta}) + \varepsilon e^{-3\hat{\tau}} \sin \hat{\theta} \equiv F(\varphi)$

with ε defined by

(12.2.13) $\varepsilon = \lambda' - \lambda.$

The constant λ will be given an arbitrary but fixed value; the quantity $2\pi/\lambda$ will then be the period of the function $\omega(\chi)$. The constant ε, and with it λ' through (12.2.13), will be fixed by the solution $\omega(\chi)$ in a manner to be indicated below, and the propagation speed U is then determined by the formula (12.2.10). As can be seen at once, the boundary condition (12.2.9) now takes the form

(12.2.14) $\theta_\psi - \lambda\theta = F(\varphi), \qquad \psi = 0.$

The reasons for writing the free surface condition in the form (12.2.14) are as follows: As remarked above, we seek a motion in the neighborhood of a uniform flow, so that ω as defined by (12.2.7) should be small in some sense. It would seem reasonable to set up an iteration procedure which starts with that solution $\omega_1(\varphi, \psi)$ of the problem which results when $F(\varphi)$, which contains the nonlinear terms in the free surface condition, vanishes identically. Afterwards the successive approximations will be inserted in $F(\varphi)$ to obtain a sequence of linear problems whose solutions ω_k converge to the desired solution of our problem.

The problem of determining $\omega_1(\varphi, \psi)$, when $F \equiv 0$, is exactly the problem posed by the linear theory which was discussed at length in Chapter 3; in fact, if $F(\varphi)$ vanishes, we know from the discussion in Chapter 3 that the only bounded conjugate harmonic functions $\theta_1(\varphi, \psi)$, $\tau_1(\varphi, \psi)$, other than $\theta_1 = \tau_1 = 0$, in the lower half plane $\psi < 0$ which satisfy the homogeneous free surface condition $\theta_{1\psi} - \lambda\theta_1 = 0$ are the functions

$$\theta_1(\varphi, \psi) = a_1 e^{\lambda\psi} \sin \lambda\varphi,$$
$$\tau_1(\varphi, \psi) = a_1 e^{\lambda\psi} \cos \lambda\varphi,$$

once φ is taken to be zero at a crest or trough of the wave. Thus the boundedness condition at ∞ and the homogeneous free surface condition lead automatically to waves which are sines or cosines of φ.

The "amplitude" a_1 is, of course, arbitrary on account of the homogeneity of the problem. The corresponding function $\omega_1(\chi)$ is then

$$\omega_1(\chi) = \theta_1 + i\tau_1 = ia_1 e^{\lambda(\psi - i\varphi)}.$$

We suppose, naturally, that the "amplitude" $|a_1|$ is small and hence that $|\omega_1|$ is also small of the same order. The basic parameter in the iteration procedure will be the quantity a_1, and the procedure will be so arranged that the quantity ε in (12.2.13) as well as the iterates ω_k will be of order $|a_1|$. It is then easily seen that $F(\varphi)$ will always be of order $|a_1|^2$, which indicates that such a scheme of iteration is reasonable. We shall show that it does indeed lead to a sequence ω_k which converges to the desired solution ω for all sufficiently small values of $|a_1|$, and that the solution ω fixes a value of ε, and hence of λ' since λ is once for all fixed, in a manner to be explained in a moment.

It might be mentioned that it is not difficult to verify that the corresponding motion furnished in the physical plane by $\chi_1(z)$ would, up to terms of first order in $a_1|$, be given by

$$\chi_1 = Uz + \frac{a_1 i}{\lambda} (e^{-i\lambda Uz} - 1),$$

and this coincides with what was found in Chapter 3 by a more direct procedure.

Iterations, as we have indicated, are to be performed, starting with the solution $\omega_1 = ia_1 e^{\lambda(\psi - i\varphi)}$ of the linearized problem, with a_1 regarded as a small parameter. This is then inserted in the right hand side of (12.2.14); a bounded harmonic function $\theta_2(\varphi, \psi)$ in the half plane $\psi < 0$ is then determined through this nonhomogeneous boundary condition and the corresponding analytic function $\omega_2(\chi) = \theta_2 + i\tau_2$ with it. In order to solve the boundary problem for θ_2 (and through it ω_2), however, it is necessary to dispose of the parameter ε in (12.2.13) appropriately. This comes about because, as we have just seen, the homogeneous linear boundary value problem for θ_1 has a non-trivial solution, $\theta_1 = a_1 e^{\lambda\psi} \sin \lambda\varphi$, and hence an orthogonality condition on $F(\varphi)$ is needed which will ensure the existence of the solution of the nonhomogeneous problem for θ_2. This condition is well known to be that the integral $\int_0^{2\pi/\lambda} F(\varphi) \sin \lambda\varphi\, d\varphi$ should vanish. It turns out that the value of ε so determined really is of the same order as a_1. Continuing the iterations in this fashion, the result is a sequence of functions $\omega_n(\chi)$, and a sequence of corresponding values ε_n of ε such that $|\omega_n|$ and $|\varepsilon_n|$ are all of order $|a_1|$. It is to be shown that both sequences converge to yield a function $\omega(\chi)$ and a number ε which solve the problem, the quantity λ' in (12.2.10) being fixed by $\varepsilon = \lim \varepsilon_n$ and the arbitrarily chosen value of λ through (12.2.13).

We observe that this whole procedure is in marked contrast with the method of solution of the problem of the solitary wave given by Friedrichs and Hyers [F.13] and explained in sec. 10.9; in the latter case the iteration procedure is quite different and it is carried out with respect to a parameter which has an entirely different significance from the parameter a_1 which is used here.

The procedure outlined here also differs from the procedure followed by Lichtenstein [L.11] in solving the same problem. Lichtenstein applies E. Schmidt's bifurcation theory to an appropriate nonlinear integral equation (essentially the counterpart of the functional equation to be used here). In this procedure, the basic idea is to modify what corresponds to the function $F(\varphi)$ in (12.2.14) in such a way that the modified problem (which is arranged to contain one or more parameters) can always be solved. Afterwards, conditions are written down to ensure that the modified problem is identical with the original problem; these conditions are called the bifurcation conditions. Such a process could have been used here in conjunction with an iteration scheme, as a substitute for the process of fixing the parameters ε_n at each stage of the iteration procedure in the manner indicated above.

Basically the method of solution of our nonlinear problem just outlined requires the solution of a sequence of linear problems. We turn next, therefore, in sec. 12.2c to the derivation of the solution of these linear problems, and afterwards, in sec. 12.2d we shall prove that an appropriate sequence of solutions of the linear problems converges to the desired solution of the nonlinear problem.

12.2c. The solution of a class of linear problems

The linear problems we have in mind to solve, in accordance with the above discussion, are problems for $\omega(\varphi, \psi) = \theta(\varphi, \psi) + i\tau(\varphi, \psi)$ when $F(\varphi)$ in (12.2.14) is regarded as a given function. That is, ω should satisfy all of the conditions formulated above, except for the free surface condition. Later on we shall begin our iteration process with a function ω_1 which has the period $2\pi/\lambda$ in φ; $F(\varphi)$, all subsequent iterates, and the solution itself, will have the same period. Since we expect the waves to be symmetrical about a crest or trough (indeed, our existence proof will yield only waves with this property), we suppose that the origin is taken at such a point, and hence that at any stage of the iteration process $\theta(\varphi, 0) = \hat{\theta}(\varphi)$ would be an odd function of φ; while $\tau(\varphi, 0) = \hat{\tau}(\varphi)$ would be an even function of φ, and both would have the period $2\pi/\lambda$. Thus $F(\varphi)$ in (12.2.14) as defined by (12.2.12) should be taken for our purposes as an odd function of φ with the real period $2\pi/\lambda$.

If we were to work at the outset with Fourier series, it follows that θ would be represented as a sine series, and $F(\varphi)$ also. However, we wish later on to carry out an iteration process in which only continuous, and not necessarily differentiable, functions of φ are employed, and in which the existence of a certain continuous periodic function $\hat{\omega}(\varphi)$ of period $2\pi/\lambda$ is first proved; this function will furnish the boundary values of the solution $\omega(\varphi, \psi)$. (Afterwards, the question of the existence of the normal derivative $\theta_\psi(\varphi, 0)$ in (12.2.14), and of other derivatives, will be dealt with separately.) In doing so, we shall have occasion to approximate such continuous periodic functions by finite Fourier series, or Fourier polynomials, a process justified by the Weierstrass approximation theorem which states that such a polynomial can always be constructed to yield a uniform approximation for all values of φ and any arbitrary degree of approximation.

Suppose, therefore, that $F(\varphi)$ in (12.2.14) had been approximated at some stage in the iteration procedure by a function $g(\varphi)$ in the form of the following finite sine series:

$$(12.2.15) \qquad g(\varphi) = \sum_{\nu=1}^{n} b_\nu \sin \nu\lambda\varphi,$$

and we seek the bounded harmonic function $\theta(\varphi, \psi)$ which satisfies the boundary condition (12.2.14) with $F \equiv g$. For this purpose we write the solution $\theta(\varphi, \psi) = \mathscr{R}e\, \omega(\varphi, \psi)$ also as a finite Fourier sum:

$$(12.2.16) \qquad \theta(\varphi, \psi) = \sum_{\nu=1}^{n} a_\nu e^{\nu\lambda\psi} \sin \nu\lambda\varphi.$$

Insertion of this sum in (12.2.14), with $F \equiv g$, leads to the following equations for the determination of the coefficients a_ν:

$$\begin{aligned} \lambda\nu a_\nu - \lambda a_\nu &= b_\nu, \\ \lambda(\nu - 1)a_\nu &= b_\nu \end{aligned} \qquad \nu = 1, 2, \ldots, n.$$

and thus to the conditions

$$(12.2.17) \qquad \begin{cases} a_1 \text{ arbitrary, } b_1 = 0 \\ a_\nu = \dfrac{b_\nu}{\lambda(\nu - 1)}, \qquad \nu = 2, 3, \ldots, n. \end{cases}$$

It is very important for the following to observe that the Fourier sine series for $F(\varphi)$ must lack the first order term: otherwise, as we have remarked above, our problem would have no solution: a term of the form $b_1 \sin \lambda\varphi$ in $F(\varphi)$ is a "resonance" term, the presence of which would preclude the existence of the solution of the nonhomogeneous problem. The unique solution for $\theta(\varphi, \psi)$ is

$$(12.2.18) \quad \theta(\varphi, \psi) = \sum_{\nu=2}^{n} \frac{b_\nu}{\lambda(\nu - 1)} e^{\nu\lambda\psi} \sin \nu\lambda\varphi + a_1 e^{\lambda\psi} \sin \lambda\varphi,$$

once a_1 is prescribed (cf. Chapter 3) and θ is assumed to be an odd function of φ. The harmonic conjugate $\tau(\varphi, \psi)$ of $\theta(\varphi, \psi)$ is obtained by integrating the Cauchy-Riemann equation $\theta_\psi = -\tau_\varphi$, with the result

$$(12.2.19) \quad \tau(\varphi, \psi) = \sum_{\nu=2}^{n} \frac{b_\nu}{\lambda(\nu - 1)} e^{\nu\lambda\psi} \cos \nu\lambda\varphi + a_1 e^{\lambda\psi} \cos \lambda\varphi.$$

(A possible additive integration constant is set equal to zero since $\tau \to 0$ when $\psi \to -\infty$.) Thus we would have for $\omega(\varphi, \psi)$ under the assumed circumstances the expression

$$(12.2.20) \qquad \omega(\varphi, \psi) = i \sum_{\nu=2}^{n} \frac{b_\nu}{\lambda(\nu - 1)} e^{\nu\lambda(\psi - i\varphi)} + i a_1 e^{\lambda(\psi - i\varphi)}.$$

In other words, if $F(\varphi)$ is given as in (12.2.15) by a finite Fourier series of sines which lacks its lowest order term, i.e. is such that

$$(12.2.21) \qquad \int_0^{2\pi/\lambda} F(\varphi) \sin \lambda\varphi \, d\varphi = 0,$$

then, as we see from (12.2.20) and the discussion preceding it, the function $\hat{\omega}(\varphi) = \omega(\varphi, 0) = \hat{\theta}(\varphi) + i\hat{\tau}(\varphi)$ given by

$$(12.2.22) \qquad \hat{\omega}(\varphi) = i \sum_{\nu=2}^{n} \frac{b_\nu}{\lambda(\nu - 1)} e^{-i\nu\lambda\varphi} + i a_1 e^{-i\nu\varphi}$$

yields the boundary values of an analytic function $\omega(\varphi, \psi)$ which would satisfy the boundary condition (12.2.14). Evidently, ω would also satisfy all of the conditions a) to e) formulated above, if the amplitude a_1 of the first order term of the Fourier series is chosen

small enough, except that the boundary condition b) is replaced by
a linear condition.

It is clear that the insertion of a function $\hat{\omega}(\varphi)$ as given by (12.2.22)
in (12.2.12) to determine a new function $F(\varphi)$ in order to continue the
iteration process would not yield in general a function representable
as a finite Fourier sum, but rather to one representable only by a
Fourier series. However, we have already stated that we wish to
carry out our iteration scheme in the nonlinear problem within the
class of continuous functions, which need not possess convergent
Fourier series. Nevertheless, the general scheme outlined above for
determining the successive iterations can still be used once it has been
extended in an appropriate way to the wider class of functions. For
this purpose, and later purposes as well, it is convenient to introduce
the terminology of functional analysis. Thus we speak of the linear
vector space of elements which are complex-valued functions
$g(\varphi) = \alpha(\varphi) + i\beta(\varphi)$, continuous for all φ and of period $2\pi/\lambda$, such that
α is an odd function and β an even function of φ. The scalars are the
real numbers. This space is made into a normed linear space by introducing as the distance from the origin to the "point" g the following
norm, written $\| g \|$:

$$\| g \| = \| \alpha + i\beta \| = \max_{\varphi} \sqrt{\alpha^2 + \beta^2} = \max | g |,$$

and as the distance between two elements or points g_1, g_2, the norm of
their difference, i.e. $\| g_1 - g_2 \|$. This space, which we shall call the
space B, is *complete*, i.e. it has the property that every Cauchy sequence in the space converges to an element in the space.* By a Cau-
chy sequence g_n we mean a sequence such that $\| g_m - g_n \| \to 0$
when $m, n \to \infty$. Since the norm is the maximum of the absolute
value of g, it follows that a sequence g_n which is a Cauchy sequence is
uniformly convergent and hence has a continuous function as a limit.
We remark also that the notion of distance thus introduced in our
space has the usual properties required for the distance function in a
metric space, i.e., the distance is positive definite:

$$\| g \| \geqq 0, \quad \text{and} \quad \| g \| = 0 \quad \text{implies} \quad g \equiv 0,$$

and the triangle inequality

$$\| g_1 + g_2 \| \leqq \| g_1 \| + \| g_2 \|$$

holds.

---

* We remark that a complete linear normed space is called a Banach space;
however, such properties of these spaces as are needed will be developed here.

We introduce next the subspace B_1 of our Banach space B which consists of all real functions $g(\varphi)$ given by finite Fourier sums of sines lacking the term of first order:

$$(12.2.23) \qquad g(\varphi) = \sum_{\nu=2}^{n} b_\nu \sin \nu\lambda\varphi, \qquad b_\nu \text{ real.}$$

With respect to this set B_1 of functions we define a transformation S as follows:

$$(12.2.24) \qquad Sg(\varphi) = i \sum_{\nu=2}^{n} \frac{b_\nu}{\lambda(\nu-1)} e^{\nu\lambda\psi} e^{-i\nu\lambda\varphi}, \qquad -\infty < \psi \leqq 0.$$

Since $\psi \leqq 0$ we have for the norm* of Sg the bound

$$\| Sg \| \leqq \sum_{\nu=2}^{n} \frac{|b_\nu|}{\lambda(\nu-1)}.$$

From Cauchy's inequality the following inequality for $\| Sg \|$ is then obtained:

$$\| Sg \| \leqq \sqrt{\sum_{\nu=2}^{n} \frac{1}{\lambda^2(\nu-1)^2}} \cdot \sqrt{\sum_{\nu=2}^{n} b_\nu^2}$$

$$\leqq \sqrt{\sum_{\nu=2}^{\infty} \frac{1}{\lambda^2(\nu-1)^2}} \cdot \sqrt{\frac{\lambda}{\pi} \int_0^{2\pi/\lambda} g^2 d\varphi},$$

the last step resulting from (12.2.23) because of the orthogonality of the functions $\sin \nu\lambda\varphi$. Since

$$\sqrt{\frac{\lambda}{\pi} \int_0^{2\pi/\lambda} g^2 d\varphi} \leqq \sqrt{2} \, |g|_{\max},$$

it follows that a constant K exists which is independent of g and n (though not of λ), such that

$$(12.2.25) \qquad \| Sg \| \leqq K \| g \|, \qquad \text{for } g \subset B_1.$$

Thus S is what is called a bounded transformation in B_1 since it transforms each element of B_1 into an element of B with a norm bounded by a constant times the norm of the original element. Clearly, S transforms a certain class of boundary data given in terms of the real function $g(\varphi)$ into an analytic function defined in the lower half plane.

We proceed next to extend the domain of definition of the transformation S in such a way that it applies to a certain set of real

---

* By the norm of a function of two variables we mean the least upper bound of its absolute value.

functions in B which contains the set B_1, i.e., to the set B_2 of *continuous real* functions g in B with vanishing first Fourier coefficients, that is, to functions such that $\int_0^{2\pi/\lambda} g(\varphi) \sin \lambda\varphi \, d\varphi = 0$; this subspace B_2 is also complete, with the same norm. The extension of the definition of S is made in the following rather natural way: Take any function g in B_2 and let g_n be a sequence of functions in B_1 (i.e., a set of real trigonometric polynomials lacking first order terms) which *approximate g uniformly*. That such a sequence exists is known from the Weierstrass approximation theorem. We then form the sequence Sg_n—which is possible since S is applicable to these functions—and observe that

$$\| Sg_m - Sg_n \| = \| S(g_m - g_n) \|$$

because S is obviously a linear transformation, and hence

$$\| Sg_m - Sg_n \| \leqq K \| g_m - g_n \|$$

from (12.2.25) since $g_m - g_n$ is an element of B_1. Thus the sequence Sg_n is a Cauchy sequence, for $\| g_m - g_n \| \to 0$ because the functions g_n are assumed to furnish uniform approximations to g; hence the sequence Sg_n has a unique continuous limit function which we define as Sg. The transformation S thus extended will be referred to by the symbol $\bar{S}$. $\bar{S}$ is easily seen to be a linear transformation and the inequality (12.2.25) holds for it with the same value of K since it holds for all the functions g_n, independent of n.

Once the definition of the transformation S is extended so that it applies to functions in B_2, it becomes possible to widen the class of functions within which a (generalized) solution of our linear problem can be sought, and at the same time to reformulate the boundary problem in terms of the the following functional equation:

$$(12.2.26) \qquad \omega(\varphi) = \bar{S}g(\varphi) + ia_1 e^{-i\lambda\varphi},$$

in which $\bar{S}g(\varphi)$ refers to the above extension of $Sg(\varphi)$, with $g(\varphi) \subset B_2$, evaluated on the boundary $\psi = 0$. By virtue of (12.2.20) and the definition of $\bar{S}g(\varphi)$ for $g(\varphi)$ in B_2, one might expect that $\hat{\omega}(\varphi)$ would yield correct boundary values for the solution $\omega(\varphi, \psi)$. We proceed to show that this is indeed the case, i.e. *that any continuous function $\hat{\omega}(\varphi)$ which is given by (12.2.26) for $g \subset B_2$ furnishes the boundary values of a function $\omega(\varphi, \psi)$ defined and continuous for $\psi \leqq 0$ which is analytic in the lower half plane, has a real part $\theta(\varphi, \psi)$ with a continuous normal derivative θ_ψ in the closed half plane, and such that the boundary condition (12.2.14) with $F(\varphi) \equiv g(\varphi)$ is satisfied.*

The regularity properties of the function $\omega(\varphi, \psi)$ on the boundary $\psi = 0$ come about because of certain smoothing properties of the transformation $\overline{S}$, which we proceed to discuss. Consider first the special case in which $g(\varphi)$ is given, as in (12.2.23), by a finite Fourier sum. The function $\overline{\alpha}(\varphi) + i\overline{\beta}(\varphi) = \overline{S}g$:

$$\overline{\alpha} + i\overline{\beta} = \overline{S}g = \sum_{\nu=2}^{n} \frac{b_\nu \sin \nu\lambda\varphi}{\lambda(\nu - 1)} + i \sum_{\nu=2}^{n} \frac{b_\nu \cos \nu\lambda\varphi}{\lambda(\nu - 1)}$$

has in this case the following property:

$$- \overline{\beta}_\varphi = g + \lambda\overline{\alpha}.$$

The validity of this formula for $\overline{S}$ follows from the fact that $Sg(\varphi) = \alpha(\varphi, \psi) + i\beta(\varphi, \psi)$ as given by (12.2.24) is an analytic function in the closed half plane $\psi \leq 0$, and that Sg satisfies the nonhomogeneous boundary condition (12.2.14) with $F \equiv g$, when $\theta(\varphi, \psi)$ is identified with $\alpha(\varphi, \psi)$. This implies, because of (12.2.26) and the triangle inequality, the inequality

$$\| \overline{\beta}_\varphi \| \leq K_1 \| g \|, \quad K_1 = \text{constant},$$

as one easily sees. If g is any function in B_2, it now can be proved that $\overline{S}g$—defined for functions g in B_2 in the manner described above—is such that its imaginary part β has a continuous derivative with respect to φ. This is done by approximating g uniformly by a sequence g_n of finite Fourier sums in B_1. The corresponding derivatives $\beta_{n\varphi}$ form a Cauchy sequence because of the above inequality and hence would converge to a continuous function. The relation $-\beta_\varphi = g + \lambda\alpha$ also would hold in the limit for the derivative β_φ; thus β_φ is again seen to be continuous. It follows, therefore, that a continuous function $\hat{\omega}(\varphi) = \hat{\theta}(\varphi) + i\hat{\tau}(\varphi)$ given by (12.2.26) has the property that $\hat{\tau}(\varphi)$ has a continuous derivative, and in addition $-\hat{\tau}_\varphi(\varphi) = g(\varphi) + \lambda\theta(\varphi)$. We observe next that $\hat{\theta} + i\hat{\tau}$ furnishes the boundary values of an analytic function $\omega(\varphi, \psi)$ defined for $-\infty < \psi < 0$ and continuous in the closed half plane: this follows again by approximating $g(\varphi)$ by functions $g_n(\varphi)$ in B_1, as in (12.2.23), defining the corresponding $\omega_n(\varphi, \psi)$ by (12.2.20), and making the passage to the limit to obtain $\omega(\varphi,\psi) = \theta(\varphi,\psi) + i\tau(\varphi,\psi)$; that the functions $\omega_n(\varphi, \psi)$ converge to a continuous function for $\psi \leq 0$ we know, and that the limit is analytic at interior points follows since it is the uniform limit of analytic functions. Since $-\tau_\varphi(\varphi, \psi) = \theta_\psi(\varphi, \psi)$ for $\psi < 0$, and since τ_φ is itself a harmonic function with continuous boundary values $\hat{\tau}_\varphi$, it follows that $\tau_\varphi(\varphi, \psi) \to \hat{\tau}_\varphi$ as $\psi \to 0$, and hence that $\hat{\theta}_\psi$ is also continuous for $\psi = 0$, i.e. $\theta_\psi(\varphi, 0) = \hat{\theta}_\psi$; hence the condition $-\hat{\tau}_\varphi = g(\varphi) + \lambda\hat{\theta}$, which we have proved above to hold becomes $\hat{\theta}_\psi - \lambda\hat{\theta} = g(\varphi)$, and this is our boundary condition.

We have therefore shown that a continuous function $\hat{\omega}(\phi)$ which is given by (12.2.26), with $g(\phi)$ any function in B_2, furnishes the boundary values of an analytic function $\omega(\phi, \psi) = \theta + i\tau$ in $\psi < 0$ which is continuous for $\psi \leq 0$, whose real part $\theta(\phi, \psi)$ has a continuous derivative θ_ψ in the closed lower half plane with $\theta_\psi(\phi, 0) - \lambda\theta(\phi, 0) = g(\phi)$ or, as we also write it: $\dot{\theta}_\psi - \lambda\hat{\theta} = g$. In the subsection immediately following we shall establish the existence of a continuous solution $\hat{\omega}(\phi)$ of (12.2.26) when $g(\phi)$ is not given a priori, but depends in a non-linear way on $\hat{\omega}(\phi) = \hat{\theta}(g) + i\hat{\tau}(\phi)$, i.e. when $g(\phi) \equiv F(\phi) = \lambda(e^{-3\hat{\tau}} \sin \hat{\theta} - \hat{\theta}) + \varepsilon e^{-3\hat{\tau}} \sin \hat{\theta}$ (cf. (12.2.12)). Assuming this to have been proved, we proceed to draw at once further conclusions regarding the properties of $\hat{\omega}(\phi)$ and its continuation $\omega(\phi, \psi)$ as an analytic function in the lower half plane $\psi \leq 0$. We show, in fact, that the solution $\hat{\omega}(\phi)$ in B of our nonlinear functional equation will not only furnish the boundary values of an analytic function $\omega(\phi, \psi)$ in $\psi < 0$, with $\hat{\omega}$ as boundary values, but that ω has continuous first derivatives in the closed half plane $\psi \leq 0$, and is as a consequence then seen actually to be analytic for $\psi = 0$. Thus, in particular, $\hat{\omega}(\phi)$ would possess a convergent Fourier series. Consider the analytic function $\tilde{F}(\chi)$ defined in the lower half plane $\psi < 0$ with boundary values $\mathscr{Re}\ \tilde{F}(\chi) = g(\varphi)$ for $\psi = 0$ and with $\tilde{F}(\chi)$ bounded at ∞. The boundary condition $\hat{\theta}_\psi - \lambda\hat{\theta} = -\hat{\tau}_\varphi - \lambda\hat{\theta} = g(\varphi)$ satisfied by our solution $\omega(\varphi, \psi)$ is also extended analytically into the lower half plane $\psi < 0$ by means of the relation $\mathscr{Re}(\omega_\psi - \lambda\omega) = \mathscr{Re}\ \tilde{F}(\chi)$ in the manner used frequently in Chapters 3 and 4; hence we have

$$i\omega_\chi - \lambda\omega = \tilde{F}(\chi), \qquad \psi < 0$$

since the imaginary parts of ω_χ and ω both vanish for $\psi = -\infty$. We have just seen above that ω has an imaginary part τ with a continuous derivative $\hat{\tau}_\varphi$ on the boundary $\psi = 0$. The fact that $\hat{\tau}_\varphi$ is continuous then makes it possible to show that $\omega(\varphi, \psi)$ is Hölder continuous for $\psi = 0$. This follows, in fact, from a classical theorem of Privaloff (Bull. Soc. Math. France, Vol. 44, 1916) which states that a function which is defined and continuous in the unit circle, analytic in the interior of the circle, and has an imaginary part which is Hölder continuous on the boundary of the circle, is itself Hölder continuous in the closed unit circle: in other words, Hölder continuity of the imaginary part brings with it the Hölder continuity of the real part of the function. This theorem is made applicable in the present case by mapping one of the period strips of the solution $\omega(\chi)$ in the χ-plane conformally on the unit circle of a ζ-plane, say: we know, in fact, that ω has the real period $2\pi/\lambda$. The part of the boundary of the strip given by $\psi = 0$ (i.e. a full period interval on the boundary) is mapped on the boundary of the unit circle. (Since $|\omega| \to 0$ as $\psi \to -\infty$, the infinity of the strip is mapped on the center of the circle.) Thus $\omega = \theta + i\tau$ has an imaginary part with a continuous derivative τ_φ on $\psi = 0$, and it follows that τ is certainly Hölder continuous for $\psi = 0$. Consequently the real part of $\omega(\zeta)$, hence $\omega(\zeta)$ itself, is Hölder continuous in the closed unit circle, since this property is not destroyed by the conformal mapping. The real part of $\tilde{F}(\chi)$ on the boundary $\psi = 0$, which is given by $g(\varphi)$, is now seen to be Hölder continuous, simply because of the way $g(\phi)$ is given in terms of $\hat{\theta}$ and $\hat{\tau}$, and the Hölder continuity of the latter functions. A second application of Privaloff's theorem, this time to $\tilde{F}(\chi)$, then leads to the Hölder continuity of $\tilde{F}(\chi)$ for $\psi \le 0$. The relation $i\omega_\chi - \lambda\omega = \tilde{F}(\chi)$ thus holds for $\psi = 0$ and it shows that ω_χ is continuous for $\psi = 0$, since both ω and $\tilde{F}$ have this property. In other words ω_φ and ω_ψ are both continuous for $\psi = 0$. Finally, once $\omega(\chi)$ is shown to have a continuous derivative with respect to χ on the boundary, we could make use of a theorem of H. Lewy [L.9] to show that $\omega(\chi)$ is actually analytic on the boundary.

12.2d. The solution of the nonlinear boundary value problem

The nonlinear problem to be solved here differs from the linear problems discussed above because of the fact that the function $g(\varphi)$, the nonhomogeneous term in the free surface condition, is not given a priori, but rather becomes known only when the solution $\omega(\varphi, \psi)$ itself is determined. On the other hand, we have seen that the equation (12.2.26) furnishes the boundary values $\hat{\omega}(\varphi)$ for $\omega(\varphi, \psi)$, in case $g(\varphi)$ is a known function in the space B_2. To solve the nonlinear problem we now reverse this process: we regard the equation (12.2.26) as a functional equation for the determination of the function $\hat{\omega}(\varphi)$ when $g(\varphi)$ is identified with the function $F(\varphi)$ in equation (12.2.12), i.e. when $g(\varphi)$ itself depends in an explicitly given way on $\hat{\omega}(\varphi)$. The discussion of the preceding subsection shows that we have to prove only that the functional equation has a solution $\hat{\omega}(\varphi)$ in the Banach space B.

The existence of the solution $\hat{\omega}(\varphi)$ of the functional equation will be carried out, as we have stated earlier, by an iteration process applied to the functional equation. To this end it is convenient to introduce a nonlinear transformation R defined for any function $g = \alpha + i\beta$ in the whole space B by means of the relation

$$(12.2.27) \qquad Rg(\varphi) = \lambda(e^{-3\beta} \sin \alpha - \alpha) + \varepsilon e^{-3\beta} \sin \alpha.$$

In order that the transformation $\bar{S}$ defined in the preceding subsection should be applicable to $Rg(\varphi)$ we require, as part of the definition of R, that ε should be so determined that $Rg(\varphi)$ lies in B_2 and thus lacks its first Fourier coefficient, i.e. such that $\int_0^{2\pi/\lambda} Rg(\varphi) \sin \lambda\varphi d\varphi = 0$; this leads to the following condition on ε:

$$(12.2.28) \qquad \varepsilon = - \frac{\int_0^{2\pi/\lambda} \lambda(e^{-3\beta} \sin \alpha - \alpha) \sin \lambda\varphi d\varphi}{\int_0^{2\pi/\lambda} e^{-3\beta} \sin \alpha \sin \lambda\varphi d\varphi}.$$

This implies that R is defined only if the denominator of (12.2.28) does not vanish. Clearly, this nonlinear transformation yields always real odd functions.

Consider now the functional equation

$$(12.2.29) \qquad \hat{\omega}(\varphi) = \bar{S}R\hat{\omega}(\varphi) + ia_1 e^{-i\lambda\varphi},$$

in which $\bar{S}g(\varphi)$ refers to the extension of $Sg(\varphi)$, as defined in the preceding subsection for functions $g(\varphi)$ in the space B_2, on the boundary $\psi = 0$ and a_1 is a given real constant. Because of (12.2.12), (12.2.14), and the discussion of the preceding paragraph, it follows that a solution $\omega(\chi)$ of the nonlinear boundary value problem will be established once a function $\hat{\omega}(\varphi)$ in B is found which satisfies the functional equation (12.2.29).

In carrying out the existence proof it is convenient to introduce a few new notations. The function $r(\varphi)$ is introduced:

$$(12.2.30) \qquad r(\varphi) = \frac{\hat{\omega}(\varphi)}{a_1} - ie^{-i\lambda\varphi},$$

and a new transformation T on r is defined by

$$(12.2.31) \qquad Tr = \frac{1}{a_1} \bar{S}R[a_1(r + ie^{-i\lambda\varphi})].$$

In other words, T is applied only to those functions r such that $a_1(r + ie^{-i\lambda\varphi})$ is in the domain of definition of R. The functional equation (12.2.29) is now seen to be equivalent to the equation

$$(12.2.32) \qquad\qquad r = Tr,$$

and we seek a solution of it in the space B.

We shall solve (12.2.32) by an iteration process which starts with a function r_1 in B such that the corresponding function $R\hat{\omega}_1$ is in B_2 (i.e. such that R is applicable to $\hat{\omega}_1$), inserts it in $r_2 = Tr_1$ etc., thus obtaining a sequence r_k with $r_k = Tr_{k-1}$. In order to make sure that (12.2.28) holds for the solution we stipulate that the parameter ε in (12.2.12) be fixed at each stage of the iteration process so that (12.2.28) is satisfied; this is done by setting

$$(12.2.33) \qquad \varepsilon_k = - \frac{\int_0^{2\pi/\lambda} \lambda(e^{-3\hat{\beta}_k} \sin \hat{\alpha}_k - \hat{\alpha}_k) \sin \lambda\varphi d\varphi}{\int_0^{2\pi/\lambda} e^{-3\hat{\beta}_k} \sin \hat{\alpha}_k \sin \lambda\varphi d\varphi}.$$

At the same time, this ensures that the transformation T is really applicable to the members of the sequence r_k. Of course, it will be necessary to show that the denominators in the equations (12.2.33) are bounded away from zero and that the sequence ε_k converges. The existence of a sequence r_k converging to a solution r of (12.2.32) will be shown by disposing properly of the arbitrary constant a_1 (the amplitude of the wave in the linearized solution of the problem), i.e. by showing that $a_1 \neq 0$ can be chosen small enough so that the sequences r_k and ε_k, each of which is a function of a_1, converges. We note in passing that if $\hat{\omega}_k = \hat{\alpha}_k + i\hat{\beta}_k$ is small of order a_1, then ε_k as given by (12.2.33) is also of order a_1—later on, we give an explicit estimate for it—so that the quantities ε_k should not turn out to be of the wrong order.

The convergence of the sequence of iterates r_k to a solution r of $r = Tr$ will be shown by proving that all of the functions r_k, for values of a_1 less than a certain fixed constant, satisfy the following conditions: for some real positive constant η and real positive $\varkappa < 1$

I) $\| r \| \leq \eta$ implies $\| Tr \| \leq \eta$,

II) $\| r_1 \|, \| r_2 \| \leq \eta$ implies $\| Tr_1 - Tr_2 \| \leq \varkappa \| r_1 - r_2 \|$,

for any pair of functions r_1, r_2. Condition I) says that the transformation T carries any function in the closed "sphere" of radius η into another function in the same "sphere", and Condition II) is a Lipschitz condition.

The iteration scheme for solving the functional equation $Tr = r$ proceeds in the following standard fashion. Take any function $r_0(\varphi)$ in B with $\| r_0 \| < \eta$ to which T is applicable and consider the iterates r_n defined by $r_n = Tr_{n-1}$. From I) we see that all such functions r have a bounded norm. We have, evidently:

$$r_{n+1} - r_n = Tr_n - Tr_{n-1}.$$

Since II) holds we may write

$$\| r_{n+1} - r_n \| = \| Tr_n - Tr_{n-1} \| \leq \varkappa \| r_n - r_{n-1} \|,$$

and hence

$$\| r_{n+1} - r_n \| \leq \varkappa^n \| r_1 - r_0 \|.$$

We consider next the norm of $r_m - r_n$, $m \geq n$:

$$\| r_m - r_n \| = \| (r_m - r_{m-1}) + (r_{m-1} - r_{m-2}) + \ldots + (r_{n+1} - r_n) \|$$

$$\leq (\varkappa^{m-1} + \varkappa^{m-2} + \ldots + \varkappa^n) \| r_1 - r_0 \| < \frac{\varkappa^n}{1 - \varkappa} \| r_1 - r_0 \|.$$

(The triangle inequality is of course used here.) Since $\varkappa < 1$ it is clear that the sequence r_n is a Cauchy sequence and hence it converges to a unique limit function r in B with norm less than η. (The uniqueness statement holds of course only for functions r with norm less than η.) That the limit function r satisfies $Tr = r$ is clear, since the sequence r_n is identical with the sequence Tr_{n-1} and hence both converge to the same limit r.

In order to establish conditions I) and II) for the functions in the sequence r_k, and hence to complete our existence proof, it is convenient to introduce certain continuous functions $F_1(N)$, $F_2(N)$, ..., which are defined for real $N \geq 0$, bounded near $N = 0$, and increasing with N.

Suppose that $r_l \subset B$ is such that $\| r_l \| \leq \eta$. We set $\hat{\omega}_l = \hat{\theta}_l + i\hat{\tau}_l$ and (cf. (12.2.30)) recall that $r_l = \dfrac{\hat{\theta}_l + i\hat{\tau}_l}{a_1} - ie^{-i\lambda\varphi}$ so that $\| \hat{\omega}_l \| \leq | a_1 | (1 + \eta) = N$. In what follows, however, we omit the circumflex over θ and τ, and we shall also omit the subscript on a.

The following inequalities hold when $| a |$ is sufficiently small:

$$(12.2.34)\quad\begin{array}{ll}1. & \|\, e^{-3\tau_l}\sin\theta_l\,\| \leq NF_1(N),\\[4pt] 2. & \|\, e^{-3\tau_l}\sin\theta_l - \theta_l\,\| \leq N^2 F_2(N),\\[4pt] 3. & \|\, e^{-3\tau_l}\sin\theta_l - e^{-3\tau_m}\sin\theta_m\,\| \leq |\,a\,|\,\|\,r_l - r_m\,\|\,F_3(N),\\[4pt] 4. & \|\, e^{-3\tau_l}\sin\theta_l - e^{-3\tau_m}\sin\theta_m - (\theta_l - \theta_m)\|\\[4pt] & \qquad\qquad \leq |\,a\,|\,\|\,r_l - r_m\,\|\,NF_4(N).\end{array}$$

These inequalities are all based on the fact that if $h(\xi) = \sum_0^\infty h_n\xi^n$ is an absolutely convergent power series for all real ξ, then $|\,\xi\,| \leq N$ implies $|\,h(\xi)\,| \leq \sum_0^\infty |\,h_n\,|\,N^n$. We derive the second and third of the above inequalities as typical cases—the others are derived in a similar way. Consider the second inequality; we write

$$\|\, e^{-3\tau}\sin\theta - \theta\,\| = \|\,\theta(e^{-3\tau} - 1) + e^{-3\tau}(\sin\theta - \theta)\,\|$$

$$\leq N(e^{3N} - 1) + e^{3N}\left|\,{-\frac{\theta^3}{3!} + \frac{\theta^5}{5!}\cdots}\,\right|$$

$$\leq 3N^2\,e^{3N} + e^{3N}\,N^3\,e^N = N^2 F_2(N),$$

with $F_2(N) = 3e^{3N} + Ne^{4N}$. Consider next the third inequality. From the mean value theorem we have

$$e^{-3\tau_l}\sin\theta_l - e^{-3\tau_m}\sin\theta_m = -3e^{-3\tau^*}\sin\theta^*(\tau_l - \tau_m) + e^{-3\tau^*}\cos\theta^*(\theta_l - \theta_m),$$

in which θ^*, τ^* are some values on the segment joining (θ_l, τ_l) and (θ_m, τ_m). From this we have

$$\|\, e^{-3\tau_l}\sin\theta_l - e^{-3\tau_m}\sin\theta_m\,\| \leq \|\,\theta_l - \theta_m\,\|\,e^{3N} + 3\,\|\,\tau_l - \tau_m\,\|\,e^{3N}$$

$$\leq |\,a\,|\,\|\,r_l - r_m\,\|\,F_3(N),$$

with $F_3(N) = 4e^{3N}$, in view of the definition of r_l, r_m given in (12.2. 30).

It is also essential to give an estimate for ε_k in (12.2.33). First we obtain a lower bound for the denominator. We have, from the definition of r and the second inequality above:

$$\int_0^{2\pi/\lambda} e^{-3\tau}\sin\theta\sin\lambda\varphi\,d\varphi$$

$$= \int_0^{2\pi/\lambda}(e^{-3\tau}\sin\theta - \theta)\sin\lambda\varphi\,d\varphi + \int_0^{2\pi/\lambda}\theta\sin\lambda\varphi\,d\varphi$$

$$\geq -\frac{2\pi}{\lambda}N^2 F_2(N) + \int_0^{2\pi/\lambda}|\,a\,|\,(\sin\lambda\varphi + \mathcal{R}e\,r)\sin\lambda\varphi\,d\varphi$$

$$\geq \frac{\pi}{\lambda}\,[\,|\,a\,|\,(1 - 2\eta) - 2N^2 F_2(N)].$$

Subscripts have been dropped in the above. Since $N = |a|(1 + \eta)$ it is clear that for $\eta \leqq \frac{1}{4}$ and $|a|$ sufficiently small, say $|a| \leqq a^{(1)}$, the resulting expression is greater than $k|a|$, with k a positive constant depending on $a^{(1)}$. Use of this fact together with the second inequality above in the definition (12.2.33) of ε leads at once to the inequality:

$$(12.2.34) \quad 5. \qquad |\varepsilon| \leqq \frac{N^2 F_5(N)}{|a|}.$$

Thus ε is of order $|a|$ if $\eta \leqq \frac{1}{4}$, since N is of order $|a|$. Thus the quantities ε_k, as defined by (12.2.33), are in fact of the correct order. In the same fashion, by using all four of the above inequalities, one obtains

$$(12.2.34) \quad 6. \qquad |\varepsilon_l - \varepsilon_m| \leqq \|r_l - r_m\| \frac{N^2 F_6(N)}{|a|}.$$

We are now in a position to show that the conditions I) and II) hold once proper choices of η and a have been made. We suppose that $\|r\| \leqq \eta \leqq \frac{1}{4}$ and choose a such that $0 < |a| \leqq a^{(1)}$; any value $\varkappa$ in the range $0 < \varkappa < 1$ is taken. As before, the norm of the function $\hat{\omega}$ defined by r satisfies $\|\hat{\omega}\| \leqq |a|(1 + \eta) = N \leqq \frac{5}{4}|a|$. Our next objective is to give an estimate for Tr as defined by (12.2.31). We have, in view of (12.2.31), and (12.2.27) and (12.2.25):

$$\|Tr\| = \frac{1}{|a|} \|\bar{S}R\hat{\omega}\| \leqq \frac{K}{|a|} \|R\hat{\omega}\| = \frac{K}{|a|} \|\lambda(e^{-3\tau}\sin\theta - \theta) + \varepsilon e^{-3\tau}\sin\theta\|,$$

and this in turn yields:

$$\|Tr\| \leqq \frac{K}{|a|} [\lambda N^2 F_2(N) + \frac{N^3}{|a|} F_1(N)F_5(N)] \leqq \bar{K}|a|F_7(N),$$

with $\bar{K}$ a fixed constant, upon using the first, second, and fifth of our inequalities, together with the fact that N is of order a. Thus if $a^{(2)} \leqq a^{(1)}$ is a positive constant such that $\bar{K}a^{(2)}F_7(\frac{5}{4}a^{(2)}) \leqq \eta$ it follows that

$$\|Tr\| \leqq \eta \text{ if } \|r\| \leqq \eta \leqq \frac{1}{4} \text{ and } a| \leqq a^{(2)}.$$

This establishes the condition I). The proof that II) holds is carried out in much the same way. Suppose that r_1, r_2 are such that $\|r_1\|$, $\|r_2\| \leqq \eta$. We have, upon using the inequalities 1. to 6.:

$$\| \, Tr_1 - Tr_2 \, \| \leqq \frac{K}{|\,a\,|} \, \| \, R\hat{\omega}_1 - R\hat{\omega}_2 \, \|$$

$$= \frac{K}{|\,a\,|} \, \| \, \lambda (e^{-3\tau_1} \sin \theta_1 - e^{-3\tau_2} \sin \theta_2 - \theta_1 + \theta_2)$$

$$+ \varepsilon_1 (e^{-3\tau_1} \sin \theta_1 - e^{-3\tau_2} \sin \theta_2) + e^{-3\tau_2} \sin \theta_2 (\varepsilon_1 - \varepsilon_2) \, \|$$

$$\leqq \frac{K}{|\,a\,|} \, \| \, r_1 - r_2 \, \| \, (\lambda \, |\,a\,| \, N F_4 + N^2 F_3 F_5 + \frac{N^3}{|\,a\,|} \, F_1 F_6)$$

$$\leqq \overline{K} \, |\,a\,| \, F_8 \, (N) \, \| \, r_1 - r_2 \, \|$$

with $\overline{K}$ a fixed positive constant. If $a^{(3)} \leqq a^{(2)}$ is a fixed positive constant such that $\overline{K} a^{(3)} F_8 (\frac{5}{4} a^{(3)}) \leqq \varkappa$, then

$$\| \, Tr_1 - Tr_2 \, \| \leqq \varkappa \, \| \, r_1 - r_2 \, \|$$

and the condition II) is verified.

It follows that an iteration process starting with an arbitrary function r_0 in B, such that $R\hat{\omega}_0$ lies in B_2, with $\| \, r_0 \, \| \leqq \eta \leqq \frac{1}{4}$ will converge to a solution r of $Tr = r$ if $0 < |\,a\,| \leqq a^{(3)}$. The function $\hat{\omega} = a(r + ie^{-i\lambda\varphi})$ is then a solution of the functional equation (12.2.29) which lies in B, and which is furthermore not the "trivial" solution $\omega = 0$ (which always exists), since $\| \, \hat{\omega} \, \| \geqq a(1 - \| \, r \, \|) \geqq \frac{3}{4}a$ since $\| \, r \, \| \leqq \frac{1}{4}$. This concludes the proof for the existence of a continuous solution $\hat{\omega}(\phi)$ of the nonlinear functional equation. Once this has been done we have seen at the end of the preceding subsection that $\hat{\omega}(\phi)$ is actually analytic in ϕ.

It is also clear that the quantities ε_k assigned to each $\hat{\omega}_k$ and r_k exist, and that they converge since the ε_k form a Cauchy sequence in view of the sixth inequality above and the fact that $\| \, r_m - r_n \, \| \to 0$. If we set $\varepsilon = \lim \varepsilon_k$, it is clear that the resulting value of λ' obtained from (12.2.13), in conjunction with the arbitrarily prescribed value of λ, yields the propagation speed U through (12.2.10) as a function of the amplitude parameter a. Since $\omega(\chi)$ has the period $2\pi/\lambda$, it follows from the discussion at the beginning of this section that the motion in the physical plane has the period, or wave length, $2\pi/\lambda U$.

Bibliography

Arthur, R. S.
[A.1]   *Revised wave forecasting graphs and procedure.* Scripps Institution of Oceanography of the University of California. Wave Report 73, 1947.
[A.2]   *Variability in direction of wave travel.* Annals of the New York Academy of Sciences, Vol. 51, Art. 3, 1949, pp. 511–522.
[A.3]   *Refraction of shallow water . . . ..* Transactions of the American Geophysical Union, Vol. 31, 1950, pp. 549–552.
[A.4]   See [M.14].
[A.5]   See [P.18].
Atkinson, F. V.
[A.6]   *On Sommerfeld's "radiation condition."* London, Dublin and Edinburgh Philosophical Magazine, Ser. 7, Vol. 40, 1949, pp. 641–651.
Abdullah, A. J.
[A.7]   *Cyclogenesis by a purely mechanical process.* Journal of Meteorology, Vol. 6, 1949, pp. 86–97.

Baird, E. G.
[B.1]   See [E.5].
Baker, B. B., and E. T.Copson
[B.2]   *The Mathematical Theory of Huygens' Principle.* The Clarendon Press, Oxford, 1939.
Bakhmeteff, B. A.
[B.3]   *Hydraulics of Open Channels.* McGraw-Hill, New York, 1932.
Barber, N. F., and F. Ursell
[B.4]   *The generation and propagation of ocean waves and swell. I. Wave periods and velocity.* Philosophical Transactions of the Royal Society of London, 240, 1948, pp. 527–560.
Bateman, H.
[B.5]   *Partial Differential Equations of Mathematical Physics.* The University Press, Cambridge (Eng.), 1932. Chapter 11, Diffraction problems.
Bates, C. C.
[B.6]   *Utilization of wave forecasting in the invasions of Normandy, Burma, and Japan.* Annals of the New York Academy of Sciences, Vol. 51, Art. 3, 1949.
Beach Erosion Board
[B.7]   Technical Report No. 2. *A summary of the theory of oscillatory waves.* U. S. Government Printing Office, 1942.
Bernard, P.
[B.8]   *Sur certaines propriétés de la houle étudiées à l'aide des enregistrements séismographes.* Bulletin Institut Océanographique, Monaco, Vol. 38, 1941.
Biesel, F.
[B.9]   *Le filtre à houle.* La Houille Blanche, Vol. 4, 1949, pp. 373–375.
[B.10] *Study of wave propagation in water of gradually varying depth.* U.S. National.Bureau of Standards. Gravity Waves. NBS Circular 521, 1952.
Birkhoff, G.
[B.11] *Hydrodynamics.* Princeton University Press, 1950.

Birkhoff, Garrett, and Jack Kotik
[B.12]   *Fourier analysis of wave trains.* Proc. of the NBS Semicentennial Symposium on Gravity Waves, 1951.
Blue, F. L., and J. W. Johnson
[B.13]   *Diffraction of waves through a breakwater gap.* Transactions of the American Geophysical Union, Vol. 30, 1949, pp. 705–718.
Bondi, H.
[B.14]   *On the problem of breakers.* Great Britain, Admiralty Computing Service, WA–2304–13, 1943.
Bouasse, H.
[B.15]   *Houle, rides, seiches et marées.* Librairie Delagrave, Paris, 1924, pp. 92–145.
Boussinesq, J.
[B.16]   *Théorie de l'intumescence liquide appelée onde solitaire ou de translation se propageant dans un canal rectangulaire.* Institut de France, Académie des Sciences, Comptes Rendus, June 19, 1871, p. 755.
[B.17]   *Essai sur la théorie des eaux courantes.* Institut de France, Académie des Sciences, Memoires présentés par divers savants, Vol. 23, 1877.
Broer, L. J. F.
[B.18]   *On the propagation of energy in linear conservative waves.* Applied Scientific Research, Vol. A 2, 1951, pp. 447–468.
Bruman, J. R.
[B.19]   *Application of the water channel—compressible gas analogy.* North American Aviation, Inc., Report NA–47–87, 1947.
Bjerknes, J., and H. Solberg
[B.20]   *Life cycle of cyclones and the polar front theory of atmospheric circulation.* Geofysiske Publikationer (Kristiania), Vol. 3, 1922, pp. 3–18.

Carr, J. H., and M. E. Stelzriede
[C.1]   *Diffraction of water waves by breakwaters.* U.S. National Bureau of Standards, Gravity Waves, NBS Circular 521, 1952.
Chrystal, G.
[C.2]   *An investigation of the seiches of the Loch Earn by the Scottish lake survey.* Transactions of the Royal Society of Edinburgh, Vol. 45, 1906, pp. 361–396.
Cooper, R. I. B., and M. S. Longuet-Higgins
[C.3]   *An experimental study of the pressure variations in standing water waves.* Proceedings of the Royal Society of London, Ser. A, Vol. 206, 1951, pp. 424–435.
Copson, E. T.
[C.4]   *On an integral equation arising in the theory of diffraction.* Quarterly Journal of Mathematics, Oxford Series, Vol. 17, 1946, pp. 19–34.
[C.5]   *The asymptotic expansion of a function defined by a definite integral or contour integral.* British Admiralty Computing Service, Report No. S.R.E./ACS 106, 1946.
   Also see [B.2]
Cornish, Vaughan
[C.6]   *Waves of the Sea, and Other Water Waves.* T. Fisher Unwin, London, 1910.
[C.7]   *Ocean Waves and Kindred Geophysical Phenomena.* Cambridge University Press, London, 1934.
Coulson, C. A.
[C.8]   *Waves.* Fourth Edition, Oliver and Boyd, Edinburgh, 1947.
Courant, R., and K. O. Friedrichs
[C.9]   *Supersonic Flow and Shock Waves.* Interscience Publishers, Inc., New York, 1948.

Courant, R., and D. Hilbert
[C.10] *Methoden der Mathematischen Physik.* 2 Volumes, Interscience Publishers, Inc., New York, 1944.
Courant, R., E. Isaacson, and M. Rees
[C.11] *On the solution of nonlinear hyperbolic differential equations by finite differences.* Communications on Pure and Applied Mathematics, Vol. 5, 1952, pp. 243–255.
Crossley, H. E., Jr.
[C.12] *The analogy between surface shock waves in a liquid and shocks in compressible gases. Experimental study of hydraulic jump interactions.* California Institute of Technology Hydrodynamics Laboratory Report No. N–54, 1, 1949.
Craya, A.
[C.13] *Calcul graphiques des régimes variables dans les canaux.* La Houille Blanche, N.S. Année 1, 1946.
[C.14] *The criterion for the possibility of roll wave formation.* U.S. National Bureau of Standards, Gravity Waves, NBS Circular 521, 1952, pp. 141–151.
Charney, J. G.
[C.15] *Dynamic forecasting by numerical process.* Compendium of Meteorology, American Meteorological Society Boston, 1951, pp. 470–482.
Carson, Rachel L.
[C.16] *The Sea Around Us.* Oxford University Press, New York, 1951.

Daily, J. W., and S. C. Stephan Jr.
[D.1] *The solitary wave.* Massachusetts Institute of Technology, Hydrodynamics Laboratory, Technical Report No. 8, 1952.
Danel, Pierre
[D.2] *On the limiting clapotis.* U.S. National Bureau of Standards, Gravity Waves, NBS Circular 521, 1952, pp. 35–38.
[D.3] *Hydraulic Research Work.* Edited by La Houille Blanche.
Darbyshire, J.
[D.4] *Identification of microseismic activity with sea waves.* Proceedings of the Royal Society of London, Vol. 202, Ser. A., 1950, pp. 439–448.
Davies, T. V.
[D.5] *Symmetrical, finite amplitude gravity waves.* U.S. National Bureau of Standards, Gravity Waves, NBS Circular 521, 1952, pp. 55–60.
Deacon, G. E. R.
[D.6] *Recent studies of waves and swell.* Annals of the New York Academy of Sciences, Vol. 51, Art. 3, 1949, pp. 475–482.
[D.7] *Analysis of sea waves.* U.S. National Bureau of Standards, Gravity Waves, NBS Circular 521, 1952, pp. 209–214.
Dean, W. R.
[D.8] *Note on waves on the surface of running water.* Proceedings of the Cambridge Philosophical Society, Vol. 43, 1947, pp. 96–99.
Deymié, P.
[D.9] *Propagation d'une intumesence allongée (problème aval).* International Congress for Applied Mechanics, 5th Proceedings, September 1938, John Wiley and Sons, New York, 1939.
Dirichlet, P. L.
[D.10] *Untersuchungen über ein Problem der Hydrodynamik.* Journal für die reine und angewandte Mathematik, Vol. 58, 1861, pp. 181–228.
Dixon, G. F.
[D.11] *Thèses présentées à la Faculté des Sciences de l'Université de Grenoble,* 1949.

Dressler, R. F.
[D.12] *Mathematical solution of the problem of roll-waves in inclined open channels.* Communications on Pure and Applied Mathematics, Vol. 2, 1949, pp. 149–194.
[D.13] *Stability of uniform flow and roll-wave formation.* U.S. National Bureau of Standards, Gravity Waves, NBS Circular 521, 1952, pp. 237–241.
Dressler, R. F., and F. V. Pohle
[D.14] *Resistance effects on hydraulic instability.* Communications on Pure and Applied Mathematics, Vol. 6, 1953, pp. 93–96.
Dubreuil-Jacotin, M. L.
[D.15] *Sur les ondes de type permanent dans les liquides·hétérogènes.* Rendiconti della Accademia Nazionale dei Lincei, Ser. 6, Vol. 15, 1932, pp. 814–819.
[D.15a] *Sur la détermination rigoureuse des ondes permanentes périodiques d'ampleur finie.* Journal de Mathématiques Pures et Appliquées, Ser. 9, Vol. 13, 1934, pp. 217–291.
Dean, W. R.
[D.16] *On some cases of the reflection of surface waves by an infinite plane barrier.* Proceedings of the Cambridge Philosophical Society, Vol. 42, 1945, pp. 24–28.
[D.17] *On the reflection of surface waves by a submerged plane barrier.* Proceedings of the Cambridge Philosophical Society, Vol. 41, 1945, pp. 231–238.

Eckart, C.
[E.1] *The ray-particle analogy.* Journal of Marine Research, Vol. 9, 1950, pp. 139–144.
[E.2] *Surface waves on water of variable depth.* Scripps Institution of Oceanography, Lecture Notes, Fall Semester 1950–51. Ref. 51–12. Wave Report No. 100, 1951.
[E.3] *The propagation of gravity waves from deep to shallow water.* U.S. National Bureau of Standards, Gravity Waves, NBS Circular 521, 1952, pp. 165–173.
Egiazarov, I.
[E.4] *On the Daily Regulation of Hdryoelectric Stations (Experimental Investigations of the Negative Wave).* Leningrad, 1931. (In Russian, with an English summary. Cited in Favre [F.2].)
Einstein, H. A., and E. G. Baird
[E.5] *Progress report of the analogy between surface shock waves on liquids and shocks in compressible gases.* California Institute of Technology, Hydrodynamics Laboratory, Report No. N–54, 1946.
Emde, F.
[E.6] See [J.1].

Favre, H.
[F.1] *Contribution à l'étude des courants liquides.* Publication du Laboratoire de Recherches hydrauliques annexé à l'Ecole Polytechnique Fédérale de Zurich, 1933.
[F.2] *Etude théorique et expérimentale des ondes de translation dans les canaux découverts.* Dunod, Paris, 1935.
Finkelstein, A.
[F.3] *The initial value problem for transient water waves.* Dissertation, New York University, 1953.
Fjeldstad, J. E.
[F.4] *Observations of the internal tidal waves.* U.S. National Bureau of Standards, Gravity Waves, NBS Circular 521, 1952.

Fleishman, B., J. J. Stoker, and L. Weliczker,
[F.5] *Floating breakwaters in shallow water.* New York University, Institute of Mathematical Sciences, Report No. IMM–192, 1953.
Forchheimer, P.
[F.6] *Hydraulik.* Encyklopädie der mathematischen Wissenschaften, Vol. IV, Teil 3, pp. 324–472, B. G. Teubner, Leipzig, 1901–1908.
Forel, F. A.
[F.7] *Le Léman, monographie limnologique.* Bibliothéque Nationale, Vol. 2, F. Rouge, Lausanne, 1895.
Frank, P., and R. v. Mises,
[F.8] *Die Differential- und Integralgleichungen der Mechanik und Physik.* F. Vieweg und Sohn, Braunschweig, 1925–27.
Freeman, J. C., Jr.
[F.9] *An analogy between the equatorial easterlies and supersonic gas flows.* Journal of Meteorology, Vol. 5, 1948, pp. 138–146.
[F.10] *The solution of nonlinear meteorological problems by the method of characteristics.* Compendium of Meteorology, American Meteorological Society, 1951, pp. 421–433.
Friedrichs, K. O.
[F.11] *On the derivation of the shallow water theory.* Appendix to *The formation of breakers and bores* by J. J. Stoker, Communications on Pure and Applied Mathematics, Vol. 1, 1948, pp. 81–85.
Friedrichs, K. O., and H. Lewy
[F.12] *The dock problem.* Communications on Pure and Applied Mathematics, Vol. 1, 1948, pp. 135–148.
Friedrichs, K. O., and D. H. Hyers
[F.13] *The existence of solitary waves.* Communications on Pure and Applied Mathematics, Vol. 7, 1954, pp. 517–550.
Friedrichs, K. O.
[F.14] *Water waves on a shallow sloping beach.* Communications on Pure and Applied Mathematics, Vol. 1, 1948, pp. 109–134.
[F.15] See [C.9].
Fleming, R. H.
[F.16] See [S.32a].

Goldstein, E., and J. B. Keller
[G.1] *Reflection of waves in shallow water.* (Unpublished.) See E. Goldstein, Master's Thesis, New York University, 1950.
Guthrie, F.
[G.2] ` *On stationary liquid waves.* London, Edinburgh and Dublin, Philosophical Magazine, Ser. 4, Vol. 50, 1875.
Gwyther, R. F.
[G.3] *The classes of progressive long waves.* London, Edinburgh and Dublin, Philosophical Magazine, Ser. 5, Vol. 1, 1900. pp. 213, 308, 349.
Goldstein, E.
[G.4] See [K.8].
Gerber, R.
[G.5] *Sur les solutions exactes des équations du mouvement avec surface libre d'un liquide pesant.* (Thesis, Univ. of Grenoble, 1955.) To appear in Journal de Mathématique.
Goldner, S. R.
[G.6] *Existence and uniqueness theorems for two systems of nonlinear hyperbolic differential equations for functions of two independent variables.* Doctor's thesis, New York University, 1949.

Hadamard, J.
[H.1]   *Sur les ondes liquides.* Institut de France. Académie des Sciences, Comptes Rendus, Vol. 150, 1910, pp. 609–611, 772–774.

Hamada, T.
[H.2]   *Breakers and beach erosion.* Report of Transportation Technical Research Institute (Report No. 1) Tokyo, 1951.

Harleman, D. R. F.
[H.3]   *Studies on the validity of the hydraulic analogy to supersonic flow.* Parts I and II. Massachusetts Institute of Technology, 1950. (Technical report to U.S. Air Force).

Haskind, M. D.
[H.4]   *Waves arising from oscillation of bodies in shallow water.* Prikladnaya Matematika i Mechanika, Vol. 10, 1946. pp. 475–480.

[H.5]   *The hydrodynamic theory of the oscillation of a ship in waves* (in Russian). Prikladnaya Matematika i Mechanika, Vol. 10, 1946, pp. 39–66. *Oscillation of a ship on a calm sea* (in Russian), Izv. Ak. Nauk, Ot. Tex. Nauk, No. 1, 1946, pp. 23–34. Translated by V. Wehausen as "Two papers on the hydrodynamic theory of heaving and pitching of a ship", Technical Research Bulletin 1–12, Society of Naval Architects and Marine Engineers, New York, 1953.

Haurwitz, B.
[H.6]   *Dynamic meteorology.* McGraw-Hill, New York, 1941.

Havelock, T. H.
[H.7]   *The wave-making resistance of ships.* Proceedings of the Royal Society of London, Ser. A, Vol. 82, 1909, pp. 276–300.

[H.8]   *Wave resistance theory and its application to ship problems.* Society of Naval Architects and Marine Engineers, New York, 1950.

[H.9]   *The propagation of groups of waves in dispersive media.* The University Press, Cambridge (Eng.), (Cambridge Tracts, No. 17), 1914.

Hayashi, T.
[H.10] *Mathematical theory of flood waves.* Japan National Congress for Applied Mechanics, First. Proceedings. 1951.

Heins, A. E.
[H.11] *Water waves over a channel of finite depth with a submerged plane barrier.* Canadian Journal of Mathematics, Vol. 2, 1950, pp. 210–222.

[H.12] *Water waves over a channel of finite depth with a dock.* American Journal of Mathematics, Vol. 70, 1948, pp. 730–748.

Henry, Marc
[H.13] *Note sur la propagation des intumescences dans un canal rectangulaire.* Communiqué à la séance du 5 nov. 1937 du comité technique de la société hydrotechnique de France.

Hinze, J. O.
[H.14] *Die Erzeugung von Ringwellen auf einer Flüssigkeitsoberfläche durch periodisch wirkende Druckkräfte.* Zeitschrift für angewandte Mathematik und Mechanik, Bd. 16, 1936.

Hogner, E.
[H.15] *A contribution to the theory of ship waves.* Arkiv för Matematik, Astronomi, och Fysik, Vol. 17, 1922.

Hopf, E.
[H.16] *On S. Bernstein's theorem on surfaces z(x, y) of nonpositive curvature.* Proceedings of the American Mathematical Society, Vol. 1, 1950, pp. 80–85.

Hyers, D. H.
See [F.13]

Hanson, E. T.
[H.17] *The theory of ship waves.* Proceedings of the Royal Society of London, Ser. A, Vol. 111, 1926, pp. 491–529.

Isaacson, E.
[I.1] *Water waves over a sloping bottom.* Communications on Pure and Applied Mathematics, Vol. 3, 1950, pp. 1–32.
[I.2] *Waves against an overhanging cliff.* Communications on Pure and Applied Mathematics, Vol. 1, 1948, pp. 201–209.
[I.3] See [C.11].
Isaacson, E., J. J. Stoker, and B. A. Troesch
[I.4] *Numerical solution of flood prediction and river regulation problems. Report 2. Numerical solution of flood problems in simplified models of the Ohio River and the junction of the Ohio and Mississippi Rivers.* New York University, Institute of Mathematical Sciences, Report No. IMM–205, 1954.
[I.4a] *Numerical solution of flood prediction and river regulation problems. Report 3. Results of the numerical prediction of the 1945 and 1948 floods in the Ohio River, of the 1947 flood through the junction of the Ohio and Mississippi Rivers, and of the floods of 1950 and 1948 through Kentucky Reservoir.* New York University, Institute of Mathematical Sciences, Report No. IMM-NYU-235, 1956.
Isaacs, J. D.
[I.5] See [J.7].
Iversen, H. W.
[I.6] *Laboratory study of breakers.* U.S. National Bureau of Standards, Gravity Waves, NBS Circular 521, 1952.

Jahnke, E., and F. Emde
[J.1] *Tables of Functions.* Dover, New York, 1945, p. 138.
Jeffreys, Harold
[J.2] *The flow of water in an inclined channel of rectangular section.* London, Edinburgh and Dublin, Philosophical Magazine, Ser. 6, Vol. 49, 1925, pp. 793–807.
Jeffreys, H. and Bertha S. Jeffreys
[J.3] *Methods of Mathematical Physics.* Cambridge (Eng.), The University Press, 1948.
John, F.
[J.4] *Waves in the presence of an inclined barrier.* Communications on Pure and Applied Mathematics, Vol. 1, 1948, pp. 149–200.
[J.5] *On the motion of floating bodies* I, II. Communications on Pure and Applied Mathematics, Vol. 2, 1949, pp. 13–57; Vol. 3, 1950, pp. 45–100.
Johnson, J. W. and M. P. O'Brien
[J.6] *The engineering aspects of waves in harbor design.* International Association of Hyd. Str. Research, Grenoble, 1949.
Johnson, J. W., M. P. O'Brien, and J. D. Isaacs
[J.7] *Graphical construction of wave refraction diagrams.* Hydrographic Office, Navy Department, Publication no. 605, 1948.
Johnson, J. W.
[J.8] See [B.13].
Johnson, M. W.
[J.9] See [S.32a].

Kampé de Feriet, J., and J. Kotik
[K.1] *Surface waves of finite energy.* Journal of Rational Mechanics and Analysis, Vol. 2, 1953, pp. 577–585.

von Kármán, T.
[K.2] *Eine praktische Anwendung der Analogie zwischen Überschallströmung in Gasen und überkritischer Strömung in offenen Gerinnen.* Zentralblatt für angewandte Mathematik und Mechanik, Vol. 18, 1939.

Karp, S. N.
[K.3] *Wiener-Hopf techniques and mixed boundary value problems.* Communications on Pure and Applied Mathematics, Vol. 3, 1950, pp. 411–426.

Keller, J. B., and P. Lax
[K.4] *Finite difference schemes for hyperbolic systems.* LAMS 1201, 1950.

Keller, J. B.
[K.5] *The solitary wave and periodic waves in shallow water.* Annals of the New York Academy of Sciences, Vol. 51, Art. 3. 1949, pp. 345–350.

[K.6] *The solitary wave and periodic waves in shallow water.* Communications on Pure and Applied Mathematics, Vol. 1, 1948, pp. 323–339.

[K.7] *Scattering of water waves treated by the variational method.* U.S. National Bureau of Standards, Gravity Waves, NBS Circular 521, 1952.

Also see [W.8]

Keller, J. B., and E. Goldstein
[K.8] *Water wave reflection due to surface tension and floating ice.* Transactions of the American Geophysical Union, Vol. 34, 1953, pp. 43–48.

Keller, J. B. and M. Weitz
[K.9] *Reflection and transmission coefficients for waves entering or leaving an icefield.* Communications on Pure and Applied Mathematics, Vol. 6, 1953, pp. 415–417.

Kellog, O. D.
[K.10] *Foundations of potential theory.* J. Springer, Berlin, 1929.

Kelvin, Lord (Sir W. Thomson)
[K.11] *On the waves produced by a single impulse in water of any depth, or in a dispersive medium.* Proceedings of the Royal Society of London, Ser. A. Vol. 42, 1887, pp. 80–85.

Keulegan, G. H.
[K.12] *The characteristics of internal solitary waves.* (Abstract only.) U.S. National Bureau of Standards, Gravity Waves, NBS Circular 521, 1952.

Keulegan, G. H., and G. W. Patterson
[K.13] *Mathematical theory of irrotational translation waves.* U.S. National Bureau of Standards Journal of Research, Vol. 24, 1940.

Kotschin, N. J., I. A. Kibel, and N. W. Rose
[K.14] *Theoretische Hydrodynamik.* Akad. Verlag, Berlin, 1954.

Korteweg, D. J., and G. de Vries
[K.15] *On the change of form of long waves advancing in a rectangular canal and on a new type of long stationary waves.* London, Dublin and Edinburgh, Philosophical Magazine, Ser. 5, Vol. 39, 1895, p. 422.

Korvin-Krukovsky, B. V., and E. V. Lewis
[K.16] *Ship motions in regular and irregular seas.* Stevens Institute of Technology, Experimental Towing Tank, Technical Memorandum no. 106, 1954.

Kotik, J.
[K.17] *Existence and uniqueness theorems for linear water waves.* Massachusetts Institute of Technology, 1952.

[K.18] See [K.1], [B.12].

Kreisel, G.
[K.19] *Surface waves.* Quarterly of Applied Mathematics, Vol. 7, 1949, pp. 21–44.

Krylov, A. N.
[K.20] *A general theory of the oscillations of a ship on waves.* Transactions of the Institute of Naval Architects and Engineers, Vol. 37, 1896.
Keldysh, M. V.
[K.21] *Remarks on certain motions of a heavy fluid.* (Russian) Tekhnich. Zametki Tsagi, no. 52, 1935.

Laitone, E. V.
[L.1] *A study of transonic gas dynamics by the hydraulic analogy.* University of California, Berkeley, California, Institute of Engineering Research; 1951.
Lamb, H.
[L.2] *On deep water waves.* Proceedings of the London Mathematical Society, Ser. 2, Vol. 2, 1904, p. 388.
[L.3] *Hydrodynamics.* Dover Publications, New York, 1945; Cambridge University Press, 1932.
Lavrentieff, M.
[L.4] *On the theory of long waves.* Akad. Nauk Ukrain. R.S.R., Zb. Proc. Inst. Math., Vol. 1, 1948. In Ukrainian.
Lax, Anneli
[L.5] *Decaying shocks.* Communications on Pure and Applied Mathematics, Vol. 1, 1948, pp. 247–257.
Lax, Peter
[L.6] See [K.4].
Levi-Civita, T.
[L.7] *Détermination rigoureuse des ondes permanentes d'ampleur finie.* Mathematische Annalen, Vol. 93, 1925, pp. 264–314.
Lewy, H.
[L.8] *Water waves on sloping beaches.* Bulletin of the American Mathematical Society, Vol. 52, 1946.
[L.9] *A note on harmonic functions and a hydrodynamical application.* Proceedings of the American Mathematical Society, Vol. 3, No. 1, 1952.
[L.9a] *On steady free surface flow in a gravity field.* Communications on Pure and Applied Mathematics, Vol. 5, 1952.
[L.10] See [F.12].
Lichtenstein, L.
[L.11] *Vorlesungen über einige Klassen nichtlinearer Integralgleichungen und Integro-Differentialgleichungen nebst Anwendungen.* J. Springer, Berlin, 1931.
[L.12] *Grundlagen der Hydromechanik.* J. Springer, Berlin, 1929.
Longuet-Higgins, M. S.
[L.13] *A theory of the generation of microseisms.* Philosophical Transactions of the Royal Society of London, Ser. A, Vol. 243, 1950, pp. 1–35.
Longuet-Higgins, M. S., and F. Ursell
[L.14] *Seawaves and microseisms.* Nature, Vol. 162, 1948.
Longuet-Higgins, M. S.
[L.15] See [C.3].
Lowell, S. C.
[L.16] *The propagation of waves in shallow water.* Communications on Pure and Applied Mathematics, Vol. 2, 1949, pp. 275–291.
[L.17] *The Propagation of Waves in Shallow Water.* Ph. D. Thesis, New York University, June 1949.
Lunde, J. K., and W. C. S. Wigley
[L.18] *Calculated and observed wave resistances for a series of forms of fuller midsection.* Quarterly Transactions of the Institute of Naval Architects, London, Vol. 90, 1948.

Lunde, J. K.
[L.19] *On the linearized theory of wave resistance for displacement ships in steady and accelerated motion.* Society of Naval Architects and Marine Engineers, 1951.
Lewis, E. V.
[L.20] See [K.16].

MacDonald, H. M.
[M.1] *A class of diffraction problems.* Proceeding of the London Mathematical Society, Ser. 2, Vol. 14, 1915, pp. 410–427.
MacRobert, T. M.
[M.2] *Functions of a Complex Variable.* MacMillan, London, 1947.
Martin, J. C., W. J. Moyce, W. G. Penney, A. T. Price, and C. K. Thornhill
[M.3] *Some gravity wave problems in the motion of perfect liquids.* Philosophical Transactions of the Royal Society of London, Ser. A, Vol. 244, 1952.
Mason, Martin A.
[M.4] *Some observations of breaking waves.* U.S. National Bureau of Standards, Gravity Waves, N. B. S. Circular 521, 1952.
Massau, J.
[M.5] *Mémoire sur l'intégration graphique des équations aux derivées partielles.* Annales des ingénieurs sortis de Gand, Tome XII, 1889.
McGowan, J.
[M.6] *On the solitary wave.* London, Dublin and Edinburgh, Philosophical Magazine, Ser. 5, Vol. 32, p. 45, 1891. *On the highest wave of permanent type.* Ser. 5, Vol. 38, p. 351, 1894.
McNown, J. S.
[M.7] *Waves and seiches in idealized ports.* U.S. National Bureau of Standards, Gravity Waves, N. B. S. Circular 521, 1952.
Miche, A.
[M.8] *Mouvements ondulatoires de la mer en profondeur constante ou décroissante.* Annales des ponts et chaussées, 1944, pp. 25–78, 131–164, 270–292, 369–406.
Michell, J. H.
[M.9] *The wave resistance of a ship.* London, Dublin and Edinburgh, Philosophical Magazine, Ser. 5, Vol. 45, 1898.
Milne-Thompson, L. M.
[M.10] *Theoretical Hydrodynamics.* MacMillan, London, 1938.
Moyce, W. J.
[M.11] See [M.3].
Munk, W. H.
[M.12] *The solitary wave theory and its application to surf problems.* Annals of the New York Academy of Sciences, Vol. 51, Art. 3, 1949, pp. 376–424.
[M.13] *Supplement to breakers and surf, principles in forecasting.* Scripps Institution Oceanography. Wave Report No. 49.
Munk, W. H., and R. S. Arthur
[M.14] *Wave intensity along a refracted ray.* U. S. National Bureau of Standards, Gravity Waves, N. B. S. Circular 521, 1952.
Munk, W. H., and M. A. Traylor
[M.15] *Height of breakers and depth of breaking.* Scripps Institution of Oceanography, Wave Report No. 47, 1945.
[M.16] *Refraction of ocean waves: a process linking underwater topography to beach erosion.* Journal of Geology, Vol. 15, 1947.
Munk, W. H.
[M.17] See [S.33], [S.34], [S.35], [S.36], [S.37].

Nekrassov, A. I.
[N.1]   *On steady waves.* Izv. Ivanovo-Voznesensk. Politekhn. In-ta, No. 3, 1921.
[N.1a] *The exact theory of steady waves on the surface of a heavy fluid.* Izdat.
        Akad. Nauk, SSSR, Moscow, 1951.
Nirenberg, L.
[N.2]   *Waves of finite amplitude created by an obstacle in a stream.* Unpublished
        manuscript (New York University).
Nakano, N., S. Unoki, and Y. Kuga
[N.3]   *On the result of wave observations at Jogashima Island and its application
        to wave forecasting.* Meteorological Research Institute, Tokyo, 1953.

O'Brien, M. P.
[O.1]   See [J.6].
[O.2]   See [J.7].

Patterson, G. W.
[P.1]   See [K.13].
Penney, W. G.
[P.2]   See [M.3]
Penney, W. G. and A. T. Price
[P.3]   *Diffraction of sea waves by breakwaters.* Directorate of Miscellaneous
        Weapons Development History No. 26–Artificial Harbors, Sec. 3D, 1944.
Peters, A. S.
[P.4]   *A new treatment of the ship wave problem.* Communications on Pure and
        Applied Mathematics, Vol. 2, 1949, pp. 123–148.
[P.5]   *The effect of a floating mat on water waves.* Communications on Pure
        and Applied Mathematics, Vol. 3, 1950, pp. 319–354.
[P.6]   *Water waves over sloping beaches and the solution of a mixed boundary
        value problem for $\Delta\varphi - k^2\varphi = 0$ in a sector.* Communications on Pure
        and Applied Mathematics, Vol. 5, 1952, pp. 87–108.
Peters, A. S., and J. J. Stoker
[P.7]   *The motion of a ship, as a floating rigid body, in a seaway.* Institute of
        Mathematical Sciences, Report No. IMM–203, New York University,
        1954.
Pierson, W. J., Jr.
[P.8]   *The interpretation of crossed orthogonals in wave refraction phenomena.*
        Technical Memorandum No. 21, Beach Erosion Board, Corps of
        Engineers.
Pierson, W. J., Jr., and M. St. Denis
[P.9]   *On the motion of ships in confused seas.* Society of Naval Architects and
        Marine Engineers, 1953.
Pierson, W. J., Jr.
[P.10] *A unified mathematical theory for the analysis, propagation, and refraction
        of storm generated ocean surface waves.* Part I, Part II, Reports of
        College of Engineering, New York University, 1952.
Pohle, F. V.
[P.11] *The Lagrangian equations of hydrodynamics: solutions which are analytic
        functions of the time.* Thesis, New York University, 1950.
[P.12] *Motions of water due to breaking of a dam, and related problems.* U. S.
        National Bureau of Standards, Gravity Waves, N. B. S. Circular 521,
        1952.
[P.13] See [D.14].
Price, A. T.
[P.14] See [P.3].
[P.15] See [M.3].

Preiswerk, E.
[P.16] *Anwendung gasdynamischer Methoden auf Wasserströmungen mit freier Oberfläche.* Mitteilungen aus dem Institut für Aerodynamik, Eidg. Technische Hochschule, Zürich, 1938.
Putman, H. J.
[P.17] *Unsteady flow in open channels.* Transactions of the American Geophysical Union, Vol. 29, 1948.
Putnam, J. A., and R. S. Arthur
[P.18] *Diffraction of water waves by breakwaters.* Transactions of the American Geophysical Union, 29 (4), 1948.
Peters, A. S., and J. J. Stoker
[P.19] *A uniqueness theorem and a new solution for Sommerfeld's and other diffraction problems.* Communications on Pure and Applied Mathematics. Vol. 7, 1954, pp. 565–585.

Rayleigh, Lord
[R.1] *On waves.* London, Dublin and Edinburgh, Philosophical Magazine, Ser. 5, Vol. 1, 1876.
[R.2] *The form of standing waves on the surface of running water.* Proceedings of the London Mathematical Society, Vol. 15, 1883–4, pp. 69–78.
[R.3] *On the theory of long waves and bores.* Proceedings of the Royal Society of London, Ser. A, Vol. 90, 1914.
[R.4] *The Theory Of Sound.* Dover Publications, New York, 1945.
Ré R.
[R.5] *Étude du lâcher instantané d'une retenue d'eau dans un canal par la méthode graphique.* La Houille Blanche N. S. Vol. 1, 1946.
Rees, M.
[R.6] See [C.11].
Rellich, F.
[R.7] *Über das asymptotische Verhalten der Lösungen von $\Delta u + u = 0$ in unendlichen Gebieten.* Jahresbericht der deutschen Mathematiker Vereinigung, Vol. 53, 1943, pp. 57–65.
Riabouchinsky, D.
[R.8] *Sur l'analogie hydraulique des mouvements d'un fluide compressible,* Institut de France, Académie des Sciences, Comptes Rendus, Vol. 195, 1932, p. 998.
Rose, N. W.
See [K.14]
Roseau, M.
[R.9] *Contribution à la théorie des ondes liquides de gravité en profondeur variable.* Publications Scientifiques et Techniques du Ministère de l'Air, No. 275, Paris, 1952.
Rouse, H.
[R.10] *Fluid mechanics for hydraulic engineers.* McGraw-Hill, New York, 1938.
Rouse, H., editor
[R.11] *Engineering Hydraulics.* John Wiley and Sons, New York, and Chapman and Hall, Ltd., London, 1950. (Especially Gilcrest, B. R., Flood routing, pp. 635–676.)
Ruellan, F., and A. Wallet
[R.12] *Trajectoires internes dans un clapotis partiel.* La Houille Blanche, Vol. 5, 1950.
Rubin, H.
[R.13] *The dock of finite extent.* Thesis, New York University, 1953.
Russell, S.
[R.14] *Report on waves.* British Association Reports, 1844.

St. Denis, M., and G. Weinblum
[S.1] *On the motions of ships at sea.* Society of Naval Architects and Marine Engineers, 1950.
[S.2] See [P.9].
Saint Venant, B. de
[S.3] *Théorie du mouvement non permanent des eaux.* Institut de France, Académie des Sciences, Comptes Rendus, Paris, July 1871, Vol. 73, pp. 147, 237.
Schönfeld, J. C.
[S.4] *Propagation of Tides and Similar Waves.* Doctor's Thesis, Delft, Holland, 1951.
[S.4a] *Distortion of long waves; equilibrium and stability.* Union Géodésique et Géophysique Internationale, Vol. 4, 1951.
Schwinger, J. S.
[S.5] *Fourier transform solution of integral equations.* M.I.T. Radiation Laboratory Report.
Scripps Institution of Oceanography
[S.6] *Waves in shallow water* (I). *Change in wave height along Scripps Institution pier with special emphasis on breaker characteristics.* Wave Report No. 13.
[S.7] *Effect of bottom slope on breaker characteristics as observed along the Scripps Institution pier.* Wave Report No. 24, 1944.
Seiwell, H. R.
[S.8] *Military oceanography in World War II.* Mil. Eng. 39 (259), 1947.
[S.9] *The principles of time series analyses applied to ocean wave data.* Proceedings of the National Academy of Sciences, Vol. 35, 1949, pp. 518–528.
[S.10] *Sea surface roughness measurements in theory and practise.* Annals of the New York Academy of Sciences, Vol. 51, Art. 3, 1949, pp. 483–501.
Sneddon, I. N.
[S.11] *Fourier Transforms.* McGraw-Hill, New York, 1951.
Solberg, H.
See [B.20]
Sommerfeld, A.
[S.12] *Mathematische Theorie der Diffraktion.* Mathematische Annalen, Bd. 47, p. 317, 1896.
[S.13] *Vorlesungen über theoretische Physik,* II, *Mechanik der deformierbaren Medien.* Akademie Verlag, Leipzig, 1949.
Southwell, R.
[S.14] *Relaxation Methods.* Oxford Press, 1940.
Sretenski, L. N.
[S.15] *Ship waves for circular courses* (in Russian). Bulletin de l'Académie des Sciences de l'URSS, 1946.
Stelzriede, M. E.
[S.16] See [C.1].
Stephan, S. C., Jr.
[S.17] See [D.1].
Stoker, J. J.
[S.18] *Surface waves in water of variable depth.* Quarterly of Applied Mathematics, Vol. 5, 1947, pp. 1–54.
[S.19] *The formation of breakers and bores.* Communications on Pure and Applied Mathematics, Vol. 1, 1948, pp. 1–87.
[S.20] *The breaking of waves in shallow water.* Annals of the New York Academy of Sciences, Vol. 51, Art. 3, 1949, pp. 360–375.
[S.21] *On radiation conditions.* Sixth Symposium on Applied Mathematics. American Mathematical Society, 1953.

[S.22] *Unsteady waves on a running stream.* Communications on Pure and Applied Mathematics, Vol. 6, 1953, pp. 471–481.

[S.23] *Numerical solution of flood prediction and river regulation problems* I: *Derivation of basic theory and formulation of numerical methods of attack.* New York University, Institute of Mathematical Sciences, Report No. IMM-200, 1953.

[S.24] *Dynamical theory for treating the motion of cold and warm fronts in the atmosphere.* Institute of Mathematical Sciences, Report No. IMM–195, 1953, New York University.

[S.25] See [F.5].

[S.26] See [I.4].

[S.27] See [P.7].

[S.27a] See [I.4a].

Stokes, G. G.

[S.28] *On the theory of oscillatory waves.* Transactions of the Cambridge Philosophical Society, Vol. 8, 1847, and Supplement. Scientific Papers, Vol. 1.

Struik, D. J.

[S.29] *Détermination rigoureuse des ondes irrotationnelles périodiques dans un canal à profondeur finie.* Mathematische Annalen, Vol. 95, 1926, pp. 595–634.

Suquet, F.

[S.30] *Remarks on graphical computation of wave refraction.* International Association for Hydraulic Research, Grenoble, 1949.

[S.31] *Etude expérimentale du déferlement de la houle.* La Houille Blanche, Vol. 5, 1950.

Sverdrup, H. U.

[S.32] *Oceanography for Meteorologists.* Prentice-Hall, Inc., N. Y., 1943.

Sverdrup, H. U., M. W. Johnson, and R. H. Fleming

[S.32a] *The Oceans.* Prentice-Hall, New York, 1946.

Sverdrup, H. U., and W. H. Munk

[S.33] *Theoretical and empirical relations in forecasting breakers and surf.* Transactions of the American Geophysical Union, Vol. 27, 1946.

[S.34] *A study of progressive oscillatory waves in water; a summary of the theory of water waves.* Technical reports No. 1, 2, Beach Erosion Board, Office of Chief of Engineering, U.S. War Department.

[S.35] *Breakers and surf.* U.S. Navy Department, Hydrographic Office, Publication No. 234, 1944.

[S.36] *Wind, sea, and swell: Theory of relations for forecasting.* U.S. Navy Department, Hydrographic Office, 1947.

[S.37] *Empirical and theoretical relations between wind, sea and swell.* Transactions of the American Geophysical Union, Vol. 27, 1946.

Sretenski, L. N.

[S.38] *Waves.* In: Mekhanika v SSSR za tridtsat'let 1917–1947. Moscow, 1950, pp. 279–299.

Thomas, H. A.

[T.1] *The propagation of stable wave configurations in steep channels.* Carnegie Institute of Technology, Pittsburgh, Pa.

[T.2] *Hydraulics of flood movements in rivers.* Carnegie Institute of Technology, Pittsburgh, Pa., 1937.

[T.3] *Propagation of waves in steep prismatic conduits.* Proceedings of Hydraulics Conference, University of Iowa, Studies in engineering, Bulletin 20, 1940.

Thorade, H. F.
  [T.4] *Probleme der Wasserwellen.* 1931, from Probleme der kosmischen Physik, Bd. 13–14.
Thornhill, C. K.
  [T.5] See [M.3].
Titchmarsh, E. C.
  [T.6] *Introduction to the Theory of Fourier Integrals.* Second Edition, The Clarendon Press, Oxford, 1948.
Traylor, M. A.
  [T.7] See [M.15].
  [T.8] See [M.16].
Troesch, B. A.
  [T.9] See [I.4].
  [T.9a] See [I.4a].
Thompson, P. D.
  [T.10] *Notes on the theory of large-scale disturbances in atmospheric flow with applications to numerical weather prediction.* Airforce Cambridge Research Center, Geophysical Research Papers, No. 16, 1952.
Tepper, M.
  [T.11] *A proposed mechanism of squall lines: the pressure jump line.* Journal of Meteorology, Vol. 7, 1950, pp. 21–29.

U.S. Navy Hydrographic Office
  [U.1] *Wind, waves and swell: Principles in forecasting.* H. O. Misc. 11, 1944.
  [U.2] *Breakers and surf, principles in forecasting.* H. O. 234, 1944.
Ursell, F.
  [U.3] *The effect of a vertical barrier on surface waves in deep water.* Proceedings of the Cambridge Philosophical Society, Vol. 43, 1947, pp. 374–382.
  [U.4] *On the waves due to the rolling of a ship.* Quarterly Journal of Mechanics and Applied Mathematics, Vol. 1, 1948, pp. 246–252.
  [U.5] *Surface waves on deep water in the presence of a submerged circular cylinder,* I, II. Cambridge Philosophical Society, Vol. 46, 1950, pp. 141–152, 153–163.
  [U.6] *On the application of harmonic analysis to ocean wave research.* Science, Vol. 111, 1950.
  [U.7] *Discrete and continuous spectra in the theory of gravity waves.* U.S. National Bureau of Standards, Gravity waves, N.B.S. Circular 521, 1952, pp. 1–5.
  [U.8] *Trapping modes in the theory of surface waves.* Proceedings of the Cambridge Philosophical Society, Vol. 47, 1951.
  [U.9] *On the rolling motion of cylinders on the surface of a fluid.* Quarterly Journal of Mechanics and Applied Mathematics, Vol. 2, 1949.
  [U.10] *On the heaving motion of a circular cylinder on the surface of a fluid.* Quarterly Journal of Mechanics and Applied Mathematics, Vol. 2, 1949.
  [U.11] *The long-wave paradox in the theory of gravity waves.* Proceedings of the Cambridge Philosophical Society, Vol. 49, 1953.
  [U.12] See [B.4].
  [U.13] See [L.14].
Unaki, G., and M. Nakano
  [U.14] *On the Cauchy-Poisson waves caused by the eruption of a submarine volcano,* (I) *and* (II). Oceanograph. Mag. 4,5; 1953.

Vedernikov, V. V.
  [V.1] *Conditions at the front of a translation wave disturbing a steady motion of a real fluid.* Doklady (Comptes rendus) Akademiya Nauk SSSR, Vol. 48, 1945.

[V.2]   *Characteristic features of a liquid flow in an open channel.* Doklady (Comptes rendus), Vol. 52, 1946, pp. 207–210.

Vergne, Henri
[V.3]   *Ondes liquides de gravité.* Actualités Scientifiques et Industrielles, Hermann et Cie., Paris, 1928.

de Vries, G.
[V.4]   See [K.15].

Wallet, A.
See [R.12]

Watson, G. N.
[W.1]   *A Treatise on the Theory of Bessel Functions.* The University Press, Cambridge, 1944.

Wilkes, M. V.
[W.2]   *Oscillations of the Earth's Atmosphere.* The University Press, Cambridge, 1949.

Weinblum, G. P.
[W.3]   *Analysis of wave resistance.* David W. Taylor Model Basin (Washington, D. C.), Report 710, 1950.

[W.4]   See [S.1].

Weinstein, A.
[W.5]   *On surface waves.* Canadian Journal of Mathematics, Vol. 1, 1949, pp. 271–278.

[W.6]   *Sur la vitesse de propagation de l'onde solitaire.* Reale Academia dei Lincei, Rendiconti, Classe di scienze fisiche, matematiche, naturali, Ser. 6, Vol. 3, 1926, pp. 463.

[W.7]   *Sur un problème aux limites dans une bande indéfinie.* Institut de France. Comptes Rendus de l'Académie des Sciences. Vol. 184, 1927, p. 497.

Weitz, M., and J. B. Keller
[W.8]   *Reflection of water waves from floating ice in water of finite depth.* Communications on Pure and Applied Mathematics. Vol. 3, 1950, pp. 305–318.

Also see [K.9]

Weliczker, L.
[W.9]   See [F.5].

Wells, Leon W. (formerly L. Weliczker)
[W.10]   *Some remarks on shallow water theory.* New York University, Institute of Mathematical Sciences, Report No. IMM–198, 1953.

Westergaard, H. M.
[W.11]   *Water pressure on dams during earthquakes.* Transactions of American Society of Civil Engineers, Vol. 93, 1933.

Whitham, G. B.
[W.12]   *Dynamics of meteorological fronts.* New York University, Institute of Mathematical Sciences, Report No. IMM–195, 1953.

Wigley, W. C. S.
[W.13]   See [L.18].

Author Index

561

Subject Index

Lightning Source UK Ltd.
Milton Keynes UK
UKHW02f0641130918
328823UK00014B/1160/P